LINES

The slope of the line through $P_1(x_1, y_1)$ and $P_2(x_2, y_2)$ is

$$m = \frac{\Delta y}{\Delta x} = \frac{y_2 - y_1}{x_2 - x_1}.$$

Point-slope form: $y - y_1 = m(x - x_1)$

Slope-intercept form: $y = mx + b$

DISTANCE FORMULA

The distance between points $P_1(x_1, y_1)$ and $P_2(x_2, y_2)$ is

$$\sqrt{(x_1 - x_2)^2 + (y_1 - y_2)^2}.$$

BINOMIAL THEOREM

$$(a + b)^n = \binom{n}{0} a^n + \binom{n}{1}a^{n-1}b + \binom{n}{2}a^{n-2}b^2 + \cdots + \binom{n}{k}a^{n-k}b^k + \cdots + \binom{n}{n}b^n,$$

where $\binom{n}{k} = \dfrac{n!}{k!(n - k)!}.$

COMBINATORICS

Permutations (arrangements)

$$P(n, k) = \frac{n!}{(n - k)!}$$

Combinations (selections without regard to order)

$$C(n, k) = \frac{n!}{k!(n - k)!}$$

GEOMETRIC FORMULAS

Area

Triangle	$A = \frac{1}{2}bh$
Square	$A = s^2$
Rectangle	$A = ab$
Circle	$A = \pi r^2$
Sphere	$A = 4\pi r^2$

Volume

Cube	$V = s^3$
Sphere	$V = \frac{4}{3}\pi r^3$

COLLEGE ALGEBRA AND TRIGONOMETRY

COLLEGE ALGEBRA AND TRIGONOMETRY

SECOND EDITION

JOHN R. DURBIN
THE UNIVERSITY OF TEXAS
AT AUSTIN

WILEY

JOHN WILEY & SONS
New York   Chichester   Brisbane   Toronto   Singapore

Cover designed by Raphael Hernandez
Text designed by Ann Marie Renzi

Library of Congress Cataloging in Publication Data:

Durbin, John R.
  College algebra and trigonometry.

  Includes index
  1. Algebra. 2. Trigonometry. I. Title.
QA154.2.D874   1988   512'.13   87-28039
ISBN 0-471-62545-0

Printed in the United States of America

10 9 8 7 6 5 4

PREFACE

This book presents the standard topics of college algebra and trigonometry. The prerequisites are one year of high school geometry and two years of high school algebra. No previous experience with trigonometry is required.

I have tried to write so that students can learn as much as possible from the book itself, and so that instructors will find the book easy to use. For students, this implies a larger than usual number of examples, explanations that are to the point but complete, and exercise sets that proceed carefully from problems that are straightforward to those requiring more thought. For instructors, helpful features include careful organization, thorough coverage, and sample assignments for each section (which are in the solutions manual, available from the publisher).

Significant changes from the first edition include a more streamlined treatment of review material (Chapter I), an expanded treatment of conic sections, and the addition of more calculator exercises throughout the book. Other changes are listed below.

Review Material   The treatment of basic algebra in Chapter I is unusually complete, though condensed from the first edition. My own experience is that many students need and appreciate such review material, especially if it cannot be covered thoroughly in lectures. The detailed presentation in the text, along with the large numbers of examples and exercises, makes it easier for the instructor to concentrate on highlights and to move more quickly into the later parts of the book.

Pace and Flexibility   The arrangement is such that the essential material in each section can be covered in one lecture. Each section is divided into (lettered) subsections to make it easy for instructors to omit topics that they do not consider essential. For example, to keep a pace of one section per lecture, it may be necessary to omit some of the subsections on applications. The use of subsections also makes it easier for students to grasp the natural organization of the material.

Exercise Sets   The exercises at the end of each section are carefully arranged so that a representative sample is obtained by choosing every third one beginning with 1, 2, or 3. Each exercise set begins with straightforward problems to reinforce the most basic ideas and to build the student's confidence. Later exercises require several ideas or more thought, or illustrate applications. Assignments can be adjusted based on such factors as the background of the students and the goals of the course. Each chapter concludes with a set of review exercises, and either the even- or the odd-numbered exercises will give a representative sample of these.

The solutions manual, available to instructors from the publisher, gives sample assignments for each section. In most cases the assignment is divided into basic exercises (usually about ten), applications, and calculator problems.

In addition, there is a choice between three options, depending on whether one wants half, all, or none of the assigned exercises to have answers in the book.

Answers An appendix gives answers to most of the odd-numbered exercises from each section, and to nearly all of the review exercises. A solutions manual for the remaining exercises is available from the publisher.

Applications Numerous applications introduce word problems covering a variety of topics, including sports, mathematics of finance, nutrition, population growth, economics, physical science, variation, number theory, and geometry. The applications generally occur at the end of a section so that they can be omitted without loss of continuity.

Calculators Although this material can be learned without a calculator, more interesting examples and exercises can be considered if a calculator is used for some of the computations. The symbol $\boxed{C}$ marks the steps where a calculator has been used in examples; it also marks those exercises for which a calculator is suggested. In examples, the results of computations marked with $\boxed{C}$ can be taken on faith. If a calculator is not available, then the exercises marked $\boxed{C}$ can be either completed without a calculator or solved with the final computations only indicated and not carried out.

Options for a Course There are a number of options for courses based on the book. By covering one section per day and omitting the sections that are marked as optional (see the Contents), a one-semester course can begin with Chapter I and run through Chapter VIII; this would cover the fundamental topics in algebra as well as polynomial, rational, exponential, logarithmic, and trigonometric functions. Later material in the book can be included by spending less time on Chapter I (Basic Algebra) or by entirely omitting material earlier in the book. It will be clear that many of the chapters are independent of one another. For example, Chapters XI–XIV are independent of Chapters IV–X, and the material on trigonometry (Chapters VI–IX) is independent of the material on polynomial, rational, exponential, and logarithmic functions (Chapters IV and V).

Remarks on the Second Edition In making changes from the first edition, I have been aided by suggestions from students, from other instructors who have taught from the book, and from independent reviewers.

- Chapter I (Basic Algebra) has been streamlined based on suggestions that the pace of the review material was too slow.
- The first edition devoted one section to conic sections. Now there is an entire chapter (Chapter XV) with three sections.
- The chapter on exponential and logarithmic functions has been revised to improve its readability. Instead of one section introducing both kinds of functions, there is now one section for each.
- More calculator exercises have been added throughout the book.
- Sample assignments have been chosen for each section of the book. (See the preceding remarks on exercise sets.)

The exposition, examples, and exercises have been changed where appropriate for updating and clarity.

It is a pleasure to acknowledge the advice I have received from the following people:

Donald Bardwell (Nicholls State University), Edward Doran (Front Range Community College), Glenn Fox (Central Virginia Community College), Joseph O. Howell (Eastern Montana College), Mark Serebransky, Luis Shapiro (Rutgers State University), A. H. Tellez, and James Washenberger (Virginia Polytechnic Institute and State University).

I am especially grateful to Patricia Hickey (Baylor University) who has helped with proofreading and checked the answers in the appendix. Finally, it is a pleasure to thank the people at John Wiley & Sons who have helped with the production of the book.

JOHN R. DURBIN

THE DURBIN PRECALCULUS SERIES

College Algebra, 2nd Ed.
A student-oriented text designed for a one-term course covering standard topics such as equations and inequalities, exponential and logarithmic functions, complex numbers, and matrices. It features flexible organization and numerous applications.

Trigonometry
A thorough introduction to trigonometry for a one-term course. This text includes basic ideas of functions and graphs, exponential and logarithmic functions, and a chapter on analytic geometry.

College Algebra and Trigonometry, 2nd Ed.
A clear, concise presentation of the standard topics of college algebra and trigonometry, featuring numerous examples and exercises, including calculator exercises. The text is flexible and is carefully organized so that sections can be covered in one lecture each.

Precalculus
A student-oriented text for a one-term course presenting a solid foundation for calculus. It features a thorough treatment of the elementary functions, flexible organization, and numerous examples and exercises.

CONTENTS

V
EXPONENTIAL AND LOGARITHMIC FUNCTIONS 194

VI
INTRODUCTION TO TRIGONOMETRY 227

VII
TRIGONOMETRIC FUNCTIONS 252

VIII
GRAPHS AND INVERSES OF TRIGONOMETRIC FUNCTIONS 289

IX
APPLICATIONS OF TRIGONOMETRY 319

BASIC ALGEBRA

This chapter introduces some of the basic ideas of algebra: real numbers, exponents, polynomials, and rational (fractional) expressions. The applications used as illustrations include simple and compound interest and examples from physical science. Section 1 is longer than other sections of the book in order to give a thorough review of the real numbers.

SECTION **1**
Real Numbers

A. SOME IMPORTANT SETS OF NUMBERS

We begin with descriptions of the following sets of numbers, which are used throughout mathematics:

> natural numbers
>
> integers
>
> rational numbers
>
> irrational numbers
>
> real numbers

Two other sets of numbers, imaginary and complex, will be described in Chapter X.

The **natural numbers** are the numbers we count with:

$$1, 2, 3, 4, 5, \ldots$$

The **integers** are the natural numbers together with their negatives and zero:

$$\ldots, -5, -4, -3, -2, -1, 0, 1, 2, 3, 4, 5, \ldots$$

The **positive integers** are the same as the natural numbers.

A number is **rational** if it can be written in the form a/b, where a and b are integers and $b \neq 0$. Examples are

$$\frac{2}{3}, \quad \frac{-43}{11}, \quad 0.5 = \frac{1}{2}, \quad 5\frac{1}{3} = \frac{16}{3}, \quad \text{and} \quad \frac{2\pi}{5\pi} = \frac{2}{5}.$$

Every integer is rational. For example,

$$-17 = \frac{-17}{1} \quad \text{and} \quad 0 = \frac{0}{1}.$$

(The word *rational* is used because of the connection between fractions and ratios. If, for instance, one object weighs $\frac{2}{3}$ as much as another, then their weights are in the *ratio* 2 to 3.)

Before defining real numbers and irrational numbers, it will be useful to look more closely at rational numbers. By division, we can represent any rational number as a decimal number. Here is the result for $\frac{2}{27}$.

$$\frac{2}{27} = 0.074074 \cdots = 0.\overline{074}$$

To indicate that a block of digits is to be repeated, we write a bar over the block, as indicated above for 074. Here are some other examples.

$$3 = 3.0 \qquad \frac{1}{3} = 0.333 \cdots = 0.\overline{3}$$

$$\frac{-7}{2} = -3.5 \qquad -\frac{103}{330} = -0.31212 \cdots = -0.3\overline{12}$$

$$\frac{15}{8} = 1.875 \qquad \frac{16}{7} = 2.\overline{285714}$$

In the three examples on the left, the decimal representation *terminates:* there is a place after which only 0's would appear. In the three examples on the right, the decimal representation *repeats:* there is a place after which only the repetitions of some block that is not all 0's would appear. These two cases—terminating and repeating—are the only ones that can arise from rational numbers:

A decimal number represents a rational number
iff*
it either terminates or repeats.

Appendix A explains why that statement is true. The statement gains interest because many decimal numbers neither terminate nor repeat. A simple example is $0.1010010001 \cdots$, in which the number of 0's between the 1's increases by one each time.

Any number that has a decimal representation is called a **real number.** The real numbers that are not rational are called **irrational numbers.** Thus a real number is irrational iff it *cannot* be written as a quotient of two integers. Moreover, from what has already been said we can draw the following conclusion.

A decimal number represents an irrational number
iff
it neither terminates nor repeats.

The example $0.1010010001 \cdots$ is irrational. Figure 1.1 summarizes the relation between the sets of numbers discussed thus far; each set contains the sets that appear beneath it.

It can be proved that $\sqrt{2}, \sqrt{3}, \sqrt{5}, \sqrt{6}, \sqrt{7}$, and $\sqrt{8}$ are irrational.† In fact, if n is a positive integer, then $\sqrt{n}$ is irrational unless n is a perfect square $(1, 4, 9, 16, 25, \ldots)$. A proof that $\sqrt{2}$ is irrational is given in Appendix A. The number π, which represents the ratio of the circumference of any circle to its

---

*We follow the practice, now widely accepted in mathematics, of using "iff" to denote "if and only if." This expression can be used only to connect statements that are equivalent. Here are some examples.

    TRUE   A number is a natural number *iff* it is a positive integer.
    TRUE   A number is a natural number *if* it is a positive integer.
    TRUE   A number is rational *if* it is an integer.
    FALSE   A number is rational *iff* it is an integer.

†Square roots are reviewed in Section 3.

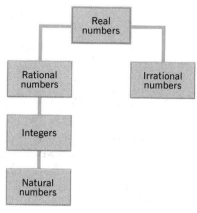

FIGURE 1.1

diameter, is also irrational (but the proof is not easy). These examples show that irrational numbers arise naturally, and not just from somewhat artificial constructions like 0.1010010001 · · ·.

Example 1.1 Which of the following are natural numbers? integers? rational numbers? irrational numbers?

$$8, \quad 1.414, \quad \frac{\pi}{2}, \quad \frac{\sqrt{25}}{3}, \quad 22.\overline{34}, \quad \sqrt{20}, \quad -6\frac{2}{3}, \quad \frac{-85}{17}$$

Solution

Natural numbers:	8
Integers:	8 (Every natural number is an integer.)
	$\dfrac{-85}{17}$ $\left(\dfrac{-85}{17} = -5\right)$
Rational numbers:	$8, \dfrac{-85}{17}$ (Every integer is a rational number.)
	1.414 (Terminating decimal.)
	$\dfrac{\sqrt{25}}{3}$ $\left(\dfrac{\sqrt{25}}{3} = \dfrac{5}{3}\right)$
	$22.\overline{34}$ (Repeating decimal.)
	$-6\dfrac{2}{3}$ $\left(-6\dfrac{2}{3} = \dfrac{-20}{3}\right)$
Irrational numbers:	$\pi/2$ (If $\pi/2$ were a fraction of integers, say $\pi/2 = a/b$, then π would also be a fraction of integers, $\pi = 2a/b$.)
	$\sqrt{20}$ (20 is not a perfect square.) ∎

Because $\sqrt{2}$ and π are irrational, no calculator can represent them exactly. A calculator with eight-digit accuracy will use the rounded-off values 1.4142136 for $\sqrt{2}$ and 3.1415927 for π. Such approximations are perfectly adequate for most practical work, but even more accurate values like 1.41421356237309504880 and 3.14159265358979323846 are not exact, since a terminating decimal cannot be exactly equal to an irrational number. Whenever the word *number* is used without qualification in this book, it will mean *real number*.

B. PROPERTIES OF THE REAL NUMBERS

This subsection summarizes the basic properties of the real numbers. The exercises at the end of this section ask you to simplify a number of expressions using these properties. In all of the properties, a, b, and c denote real numbers.

Commutative properties

$a + b = b + a$ for all a and b.
$ab = ba$ for all a and b.

Associative properties

$a + (b + c) = (a + b) + c$ for all a, b, and c.
$a(bc) = (ab)c$ for all a, b, and c.

Distributive properties

$a(b + c) = ab + ac$ for all a, b, and c.
$(a + b)c = ac + bc$ for all a, b, and c.

Zero and one

$a + 0 = 0 + a = a$ for every a.
$a \cdot 1 = 1 \cdot a = a$ for every a.

Negatives

For each number a there is a unique number denoted $-a$, and called the **negative** (or **additive inverse**) of a, such that

$$a + (-a) = (-a) + a = 0.$$

Reciprocals

For each number a except 0 there is a unique number denoted $1/a$, and called the **reciprocal** (or **multiplicative inverse**) of a, such that

$$a \cdot \frac{1}{a} = \frac{1}{a} \cdot a = 1.$$

The preceding properties are called the **field axioms** (because any system that satisfies them, like the real numbers, is called a **field**). We also need the following properties of equality.

Equality

$a = a$ for every a.　　*Reflexive property*

If $a = b$, then $b = a$.　　*Symmetric property*

If $a = b$ and $b = c$, then $a = c$.　　*Transitive property*

If $a = b$ and $c = d$, then

$\quad a + c = b + d$ and $ac = bd$.　　*Substitution property*

The remaining properties in this subsection can be proved from the field axioms and the preceding properties of equality.

Other properties of equality

If $a = b$, then $a + c = b + c$ and $ac = bc$ for every c.

If $a + c = b + c$, then $a = b$.　　*Cancellation for addition*

If $ac = bc$ and $c \neq 0$, then $a = b$.　　*Cancellation for multiplication*

Properties of zero

$a \cdot 0 = 0 \cdot a = 0$ for every a.

$ab = 0$ iff either $a = 0$ or $b = 0$.

Subtraction

This property is defined in terms of addition and negatives, as follows:

$$a - b = a + (-b) \qquad \text{for all } a \text{ and } b.$$

Other properties of negatives

For all real numbers a, b, and c,

$$-(-a) = a$$

$$-(a + b) = (-a) + (-b)$$

$$(-a)b = a(-b) = -(ab)$$

$$(-a)(-b) = ab$$

$$a(b - c) = ab - ac.$$

Division

This property is defined in terms of multiplication and reciprocals, as follows: For all real numbers a and b with $b \neq 0$,

$$\frac{a}{b} = a \cdot \frac{1}{b} = \frac{1}{b} \cdot a.$$

The number a/b is called a **fraction** with **numerator** a and **denominator** b. Notice that division by zero is undefined.

Properties of fractions

For all real numbers a, b, c, and d with $b \neq 0$ and $d \neq 0$,

$$\frac{-a}{b} = \frac{a}{-b} = -\frac{a}{b}$$

$$\frac{-a}{-b} = \frac{a}{b}$$

$$\frac{a}{b} = \frac{c}{d} \quad \text{iff} \quad ad = bc$$

$$\frac{a}{b} \cdot \frac{c}{d} = \frac{ac}{bd}$$

$$\frac{ac}{bc} = \frac{a}{b} \quad (c \neq 0)$$

$$\frac{\dfrac{a}{b}}{\dfrac{c}{d}} = \frac{a}{b} \cdot \frac{d}{c} = \frac{ad}{bc} \quad (c \neq 0)$$

$$\frac{a}{d} + \frac{b}{d} = \frac{a+b}{d}$$

$$\frac{a}{d} - \frac{b}{d} = \frac{a-b}{d}.$$

C. REAL LINES

Much of the importance of the real numbers derives from their use for measurement. This, in turn, is based on the assumption of a one-to-one correspondence between the set of real numbers and the set of points on a line. To establish such a correspondence, we first choose two points on a straight line and label them 0 and 1 (Figure 1.2). The line is made into a *directed line* by taking the direction from 0 toward 1 to be positive. The real numbers are then made to correspond to the points on the line in the way that will be familiar from previous experience with algebra.

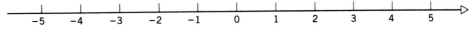

FIGURE 1.2

A line together with such a correspondence is called a **real line** (or **number line** or **coordinate line**). The number associated with a point on a real line is called the **coordinate** of the point. Figure 1.3 shows several examples. Points with irrational coordinates are located using decimal or fractional approximations. For example, 3.14 will usually suffice as an approximate decimal coordinate for π.

FIGURE 1.3

Instead of saying "the point with coordinate c" when referring to a point on a real line, we generally just say "the point c" if there is no chance of confusion.

Consider a horizontal real line directed to the right. If a and b are real numbers, and a is to the *left* of b on the line, then we say that a is **less than** b, and we write $a < b$. Thus

$$3 < 6, \ -17 < 2, \ -20 < -17, \text{ and } 0 < \pi.$$

The notation $b > a$ means the same as $a < b$. Thus

$$5 > 2, 5 > -2, 2 > -5, \text{ and } -2 > -5.$$

We say that b is **greater than** a if $b > a$.

The notation $a \leq b$ means that either $a = b$ or $a < b$. Similarly for $b \geq a$. Thus

$$4 \leq 9, 9 \leq 9, -1 \geq -3, \text{ and } -1 \geq -1.$$

The notation $a \leq b$ is read "a is less than or equal to b." The notation $b \geq a$ is read "b is greater than or equal to a."

A number c is **positive** if $c > 0$, and **negative** if $c < 0$. Statements involving $<$, $>$, $\leq$, or $\geq$ are called **inequalities.** More will be said about inequalities in Section 10.

D. ABSOLUTE VALUE

The **absolute value** of a real number a, denoted $|a|$, is the distance between the points with coordinates 0 and a on a real line, without regard to direction.

Example 1.2 (a) $|5| = 5$ and $|-5| = 5$. (See Figure 1.4.)

(b) $|\pi| = \pi$ and $|-\pi| = \pi$.

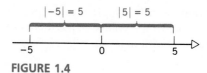

FIGURE 1.4

Here are three properties of absolute value.

$|0| = 0$.

If $a \neq 0$, then $|a| > 0$.

$|a| = |-a|$ for every a.

If a is positive, then $|a| = a$. If a is negative, however, then $|a| = -a$. For instance, $|-5| = 5 = -(-5)$. The point is that if a is negative, then $-a$ is positive. This allows us to make the following equivalent definition of $|a|$.

$$|a| = \begin{cases} a & \text{if } a \geq 0 \\ -a & \text{if } a < 0 \end{cases}$$

If a and b are real numbers, then the **distance** between a and b on a real line, without regard to direction, is $|a - b|$. (Strictly speaking, $|a - b|$ is the distance between the *points* with *coordinates* a and b.) Also, $|a - b|$ is the distance between $a - b$ and 0. Notice that $|a - b| = |b - a|$.

Example 1.3 (See Figure 1.5.)

(a) The distance between 8 and 5 is $|8 - 5| = |3| = 3$ or $|5 - 8| = |-3| = 3$.

(b) The distance between 3.5 and -4 is

$$|3.5 - (-4)| = |7.5| = 7.5 \quad \text{or} \quad |-4 - 3.5| = |-7.5| = 7.5. \quad \blacksquare$$

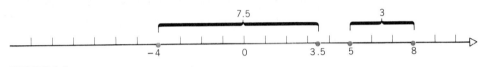

FIGURE 1.5

E. APPLICATION: PERCENTAGE AND SIMPLE INTEREST

Recall that **percentage** (or **percent**) means parts per hundred. For example, $25\% = \frac{25}{100} = 0.25$, $100\% = \frac{100}{100} = 1$, $110\% = \frac{110}{100} = 1.1$, and $\frac{1}{2}\% = 0.5\% = 0.005$. (To remove the $\%$ sign from a decimal number, move the decimal point two places to the left.)

Example 1.4 (a) 5% of 80 is 0.05×80, which is 4. (Notice that *of* is here a signal to multiply.)

(b) If 18 is 75% of x, then $0.75x = 18$, and $x = 18/0.75 = 24$.

(c) If 300 is increased by 15%, the result is

$$300 + 0.15(300) = 300 + 45 = 345. \quad \blacksquare$$

Example 1.5 The population of the U.S. increased from 203,200,000 in 1970 to 226,500,000 in 1980 (each figure to the nearest 100,000). Find the percentage increase.

Solution The absolute increase was $226,500,000 - 203,200,000 = 23,300,000$. Thus the percentage increase was

$$\frac{23,300,000}{203,200,000} \approx 11.5\% \qquad \boxed{C}$$

Reminder $\approx$ means *approximately equal*. $\blacksquare$

Example 1.6 Suppose that a salesman works on an 8% rate of commission. What must his sales be if he is to earn $1200?

Solution Let A denote the total amount of sales. We must determine A such that 8% of A is $1200:

$$8\% \text{ of } A \text{ is } 1200$$

gives

$$0.08A = 1200$$

and

$$A = 1200/0.08 = \$15,000.$$ ■

The formula for **simple interest** is

$$I = Prt, \tag{1.1}$$

where P is the **principal** (amount invested), r is the **rate** per year (as a percentage expressed in decimal form), and t is the **time** of the investment in years. (In practice, most interest is not *simple* interest but *compound* interest, which is discussed in Section 2.)

Example 1.7   What is the interest on \$500 invested at 9% per year simple interest for 4 years?

Solution   Use $P = \$500$, $r = 0.09$, and $t = 4$ in Equation (1.1):

$$I = 500(0.09)(4) = \$180.$$ ■

Example 1.8   What amount must be invested at 8% simple interest so that \$50 interest is earned at the end of 6 months?

Solution   We first divide both sides of Equation (1.1) by rt, since we want to compute P. Then use $I = 50$, $r = 0.08$, $t = \frac{1}{2}$ (6 months = $\frac{1}{2}$ year).

$$P = I/rt$$

$$P = 50/[(0.08)\,(\tfrac{1}{2})]$$

$$P = 50/0.04$$

$$P = \$1250$$ ■

F. CALCULATORS AND APPROXIMATE NUMBERS

For some of the exercises in this book you will be asked to use a calculator. The sequence of steps required for a calculation often depends on the type of calculator. To ensure that you know how to handle the basic operations and square roots ($+$, $-$, $\times$, $\div$, and $\sqrt{\ }$) when they are combined in different orders, use a calculator to verify each computation in the following example. Exercises 16–48 can be used for further practice.

Example 1.9   (a) $13 + 2 \times 6 = 25$ Ⓒ
(b) $(13 + 2) \times 6 = 90$
(c) $6 - (7 - 2) = 1$
(d) $6 - 7 - 2 = -3$
(e) $4.20 \times (-1.85) \div 13.96 \approx -0.557$
(f) $(4 + 6) \div 5 = 2$
(g) $4 + 6 \div 5 = 5.2$

(h) $\dfrac{19.3 - 27.5}{4.28 + 0.913} \approx -1.58$

(i) $\dfrac{7\frac{1}{3} + 2\frac{1}{4}}{3\frac{1}{7}} \approx 3.049$ (exact answer: 805/264)

(j) $\sqrt{7} + 2 \approx 4.6458$

(k) $\sqrt{14} - \sqrt{19} \approx -0.6172$ ■

In applying mathematics it is important to distinguish between *exact* numbers (numbers that measure something exactly) and *approximate* numbers (numbers that measure something approximately). Here are two examples.

There are 100 centimeters in a meter. (*100 is exact.*)

The world population in 1987 was 5 billion. (*5 billion is approximate, although accurate at least to the nearest billion.*)

Approximate numbers generally arise either from measurement or from decimal approximations of irrational numbers like π and $\sqrt{2}$. In particular, no measurement can be more accurate than the instrument used to obtain it.

For dealing with approximate numbers it is convenient to use the notion of significant figures, which can be thought of as the digits in a number about whose accuracy we feel reasonably certain. The significant figures in a number sometimes must simply be inferred. If the airline distance from New York to San Francisco is given as 2570 miles, for example, the digits 2, 5, and 7 should be thought of as significant. Unless we are certain that a zero on the end of a number is a significant figure, we should assume that it is not.

The guiding principle in working with approximate numbers is that we must not assume a final answer to be more accurate than the numbers with which we begin. Here are two rules that result from that principle. Examples follow the rules.

Rule. The result of *adding* or *subtracting* approximate numbers should not be given to more *decimal places* than any of the numbers with which we begin.

Rule. The result of *multiplying* or *dividing* approximate numbers should not be given with more *significant figures* than any of the numbers with which we begin.

These rules are really only thinly disguised common sense. If one edge of a rectangular field is measured only to the nearest meter, for instance, but an adjacent edge is measured to the nearest centimeter, it would be foolish to carry out the computation of the perimeter beyond the nearest meter.

Example 1.10   (a) Give the sum of the approximate numbers 4.321 and 91.3 as 95.6, not 95.621.

(b) Give the product of the approximate numbers 43.21 (four significant figures) and 123 (three significant figures) as 5310 (three significant figures), not 5314.83 (six significant figures), or even 5315 (four significant figures).   ■

Exact numbers put no restriction on the accuracy of an answer. For example, the accuracy of an answer from using the formula $P = 4s$ for the perimeter of a square is restricted only by the accuracy of s (the length of a side); the number 4 is exact in this formula. In Example 1.10(b), if the number 123 were exact rather than approximate, then the product would be given as 5315 (four significant figures).

EXERCISES FOR SECTION 1

1. Each of the following real numbers is rational. Give a reason in each case.

(a) $16\frac{2}{3}$ (b) -3.14

(c) $4.\overline{07}$ (d) $-\sqrt{25}/9$

2. Each of the following real numbers is irrational. Give a reason in each case.

(a) $\sqrt{10}$ (b) $5\sqrt{3} - 4\sqrt{3}$

(c) $0.1001000010000001 \cdots$ (d) $-\pi$

3. Indicate whether each of the following statements is true or false.

(a) Every negative integer is rational.

(b) A number is irrational if it is real but not an integer.

(c) Every positive rational number is a natural number.

(d) If the decimal representation of a real number does not terminate, then the number is irrational.

In each of Exercises 4–6, draw a real line and label the points with the given numbers as coordinates.

4. $4\frac{1}{2}$, $-2.\overline{3}$, π, $-2\sqrt{2}$, $\frac{5}{7}$, -2.3

5. $-\frac{3}{2}$, 2π, $\sqrt{2}/2$, 1.6, $-3.\overline{6}$, -3.7

6. $\pi/2$, 1.3, $-\sqrt{2}/3$, $\frac{-7}{4}$, 0.45, $1.\overline{3}$

Replace each □ with the appropriate sign, either $<$ or $>$.

7. $-2 \; \square \; 5$

8. $0.7 \; \square \; \frac{3}{4}$

9. $\frac{1}{5} \; \square \; \frac{1}{6}$

10. $-0.2 \; \square \; -0.19$

11. $0.3 \; \square \; 0.03$

12. $|-0.2| \; \square \; -20$

13. $-|-0.2| \; \square \; -0.\overline{2}$

14. $-|\frac{1}{4}| \; \square \; -|-\frac{1}{5}|$

15. $1/0.2 \; \square \; 1/0.1$

In each of Exercises 16–18, use a calculator to help arrange the four given numbers in increasing order. (If you do not have a $\boxed{\pi}$ key, use 3.14159 for π.)

$\boxed{C}$

16. $\sqrt{2}$, $41/29$, $4\sqrt{21}/13$, $1\frac{29}{70}$

17. π, $22/7$, $99\sqrt{10}/100$, $3\frac{10}{71}$

18. $(1 + \sqrt{5})/2$, $21/13$, $13\pi/25$, $1\frac{34}{55}$

Simplify. Suggestion: Do without a calculator and then check with the help of a calculator.

19. $-2(3 - 5)$ **20.** $-(-7)(5)$

21. $(1 - 6)(-3)$ **22.** $(-45) \div 5$

23. $0(0 - 12)$ **24.** $-[-(-5)]$

25. $4 \cdot 5 + |-7|$ **26.** $3 - 2|-9|$

27. $-6 + |-2|5$

28. $(-6)(-8) \div (-12)$

29. $121 \div (-11)$

30. $(12 + 3) \div (-5)$

31. $(-4) \div |-4 - 1|$

32. $(|-12| - |0|)/|3|$

33. $0 \div |-6|$

34. $-0(14 \div 2)$

35. $(0.08)(-12.5)$

36. $(-0.39) \div 1.3$

37. $(-1.21) \div (1.1)$

38. $0.5[(0.27) \div 0.3]$

39. $1 - 3[5 - 7(11 - 13)]$

40. $2.4[2 - 3(4 - 5)] \div (-0.5)$

41. $3.5[1 - 3(5 - 2)] \div (-0.4)$

42. $\{8[5 - 3(2 - 1)] \div [4(3 - 4)]\} \div 2$

43. $\{-4[-5 - 2(-3)] \div 2\} \div |2 - 4|$

44. $(-15.41) \div (-0.67)$

45. $2|19.51 - 73.3| \div 0.2$

46. $(1.234 - 3.21) \div 0.04$

47. $[(6.594) \div (-0.21)] \div 3.14$

48. $(4.192 - 5.82 - 1.7074) \div 0.9$

Simplify each of the following fractions, writing the final answer in the form a/b, *where* a *and* b *are integers.*

49. $(\frac{1}{2})(\frac{4}{3})$

50. $(\frac{5}{2})(-2\frac{1}{3})$

51. $(1\frac{1}{5})(1\frac{2}{3})$

52. $(-6\frac{3}{7})(\frac{14}{-5})$

53. $(-\frac{7}{10})(\frac{5}{6})(\frac{-9}{-14})$

54. $(\frac{-3}{28})(1\frac{2}{5})(\frac{10}{-9})$

55. $\frac{1}{3} + \frac{1}{4}$

56. $\frac{2}{5} - \frac{1}{3}$

57. $\frac{5}{6} - \frac{1}{4}$

58. $\frac{1}{3} - \frac{1}{6} + \frac{1}{9}$

59. $1\frac{1}{2} + 2\frac{1}{3} - 3\frac{1}{4}$

60. $1\frac{3}{8} - 2\frac{5}{6} - 1$

61. $\frac{-8}{5}/\frac{15}{2}$

62. $2\frac{1}{2} \div (\frac{1}{3})$

63. $1 \div (-4\frac{1}{7})$

64. $(-4\frac{2}{3})/(-3)$

65. $(-1\frac{2}{7})/3$

66. $(6\frac{2}{3})/(-5)$

67. $(\frac{3}{2}) \div (\frac{7}{4} - 3)$

68. $(-\frac{1}{4}) \div (2 - \frac{4}{3})$

69. $(\frac{-12}{5}) \div (\frac{1}{3} - \frac{7}{3})$

70. $(4\frac{1}{3} + 1\frac{2}{5}) \div (2\frac{1}{2})$

71. $(\frac{7}{50} - \frac{7}{30}) \div (\frac{1}{6} + \frac{1}{5})$

72. $(3 + \frac{1}{2}) \div (2 - \frac{1}{3})$

Rewrite each expression without using parentheses and simplify.

73. $-(2a - b)$

74. $-2(-a + b)$

75. $-(-a - b)$

76. $3[-(2b + c) + (c - b) - b]$

77. $-5[a - 2(a - b) - (2a + b)]$

78. $-[-2(a + c) - (b - c) + 3(-a + c)]$

79. $-\{-2[-(a - b) + 2(-a + b)] + a\}$

80. $4\{-[(a - b) + 2(b - a)] - (a + b)\}$

81. $-3\{4(-a + b) - 3[-(2a - b) - 5(-a + b)]\}$

Determine the distance on a real line between the points with the given coordinates.

82. 12 and -2

83. -5.1 and -7

84. $16\frac{1}{2}$ and 4.2

85. -3π and $-5\pi/2$

86. $\frac{1}{7}$ and $-\frac{1}{8}$

87. $1/\pi$ and $-3/(2\pi)$

88. If 720 is 0.6% of x, what is x?

89. If 15 is $\frac{1}{4}$% of y, what is y?

90. If u is 175% of v, what percent of u is v?

91. What is the percentage profit on an item that is bought for $20.00 and sold for $22.50?

92. What total sales will produce a $500 commission if the commission rate is 20%?

93. If a woman must pay 17% tax on her taxable income, what taxable income will produce an after-tax income of $12,500? $\boxed{C}$

94. What is the interest on $200 invested at 8% per year simple interest for 2 years?

95. What is the interest on $1000 invested at 9% per year simple interest for 3 years?

96. What is the interest on $500 invested at 12% per year simple interest for 6 months?

97. What amount must be invested at 6% per year simple interest so that $40 interest is earned at the end of 18 months?

98. What amount must be invested at 10% per year simple interest so that $250 interest is earned at the end of 30 months?

99. What amount must be invested at 8% per year simple interest so that $200 interest is earned at the end of 5 years?

100. The population of the U.S.S.R. in 1980 was 266 million. The population increased by 4.5% between 1980 and 1985. What was the population in 1985?

101. In 1975, 25% of the population of Northern America (which consists primarily of the U.S. and Canada) was less than 15 years old, and 42% of the population of Latin America was less than 15 years old. The populations of Northern America and Latin America were approximately 237 million and 324 million, respectively. What percent of the total population of Northern and Latin America was less than 15 years old? $\boxed{C}$

102. In 1980, 76.0% of the U.S. population of 227 million was metropolitan and the remainder was nonmetropolitan. Of the total metropolitan population, 5.8% lived in New England. Of the total nonmetropolitan population 4.3% lived in New England. Determine the 1980 population of New England to the nearest 100,000. $\boxed{C}$

SECTION **2**
Integral Exponents

A. DEFINITIONS AND PROPERTIES

To review the laws (properties) of integral exponents, we begin by recalling the definitions of a^n for n positive, zero, and negative, respectively.

If n is a positive integer and a is a real number, then

$$\overbrace{a^n = a \cdot a \cdot \cdots \cdot a.}^{n \text{ factors}} \qquad (2.1)$$

In particular, $a^1 = a$. We call a the **base,** n the **exponent,** and a^n the **nth power** of a.

Example 2.1 $2^3 = 2 \cdot 2 \cdot 2 = 8$ $(-5)^4 = (-5)(-5)(-5)(-5) = 625$

If a is any nonzero real number, then
$$a^0 = 1. \qquad (2.2)$$

Notice that a is nonzero for (2.2); 0^0 is undefined.

Example 2.2 $6^0 = 1$ $\left(\dfrac{2}{3}\right)^0 = 1$ $\dfrac{2}{3^0} = \dfrac{2}{1} = 2$ $(-5)^0 = 1$ $-5^0 = -1$

If a is any nonzero real number and n is a positive integer, then

$$a^{-n} = \frac{1}{a^n}. \qquad (2.3)$$

Example 2.3 (a) $a^{-1} = 1/a$
(b) $2^{-3} = 1/2^3 = 1/8$
(c) $(-3)^{-2} = 1/(-3)^2 = 1/9$ but $-3^{-2} = -(1/3^2) = -1/9$
(d) $(-0.2)^{-1} = 1/(-0.2) = -5$

Notice that Equation (2.3) is valid for $n < 0$ as well as $n > 0$, because if

$n < 0$, then $-n > 0$, so

$$a^{-n} = \frac{1}{1/a^{-n}} \qquad \text{because } u = \frac{1}{1/u} \text{ for } u \neq 0$$

$$= \frac{1}{a^n} \qquad \text{by (2.3) with } n \text{ replaced by } -n.$$

For instance,

$$4^2 = 4^{-(-2)} = \frac{1}{4^{-2}}.$$

The following display summarizes the laws of integral exponents. It is assumed that you have seen these laws explained, and worked with them, in previous algebra courses. The examples and exercises provide a review.

Laws of integral exponents

If a and b are nonzero real numbers, and m and n are integers, then

$$a^m a^n = a^{m+n} \tag{2.4}$$

$$(a^m)^n = a^{mn} \tag{2.5}$$

$$(ab)^m = a^m b^m \tag{2.6}$$

$$\frac{a^m}{b^m} = \left(\frac{a}{b}\right)^m \tag{2.7}$$

$$(-a)^n = \begin{cases} a^n & \text{if } n \text{ is even} \\ -a^n & \text{if } n \text{ is odd} \end{cases} \tag{2.8}$$

$$\frac{a^m}{a^n} = a^{m-n}. \tag{2.9}$$

Example 2.4 This example illustrates Properties (2.4)–(2.9) and will also serve as a reminder of why each property is true.

(a) $3^4 \cdot 3^2 = (3 \cdot 3 \cdot 3 \cdot 3)(3 \cdot 3) = 3^{4+2} = 3^6 = 729$

(b) $(x^4)^3 = (x \cdot x \cdot x \cdot x)(x \cdot x \cdot x \cdot x)(x \cdot x \cdot x \cdot x) = x^{4 \cdot 3} = x^{12}$

(c) $(ab)^3 = ab \cdot ab \cdot ab = a \cdot a \cdot a \cdot b \cdot b \cdot b = a^3 b^3$

(d) $\dfrac{y^4}{z^4} = \dfrac{y \cdot y \cdot y \cdot y}{z \cdot z \cdot z \cdot z} = \dfrac{y}{z} \cdot \dfrac{y}{z} \cdot \dfrac{y}{z} \cdot \dfrac{y}{z} = \left(\dfrac{y}{z}\right)^4$

(e) $(-2)^4 = 16 = 2^4$ but $-2^4 = -16$

(f) $(-2)^5 = -32 = -2^5$

(g) $\dfrac{a^5}{a^3} = \dfrac{a \cdot a \cdot a \cdot a \cdot a}{a \cdot a \cdot a} = a^{5-3} = a^2$

(h) $\dfrac{a^3}{a^3} = \dfrac{a \cdot a \cdot a}{a \cdot a \cdot a} = a^{3-3} = a^0 = 1$

(i) $\dfrac{a^3}{a^5} = \dfrac{a \cdot a \cdot a}{a \cdot a \cdot a \cdot a \cdot a} = \dfrac{1}{a^{5-3}} = \dfrac{1}{a^2} = a^{3-5} = a^{-2}$

The next example illustrates that *if the numerator and denominator of a fraction are products,* then a factor of the numerator can be moved to the denominator if we change the sign of the exponent on the factor, and a factor of the denominator can be moved to the numerator if we change the sign of the exponent on the factor. This follows from multiplying the fraction by 1 in a conveniently chosen form.

Example 2.5 (a) $\dfrac{a^{-3}b}{c} = \dfrac{b}{a^3c}$ because $\dfrac{a^{-3}b}{c} = \dfrac{a^3}{a^3} \cdot \dfrac{a^{-3}b}{c} = \dfrac{b}{a^3c}.$

(b) $\dfrac{u}{v^{-2}} = uv^2$ because $\dfrac{u}{v^{-2}} = \dfrac{u}{v^{-2}} \cdot \dfrac{v^2}{v^2} = uv^2.$

(c) $\dfrac{xy^{-2}}{x^{-3}y} = \dfrac{xx^3}{yy^2} = \dfrac{x^4}{y^3}$

(d) $\dfrac{1 + a^{-2}b}{c} \neq \dfrac{1+b}{a^2c}$. We cannot move a^{-2} to the denominator as a^2 because a^{-2} is not a factor of the *whole* numerator. In fact,

$$\frac{1 + a^{-2}b}{c} = \frac{1 + \dfrac{b}{a^2}}{c} = \frac{\dfrac{a^2 + b}{a^2}}{\dfrac{c}{1}} = \frac{a^2 + b}{a^2c}.$$

Example 2.6 The following illustrations show how to simplify expressions so that only positive exponents appear in the final answer. Assume that n is a positive integer in part (d).

(a) $(a^{-2})^{-5} = a^{(-2)(-5)} = a^{10}$

(b) $\left(\dfrac{u^2}{v^3}\right)^{-4} = \dfrac{(u^2)^{-4}}{(v^3)^{-4}} = \dfrac{u^{-8}}{v^{-12}} = \dfrac{v^{12}}{u^8}$

(c) $\dfrac{(x^2y^{-3})^4}{(xy^2)^{-2}} = \dfrac{(x^2)^4(y^{-3})^4}{x^{-2}(y^2)^{-2}} = \dfrac{x^8y^{-12}}{x^{-2}y^{-4}}$

$$= x^8y^{-12} \cdot x^2y^4 = x^{10}y^{-8} = \frac{x^{10}}{y^8}$$

(d) $\left(\dfrac{a^{-1}b^5}{a^3b^{-2}}\right)^{-3n} = \left(\dfrac{b^5b^2}{a^3a^1}\right)^{-3n} = \left(\dfrac{b^7}{a^4}\right)^{-3n} = \dfrac{(b^7)^{-3n}}{(a^4)^{-3n}}$

$$= \frac{b^{-21n}}{a^{-12n}} = \frac{a^{12n}}{b^{21n}}$$

B. APPLICATION: SCIENTIFIC NOTATION

One useful application of exponents begins from the following observation: To multiply a decimal number by 10^n we simply move the decimal point n places to the right if $n > 0$, and $|n|$ places to the left if $n < 0$. For example,

$$1.47 \times 10^2 = 147 \quad \text{and} \quad 3892 \times 10^{-5} = 0.03892.$$

This means that we can move the decimal point wherever we like provided we compensate by multiplying by an appropriate power of 10. If the decimal point is moved just to the right of the leftmost nonzero digit, then the resulting form is called **scientific notation.** Thus the form of a positive number written in scientific notation is

$$a \times 10^n, \text{ where } n \text{ is an integer and } 1 \leq a < 10.$$

Example 2.7 (a) To write 147 in scientific notation, first replace 147 by 1.47; then multiply 1.47 by 10^2 to compensate:

$$147 = 1.47 \times 10^2.$$

(b) To write 0.03892 in scientific notation, first replace 0.03892 by 3.892; then multiply by 10^{-2} to compensate:

$$0.03892 = 3.892 \times 10^{-2}.$$

Example 2.8 Scientific notation is especially convenient when we are faced with extremely large or extremely small numbers.

The earth is approximately $93,000,000 = 9.3 \times 10^7$ miles from the sun.

A light-year (the distance light travels in one year) is approximately $5,880,000,000,000 = 5.88 \times 10^{12}$ miles.

The mass of a hydrogen atom has been found to be approximately $0.0000000000000000000001675 = 1.675 \times 10^{-24}$ grams.

Calculators display extremely large or extremely small numbers with scientific notation in something like this form:

$$\boxed{2.5344 \quad 11} \text{ for } 2.5344 \times 10^{11}$$

$$\boxed{3.9457 \ -12} \text{ for } 3.9457 \times 10^{-12}.$$

The next example shows how to use scientific notation to keep track of decimal points in computations. Given a product or quotient, we can first write each factor in scientific notation and then determine the proper placement of the decimal point in the final answer merely by working with powers of 10.

Example 2.9 (a) $2,000,000 \times 0.03 \times 160 = (2 \times 10^6) \times (3 \times 10^{-2}) \times (1.6 \times 10^2) = (2 \times 3 \times 1.6) \times 10^{6-2+2} = 9.6 \times 10^6 = 9,600,000$

(b) $\dfrac{70 \times 0.0032}{0.001 \times 800} = \dfrac{(7 \times 10) \times (3.2 \times 10^{-3})}{(1 \times 10^{-3}) \times (8 \times 10^2)}$

$$= \frac{7 \times 3.2}{8} \times 10^{1-3+3-2}$$

$$= \frac{22.4}{8} \times 10^{-1}$$

$$= 2.8 \times 10^{-1}$$

$$= 0.28$$

C. APPLICATION: COMPOUND INTEREST

The two basic kinds of interest on invested money are *simple* interest and *compound* interest. In each case an amount of interest is paid at the end of regular time periods. The difference is that for simple interest the amount of interest paid is the same for each period, being computed solely on the original principal (amount invested); for compound interest the amount of interest paid increases from period to period, because the interest paid during each period is added to the principal to create a new principal, and interest for the following period is computed on this new (increased) principal. Simple interest was discussed in Section 1. Here we look at compound interest.

Compound Amount Formula

The time periods over which compound interest is computed are called **conversion periods.** Interest rates are conventionally given as annual rates and then converted to rates per conversion period. For example, a rate of 10% compounded quarterly yields a rate of $\frac{1}{4} \cdot 10\% = 2.5\%$ per quarter (the conversion period).

Example 2.10 Suppose that $1000 is invested at an annual rate of 6%, compounded semiannually. How much (principal plus interest) will have accumulated at the end of 1 year?

Solution The rate per conversion period is $\frac{1}{2} \cdot 6\% = 3\%$. The amount of interest earned during the first period is

$$1000 \times 0.03 = \$30.00.$$

Therefore, the principal for the second period is $1030 (Figure 2.1).
The interest earned during the second period is

$$1030 \times 0.03 = \$30.90.$$

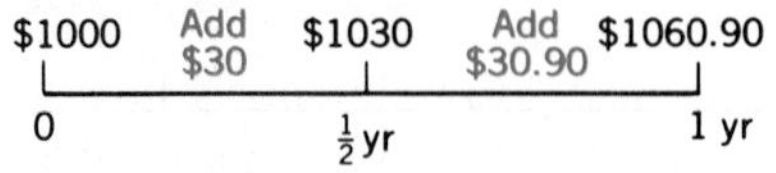

$$\begin{array}{cccccc} \$1000 & \text{Add} & \$1030 & \text{Add} & \$1060.90 \\ & \$30 & & \$30.90 \\ \rule{0pt}{0pt} & & & & \\ 0 & & \frac{1}{2}\,\text{yr} & & 1\,\text{yr} \end{array}$$

FIGURE 2.1

The amount accumulated at the end of 1 year is, therefore,

$$\$1030.00 + \$30.90 = \$1060.90.$$

(At 6% per year *simple* interest, $1060.00 would have accumulated.)

By generalizing the preceding example we can derive a formula for **compound amount**—that is, the amount accumulated (principal plus interest) in any compound interest problem. Let

$P =$ original principal

$r =$ annual interest rate expressed in decimal form

$n =$ number of conversion periods per year

$S_k =$ compound amount at the end of k periods.

The interest rate per conversion period is r/n. Therefore, the compound amount at the end of the first period will be

$$S_1 = P + P\frac{r}{n} = P\left(1 + \frac{r}{n}\right).$$

The principal for the second period will be S_1, so that

$$S_2 = S_1 + S_1\left(\frac{r}{n}\right) = P\left(1 + \frac{r}{n}\right) + P\left(1 + \frac{r}{n}\right)\frac{r}{n}$$

$$= P\left(1 + \frac{r}{n}\right)\left(1 + \frac{r}{n}\right)$$

$$= P\left(1 + \frac{r}{n}\right)^2.$$

In the same way,

$$S_3 = S_2 + S_2\left(\frac{r}{n}\right) = P\left(1 + \frac{r}{n}\right)^2 + P\left(1 + \frac{r}{n}\right)^2\left(\frac{r}{n}\right) = P\left(1 + \frac{r}{n}\right)^3.$$

Continuing, we see that in general

$$S_k = P\left(1 + \frac{r}{n}\right)^k. \tag{2.10}$$

Over t years the number k of conversion periods will be nt (the number of conversion periods per year times the number of years). Therefore, Equation (2.10) justifies the following conclusion.

Compound amount

> If a principal amount P is invested at an annual interest rate r compounded n times per year for t years, then the compound amount S is given by
>
> $$S = P\left(1 + \frac{r}{n}\right)^{nt}. \tag{2.11}$$

Example 2.11 What is the compound amount on $500 invested for 2 years at 12% compounded quarterly?

Solution Here $P = \$500$, $n = 4$(quarterly compounding), $r/n = 0.12/4 = 0.03$, and $t = 2$. Equation (2.11) gives

$$S = 500(1.03)^8 = \$633.39.$$

Throughout this section and its exercises, the rates are annual rates, like r in Equation (2.11).

Present Value Formula

Suppose we want to know how much to invest at a specified rate and compounding period, and for a specified time, in order to accumulate a specified amount. Then r, n, t, and S are known, and we can find P by multiplying both sides of Equation (2.11) by $[1 + (r/n)]^{-nt}$. This amount P is called the **present value** of S (for the given r, n, and t).

Present value

$$P = S\left(1 + \frac{r}{n}\right)^{-nt} \qquad (2.12)$$

Example 2.12 How much must be invested at 10% compounded semiannually so that $1000 will be accumulated at the end of 2 years?

Solution The problem asks for the present value of $1000 under the stated interest conditions. In this case $r = 0.10$, $n = 2$, $r/n = 0.05$, $t = 2$, and $S = 1000$. Equation (2.12) gives

$$P = 1000(1.05)^{-4} = 1000/1.05^4 = \$822.70. \qquad \boxed{C} \ \blacksquare$$

EXERCISES FOR SECTION 2

Compute each power and simplify.

1. $(-2)^3/9^2$

2. $-3^2/(-4)^3$

3. $7^2/(-5)^3$

4. $(2^{-3})^2$

5. $(1/2)^{-5}$

6. $(3^{-2})^{-3}$

7. $(-0.02)^{-1}$

8. -0.2^{-2}

9. $(-0.1)^{-3}$

Simplify, using no negative exponents in the final answers. Assume that n is a positive integer.

10. $(a^2)^4 a^3$

11. $(t^3 t)^2$

12. $(x^2 x^4 x^3)^5$

13. $(-1.5u^3)^2$

14. $(-3v^2)^3$

15. $(-0.5w^4)^3$

16. $(3b^{-3})^2$

17. $\left(\dfrac{x^2}{2y}\right)^{-1}$

18. $[(2x)^{-2}y]^{-3}$

19. $\dfrac{(-a^{-2}b)^3}{(ab^{-1})^2}$

20. $\dfrac{x^{-3}y^4}{(-x^2y^2)^{-3}}$

21. $-\left(\dfrac{-ut^{-1}}{3u^{-2}t}\right)^4$

22. $\dfrac{-3a^0b}{(-2ab)^0}$

23. $\dfrac{(-6x)^0}{(2x^2)^{-1}}$

24. $\dfrac{5t^0}{(-10t)^{-1}}$

25. $\left(\dfrac{5a^{-1}}{b^2}\right)^2$

26. $\dfrac{(-0.3x)^2}{(10xy)^{-3}}$

27. $\dfrac{(2ab)^{-2}}{a^2b^{-1}}$

28. $\dfrac{0.4a}{(5a^3)^{-2}}$

29. $ab\left(\dfrac{-0.5a^2}{b^3}\right)^{-4}$

30. $\left(\dfrac{4a}{b^2}\right)^{-2}\left(\dfrac{2b}{a^2}\right)^4$

31. $\dfrac{a^n b^{2n}}{a^{2n}b^{-n}}$

32. $(u^2v)^{-2n}(uv^{-1})^{5n}$

33. $\left(\dfrac{x}{y^2}\right)^{-3n}\left(\dfrac{y^3}{x}\right)^{-n}$

34. $\dfrac{(5u^{-2}vw^3)^{3n}}{(5u^{-1}v^4w^{-2})^{2n}}$

35. $\left(\dfrac{x}{y}\right)^{-2n}\left(\dfrac{2y}{z}\right)^{-n}\dfrac{(2x)^{3n}}{z^{3n}}$

36. $\dfrac{[(r^{-1}s)^{-2}t]^{-3n}}{[(rs)^2t]^{3n}}$

Write each of the following numbers in scientific notation.

37. 0.00000001 (The length in centimeters of an angstrom, a unit used in expressing the length of light waves.)

38. 1,454 (The height in feet of the Sears Tower in Chicago.)

39. 3,212 (The drop in feet of Angel Falls in Venezuela.)

40. 29,028 (The height in feet of Mount Everest.)

41. 36,000 (The approximate depth in feet of the deepest part of the Pacific Ocean, in the Mariana trench.)

42. 25,200,000,000,000 (The approximate distance in miles to *Proxima Centauri,* the nearest star.)

43. 0.0000000033 (The approximate time in seconds for light to travel one meter in a vacuum.)

44. 0.00067 (The approximate time in seconds for sound to travel one meter in fresh water at room temperature.)

45. 0.0029 (The approximate time in seconds for sound to travel one meter in a vacuum at room temperature.)

Perform the following calculations using scientific notation to handle decimal points, as in Example 2.9. Show each step of the solution.

46. 890×0.0004

47. $400 \times 0.0008 \times 2000$

48. $0.12/0.000006$

49. $14 \times 0.0005/0.007$

50. $(120,000 \times 0.0003)/(90 \times 0.02)$

51. $(0.03 \times 5400)/(900 \times 0.0006)$

In Exercises 52–54, use a calculator with a $\boxed{y^x}$ *key.*

52. Compute $\left(1 + \dfrac{1}{x}\right)^x$ for $x = 10$, $x = 1000$, and $x = 100,000$. $\boxed{C}$

53. Compute $\dfrac{(2 + h)^3 - 2^3}{h}$ for $h = 0.1, h = 0.001$, and $h = 0.00001$. $\boxed{C}$

54. Compute n^n for $n = 10$, $n = 20$, $n = 30$, and so on, until you exceed the capacity of your calculator. Now try other values to find the largest integer n for which your calculator will compute n^n. What does it display for n^n in this case? $\boxed{C}$

In Exercises 55–57 determine the compound amount for the given principal and investment conditions. *

55. $100 for 2 years at 12% compounded quarterly.

56. $200 for 1 year at 8% compounded quarterly.

*The integral and fractional powers needed for Exercises 55–63 can be computed with a calculator or taken from the list at the end of the exercises.

57. $500 for 3 years at 10% compounded semiannually.

58. How much must be invested at 6% compounded monthly so that $500 accumulates at the end of 18 months?

59. How much must be invested at 12% compounded monthly so that $500 accumulates at the end of one year?

60. In $3\frac{1}{2}$ years you will need $3000 and you want to take care of this by investing now in an account that pays 10% compounded monthly. How much should you invest?

61. (a) Which would be worth more at the end of 10 years, $100 invested now at 8% compounded annually or $200 invested now at 4% compounded annually?

(b) Repeat part (a) with 10 years replaced by 20 years.

The next two exercises can be solved with appropriate extensions of the ideas and formulas in Subsection C.

62. Suppose the cost of living increases 5% per year over the next 20 years. Compute the cost at the end of that period of an item that costs $1.00 now. $\boxed{C}$

63. The Consumer Price Index rose from 100.0 to 104.2 from 1967 to 1968. What would the index have been in 1977 if it had increased at that same annual rate from 1967 to 1977? (The 1977 index was actually 181.5.) $\boxed{C}$

The following are five-place approximations for the powers needed in Exercises 55–63. (The order here is more or less random.)

$1.02^4 = 1.08243$

$1.08^{10} = 2.15892$

$1.04^{20} = 2.19112$

$1.04^{10} = 1.48024$

$1.05^6 = 1.34010$

$1.03^8 = 1.26677$

$1.05^{20} = 2.65330$

$1.008\overline{3}^{-42} = 0.70571$

$1.08^{20} = 4.66096$

$1.042^{10} = 1.50896$

$1.005^{-18} = 0.91414$

$1.01^{-12} = 0.88745$

SECTION **3**

Rational Exponents. Radicals

A. RATIONAL EXPONENTS DEFINED

Positive integral exponents relate to repeated multiplication, as in $a^2 = a \cdot a$. *Negative integral* exponents relate to multiplication and inversion, such as $a^{-2} = 1/(a \cdot a)$. We now consider *rational exponents*. The definition will be given in three steps, in Equations (3.1)–(3.3). This will be done in such a way that, with some slight restrictions, the laws of integral exponents (2.4)–(2.9) apply when the exponents are any rational numbers, as well as when they are integers.

Let a denote a real number and n a positive integer. It can be proved that if $a \geq 0$, then there is precisely one real number $b \geq 0$ such that $b^n = a$. This number b is called the **principal nth root** of a and is denoted $a^{1/n}$. That is:

For $a \geq 0$, $b \geq 0$, and n a positive integer,

$$b = a^{1/n} \quad \text{iff} \quad b^n = a. \tag{3.1}$$

Example 3.1 (a) $10 = 1000^{1/3}$ because $10^3 = 1000.$

(b) $2 = 32^{1/5}$ because $2^5 = 32.$

(c) $\frac{1}{3} = \left(\frac{1}{81}\right)^{1/4}$ because $\left(\frac{1}{3}\right)^4 = \frac{1}{81}.$

To define $a^{1/n}$ when $a < 0$, we must consider two cases, depending on whether the integer n is even or odd. If $a < 0$ and n is even, then there is *no* real number b such that $b^n = a$, because if n is even then $b^n \geq 0$ for every real number b. [For example, $(-3)^2 = 9 > 0$.] However, if $a < 0$ and n is odd, then there is precisely one real number $b < 0$ such that $b^n = a$. This number b is called the **principal nth root** of a and is denoted $a^{1/n}$. Thus:

For $a < 0$, $b < 0$, and n an odd positive integer,

$$b = a^{1/n} \quad \text{iff} \quad b^n = a. \tag{3.2}$$

Example 3.2 (a) $-2 = (-8)^{1/3}$ because $(-2)^3 = -8.$

(b) $-0.1 = (-0.00001)^{1/5}$ because $(-0.1)^5 = -0.00001.$

(c) $-\frac{3}{4} = \left(-\frac{27}{64}\right)^{1/3}$ because $\left(-\frac{3}{4}\right)^3 = -\frac{27}{64}.$

Notice that if n is even and $a > 0$ then there are two real numbers b such that $b^n = a$, but $a^{1/n}$ denotes only the positive one of these. For example, $625^{1/4} = 5$, even though $(-5)^4 = 625$ as well as $5^4 = 625$. In general, for n even and $a > 0$, both $a^{1/n}$ and $-a^{1/n}$ are called **nth roots** of a, but the *principal* nth root is $a^{1/n}$. The notation $\pm a^{1/n}$ represents the pair, $a^{1/n}$ and $-a^{1/n}$.

The definition of $a^{1/n}$ provided by (3.1) and (3.2) is just what it must be to make the law $(a^m)^n = a^{mn}$ for integral exponents carry over for $1/n$ in place of m. Specifically, from

$$(a^{1/n})^n = a^{(1/n)n} = a^1 = a$$

we see that the nth power of $a^{1/n}$ must be a. In the same way, the next step is just what it must be to make the laws of integral exponents carry over to all rational exponents.

Suppose that a is a nonzero real number and that m and n are integers such that $n > 0$ and m/n is reduced to lowest terms. Then $a^{m/n}$ is defined by

$$a^{m/n} = (a^{1/n})^m, \tag{3.3}$$

provided $a^{1/n}$ exists (that is, provided $a > 0$, or $a < 0$ and n is odd). If m/n is a positive rational number then $0^{m/n} = 0$.

Example 3.3 (a) $25^{3/2} = (25^{1/2})^3 = 5^3 = 125$

(b) $0.00001^{3/5} = (0.00001^{1/5})^3 = 0.1^3 = 0.001$

(c) $81^{0.75} = 81^{3/4} = (81^{1/4})^3 = 3^3 = 27$

The idea in part (c) can be used for any decimal exponent that is also rational. Irrational exponents will be discussed in Section 24.

To compute $a^{m/n}$ when m/n is negative, we can assume $m < 0$ and $n > 0$ [for example, $-(2/3) = (-2)/3$]. Then Equation (3.3) will apply, as in the following special cases.

Example 3.4 (a) $8^{-2/3} = (8^{1/3})^{-2} = 2^{-2} = 1/2^2 = 1/4$

(b) $0.04^{-3/2} = (0.04^{1/2})^{-3} = 0.2^{-3} = 1/0.2^3 = (1/0.2)^3 = 5^3 = 125$

To compute $a^{m/n}$ always reduce the fraction in the exponent to lowest terms if it is not already so reduced. For example, $(-1)^{6/2}$ is not defined by the equation in (3.3), because no real number is equal to $(-1)^{1/2}$. However, if we reduce $6/2$ to 3 then we have $(-1)^{6/2} = (-1)^3 = -1$.

Powers of numbers can be computed quickly on a calculator having a $\boxed{y^x}$ key. This does not lessen the importance of knowing how to use the laws of exponents. In manipulating algebraic expressions, for example, we are often faced not with specific numbers but with letters that represent numbers.

B. PROPERTIES

It can be proved that the laws for integral exponents (Section 2A) carry over to rational exponents as follows.

Laws of rational exponents

Each of the following equations is valid for all real numbers a and b and all rational numbers r and s for which both sides of the equation exist.

$$a^r a^s = a^{r+s} \tag{3.4}$$

$$(a^r)^s = a^{rs} \ (a > 0) \tag{3.5}$$

$$(ab)^r = a^r b^r \tag{3.6}$$

$$\frac{a^r}{b^r} = \left(\frac{a}{b}\right)^r \tag{3.7}$$

$$\frac{a^r}{a^s} = a^{r-s} \tag{3.8}$$

Example 3.5 Simplify, using no negative exponents in the final answer. Each letter represents a positive real number.

$$\frac{(x^{-1/2}y)^3 \, x^{-5/2} \, y^4}{(xy^2)^{-2}} = \frac{(x^{-1/2})^3 \, y^3 \, x^{-5/2} \, y^4}{x^{-2}(y^2)^{-2}}$$

$$= \frac{x^{-3/2} \, y^3 \, x^{-5/2} \, y^4}{x^{-2} \, y^{-4}}$$

$$= \frac{x^{-4} \, y^7}{x^{-2} \, y^{-4}}$$

$$= \frac{y^{7+4}}{x^{4-2}}$$

$$= \frac{y^{11}}{x^2}$$

C. RADICALS DEFINED

If a is a real number and n is a positive integer, then $\sqrt[n]{a}$ is defined by

$$\boxed{\sqrt[n]{a} = a^{1/n}} \tag{3.9}$$

provided $a^{1/n}$ exists. In the notation $\sqrt[n]{a}$, the integer n is called the **index,** a is called the **radicand,** and $\sqrt[n]{}$ is called the **radical.** Observe that $\sqrt[n]{a}$ is the *principal nth root* of a. If $a > 0$, then $\sqrt{a}$, which is the same as $\sqrt[2]{a}$, is the principal square root of a. Both $\sqrt{a}$ and $-\sqrt{a}$ are square roots of a.

Example 3.6 (a) $\sqrt[3]{-64} = -4$

(b) The *principal square root* of 49 is $\sqrt{49} = 7$.
The *square roots* of 49 are $\pm \sqrt{49} = \pm 7$, that is, 7 and -7.

D. PROPERTIES

Notice that $\sqrt[n]{a}$ is not a new concept; it is merely different notation for $a^{1/n}$. Thus the properties of radicals, which follow, are just special cases of properties of exponents.

Laws of exponents and radicals

Each of the following equations is valid for all values of m, n, a, and b for which the roots in the equation exist.

$$(\sqrt[n]{a})^n = a \tag{3.10}$$

$$\text{If } n \text{ is even, then } \sqrt[n]{a^n} = |a|. \tag{3.11}$$

$$\text{If } n \text{ is odd, then } \sqrt[n]{a^n} = a. \tag{3.12}$$

$$\sqrt[n]{ab} = \sqrt[n]{a}\,\sqrt[n]{b} \tag{3.13}$$

$$\sqrt[n]{a/b} = (\sqrt[n]{a})/(\sqrt[n]{b}) \tag{3.14}$$

$$\sqrt[m]{\sqrt[n]{a}} = \sqrt[mn]{a} \tag{3.15}$$

$$\sqrt[n]{a^m} = (\sqrt[n]{a})^m \tag{3.16}$$

The following special case of (3.11) is worth emphasizing.

$$\sqrt{a^2} = |a|$$

For example,

$$\sqrt{5^2} = \sqrt{25} = 5 \text{ and } \sqrt{(-5)^2} = \sqrt{25} = 5.$$

Example 3.7 Here are illustrations, in order, of Laws (3.10)–(3.16).

(a) $(\sqrt{5})^2 = 5$ and $(\sqrt[3]{-8})^3 = (-2)^3 = -8$

(b) $\sqrt[4]{2^4} = \sqrt[4]{16} = 2 = |2|$ and $\sqrt[4]{(-2)^4} = \sqrt[4]{16} = 2 = |-2|$

(c) $\sqrt[3]{2^3} = \sqrt[3]{8} = 2$ and $\sqrt[3]{(-2)^3} = \sqrt[3]{-8} = -2$

(d) $\sqrt{4 \cdot 9} = \sqrt{36} = 6 = 2 \cdot 3 = \sqrt{4}\,\sqrt{9}$

(e) $\sqrt[3]{(-27/64)} = -3/4 = (\sqrt[3]{-27})/(\sqrt[3]{64})$

(f) $\sqrt[3]{\sqrt[2]{64}} = \sqrt[3]{8} = 2 = \sqrt[6]{64}$

(g) $\sqrt[5]{(-32)^2} = \sqrt[5]{1024} = 4 = (-2)^2 = (\sqrt[5]{-32})^2$

Following are remarks on the proofs of Laws (3.10)–(3.16). The laws are obviously true for $a = 0$, so we assume throughout that $a \neq 0$. Law (3.10) simply restates the meaning of $\sqrt[n]{a}$.

For Law (3.11), first remember that, since n is even, $\sqrt[n]{a^n}$ denotes the unique positive real number b such that $b^n = a^n$. If $a > 0$, it follows from the uniqueness of b that $b = a$. If $a < 0$, then, since $-a > 0$ and $(-a)^n = a^n$, it follows from the uniqueness of b that $b = -a$. So in either case, $b = |a|$, as required. (Remember that $|a| = -a$ if $a < 0$.) Law (3.12) is similar to Law (3.11).

Laws (3.13)–(3.16) are consequences of the laws of exponents in Subsection 3B. For example,

$$\sqrt[m]{\sqrt[n]{a}} = (a^{1/n})^{1/m} = a^{(1/n)(1/m)} = a^{1/(mn)} = \sqrt[mn]{a}.$$

E. MORE EXAMPLES

Here are some things to keep in mind when simplifying expressions that involve radicals.

- Multiples of radicals with the same index and the same radicand can be combined through addition and subtraction. Thus, for example, by a distributive law, $6\sqrt{5} - 4\sqrt{5} = (6 - 4)\sqrt{5} = 2\sqrt{5}$.
- If $a \neq 0$, $b \neq 0$, and $n > 1$, then $\sqrt[n]{a + b} \neq \sqrt[n]{a} + \sqrt[n]{b}$. For example, $\sqrt{4 + 9} = \sqrt{13}$, but $\sqrt{4} + \sqrt{9} = 2 + 3 = 5$; thus $\sqrt{4 + 9} \neq \sqrt{4} + \sqrt{9}$.
- Try to use $\sqrt[n]{a^m} = (\sqrt[n]{a})^m$ when faced with computing $\sqrt[n]{a^m}$, because roots are easier to recognize for small numbers than for large ones. For example, $\sqrt[3]{27^4} = \sqrt[3]{531{,}441}$, and we are not likely to recognize 531,441 as a perfect cube. However, $\sqrt[3]{27^4} = (\sqrt[3]{27})^4 = 3^4 = 81$.
- Many expressions that contain radicals can be simplified most easily by converting to rational exponents. For example, $\sqrt[5]{a^{10}} = a^{10/5} = a^2$.

To simplify an expression with a radical we write it so that the radicand contains no factors with exponents greater than or equal to the index of the radical.

Example 3.8 Simplify. Assume that each variable represents a positive real number.

(a) $\sqrt{a^3} = \sqrt{a^2 \cdot a} = \sqrt{a^2}\sqrt{a} = a\sqrt{a}$

(b) $(\sqrt[3]{x})^{12} = [(\sqrt[3]{x})^3]^4 = x^4$

(c) $\sqrt[5]{x^{11}} = \sqrt[5]{x^{10}x} = \sqrt[5]{x^{10}}\sqrt[5]{x} = x^2\sqrt[5]{x}$

(d) $\sqrt{8} - \sqrt{18} = \sqrt{4 \cdot 2} - \sqrt{9 \cdot 2} = \sqrt{4}\sqrt{2} - \sqrt{9}\sqrt{2} = 2\sqrt{2} - 3\sqrt{2} = -\sqrt{2}$

(e) $6\sqrt{3a} - \sqrt{12a} + 2\sqrt{75a^3} = 6\sqrt{3a} - \sqrt{4}\sqrt{3a} + 2\sqrt{25a^2}\sqrt{3a} = 6\sqrt{3a} - 2\sqrt{3a} + 10a\sqrt{3a} = (4 + 10a)\sqrt{3a}$

Example 3.9 Simplify. The variables represent positive real numbers.

(a) $\sqrt{t}/\sqrt[3]{t} = t^{1/2}/t^{1/3} = t^{(1/2)-(1/3)} = t^{1/6} = \sqrt[6]{t}$

(b) $\sqrt[3]{25}\,\sqrt[3]{5^4} = (5^2)^{1/3} \cdot 5^{4/3} = 5^{2/3} \cdot 5^{4/3} = 5^{(2/3)+(4/3)} = 5^2 = 25$

(c) $(27a^6)^{2/3} = 27^{2/3}(a^6)^{2/3} = (\sqrt[3]{27})^2 a^{6(2/3)} = 3^2 a^{12/3} = 9a^4$

(d) $t^{-3}\sqrt[2]{\sqrt[3]{t^{12}}} = t^{-3}[(t^{12})^{1/3}]^{1/2} = t^{-3}(t^4)^{1/2} = t^{-3}t^2 = t^{-1} = 1/t$

(e) $\sqrt{\dfrac{x^{m-2}}{x^{4-m}}} = [x^{(m-2)-(4-m)}]^{1/2} = [x^{2m-6}]^{1/2} = x^{(2m-6)/2} = x^{m-3}$

If a variable in a radicand is not known to be nonnegative, then we must take that into account. For example, unless we know that $a \geq 0$ then we should write $\sqrt{a^2b} = |a|\sqrt{b}$, by (3.11), and not $\sqrt{a^2b} = a\sqrt{b}$. Of course, $b \geq 0$ here or the expressions are undefined.

F. RATIONALIZING DENOMINATORS

We sometimes encounter fractions with one or more radicals in their denominators. To **rationalize the denominator** of such a fraction is to rewrite the fraction in a form without a radical in the denominator. Most cases can be handled by one of the two techniques that follow. Each technique depends on multiplying the fraction by 1 in an appropriately chosen form.

> If the denominator has the form $a\sqrt{b}$, multiply the fraction by $\sqrt{b}/\sqrt{b}$ and simplify.

Example 3.10 Rationalize each denominator.

(a) $\dfrac{1}{\sqrt{2}} = \dfrac{1}{\sqrt{2}} \cdot \dfrac{\sqrt{2}}{\sqrt{2}} = \dfrac{\sqrt{2}}{2}$

(b) $\dfrac{\sqrt{3}}{2\sqrt{5}} = \dfrac{\sqrt{3}}{2\sqrt{5}} \cdot \dfrac{\sqrt{5}}{\sqrt{5}} = \dfrac{\sqrt{15}}{10}$

> If the denominator has the form $a + \sqrt{b}$, multiply the fraction by $(a - \sqrt{b})/(a - \sqrt{b})$ and simplify. The expression $a - \sqrt{b}$ is called the **conjugate** of $a + \sqrt{b}$. Similarly, for $\sqrt{a} + \sqrt{b}$ use the conjugate $\sqrt{a} - \sqrt{b}$, and for $c\sqrt{a} + d\sqrt{b}$ use the conjugate $c\sqrt{a} - d\sqrt{b}$.

These ideas work because $(u + v)(u - v) = u^2 - v^2$. In particular,

$$(a + \sqrt{b})(a - \sqrt{b}) = a^2 - (\sqrt{b})^2 = a^2 - b$$

and

$$(\sqrt{a} + \sqrt{b})(\sqrt{a} - \sqrt{b}) = (\sqrt{a})^2 - (\sqrt{b})^2 = a - b.$$

Example 3.11 Rationalize each denominator.

(a) $\dfrac{1}{2 + \sqrt{3}} = \dfrac{1}{2 + \sqrt{3}} \cdot \dfrac{2 - \sqrt{3}}{2 - \sqrt{3}} = \dfrac{2 - \sqrt{3}}{4 - 3} = 2 - \sqrt{3}$

(b) $\dfrac{\sqrt{r}}{\sqrt{r} - 2\sqrt{s}} = \dfrac{\sqrt{r}}{\sqrt{r} - 2\sqrt{s}} \cdot \dfrac{\sqrt{r} + 2\sqrt{s}}{\sqrt{r} + 2\sqrt{s}} = \dfrac{r + 2\sqrt{rs}}{r - 4s}$

(c) $\dfrac{1}{1 - \sqrt{x + 1}} = \dfrac{1}{1 - \sqrt{x + 1}} \cdot \dfrac{1 + \sqrt{x + 1}}{1 + \sqrt{x + 1}} = \dfrac{1 + \sqrt{x + 1}}{1 - (x + 1)}$

$$= -\dfrac{1 + \sqrt{x + 1}}{x}$$

The next example illustrates how to handle roots other than square roots.

Example 3.12 Rationalize the denominator.

$$\frac{2}{\sqrt[3]{5}} = \frac{2}{\sqrt[3]{5}} \cdot \frac{\sqrt[3]{5^2}}{\sqrt[3]{5^2}} = \frac{2\sqrt[3]{5^2}}{\sqrt[3]{5^3}} = \frac{2\sqrt[3]{25}}{5}$$

Sometimes there is a good reason to rationalize the *numerator* of a fraction. The following example arises in calculus; it uses an obvious variation on one of the techniques for rationalizing denominators.

Example 3.13 Rationalize the numerator.

$$\frac{\sqrt{x+h} - \sqrt{x}}{h} = \frac{\sqrt{x+h} - \sqrt{x}}{h} \cdot \frac{\sqrt{x+h} + \sqrt{x}}{\sqrt{x+h} + \sqrt{x}}$$

$$= \frac{(x+h) - x}{h(\sqrt{x+h} + \sqrt{x})}$$

$$= \frac{1}{\sqrt{x+h} + \sqrt{x}}$$

EXERCISES FOR SECTION 3

Compute. (Do not use a calculator.)

1. $16^{1/4}$

2. $10,000^{1/4}$

3. $64^{1/3}$

4. $4^{5/2}$

5. $8^{2/3}$

6. $9^{3/2}$

7. $9^{-1/2}$

8. $4^{-7/2}$

9. $25^{-3/2}$

10. $0.001^{4/3}$

11. $0.25^{5/2}$

12. $0.027^{2/3}$

13. $(-1000)^{-1/3}$

14. $(-32)^{-3/5}$

15. $(-64)^{-2/3}$

16. $(-27,000)^{2/3}$

17. $(-125)^{5/3}$

18. $81^{-3/4}$

19. $25^{2.5}$

20. $49^{1.5}$

21. $(-32)^{1.4}$

22. $(16^{0.75})^{-3}$

23. $(10^8)^{-0.25}$

24. $(10^{-9})^{0.\overline{3}}$

Simplify. The letters represent positive real numbers.

25. $\dfrac{x^{1/3}x^{7/3}}{x^{2/3}}$

26. $\dfrac{y^{-1/3}y^{2/3}}{y^{-7/3}}$

27. $\dfrac{z^2 z^{1/2}}{z^{-1/2}}$

28. $(125x^3)^{2/3}$

29. $(16y^6)^{1/2}$

30. $(32z^{10})^{3/5}$

31. $\dfrac{a^2(a^3b^{1/2})^{-4}}{(a^3b)^2}$

32. $\dfrac{(c^{-3}d^{2/3})^{-1}d}{cd^{4/3}}$

33. $\dfrac{(mn^2)^{-1/2}(m^{1/2}n)^{-1}}{(mn)^{-2}}$

Find each root.

34. $\sqrt[4]{81}$

35. $\sqrt[5]{-32}$

36. $\sqrt[5]{100,000}$

37. $\sqrt[3]{-125}$

38. $\sqrt[3]{8000}$

39. $\sqrt[3]{-1000}$

40. $\sqrt[4]{0.0001}$

41. $\sqrt[3]{-0.008}$

42. $\sqrt[3]{0.125}$

43. $\sqrt[5]{-1/32}$

44. $\sqrt[4]{81/16}$

45. $\sqrt[3]{-8/27}$

Simplify. Use Laws (3.10)–(3.16) as much as possible. Lengthy calculations should be unnecessary. The letters represent positive real numbers.

46. $\sqrt[4]{(-7)^4}$

47. $(\sqrt[4]{20})^4$

48. $(\sqrt[5]{100})^5$

49. $\sqrt[3]{(-13)^3}$

50. $\sqrt[10]{(-1/5)^{10}}$

51. $\sqrt[7]{(-1/12)^7}$

52. $(\sqrt[6]{25})^6$

53. $\sqrt[5]{(-1.7)^5}$

54. $\sqrt[6]{(-3)^6}$

55. $(\sqrt{0.09})^3$

56. $(\sqrt[3]{0.064})^2$

57. $(\sqrt{0.81})^3$

58. $\sqrt{5} + \sqrt{20}$

59. $\sqrt{12} - \sqrt{3}$

60. $\sqrt{50} + \sqrt{18}$

61. $3\sqrt{200} - 5\sqrt{72}$

62. $2\sqrt{40} + \sqrt{90}$

63. $2\sqrt{48} - 5\sqrt{27}$

64. $\sqrt{12\sqrt[3]{27}}$

65. $\sqrt[3]{270\sqrt{10000}}$

66. $\sqrt{5\sqrt[3]{8000}}$

67. $(\sqrt[3]{a})^6$

68. $(\sqrt{b})^4$

69. $(\sqrt[5]{c})^{10}$

70. $\sqrt[6]{x^{19}}$

71. $\sqrt[3]{y^{10}}$

72. $\sqrt{z^7}$

73. $2\sqrt{9u} - \sqrt{25u}$

74. $6\sqrt{v} - \sqrt{100v}$

75. $\sqrt{w} + \sqrt{49w}$

76. $2\sqrt{r^3} - r(\sqrt{r} - \sqrt{9r})$

77. $s\sqrt{2s} - \sqrt{18s^3} + s\sqrt{50s}$

78. $\sqrt[3]{t^4} - 2t(\sqrt[3]{t} - \sqrt[3]{8t})$

79. $\sqrt[3]{a}/\sqrt[4]{a}$ **80.** $\sqrt[4]{b}/\sqrt[3]{b}$

81. $\sqrt[5]{c}/\sqrt[4]{c}$ **82.** $\sqrt[3]{4}/\sqrt[3]{2^4}$

83. $\sqrt{27}\,\sqrt{3^5}$ **84.** $\sqrt[4]{8}\,\sqrt{2^5}$

85. $\sqrt[3]{\sqrt{\sqrt{r^{-12}}}}$ **86.** $\sqrt{\sqrt{\sqrt{s^{20}}}}$

87. $(\sqrt{\sqrt[3]{t^2}})^9$

88. $\dfrac{\sqrt[3]{3a^5bc^2}\,\sqrt[3]{36ab^4c}}{\sqrt[3]{4b^2c^3}}$

89. $\dfrac{u\sqrt{v\sqrt{w}}\,v\sqrt{w\sqrt{u}}}{w\sqrt{vw\sqrt{uw}}}$

90. $\dfrac{\sqrt[4]{\dfrac{x^3z}{y^2}}\,\sqrt[4]{\dfrac{x^5z}{y}}}{\sqrt[4]{\dfrac{y}{z^2}}}$ **91.** $\sqrt{\dfrac{a^{2m+1}}{a^{-3}}}$

92. $\sqrt{\dfrac{b^{4n+1}}{b^{2n-1}}}$ **93.** $\sqrt[3]{\dfrac{c^{2n}}{c^{6-n}}}$

Compute with a calculator having a $\boxed{y^x}$ *key.* $\boxed{C}$

94. $10^{1/5}$ **95.** $5^{3/10}$

96. $6^{-2/3}$ **97.** $(\sqrt[5]{7.06})^{-12}$

98. $\sqrt[4]{12.8^{-3}}$ **99.** $(\sqrt[8]{6.1413})^7$

Rationalize each denominator and simplify.

100. $2/\sqrt{7}$ **101.** $10/\sqrt{10}$

102. $1/\sqrt{3}$ **103.** $1/(3 + \sqrt{2})$

104. $2/(1 - \sqrt{3})$ **105.** $5/(1 - \sqrt{6})$

106. $1/(\sqrt{x} - \sqrt{y})$ **107.** $1/(\sqrt{x + 1} - \sqrt{x})$

108. $1/(\sqrt{a} + 1)$ **109.** $3/\sqrt[3]{2}$

110. $-5/\sqrt[3]{10}$ **111.** $6/\sqrt[3]{8}$

Rationalize each numerator and simplify.

112. $(1 - \sqrt{2})/3$

113. $(2 - \sqrt{3})/(2 + \sqrt{3})$

114. $(\sqrt{5} + \sqrt{2})/3$

115. $(\sqrt{x - h} - \sqrt{x})/h$

116. $(\sqrt{x + 1 + h} - \sqrt{x + 1})/h$

117. $(\sqrt{1 - x + h} - \sqrt{1 - x})/h$

118. Compute $\dfrac{\sqrt{2 + h} - \sqrt{2}}{h}$ for $h = 0.1$, $h =$ 0.001, and $h = 0.00001$. $\boxed{C}$

119. Compute $\dfrac{\sqrt{(5 + h)^3} - \sqrt{5^3}}{h}$ for $h = 0.5$, $h =$ 0.005, and $h = 0.00005$. $\boxed{C}$

120. Compute $\dfrac{-1}{3\sqrt{3 + h} + \sqrt{3}\,(3 + h)}$ for $h =$ 0.1, $h = 0.001$, and $h = 0.00001$. $\boxed{C}$

Let R *denote the average distance of a planet from the sun, and let* T *denote the period of the planet (the time required for one complete orbital revolution about the sun) measured in Earth years, so that* T = *1 for the Earth. Then Kepler's third law of planetary motion states that* R³/T² *is the same for all planets. If this constant value is denoted by* K, *then*

$$R^3/T^2 = K. \qquad (3.17)$$

121. The average distance between the Earth and the sun is approximately 93 million miles. Use this (with $T = 1$) to compute K in Equation (3.17). $\boxed{C}$

122. (a) Prove that Equation (3.17) implies $R = \sqrt[3]{KT^2}$.

(b) Prove that Equation (3.17) implies $T = \sqrt{R^3/K}$.

123. The period of Mars is approximately 1.88 Earth years. What is the average distance between Mars and the sun? [Use Exercise 122(a) and K from the answer to Exercise 121.]

SECTION **4**

Polynomials

A. TERMINOLOGY

A **variable** (or, more precisely, a **real variable**) is a letter representing an unspecified real number. In the formula $V = x^3$ for the volume of a cube, both V and x are variables. A **monomial** in a variable x is a product of a real number, called the **coefficient,** and a power x^n, where n is a nonnegative integer.* The integer n is called the **degree** of the monomial, except that a monomial with coefficient 0 is not assigned a degree. We write x^n for $1 \cdot x^n$.

Example 4.1

	Monomial	Coefficient	Degree
(a)	x^3	1	3
(b)	$-7x^5$	-7	5
(c)	$\frac{1}{2}x^0 = \frac{1}{2}$	$\frac{1}{2}$	0
(d)	$0x = 0$	0	undefined

A **polynomial** in a variable x is a finite sum of monomials in x. Notice that *sum* includes *difference* since we allow negative coefficients: for example, $6x^2 - 3x + 1 = 6x^2 + (-3)x + 1$. The individual monomials are called the **terms** of the polynomial: the terms of $6x^2 - 3x + 1$ are $6x^2$, $-3x$, and 1. The largest degree from all of the terms is the **degree** of the polynomial, and the corresponding coefficient is the **leading coefficient** of the polynomial. The zero polynomial (the polynomial having only zero coefficients) is not assigned a degree. A term of degree 0 is called a **constant term.** A polynomial with no term of degree 0 is said to have constant term 0.

Example 4.2

	Polynomial	Degree	Leading coefficient
(a)	$6x^2 - 3x + 1$	2	6
(b)	$x^3 - 5$	3	1
(c)	$-\sqrt{2}x^9 + 4x^{25} - 7x$	25	4
(d)	πx^2	2	π
(e)	-17	0	-17

Observe that we include monomials as polynomials: think of πx^2 and -17 as sums with one term, for instance.

Unless there is a special reason to do otherwise, it is generally best to arrange the terms of a polynomial so that their degrees are either in descending order or in ascending order; we usually choose the former. For example, we would usually write the polynomial in Example 4.2(c) as $4x^{25} - \sqrt{2}x^9 - 7x$.

The general form for a polynomial of degree n in x is

$$a_n x^n + a_{n-1}x^{n-1} + \cdots + a_1 x + a_0$$

where $a_n, a_{n-1}, \ldots, a_1$, and a_0 denote the coefficients and $a_n \neq 0$. The variable in a polynomial need not be x. For example, $16t^2 - 32t + 48$ is a polynomial in the variable t.

Polynomials in more than one variable are defined as follows. First, a **monomial** in several variables is a product consisting of a real number, called the **coefficient,** and nonnegative integral powers of the variables. Thus $3xy$ and $-\sqrt{2}x^5y$ are monomials in x and y. The **degree** of a monomial in several variables is the sum of the degrees on its individual variables. For example, $3xy$ has degree 2, and $-\sqrt{2}x^5y$ has degree 6. As in the case of one variable, a **polynomial** in several variables is a finite sum of monomials.

---

*The phrase "monomial *in* a variable x" may be confusing at first, but it is the accepted terminology. A better phrase would be "monomial *with* x as variable."

Example 4.3 (a) Let x and y denote the lengths of the sides of a rectangle. Then the area of the rectangle is given by xy, which is a monomial (and a polynomial) in x and y.

(b) $u^2 - uv + v^2$ is a polynomial in u and v. ◼

A sum of two monomials (such as $x^2 + y^2$) is called a **binomial.** A sum of three monomials (such as $u^2 - uv + v^2$) is called a **trinomial.**

B. OPERATIONS WITH POLYNOMIALS

Since the variables in polynomials represent real numbers, we can add, subtract, multiply, and divide polynomials by using the properties of real numbers from Section 1 (commutativity, associativity, and so on). Addition, subtraction, and multiplication will be discussed here, and division will be discussed in Section 18. Sums and differences of polynomials can be simplified by combining like terms (monomials of equal degree) as in the next two examples.

Example 4.4 Add $2x^2 - 3x + 6$ and $x^2 + x - 4$.

$$\begin{array}{r} 2x^2 - 3x + 6 \\ + \; x^2 + \; x - 4 \\ \hline 3x^2 - 2x + 2 \end{array}$$ ◼

We usually carry out such operations without writing one polynomial below the other. Also, the intermediate steps in examples like the following can often be done mentally.

Example 4.5 (a) $(3x - y + 4) + (2x + 2y - 3) = (3 + 2)x + (-1 + 2)y + (4 - 3)$
$= 5x + y + 1$

(b) $(x^2 + 4xy + y^2) - (2x^2 - 3xy + y^2)$
$= (1 - 2)x^2 + (4 + 3)xy + (1 - 1)y^2 = -x^2 + 7xy + 0y^2$
$= -x^2 + 7xy$ ◼

Example 4.6 To multiply monomials, multiply their coefficients and variables separately.

(a) $(2x)(x^3) = 2x^4$

(b) $\frac{1}{4}(8x^6) = 2x^6$

(c) $(3xy^2z)(-5x^2y^2) = [3 \cdot (-5)](x \cdot x^2)(y^2 \cdot y^2)z = -15x^3y^4z$ ◼

Example 4.7 To multiply a monomial times a polynomial with more than one term, multiply the monomial times each term of the polynomial and then simplify. This amounts to using one of the distributive laws (Section 1).

(a) $2x(5x + 4) = 2x(5x) + 2x(4) = 10x^2 + 8x$

(b) $uv(u^2 - uv + v^2) = u^3v - u^2v^2 + uv^3$ ◼

The next two examples show two ways to compute more complicated products. By the distributive laws (Section 1B), we must multiply each term in one of the polynomials by each term in the other polynomial. Then we simplify.

Example 4.8 To compute $(3x + 2)(x^2 - 4x + 1)$, we must compute $3x(x^2 - 4x + 1) + 2(x^2 - 4x + 1)$. The work can be arranged as follows (although the shorter method in the next example is usually better).

$$
\begin{array}{rl}
& x^2 - 4x + 1 \\
\times & 3x + 2 \\
\hline
3x^3 - 12x^2 + 3x & = 3x(x^2 - 4x + 1) \\
2x^2 - 8x + 2 & = \underline{2(x^2 - 4x + 1)} \text{ add} \\
\hline
3x^3 - 10x^2 - 5x + 2 & = (3x + 2)(x^2 - 4x + 1)
\end{array}
$$

This shows what each line to the left represents, and can be omitted.

Example 4.9 In this example, we multiply the second polynomial first by x, then by $-y^2$, and then add and simplify.

$$(x - y^2)(x^2 + xy^2 + y^4) = x(x^2 + xy^2 + y^4) - y^2(x^2 + xy^2 + y^4)$$
$$= x^3 + x^2y^2 + xy^4 - x^2y^2 - xy^4 - y^6$$
$$= x^3 - y^6$$

C. SOME SPECIAL PRODUCTS

Any product of two binomials has the following form.

$$(p + q)(r + s) = p(r + s) + q(r + s) = pr + ps + qr + qs \quad (4.1)$$

The four products on the right can be visualized and remembered as follows.

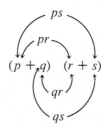

In any special case we simply make appropriate substitutions for p, q, r, and s. For example, with

$$p = u, q = v, r = u, s = -v,$$

we obtain

$$(u + v)(u - v) = u^2 - uv + vu - v^2 = u^2 - v^2.$$

This justifies the first of the three products that follow; the others can be justified in the same way. All three products should be memorized.

Difference of two squares

$$(u + v)(u - v) = u^2 - v^2 \quad (4.2)$$

Square of a binomial sum

$$(u + v)^2 = u^2 + 2uv + v^2 \quad (4.3)$$

Square of a binomial difference

$$(u - v)^2 = u^2 - 2uv + v^2 \qquad (4.4)$$

Example 4.10 (a) $(x + 2)(x - 2) = x^2 - 2^2 = x^2 - 4$

(b) $(t + 3)^2 = t^2 + 2(3)t + 3^2 = t^2 + 6t + 9$

(c) $(5y - \sqrt{2})^2 = (5y)^2 - 2(5y)(\sqrt{2}) + (\sqrt{2})^2 = 25y^2 - 10\sqrt{2}y + 2$ ■

The coefficients of the product $(ax + by)(cx + dy)$ are indicated by the following arrows.

$(ax + by)(cx + dy)$ ac is the coefficient of x^2

$(ax + by)(cx + dy)$ $ad + bc$ is the coefficient of xy

$(ax + by)(cx + dy)$ bd is the coefficient of y^2

Thus

$$(ax + by)(cx + dy) = acx^2 + (ad + bc)xy + bdy^2.$$

Example 4.11 (a) $(x + 2y)(3x + 4y) = 1 \cdot 3x^2 + (1 \cdot 4 + 2 \cdot 3)xy + 2 \cdot 4y^2 = 3x^2 + 10xy + 8y^2$

(b) $(u - 2v)(3u + 4v) = 3u^2 + (4 - 6)uv - 8v^2 = 3u^2 - 2uv - 8v^2$

(c) $(0.1m + 0.3n)(0.2m + 0.4n) = 0.02m^2 + (0.04 + 0.06)mn + 0.12n^2 = 0.02m^2 + 0.1mn + 0.12n^2$ ■

D. INTRODUCTION TO FACTORING

To **factor** a polynomial is to write it as a product of other polynomials; these other polynomials are then called **factors** of the original polynomial. For example, $x^2 - 2x = x(x - 2)$, so that x and $x - 2$ are factors of $x^2 - 2x$. Factoring is essential in simplifying algebraic expressions and in solving equations. In this section, we consider the problem of factoring **quadratics,** that is, second-degree polynomials. In the next section we consider general factoring.

Example 4.12 A second-degree polynomial with one variable and no constant term is easy to factor. By a distributive law, the variable itself is a factor.

(a) $x^2 - 2x = x(x - 2)$

(b) $3t^2 + 15t = 3t(t + 5)$ ■

Notice that $x^2 - 2x$ could also be factored as $2x(\frac{1}{2}x - 1)$. However, it is better to avoid fractions if possible. On the other hand, integral common factors, such as 3 in Example 4.12(b), generally should be removed from each term.

The next simplest cases are those that fit one of the standard forms in Equations (4.2)–(4.4). To use these equations for factoring, read each equation from right to left rather than from left to right. We can extend the applicability of these forms by substituting freely for u and v. It is the *form* that's important, not the specific variables or numbers. In Equation (4.2), for example, the form is revealed by the name, a *difference of two squares*.

Example 4.13 Factor.

(a) $x^2 - 9 = (x + 3)(x - 3)$ [Equation (4.2) with $u = x$, $v = 3$.]
(b) $y^2 + 10y + 25 = (y + 5)^2$ [Equation (4.3) with $u = y$, $v = 5$.]
(c) $z^2 - 14z + 49 = (z - 7)^2$ [Equation (4.4) with $u = z$, $v = 7$.] ■

The next example shows still more variety in the possibilities from Equations (4.2)–(4.4).

Example 4.14 Factor.

(a) $4m^2 - 9n^2 = (2m)^2 - (3n)^2 = (2m + 3n)(2m - 3n)$
(b) $9t^2 + 6t + 1 = (3t)^2 + 2(1)(3t) + 1^2 = (3t + 1)^2$
(c) $0.01x^2 - 0.2xy + y^2 = (0.1x)^2 - 2(0.1x)y + y^2 = (0.1x - y)^2$ ■

The most general form for a factored second-degree polynomial in x is as follows.

$$acx^2 + (ad + bc)x + bd = (ax + b)(cx + d) \qquad (4.5)$$

The choices $a = 1$ and $c = 1$ give the following important special case.

$$x^2 + (b + d)x + bd = (x + b)(x + d) \qquad (4.6)$$

Example 4.15 Factor $x^2 + x - 2$.

Solution Comparing Equation (4.6), we see that we want factors of the form $x + b$ and $x + d$ with

$$b + d = 1 \text{ and } bd = -2.$$

The appropriate choices are $b = 2$ and $d = -1$. That is,

$$x^2 + x - 2 = (x + 2)(x - 1).$$ ■

Example 4.16 Factor $3x^2 + 7x + 2$.

Solution Comparing Equation (4.5), we want

$$3x^2 + 7x + 2 = (ax + b)(cx + d).$$

Thus

$$ac = 3 \text{ and } bd = 2.$$

The choices $a = 3$, $c = 1$, $b = 2$, and $d = 1$ will not work:

$$(3x + 2)(x + 1) = 3x^2 + 5x + 2.$$

(The coefficient on x must be 7, not 5.) But the choices $a = 3$, $c = 1$, $b = 1$, and $d = 2$ will work:

$$(3x + 1)(x + 2) = 3x^2 + 7x + 2.$$ ■

Example 4.17 To factor $12t^2 - t - 6$, we need

$$12t^2 - t - 6 = (at + b)(ct + d).$$

In particular, we must have $ac = 12$ and $bd = -6$. The first condition can be met in several ways: $12 = 12 \cdot 1 = 6 \cdot 2 = 4 \cdot 3 = 3 \cdot 4 = 2 \cdot 6 = 1 \cdot 12$. (Both factors of 12 could also be negative, but it is sufficient to consider positive factors here.) The condition $-6 = bd$ can be met with b and d chosen appropriately form $\pm 1, \pm 2, \pm 3$, and ± 6. Trial and error leads to

$$12t^2 - t - 6 = (4t - 3)(3t + 2). \qquad \blacksquare$$

EXERCISES FOR SECTION 4

Carry out the indicated operations and simplify.

1. $(x^2 - x + 3) + (3x^2 - x - 5)$
2. $(y^3 + 2y - 5) - (y^3 + y^2 - 1)$
3. $2(t^2 + 1) - (t^2 - 1)$
4. $2(z^2 - 1) + (z + 4) - (z^2 - z - 1)$
5. $3(x^3 - x) - (x^3 - x^2 + x) + (x^2 + 2x - 1)$
6. $(v^2 - v + 1) - (2v^2 - v) + 3(v^2 + v - 1)$
7. $(0.5u^2v)(-2uv^3)$
8. $(-5a^2b^3)(-7ab^2)$
9. $(-\frac{1}{3}xyz)(12xz^2)$
10. $\frac{1}{2}x(2x^2 - 6x + 1)$
11. $3u^2(\frac{1}{3}u^2 - \frac{1}{2}u)$
12. $\frac{1}{2}x^4(x - \frac{1}{3})$
13. $(3x - 1)(x^2 + x - 2)$
14. $(t + 2)(t^2 - 3t + 1)$
15. $(4y - 1)(y^2 - y - 3)$
16. $2t + 3[t - 5(t + 1)]$
17. $0.5x - 2[0.3x - 0.1(x - 2)]$
18. $\frac{2}{3}[3(x - \frac{1}{2}) - 2(\frac{1}{3} + x)]$
19. $xy + y[2(x - 1) - (x + 3)]$
20. $3u[v + \frac{5}{2}(1 - v)] - v[3u + \frac{1}{2}(1 - u)]$
21. $m(1 + n + n^2) - mn(1 + m + n) - n(2 - m - m^2)$
22. $1.5[a + b(u + v)] - 0.5[a + b(u - v)]$
23. $\frac{1}{2}(c - d)m - \frac{1}{2}(c + d)m$
24. $0.2[u - 0.6(t - 1)] + 0.4[t - 0.3(u - 1]$

Compute each product and simplify.

25. $(x + 1)(x + 2)(x + 3)$
26. $(y - 1)(y^3 + y^2 + y + 1)$
27. $(z + 1)(z^3 - z^2 + z - 1)$
28. $(x + y)(2x + y)$
29. $(m - n)(2m + n)$
30. $(3u + v)(u + 3v)$
31. $(0.1y + 0.2z)(0.3y - 0.4z)$
32. $(1.5a - 0.5b)(a + b)$
33. $(0.1x + 0.02y)(0.01x + 0.2y)$
34. $(\frac{1}{2}u + \frac{1}{3}v)(\frac{1}{2}u - \frac{1}{3}v)$
35. $(y - \sqrt{2})^2$
36. $(m - \frac{1}{4}n)^2$
37. $(x + \sqrt{5})^2$
38. $(x + 1)(x - 1)$
39. $(z + \sqrt{3})(z - \sqrt{3})$
40. $(t - c)^2$
41. $(u + \frac{1}{2})^2$
42. $(a + 2b)^2$
43. $(x^{1/2} + y^{1/2})(x^{1/2} - y^{1/2})$
44. $(\sqrt{2a} - \sqrt{b})(\sqrt{2a} + \sqrt{b})$
45. $(5^{1/2} + 3^{1/2})(5^{1/2} - 3^{1/2})$
46. $(a^m - b^n)(a^m + b^n)$
47. $(x^p + x^q)^2$
48. $(y^r - 2)(y^r + 5)$

Factor, using Equations (4.2)–(4.4).

49. $x^2 - 25$
50. $u^2 + 12u + 36$
51. $x^2 - 2x + 1$
52. $t^2 - 6t + 9$
53. $x^2 - 10x + 25$
54. $m^2 + 14m + 49$

55. $y^2 + 8y + 16$

56. $t^2 - 3$

57. $25w^2 + 10w + 1$

58. $4x^2 + 4x + 1$

59. $v^2 + v + \frac{1}{4}$

60. $2y^2 + 2\sqrt{2}y + 1$

61. $u^2 + 2\sqrt{3}u + 3$

62. $9y^2 - 6y + 1$

63. $x^2 - \frac{2}{3}x + \frac{1}{9}$

64. $9z^2 - 0.01$

65. $0.04a^2 - 0.2ab + 0.25b^2$

66. $w^2 + 1.4w + 0.49$

Factor.

67. $2y^2 + y - 1$

68. $5x^2 + 6x + 1$

69. $3w^2 + 4w + 1$

70. $9m^2 + 12m + 4$

71. $4y^2 - 2y - 6$

72. $4t^2 - 20t + 25$

73. $6v^2 + 5v - 6$

74. $4p^2 - 12p + 9$

75. $4n^2 - 13n + 3$

An ordered triple (a,b,c) *of positive integers is called a* **Pythagorean triple** *if* $a^2 + b^2 = c^2$. *For example,* (3, 4, 5) *is a Pythagorean triple because* $3^2 + 4^2 = 5^2$. *By the Pythagorean Theorem, the Pythagorean triples are the triples of positive integers that can serve as the lengths of the sides of a right triangle* (*relative to a common unit of length*).

76. Prove that if (a,b,c) is a Pythagorean triple and k is a positive integer, then (ka, kb, kc) is a Pythagorean triple. Using $(a,b,c) = (3,4,5)$ and different values for k, find four other Pythagorean triples.

77. Prove that if m and n are positive integers with $m > n$, and if

$$a = m^2 - n^2$$
$$b = 2mn$$
$$c = m^2 + n^2,$$

then (a,b,c) is a Pythagorean triple.

78. An ordered quadruple (x,y,z,w) of positive integers such that $x^2 + y^2 + z^2 = w^2$ is called a **Pythagorean quadruple.**

(a) Verify that $(3,4,12,13)$ is a Pythagorean quadruple.

(b) Prove that if (x,y,a) and (a,z,w) are both Pythagorean triples, then (x,y,z,w) is a Pythagorean quadruple. [For example, the triples $(3,4,5)$ and $(5,12,13)$ produce the quadruple $(3,4,12,13)$.]

79. Use $m = 19$ and $n = 1$ in Exercise 77 to find a Pythagorean triple of the form $(a,38,c)$. Check that $a^2 + 38^2 = c^2$ for the triple. $\boxed{C}$

80. Use Exercise 77 to find three Pythagorean triples of the form $(a,100,c)$. [Suggestion: Choose m and n such that $2mn = 100$, that is, $mn = 50$.] $\boxed{C}$

81. Use Exercise 78(b) and the Pythagorean triples $(9,12,15)$ and $(15,112,113)$ to produce a Pythagorean quadruple. Check that your answer is indeed a Pythagorean quadruple. $\boxed{C}$

SECTION 5

Factoring

A. FACTORING QUADRATICS: REVIEW

The following three formulas for factoring quadratics (second-degree polynomials) were illustrated in Section 4.

$$u^2 - v^2 = (u + v)(u - v) \tag{5.1}$$

$$u^2 + 2uv + v^2 = (u + v)^2 \tag{5.2}$$

$$u^2 - 2uv + v^2 = (u - v)^2 \tag{5.3}$$

The next example will emphasize that squares can appear in disguise. For instance,

$$7 = (\sqrt{7})^2, \quad x^4 = (x^2)^2, \quad \text{and} \quad a^{2n} = (a^n)^2.$$

Example 5.1 Factor.

(a) $4x^2 - 7 = (2x)^2 - (\sqrt{7})^2 = (2x + \sqrt{7})(2x - \sqrt{7})$

(b) $x^4 + 2x^2y^2 + y^4 = (x^2)^2 + 2x^2y^2 + (y^2)^2 = (x^2 + y^2)^2$

(c) $a^{2n} - 2a^n + 1 = (a^n)^2 - 2a^n + 1^2 = (a^n - 1)^2$ ◼

Notice that in Example 5.1(a) we used the irrational coefficient $\sqrt{7}$ in two of the factors. That brings out the point that the coefficients in our polynomials can be any real numbers. If only integral coefficients were allowed, then $4x^2 - 7$ could not be factored. Although any real coefficients will be allowed, most of the polynomials in our examples and exercises can be factored with integral coefficients.

B. CUBES

Here are the first two of four special products involving cubes (third powers). The exercises ask you to verify that these are correct. Make a special note of the signs on both sides of these equations.

Sum of two cubes

$$u^3 + v^3 = (u + v)(u^2 - uv + v^2) \tag{5.4}$$

Difference of two cubes

$$u^3 - v^3 = (u - v)(u^2 + uv + v^2) \tag{5.5}$$

Example 5.2 Factor.

(a) $u^3 + 1$ (b) $a^3 - 8b^3$ (c) $x^{3n} - y^{3n}$

Solution (a) Use Equation (5.4) with 1 in place of v.

$$u^3 + 1 = u^3 + 1^3 = (u + 1)(u^2 - u + 1)$$

(b) Use Equation (5.5) with a in place of u, and $2b$ in place of v.

$$a^3 - 8b^3 = a^3 - (2b)^3$$
$$= (a - 2b)[a^2 + a(2b) + (2b)^2]$$
$$= (a - 2b)(a^2 + 2ab + 4b^2)$$

(c) Use Equation (5.5) with x^n in place of u, and y^n in place of v.

$$x^{3n} - y^{3n} = (x^n)^3 - (y^n)^3$$
$$= (x^n - y^n)[(x^n)^2 + x^ny^n + (y^n)^2]$$
$$= (x^n - y^n)(x^{2n} + x^ny^n + y^{2n})$$ ◼

Equations (5.4) and (5.5) concern sums and differences of cubes; the next equations concern cubes of sums and differences. Contrast $u^3 + v^3$ with $(u + v)^3$, for example. Equations (5.4)–(5.7) should all be memorized.

| Cube of a binomial sum | $(u + v)^3 = u^3 + 3u^2v + 3uv^2 + v^3$ | (5.6) |

| Cube of a binomial difference | $(u - v)^3 = u^3 - 3u^2v + 3uv^2 - v^3$ | (5.7) |

Although the preceding two identities are included in this section on factoring, we are more likely to need them for expanding each left side than for factoring each right side. The next two examples illustrate both uses.

Example 5.3 Expand and simplify.

(a) $(a + 2b)^3 = a^3 + 3a^2(2b) + 3a(2b)^2 + (2b)^3$
$$= a^3 + 6a^2b + 12ab^2 + 8b^3$$

(b) $(u - \tfrac{1}{2}t)^3 = u^3 - 3u^2(\tfrac{1}{2}t) + 3u(\tfrac{1}{2}t)^2 - (\tfrac{1}{2}t)^3$
$$= u^3 - \tfrac{3}{2}u^2t + \tfrac{3}{4}ut^2 - \tfrac{1}{8}t^3$$

Example 5.4 Factor.

(a) $m^6 + 3m^4n^2 + 3m^2n^4 + n^6$
$$= (m^2)^3 + 3(m^2)^2(n^2) + 3(m^2)(n^2)^2 + (n^2)^3$$
$$= (m^2 + n^2)^3$$

(b) $8 - 12z + 6z^2 - z^3 = 2^3 - 3 \cdot 2^2z + 3 \cdot 2z^2 - z^3$
$$= (2 - z)^3$$

C. COMMON FACTORS. REGROUPING

The first step in factoring should be to factor out all *common factors*. These include the highest power of each variable that appears in every term. If the coefficients are all integers, also remove the greatest integral common factor.

Example 5.5 Factor.

(a) $10x^3 - 15x^2 = 5x^2(2x - 3)$

(b) $2at^2 + 4at + 2a = 2a(t^2 + 2t + 1) = 2a(t + 1)^2$

(c) $q^4 + 8q = q(q^3 + 8) = q(q + 2)(q^2 - 2q + 4)$

The factor $q^2 - 2q + 4$ in (c) cannot be factored; it is *irreducible,* in the sense defined in the next subsection.

Some expressions can be factored more easily after the terms have been rearranged or regrouped.

Example 5.6 Factor.

(a) $2x + 2y + ax + ay = 2(x + y) + a(x + y)$
$$= (2 + a)(x + y)$$

(b) $a^2 - 2ab + b^2 - 1 = (a^2 - 2ab + b^2) - 1$
$$= (a - b)^2 - 1^2 \quad [\text{difference of two squares}]$$
$$= [(a - b) + 1][(a - b) - 1]$$
$$= (a - b + 1)(a - b - 1)$$

(c) $rt - su - ru + st = rt - ru + st - su$
$$= r(t - u) + s(t - u)$$
$$= (r + s)(t - u)$$ ■

D. COMPLETE FACTORIZATION

A polynomial is **reducible** if it can be written as a product of two other polynomials that are both of positive degree. A polynomial that is not reducible is said to be **irreducible.**

Example 5.7 (a) $x^2 + x$ is reducible because $x^2 + x = x(x + 1)$ and both x and $x + 1$ have positive degree.

(b) $y^3 + 0.001$ is reducible because
$$y^3 + 0.001 = (y + 0.1)(y^2 - 0.1y + 0.01).$$

(c) $2x^2 + 2$ is irreducible. Although we can write
$$2x^2 + 2 = 2(x^2 + 1),$$

we cannot write $2x^2 + 2$ as a product of two polynomials of degree one, which would be the case if $2x^2 + 2$ were reducible. ■

Any first-degree polynomial is irreducible. Example 5.7 shows that some second-degree polynomials are irreducible and some are not. Each polynomial of degree greater than two is reducible, but it may not be easy to find factors.

We say that a polynomial has been **factored completely** when it has been written as a product of irreducible polynomials. Although factoring can be tricky, many polynomials can be handled by applying the following suggestions. (Practice helps, too.)

- Factor out all common factors.
- Look for the standard forms in Equations (5.1), (5.2), and (5.3).
- If the polynomial is quadratic and (5.1), (5.2), and (5.3) do not apply, try to factor it as a general quadratic (Section 4D).
- Look for the standard forms in Equations (5.4) and (5.5), and then in (5.6) and (5.7).
- Attempt to regroup the terms so that one of the previous suggestions applies.

Example 5.8 Factor completely.

(a) $x^3 - 6x^2 + 9x = x(x^2 - 6x + 9) = x(x - 3)^2$

(b) $15 - 3t^2 = 3(5 - t^2) = 3(\sqrt{5} + t)(\sqrt{5} - t)$

(c) $x^4 - 1 = (x^2)^2 - 1 = (x^2 + 1)(x^2 - 1)$
$$= (x^2 + 1)(x + 1)(x - 1)$$

(d) $y^6 + y^5 + \frac{1}{4}y^4 = y^4(y^2 + y + \frac{1}{4}) = y^4(y + \frac{1}{2})^2$

(e) $ac^2 - bd^2 - ad^2 + bc^2 = ac^2 - ad^2 + bc^2 - bd^2$

$$= a(c^2 - d^2) + b(c^2 - d^2)$$
$$= (a + b)(c^2 - d^2)$$
$$= (a + b)(c + d)(c - d)$$

E. LEAST COMMON MULTIPLES

One polynomial is said to be a **multiple** of another polynomial if the second is a factor of the first. For example, $x^2 + x$ is a multiple of $x + 1$ because $x^2 + x = x(x + 1)$. A **least common multiple (l.c.m.)** of a set of polynomials is a polynomial of smallest degree that is a multiple of each polynomial in the set.

The l.c.m.'s of some sets of polynomials can be computed by inspection. Here is a systematic scheme (followed by an example) for more difficult cases.

Step I. Factor each polynomial in the set into powers of distinct irreducible polynomials.

Step II. Form the product of the irreducible polynomials that appear in Step I, and on each irreducible factor place the greatest exponent that appears on that factor in any of the factorizations in Step I. This product is an l.c.m. of the original set.

Example 5.9 Find an l.c.m. of $x^2 - y^2$, $x^2 + 2xy + y^2$, and $x^2 - xy$.

Solution First, factor each polynomial completely.

$$x^2 - y^2 = (x + y)(x - y)$$
$$x^2 + 2xy + y^2 = (x + y)^2$$
$$x^2 - xy = x(x - y)$$

Now, form the product of the irreducible polynomials that appear in these factorizations, and on each irreducible factor place the greatest exponent that appears on that factor in any of the above factorizations. In this case, we use x, $(x + y)^2$, and $x - y$. That is, an l.c.m. is $x(x + y)^2(x - y)$.

EXERCISES FOR SECTION 5

Factor completely.

1. $x^2 + 12x + 36$
2. $y^2 - 6y + 9$
3. $z^2 - 1$
4. $y^3 - y$
5. $z^4 + 2z^3 + z^2$
6. $x^3 - 4x^2 + 4x$
7. $4z^2 - 4z + 1$
8. $9x^2 - 25$
9. $9y^2 + 6y + 1$
10. $u^3 - 27$
11. $v^3 - 1$
12. $w^3 - 8$
13. $v^4 - 4v^3 + 3v^2$
14. $2w - 3w^2 + w^3$
15. $u^5 + 7u^4 + 6u^3$
16. $w^3 + u^3$
17. $u^3 - 1000v^3$
18. $8v^3 + w^3$

19. $r^4 - 125r$
20. $27s^5 + s^2$
21. $64t - t^4$
22. $s^4 - t^4$
23. $t^4 - 2t^2 + 1$
24. $1 - r^4$
25. $t^3 - 0.001$
26. $0.027 + r^3$
27. $s^3 + 0.008$
28. $a^{2n} + 2a^n + 1$
29. $b^{2n} - 1$
30. $4 + 4c^n + c^{2n}$
31. $b^3 + 3b^2 + 3b + 1$
32. $c^3 - 3c^2 + 3c - 1$
33. $a^3 - 6a^2 + 12a - 8$

34. $c^3 - 3c^2a + 3ca^2 - a^3$

35. $a^3 + 9a^2b + 27ab^2 + 27b^3$

36. $b^3 + 0.3b^2 + 0.03b + 0.001$

37. $x^3y - 10x^2y^2 + 25xy^3$

38. $4u^3v - 9uv^3$

39. $r^3s - 6r^2s^2 + 9rs^3$

40. $m^3 + \frac{3}{2}m^2 + \frac{3}{4}m + \frac{1}{8}$

41. $\frac{1}{27} - \frac{1}{3}n + n^2 - n^3$

42. $p^3 + \frac{3}{2}p^2q + \frac{3}{4}pq^2 + \frac{1}{8}q^3$

43. $ax + ay - bx - by$

44. $ax - ay + bx - by$

45. $ax + by + ay + bx$

46. $ax^3 - by^3 - bx^3 + ay^3$

47. $cu^2 - dv^2 + du^2 - cv^2$

48. $ap^3 + aq^3 - bp^3 - bq^3$

49. $a^6 - b^6$

50. $c^6 + 8d^6$

51. $x^8 - y^8$

52. $x^2 - y^2 - 4y - 4$

53. $4u^2 - v^2 + 6v - 9$

54. $a^2 - 4ab + 4b^2 - c^2$

55. $4x^{2n+2} - 5x^{n+2} + x^2$

56. $y^{4n}z^n - y^nz^{4n}$

57. $x^nz^{2n} + y^nw^{2n} + x^nw^{2n} + y^nz^{2n}$

58. $a^{4n} - b^{6n}$

59. $x^{2p+2}y^2 + 2x^{p+2}y^{q+2} + x^2y^{2q+2}$

60. $u^9v - 6u^5v^4 + 12u^3v^7 - 8uv^{10}$

Expand and simplify.

61. $(t + 1)^3$

62. $(u - 2)^3$

63. $(v + 5)^3$

64. $(y - 0.5)^3$

65. $(x + \frac{1}{2})^3$

66. $(a^2 - b^2)^3$

67. $(\frac{1}{2}x^2 + 1)^3$

68. $(z^2 + 0.1)^3$

69. $(y - \frac{2}{3})^3$

70. $(a^{1/3} - b^{1/3})(a^{2/3} + a^{1/3}b^{1/3} + b^{2/3})$

71. $(c^{1/3} + d^{1/3})(c^{2/3} - c^{1/3}d^{1/3} + d^{2/3})$

72. $(x^{1/4} + y^{1/4})(x^{1/4} - y^{1/4})(x^{1/2} + y^{1/2})$

Expand with the help of a calculator. Round each coefficient to three significant figures.

73. $(1.42a - 1.35b)^3$ C

74. $(314x + 159y)^3$ C

75. $(69.3u + 14.7v^2)^3$ C

76. Verify Equation (5.4) by showing that when the two factors on the right side are multiplied, the result simplifies to the left side.

77. Same as Exercise 76, with Equation (5.5) in place of Equation (5.4).

78. Verify Equation (5.6) by multiplying the right side of $(u + v)^2 = u^2 + 2uv + v^2$ by $u + v$ and showing that the result simplifies to the right side of Equation (5.6).

79. Verify Equation (5.7) by multiplying the right side of $(u - v)^2 = u^2 - 2uv + v^2$ by $u - v$ and showing that the result simplifies to the right side of Equation (5.7).

80. Verify that $(u + v)^4 = u^4 + 4u^3v + 6u^2v^2 + 4uv^3 + v^4$.

81. Verify that $(u - v)^4 = u^4 - 4u^3v + 6u^2v^2 - 4uv^3 + v^4$.

Find an l.c.m. for each set of polynomials. Write the answers in factored form.

82. $x^2, x^2 + x, x - 1$

83. $x^3 - 1, x + 1$

84. $x^2 + 9x + 20, (x + 5)^2$

85. $a^2 - b^2, (a - b)^2$

86. $a^3 - ab^2, a^2b^2, b^2 - ab$

87. $a - b, a^2 + b^2, a^4 - b^4$

SECTION **6**

Rational Expressions

A. INTRODUCTION

Any polynomial, or result obtained from combining polynomials through a finite number of additions, subtractions, multiplications, divisions, or extractions of roots, is called an **algebraic expression.** An algebraic expression is called a

rational expression if it can be written as a fraction in which both the numerator and nonzero denominator are polynomials. (Recall that a *number* is *rational* if it can be written as a fraction in which both the numerator and nonzero denominator are integers.)

Example 6.1 Each of the following is a rational expression.

(a) $\dfrac{x^3 - 3x + 2}{x - 2}$

(b) $\dfrac{x^2y^2}{x^2 + y^2}$

(c) $\dfrac{x}{2} + \dfrac{2}{x} = \dfrac{x^2 + 4}{2x}$

∎

Also, any polynomial is a rational expression. For example, $x^2 + 1 = (x^2 + 1)/1$ is a rational expression, just as $3 = \frac{3}{1}$ is a rational number. The expression $\sqrt{x} + 1$ is not rational. In later sections we'll meet other examples of expressions that are not rational.

This section is concerned with simplifying rational and other fractional expressions. In general, any expression that results from combining polynomials through addition, subtraction, multiplication, or division is a rational expression; by simplification, if necessary, it can be written as a single fraction in which both the numerator and denominator are polynomials.

Since the variables in rational expressions represent real numbers, we can manipulate rational expressions just like ordinary fractions. Remember, however, that a fraction is undefined when its denominator is zero. Therefore, we must always exclude values that make a denominator zero. When we work with $1/(x^2 - 4)$, for example, we must exclude both $x = 2$ and $x = -2$. Such restrictions will always be assumed, whether explicitly stated or not.

Following is a summary of the basic laws about fractions, which were stated in Section 1. The first law concerns equality.

$$\frac{a}{b} = \frac{c}{d} \quad \text{iff} \quad ad = bc. \tag{6.1}$$

Next is the *fundamental principle of fractions*.

$$\text{If } c \neq 0, \text{ then } \frac{ac}{bc} = \frac{a}{b}. \tag{6.2}$$

Property (6.1) implies (6.2) because $acb = bca$.

Example 6.2 Simplify.

(a) $\dfrac{3x^2y^2}{9y^2z} = \dfrac{x^2}{3z}$

(b) $\dfrac{4 - 2x}{x - 2} = \dfrac{-2(x - 2)}{x - 2} = -2$

(c) $\dfrac{x^2 - 1}{x - 1} = \dfrac{(x + 1)(x - 1)}{x - 1} = x + 1$

Notice that the last equality is valid only for $x \neq 1$, because if $x = 1$ then $(x^2 - 1)/(x - 1)$ is undefined. A similar remark applies to the other examples. Fractions are multiplied and divided as follows.

$$\frac{a}{b} \cdot \frac{c}{d} = \frac{ac}{bd} \tag{6.3}$$

$$\frac{\dfrac{a}{b}}{\dfrac{c}{d}} = \frac{ad}{bc} \tag{6.4}$$

Example 6.3 Simplify.

(a) $\dfrac{3x}{y} \cdot \dfrac{x}{2z} = \dfrac{3x^2}{2yz}$

(b) $\dfrac{p^2q}{r^2} \cdot \dfrac{r^3}{pq^2} = \dfrac{p^2qr^3}{r^2pq^2} = \dfrac{pr}{q}$

(c) $\dfrac{x^2}{x + 1} \cdot \dfrac{x^2 + 2x + 1}{x} = \dfrac{x^2(x + 1)^2}{(x + 1)x} = x(x + 1)$

Example 6.4 Simplify.

(a) $\dfrac{6x/y^2}{10x^2/y} = \dfrac{6x}{y^2} \cdot \dfrac{y}{10x^2} = \dfrac{3}{5xy}$

(b) $\dfrac{(ab^2)/c}{(a^2b)/c^2} = \dfrac{ab^2}{c} \cdot \dfrac{c^2}{a^2b} = \dfrac{ab^2c^2}{a^2bc} = \dfrac{bc}{a}$

(c) $\dfrac{t/(t + 2)}{t^2/(t^2 - 4)} = \dfrac{t(t^2 - 4)}{(t + 2)t^2} = \dfrac{t(t + 2)(t - 2)}{t^2(t + 2)} = \dfrac{t - 2}{t}$

(d) $\dfrac{(u^3 - v^3)/u^2v^2}{(v - u)/uv} = \dfrac{(u^3 - v^3)uv}{u^2v^2(v - u)}$

$\qquad = \dfrac{(u - v)(u^2 + uv + v^2)uv}{-u^2v^2(u - v)} = \dfrac{-(u^2 + uv + v^2)}{uv}$

Fractions with common (that is, equal) denominators are easy to add or subtract:

$$\frac{a}{d} + \frac{b}{d} = \frac{a+b}{d}, \qquad \frac{a}{d} - \frac{b}{d} = \frac{a-b}{d}. \tag{6.5}$$

Example 6.5 Simplify.

(a) $\dfrac{x-1}{x} + \dfrac{1}{x} = \dfrac{(x-1)+1}{x} = \dfrac{x}{x} = 1$

(b) $\dfrac{u-v}{t} - \dfrac{2u-v}{t} + \dfrac{u}{t} = \dfrac{(u-v)-(2u-v)+u}{t} = \dfrac{0}{t} = 0$ ◾

Example 6.6 In the following two cases we add after converting to common denominators.

(a) $\dfrac{1}{u} + \dfrac{1}{v} = \dfrac{1}{u} \cdot \dfrac{v}{v} + \dfrac{1}{v} \cdot \dfrac{u}{u} = \dfrac{v+u}{uv}$

(b) $\dfrac{a}{x} + \dfrac{b}{y} + \dfrac{c}{z} = \dfrac{a}{x} \cdot \dfrac{yz}{yz} + \dfrac{b}{y} \cdot \dfrac{xz}{xz} + \dfrac{c}{z} \cdot \dfrac{xy}{xy} = \dfrac{ayz + bxz + cxy}{xyz}$ ◾

B. LEAST COMMON DENOMINATORS

The simplest denominator to use when adding two or more fractions or rational expressions is usually a *least common denominator*. A **least common denominator (l.c.d.)** of a set of fractions or rational expressions is a least common multiple (l.c.m., as in Section 5E) of the set of denominators of the fractions or expressions.

Example 6.7 Perform the indicated operations and simplify.

(a) $\dfrac{1}{2xy^2} + \dfrac{1}{6x^2y}$ (b) $\dfrac{x}{x^2 - y^2} - \dfrac{y}{x^2 + 2xy + y^2}$

Solution (a) An l.c.m. of $2xy^2$ and $6x^2y$ is $6x^2y^2$. To use this l.c.m. as a common denominator, we multiply the first fraction by $3x/3x$ and the second by y/y.

$$\frac{1}{2xy^2} + \frac{1}{6x^2y} = \frac{1}{2xy^2} \cdot \frac{3x}{3x} + \frac{1}{6x^2y} \cdot \frac{y}{y}$$

$$= \frac{3x + y}{6x^2y^2}.$$

(b) An l.c.m. of $x^2 - y^2$ and $x^2 + 2xy + y^2$ is $(x+y)^2(x-y)$. Therefore, we have

$$\frac{x}{x^2 - y^2} - \frac{y}{x^2 + 2xy + y^2}$$

$$= \frac{x}{(x+y)(x-y)} - \frac{y}{(x+y)^2} \qquad \text{Factor denominators to determine an l.c.d.}$$

$$= \frac{x}{(x+y)(x-y)} \cdot \frac{x+y}{x+y}$$

$$- \frac{y}{(x + y)^2} \cdot \frac{x - y}{x - y} \qquad \text{Convert to common denominators.}$$

$$= \frac{x(x + y) - y(x - y)}{(x - y)(x + y)^2} \qquad \text{Subtract.}$$

$$= \frac{x^2 + xy - xy + y^2}{(x - y)(x + y)^2} \qquad \text{Simplify.}$$

$$= \frac{x^2 + y^2}{(x - y)(x + y)^2}. \qquad \blacksquare$$

Although a least common denominator is usually the simplest denominator to use when adding rational expressions, the product of the denominators (which may or may not be a *least* common denominator) can also be used. The main idea is to use some common denominator and then simplify in the end.

C. COMPLEX FRACTIONS

A fraction or rational expression in which the numerator or denominator (or both) contain fractions or rational expressions is called a **complex fraction.** Complex fractions can be simplified by first simplifying the numerator and denominator separately, and then inverting the denominator and multiplying [Equation (6.4)].

Example 6.8 Simplify.

(a) $$\frac{4 + \dfrac{1}{2}}{\dfrac{1}{3} - 1} = \frac{\dfrac{8 + 1}{2}}{\dfrac{1 - 3}{3}} = \frac{\dfrac{9}{2}}{\dfrac{-2}{3}} = \frac{9}{2} \cdot \frac{3}{-2} = -\frac{27}{4}$$

(b) $$1 - \cfrac{1}{1 - \cfrac{1}{2}} = 1 - \cfrac{1}{\cfrac{2 - 1}{2}} = 1 - \cfrac{1}{\cfrac{1}{2}} = 1 - 2 = -1 \qquad \blacksquare$$

Example 6.9 Simplify.

$$\frac{x - \dfrac{x}{x + 2}}{x + \dfrac{x}{x + 1}} = \frac{\dfrac{x(x + 2)}{1(x + 2)} - \dfrac{x}{x + 2}}{\dfrac{x(x + 1)}{1(x + 1)} + \dfrac{x}{x + 1}}$$

$$= \frac{\dfrac{x(x + 2) - x}{x + 2}}{\dfrac{x(x + 1) + x}{x + 1}} = \frac{\dfrac{x^2 + x}{x + 2}}{\dfrac{x^2 + 2x}{x + 1}}$$

$$= \frac{x^2 + x}{x + 2} \cdot \frac{x + 1}{x^2 + 2x}$$

$$= \frac{x(x + 1)^2}{x(x + 2)^2}$$

$$= \frac{(x + 1)^2}{(x + 2)^2}$$

Example 6.10 Simplify.

$$\frac{\dfrac{1}{x} - \dfrac{1}{y}}{x - y} = \frac{\dfrac{1}{x} \cdot \dfrac{y}{y} - \dfrac{1}{y} \cdot \dfrac{x}{x}}{x - y} = \frac{\dfrac{y - x}{xy}}{\dfrac{x - y}{1}}$$

$$= \frac{y - x}{xy} \cdot \frac{1}{x - y} = \frac{-(x - y)}{xy(x - y)} = -\frac{1}{xy}$$

Example 6.11 Simplify.

$$\frac{(x^2 + 1)^{1/2} \cdot 1 - x \cdot \frac{1}{2}(x^2 + 1)^{-1/2} \cdot 2x}{[(x^2 + 1)^{1/2}]^2}$$

$$= \frac{(x^2 + 1)^{1/2} - \dfrac{x^2}{(x^2 + 1)^{1/2}}}{x^2 + 1} = \frac{\dfrac{(x^2 + 1) - x^2}{(x^2 + 1)^{1/2}}}{x^2 + 1}$$

$$= \frac{1}{(x^2 + 1)^{1/2}} \cdot \frac{1}{(x^2 + 1)} = \frac{1}{(x^2 + 1)^{3/2}}$$

Another method for simplifying complex fractions is to multiply both the numerator and denominator by the l.c.d. of all the fractions in both the numerator and denominator, as in the following examples.

Example 6.12 The l.c.d. of $\frac{5}{2}$, $-\frac{2}{3}$, $\frac{3}{4}$, and $\frac{1}{6}$ is 12, so we multiply by $\frac{12}{12}$ in the following case.

$$\frac{\dfrac{5}{2} - \dfrac{2}{3}}{\dfrac{3}{4} + \dfrac{1}{6}} = \frac{\dfrac{5}{2} - \dfrac{2}{3}}{\dfrac{3}{4} + \dfrac{1}{6}} \cdot \frac{12}{12} = \frac{\left(\dfrac{5}{2}\right)12 - \left(\dfrac{2}{3}\right)12}{\left(\dfrac{3}{4}\right)12 + \left(\dfrac{1}{6}\right)12}$$

$$= \frac{30 - 8}{9 + 2}$$

$$= \frac{22}{11}$$

$$= 2.$$

Example 6.13 An l.c.m. of $x + 1$ and $x - 1$ is $(x + 1)(x - 1)$. Therefore, in the following simplification we multiply by $\dfrac{(x + 1)(x - 1)}{(x + 1)(x - 1)}$.

$$\frac{\dfrac{1}{x + 1} - \dfrac{2}{x - 1}}{\dfrac{3}{x + 1} + \dfrac{4}{x - 1}} = \frac{(x + 1)(x - 1)}{(x + 1)(x - 1)} \cdot \frac{\dfrac{1}{x + 1} - \dfrac{2}{x - 1}}{\dfrac{3}{x + 1} + \dfrac{4}{x - 1}}$$

$$= \frac{(x - 1) - 2(x + 1)}{3(x - 1) + 4(x + 1)} = \frac{-x - 3}{7x + 1}$$

D. APPLICATION: UNITS

Many problems, especially those involving a change of units, can be solved most easily by using the convenient fact that units can be manipulated much like numbers and variables.

Example 6.14 The basic formula for uniform motion problems is

$$\text{distance} = \text{rate} \times \text{time}$$

$$D = RT. \tag{6.6}$$

If D is measured in miles and T is measured in hours, then R will be given in miles per hour, which we write miles/hour. Compare what happens if we replace the symbols in Equation (6.6) by the units with which they are measured:

$$\text{miles} = \frac{\text{miles}}{\text{hour}} \times \text{hours}.$$

The *hours* cancel on the right to leave *miles,* which is what is on the left. Units can always be canceled and multiplied in this way. ∎

More examples will follow these lists of frequently used abbreviations, prefixes, and conversion factors.

ABBREVIATIONS

cm	centimeter	in.	inch	mi	mile
cu	cubic	kg	kilogram	qt	quart
ft	feet	km	kilometer	sec	second
g	gram	l	liter	sq	square
gal	gallon	m	meter	yd	yard
hr	hour				

PREFIXES

Definitions	*Examples*
milli means multiply by 10^{-3}	1 milliliter = 0.001 liter
centi means multiply by 10^{-2}	1 centimeter = 0.01 meter
deci means multiply by 10^{-1}	1 decigram = 0.1 gram
deka means multiply by 10	1 dekaliter = 10 liters
hecto means multiply by 10^2	1 hectometer = 100 meters
kilo means multiply by 10^3	1 kilometer = 1000 meters

CONVERSION FACTORS

Length	*Volume*
1 in. = 2.54 cm	1 qt (liquid, U.S.) = 0.946 l
1 yd = 0.9144 m	4 qt = 1 gal
1 m = 39.37 in.	*Weight*
1 mi = 1760 yd = 5280 ft	
1 km = 0.621 mi	454 g = 1 lb
1 mi = 1.61 km	1 kg = 2.205 lb

Square and cubic factors are also written with exponents. For example, sq ft = ft² and cu cm = cm³.

For problems that involve units, look for conversion factors that will lead *from* the units of quantities that are given *to* the units of quantities that are to be found.

Example 6.15 Convert 440 yards to meters. Treat 440 as an exact number.

Solution To convert yards to meters, we look for a conversion factor with yards in the denominator and meters in the numerator; then yards will cancel, leaving us with meters. The conversion factor connecting yards and meters is 1 yd = 0.9144 m. This can be rewritten as either

$$\frac{1 \text{ yd}}{0.9144 \text{ m}} = 1 \quad \text{or} \quad \frac{0.9144 \text{ m}}{1 \text{ yd}} = 1.$$

In this case we need the second choice:

$$440 \text{ yd} = 440 \text{ yd} \times \frac{0.9144 \text{ m}}{1 \text{ yd}} = 440 \times 0.9144 \text{ m} \approx 402.3 \text{ m}.$$

⦿

This shows that a 440 yard dash is approximately 2⅓ meters longer than a 400 meter dash.

■

Example 6.16 Express 60 mi/hr in ft/sec.

Solution Here we must convert both length and time. To convert length, we need a factor for ft/mi, so that *mi* will cancel and *ft* will remain. From 1 mi = 5280 ft, we can use 5280 ft/mi = 1. Thus

$$60 \frac{\text{mi}}{\text{hr}} = \left(60 \frac{\text{mi}}{\text{hr}} \right) \times \left(5280 \frac{\text{ft}}{\text{mi}} \right) = 60 \times 5280 \frac{\text{ft}}{\text{hr}}.$$

To convert time, we need a factor for hr/sec. From 1 hr = 60 min and 1 min = 60 sec, we have 1 hr = 3600 sec, or (1/3600) hr/sec = 1. Thus

$$60 \frac{\text{mi}}{\text{hr}} = (60 \times 5280) \frac{\text{ft}}{\text{hr}} \times \left(\frac{1}{3600} \right) \frac{\text{hr}}{\text{sec}}$$

$$= \left(60 \times \frac{5280}{3600} \right) \frac{\text{ft}}{\text{sec}}$$

$$= 88 \frac{\text{ft}}{\text{sec}}.$$

■

Many problems that involve units can be solved by forcing the units to turn out right. Here is an example that concerns density.

Example 6.17 The density of gold (at 20°C) is 19.3 g/cm³. How many cubic centimeters are in 100 grams of gold?

Solution We are given a quantity in g and a density in g/cm³, and we need an answer in cm³. To obtain an answer with cm³, working from g and g/cm³, we must divide:

$$g \div (g/cm^3) = g \times (cm^3/g) = cm^3.$$

Therefore, the answer is

$$(100 \text{ g}) \div (19.3 \text{ g/cm}^3) = (100/19.3) \text{ cm}^3 \approx 5.18 \text{ cm}^3.$$

■

EXERCISES FOR SECTION 6

Simplify.

1. $\dfrac{x + 2}{5x + 10}$

2. $\dfrac{6x - 3}{2x - 1}$

3. $\dfrac{8x^2 + 4}{2x^2 + 1}$

4. $\dfrac{y^2 + 2y + 1}{y^2 - 1}$

5. $\dfrac{2y - 6}{y^2 - 6y + 9}$

6. $\dfrac{y^2 + 2y}{y^2 - 4}$

7. $\dfrac{z^3 - 1}{z^2 - 1}$

8. $\dfrac{z^3 + 8}{z + 2}$

9. $\dfrac{(z + 1)^3}{z^3 + 1}$

Perform the indicated operations and simplify.

10. $\dfrac{x + 5}{x} \cdot \dfrac{x^2}{(x + 5)^2}$

11. $\dfrac{4y}{(y + 1)^2} \cdot \dfrac{y + 1}{2y^2}$

12. $\dfrac{x}{y} \cdot \dfrac{xy + y^2}{x^2 + xy}$

13. $\dfrac{a^2/(a - 1)}{a/(a - 1)^2}$

14. $\dfrac{1/b}{2/(b - 1)}$

15. $\dfrac{(c - 1)/(c + 1)}{1/(c + 1)}$

16. $\dfrac{(r^2 - 10r + 25)/r}{r - 5}$

17. $\dfrac{(1 - s^3)/(1 + s)}{(1 - s)/(1 + s^3)}$

18. $\dfrac{(t^2 + t + 1)/(t + 1)}{t^3 - 1}$

19. $\dfrac{(2u + v)/u^2v}{(8u^3 + v^3)/uv^2}$

20. $\dfrac{w^3 - 125}{(w - 5)/w}$

21. $\dfrac{(9v^2 - 4w^2)/(3v - 2w)}{(6v + 4w)/10}$

22. $\dfrac{y - 2}{y} - \dfrac{y + 2}{y}$

23. $\dfrac{2y + 1}{yz} + \dfrac{y - 1}{yz}$

24. $\dfrac{z^3 + 1}{z^2} - \dfrac{z^2 + 1}{z^2}$

25. $\dfrac{1}{a + 1} + \dfrac{a^3}{a + 1}$

26. $\dfrac{2b^2}{b + 1} - \dfrac{2}{b + 1}$

27. $\dfrac{1000}{100 - c^2} - \dfrac{c^3}{100 - c^2}$

28. $\dfrac{1}{x + h} - \dfrac{1}{x}$

29. $\dfrac{1}{(x + h)^2} - \dfrac{1}{x^2}$

30. $\dfrac{1}{(x + h)^3} - \dfrac{1}{x^3}$

31. $\dfrac{1}{a - b} - \dfrac{b}{a^2 - b^2}$

32. $\dfrac{1}{x + y} + \dfrac{y}{x^2 + xy}$

33. $\dfrac{c}{c - d} + \dfrac{d}{d - c}$

34. $\dfrac{2}{x^2 + x} + \dfrac{1}{x - 1} - \dfrac{1}{x + 1}$

35. $\dfrac{1}{c - d} \cdot \dfrac{c}{c^2 - d^2} + \dfrac{1}{2c + 2d}$

36. $\dfrac{1}{a + 5} - \dfrac{5}{a^2 + 10a + 25}$

37. $\dfrac{1 + \dfrac{1}{x - 1}}{1 - \dfrac{1}{x - 1}}$

38. $\dfrac{1 + \dfrac{1}{y}}{y - \dfrac{1}{y}}$

39. $\dfrac{2 + \dfrac{2}{z + 1}}{\dfrac{2z}{z + 1} - 2}$

40. $\dfrac{y + 1}{y - \dfrac{1}{y}}$

41. $\dfrac{z^2 - \dfrac{8}{z}}{z^2 - 4}$

42. $\dfrac{25 + 10y + y^2}{1 + \dfrac{5}{y}}$

43. $\dfrac{z - \dfrac{1}{z^2}}{1 - \dfrac{1}{z^2}}$

44. $\dfrac{x + 2}{\dfrac{x}{x + 1} + \dfrac{2}{x + 1}}$

45. $\dfrac{\dfrac{1}{y-1} - \dfrac{1}{y+1}}{\dfrac{1}{y-1} - \dfrac{1}{y}}$

46. $\dfrac{\dfrac{1}{a} + \dfrac{1}{b}}{a+b}$

47. $\dfrac{\dfrac{2}{c} - \dfrac{2}{d}}{c-d}$

48. $\dfrac{m+n}{\dfrac{1}{m^2} - \dfrac{1}{n^2}}$

49. $\dfrac{\dfrac{1}{x+h} - \dfrac{1}{x}}{h}$

50. $\dfrac{\dfrac{1}{(x+h)^2} - \dfrac{1}{x^2}}{h}$

51. $\dfrac{\dfrac{1}{(x+h)^3} - \dfrac{1}{x^3}}{h}$

52. $\dfrac{x \cdot \frac{1}{2}(x^2+1)^{-1/2} \cdot 2x - (x^2+1)^{1/2} \cdot 1}{x^2}$

53. $\dfrac{(x+1)^{3/2} \cdot 2x - x^2 \cdot \frac{3}{2}(x+1)^{1/2}}{[(x+1)^{3/2}]^2}$

54. $\dfrac{(x^2+1)^{1/3} \cdot 2x - x^2 \cdot \frac{1}{3}(x^2+1)^{-2/3} \cdot 2x}{[(x^2+1)^{1/3}]^2}$

55. $\left(\dfrac{a^{n^2+1}}{a^{1-n}}\right)^{1/(n+1)}$

56. $\left(\dfrac{b^{n-1}}{b^{n-n^3}}\right)^{1/(n^2+n+1)}$

57. $\left(\dfrac{c^{n^2-2}}{c^{n+4}}\right)^{1/(n+2)}$

58. $(6x^{3n} - 4x^{2n}) \div (-2x^n)$

59. $(x^{4n} - 1) \div (x^{2n} + 1)$

60. $(x^{4n}y^n - x^ny^{4n}) \div (x^ny^{3n} - x^{3n}y^n)$

61. Convert 200 acres to square miles given that 1 square mile = 640 acres.

62. Convert 18 drams to tablespoons given that 1 tablespoon = 4 fluid drams.

63. Convert $9\frac{1}{2}$ furlongs to miles given that 1 mile = 8 furlongs.

64. Express 5 cm/sec in m/min.

65. Express 20 ft/min in yd/hr.

66. Express 10 yd/min in m/sec.

67. The density of lead (at 20°C) is 11.34 g/cm³. How many cubic centimeters are in 20 grams of lead? [C]

68. The density of aluminum (at 20°C) is 2.7 g/cm³. How many grams are in 50 cubic centimeters of aluminum? [C]

69. The density of mercury (at 20°C) is 13.55 g/cm³. How many grams are in 20 cubic centimeters of mercury? [C]

70. In 1980 the population density of the U.S. was 64.0 persons per square mile (of land area). The land area was 3,540,023 square miles. What was the population? How many persons were there per square kilometer? [C]

71. In 1980 the population density of the earth was 86 persons per square mile (of land area). How many persons were there per square kilometer? [C]

72. Assume that 1 dollar = 5 francs and that one liter of gasoline costs 2.45 francs. What is the cost in dollars per gallon? [C]

REVIEW EXERCISES FOR CHAPTER I*

1. Which of the following numbers are natural numbers? integers? rational numbers? irrational numbers?

$$20.\overline{1}, \quad 3\pi, \quad 0/\sqrt{2}, \quad \sqrt{5},$$
$$-12/\sqrt{9}, \quad \sqrt{36}, \quad -3.14,$$
$$-\sqrt{2}\,\sqrt{2}$$

2. Define each of the following sets: natural numbers, integers, rational numbers, irrational numbers, and real numbers.

Simplify.

3. $(-8)(-0.025)$

4. $(0.16)(-5.5)$

5. $(-0.048) \div (0.12)$

6. $(-16.4) \div (-41)$

7. $(9.2 - 6.56) \div 4$

8. $(10.1 - 12.25) \div 5$

9. $4[1 - 2(3+4)] - 2(3-6)$

10. $5 - 2[1 - 6(7-9)]$

11. $-2\{5(b-a) - 2[a + (-3a - b)]\}$

12. $a - c - \{-(a-b) + 2(b-c) - [(a-b) - c]\}$

*Answers to all of the review exercises are at the back of the book.

Write each expression as a single fraction and simplify.

13. $\frac{5}{6} - \frac{3}{10}$

14. $\frac{1}{14} - \frac{1}{6}$

15. $-(2 + \frac{2}{3}) \div (-\frac{2}{3})$

16. $(-\frac{1}{5}) \div (\frac{1}{5} - 1)$

17. $\left(1 + \frac{1}{x}\right) \div \left(1 - \frac{1}{x}\right)$

18. $2 + \dfrac{1 - \dfrac{2}{y}}{1 + \dfrac{2}{y}}$

19. $\left(\dfrac{2}{a} - \dfrac{1}{a^2}\right) \div \left(\dfrac{1}{a} - 2\right)$

20. $\left(\dfrac{2}{b} + \dfrac{2}{c}\right) \div \left(\dfrac{1}{bc^2} + \dfrac{1}{b^2c}\right)$

Rewrite without absolute value signs and simplify.

21. $-|12| \div (|-2| - |5|)$

22. $(|-2.5\pi| - |\pi/2|) \div (-|1.5\pi|)$

Determine the distance on a coordinate line between the points with the given coordinates.

23. 5.1 and 3.11

24. $4.\overline{3}$ and $-5\frac{2}{3}$

Simplify.

25. $(-\sqrt{7})^4$

26. $-(-1/\sqrt{10})^6$

27. -0.3^{-2}

28. $(-0.2)^{-3}$

Simplify, using no negative exponents or square root signs in the final answers. Assume that n is a positive integer.

29. $[x(3y)^{-1}]^3$

30. $(2a^2b^{-1})^5$

31. $\left(\dfrac{(-6a)^2}{(2b)^0}\right)^{-1}$

32. $\left(\dfrac{(15y)^0}{(-y/2)^4}\right)^{-2}$

33. $(\sqrt{100a^4b^2})^3$

34. $(\sqrt{25x^6y^{10}})^3$

35. $\left(\dfrac{a^2}{b}\right)^n \left(\dfrac{b^2}{a}\right)^{-4n}$

36. $\dfrac{(\sqrt{2}u)^{6n}}{(\sqrt{6}uv^{-1})^{4n}}$

Write each of these numbers in scientific notation.

37. $772,000,000$

38. 0.0000048

Simplify. Assume that the variables represent positive real numbers.

39. $\sqrt[3]{8a^6}$

40. $\sqrt[4]{16a^8b^{12}}$

41. $3\sqrt{x^4y} - \sqrt{x^4y} + x^2\sqrt{y}$

42. $(\sqrt{x} - 2\sqrt{y})(\sqrt{x} + \sqrt{4y})$

43. $\dfrac{32^{3/5}}{32^{-3/5}}$

44. $\left(\dfrac{-27}{8}\right)^{2/3}$

45. $\left(\dfrac{4a^4b^2}{25a^{-2}b^6}\right)^{-3/2}$

46. $\left(\dfrac{c^{3/2}}{c^{2/3}}\right)^{12}$

47. $\sqrt{\dfrac{a^{3m+3}}{a^{1-m}}}$

48. $\sqrt[3]{\dfrac{b^{4m-2}}{b^{m-5}}}$

Carry out the indicated operations and simplify.

49. $(2x - 1)(x^2 + x + 1)$

50. $(y - 3)(y^2 + 2y - 2)$

Expand and simplify.

51. $(a - \frac{1}{3})^2$ **52.** $(b + 0.2)^2$

53. $(a + 2b)^3$ **54.** $(c^2 - d^2)^3$

Factor completely.

55. $t^2 - 9$ **56.** $u^2 - 22u + 121$

57. $v^2 + 12v + 36$ **58.** $4w^2 - 1$

59. $10x^2 - x - 2$ **60.** $9y^2 + 3y - 2$

61. $9x^3 - 6x^2 + x$ **62.** $y^5 + y^2$

63. $15a^2 + 4ab - 4b^2$

64. $c^4 - d^4$

65. $m^3 - 3m^2n + 3mn^2 - n^3$

66. $u^2 + 2uv + v^2 - 4$

67. $x^{2n+1} - xy^{2n} + x^{2n}y - y^{2n+1}$

68. $x^{3n} - y^{3n}$

Find an l.c.m. of each set of polynomials. Write the answer in factored form.

69. $x^2 + x, x^2 - 1, x^3 + 1$

70. $y^3 - 8, (y - 2)^2, y^3 + 2y^2 + 4y$

Perform the indicated operations and simplify.

71. $\dfrac{m^2 + 2m + 1}{3} \cdot \dfrac{6}{m + 1}$

72. $\dfrac{(u - v)}{uv} \cdot \dfrac{v^2}{u^2 + uv - 2v^2}$

73. $\dfrac{x^2 + 3x + 2}{2x} \div \dfrac{2x + 4}{x}$

74. $\dfrac{2x + 2}{x} \div \dfrac{x^2 - 1}{2x^2}$

75. $\dfrac{x}{x^2 - 1} - \dfrac{1}{x + 1}$

76. $\dfrac{1}{x^2 - x - 2} - \dfrac{1}{x^2 + x}$

77. $\dfrac{a - b}{\dfrac{1}{a} - \dfrac{1}{b}}$

78. $\dfrac{1 - \dfrac{c^2}{d^2}}{1 - \dfrac{c}{d}}$

Rationalize each denominator and simplify.

79. $\dfrac{5}{\sqrt{3}}$

80. $\dfrac{1}{a^{1/2} - b^{1/2}}$

81. What is 1.8% of 75?

82. What is 210% of 70?

83. The world population increased 20.21% from 1970 to 1980, and 8.77% from 1980 to 1985. The 1970 population was 3.721 billion. What was the 1985 population? Ⓒ

84. In 1976, Spanish was estimated to be the native language of approximately 210 million persons, and that was 76.4% of the number of native speakers of English. For how many was English the native language? Ⓒ

85. What is the interest on $2000 invested at 7% per year simple interest for 18 months?

86. What amount must be invested at 10% per year simple interest so that $60 interest is earned at the end of 4 years?

For Exercises 87–90, if you do not have a calculator to evaluate the powers that arise, simply write an expression that will give the answer but do not carry out the computations.

87. What is the compound amount on $2000 invested for 3 years at 8% compounded quarterly?

88. What is the compound amount on $5000 invested for $2\frac{1}{2}$ years at 6% compounded monthly?

89. How much must be invested at 8% compounded quarterly so that $10,000 will be accumulated at the end of 10 years?

90. How much must be invested at 10% compounded quarterly so that $2000 will be accumulated at the end of 4 years?

91. Express 1000 yd/hr in ft/min.

92. Express 5000 gal/hr in l/min.

93. The density of gold (at 20°C) is 19.3 g/cm³. How many grams are in 4 cubic centimeters of gold? Ⓒ

94. The density of mercury (at 20°C) is 13.55 g/cm³. How many cubic centimeters are in 50 grams of mercury? Ⓒ

95. Compute $\dfrac{\sqrt{3 + h} - \sqrt{3}}{h}$ for $h = 0.5$ and $h = 0.005$. Ⓒ

96. In the seventeenth century the English mathematician John Wallis represented $4/\pi$ as the infinite product

$$\frac{3 \cdot 3 \cdot 5 \cdot 5 \cdot 7 \cdot 7 \cdots}{2 \cdot 4 \cdot 4 \cdot 6 \cdot 6 \cdot 8 \cdots}.$$

By using the same number of factors in both the numerator and the denominator, the fraction approximates $4/\pi$, and the approximation becomes more accurate by increasing the number of factors.

(a) Compute an approximate value of π by using ten factors in both the numerator and the denominator.

(b) Repeat part (a) with twenty-one factors in place of ten.

[By 1987, the value of π had been computed to 134 million decimal places (but not by this method).] Ⓒ

CHAPTER II

EQUATIONS AND INEQUALITIES

In using mathematics we are continually concerned with equations and inequalities that contain variables. In this chapter we consider some of the most common types of equations and inequalities with one variable.

SECTION 7
Linear Equations

A. EQUATIONS WITH VARIABLES

We begin with some general remarks about expressions and equations that contain variables.

The **domain of an expression** containing one variable is the set of all real numbers that can be substituted for the variable so that the expression is also equal to a real number.* Numbers that are excluded from the domain include those that make a denominator zero and those that make the number under a square root sign negative. For example, 0 is excluded from the domain of $1/x$. And negative numbers are excluded from the domain of $\sqrt{x}$.

Domains can be recorded concisely with the set-builder notation

$$\{x: \cdots\},$$

which means

$$\text{the set of all } x \text{ such that } \cdots.$$

For example,

$$\{x: x < 5\}$$

represents

$$\text{the set of all } x \text{ such that } x < 5.$$

(From the context, x is assumed to represent real numbers in this case. For more examples with set-builder notation, see Appendix B.)

Example 7.1   (a) The domain of $\sqrt{x - 2}$ is

$$\{x: x \geq 2\}.$$

Each real number less than 2 is excluded from the domain because $x - 2 < 0$ if $x < 2$, and the square root of a negative number is not a real number.

(b) The domain of $1/(x - 3)$ is

$$\{x: x \neq 3\}.$$

The number 3 is excluded because it would lead to division by 0.

(c) The domain of any polynomial with one variable is the set of all real numbers. ∎

A number is in the **domain of an equation** iff it is in the domain of *both* sides of the equation.

---

*Thus far we have considered only *algebraic expressions*, as defined in Section 6. Other types of expressions (exponential, logarithmic, and trigonometric) will occur later in the book.

Example 7.2 The domain of

$$\sqrt{x - 2} = \frac{1}{x - 3}$$

is

$$\{x: x \geq 2 \text{ and } x \neq 3\}.$$

Consider an equation with one variable. If the variable in any such equation is replaced by a real number that is in the domain of the equation, then each side of the equation will also be equal to a real number. If these numbers on the two sides are equal, then the number that replaced the variable is called a **solution,** or **root,** of the equation; we also say in this case that the number **satisfies** the equation. To check whether a number is a solution of an equation, simply substitute the number for the variable in the equation and then simplify to see if the two sides are equal. Here are some examples. (We are not concerned yet with *finding* solutions; only with *checking* possibilities.)

Example 7.3 Is $\frac{2}{3}$ a solution of $3x + 2 = 4$?

Left side   *Right side*

$$3\left(\frac{2}{3}\right) + 2 \overset{?}{=} 4$$

$$4 = 4$$

The answer is Yes.

Example 7.4 (a) Is 5 a solution of $\sqrt{x + 4} = x - 2$?

Left side   *Right side*

$$\sqrt{5 + 4} \overset{?}{=} 5 - 2$$

$$\sqrt{9} \overset{?}{=} 3$$

$$3 = 3$$

The answer is Yes.

(b) Is 0 a solution of $\sqrt{x + 4} = x - 2$?

Left side   *Right side*

$$\sqrt{0 + 4} \overset{?}{=} 0 - 2$$

$$2 \neq -2$$

The answer is No.

Example 7.5 The equation $y = y + 1$ has no solution, since the right side will be one more than the left side no matter what number y represents.

Example 7.6 Every real number is a solution of

$$(t + 1)^2 = t^2 + 2t + 1$$

since the left side is just the factored form of the right side.

The equation in Example 7.6 is an *identity:* An equation is an **identity** if every real number in the domain of the equation is a solution of the equation. An equation that contains a variable but is *not* an identity is called a **conditional equation.** To **solve** an equation means to find its solutions. Special techniques are required to solve some types of equations, but others can be solved just by using the following general ideas.

Two equations with the same variable are called **equivalent** if they have the same solutions. The most basic plan for solving an equation is to replace it by simpler and simpler equivalent equations until the solutions (if there are any) become obvious. In doing this we make frequent use of the following properties of equality, which were stated in Section 1B.

$$\text{If } a = b, \text{ then } a + c = b + c \text{ and } ac = bc \text{ for every } c. \tag{7.1}$$

$$\text{If } a + c = b + c, \text{ then } a = b. \tag{7.2}$$

$$\text{If } ac = bc \text{ and } c \neq 0, \text{ then } a = b. \tag{7.3}$$

Example 7.7 Solve $2x - 3 = 4$.

Solution

$$2x - 3 = 4$$

$2x - 3 + 3 = 4 + 3$ Add 3 to both sides, using (7.1).

$2x = 7$ Simplify.

$x = \dfrac{7}{2}$ Divide both sides by 2 $\left(\text{multiply by } \dfrac{1}{2}\right)$, using (7.3).

We have transformed the original equation into the equivalent equation $x = \frac{7}{2}$, and the solution to the latter is obviously the number $\frac{7}{2}$.

Sometimes we apply (7.1)–(7.3) with c replaced by an expression that involves the variable. Notice that (7.3) insists that we multiply or divide only by *nonzero* expressions. Whenever we multiply or divide by an expression involving the variable, we run a risk of violating this restriction and creating an inequivalent equation. Here is an example to show why.

Example 7.8 We can transform $x^2 = x$ into $x = 1$ by *dividing* both sides by x. Conversely, we can transform $x = 1$ into $x^2 = x$ by *multiplying* both sides by x. But the equations $x^2 = x$ and $x = 1$ are not equivalent, because 0 is a solution of $x^2 = x$ but is not a solution of $x = 1$.

The point to remember is that whenever an equation is solved by changing its form, whether by using (7.1)–(7.3) or any other operations, unless you are certain that a final equation is equivalent to the original equation, then you must check the answer by substitution in the original equation.

B. LINEAR EQUATIONS

A **linear equation** in a variable x is an equation that can be written in the form

$$ax + b = 0,$$

where a and b are real numbers with $a \neq 0$. Linear equations in other variables are defined similarly. The equations in Examples 7.7, 7.9, and 7.10 are linear. (The reason these are called *linear* equations will become clear in Chapter III.)

Following are more examples of how to solve linear equations. The idea is to transform each equation using Properties (7.1)–(7.3) until the variable is isolated on one side and only a number (the solution) remains on the other. Each answer can be checked by substitution.

Example 7.9

$$3(2t - 2) = 2(9t + 1)$$

$$6t - 6 = 18t + 2 \qquad \text{Distributive property.}$$

$$6t = 18t + 8 \qquad \begin{array}{l}\text{Add 6 to both sides}\\ \text{and simplify.}\end{array}$$

$$-12t = 8 \qquad \begin{array}{l}\text{Subtract } 18t \text{ from both}\\ \text{sides and simplify.}\end{array}$$

$$t = -\frac{2}{3} \qquad \begin{array}{l}\text{Divide by } -12 \text{ and}\\ \text{simplify.}\end{array}$$

Example 7.10

$$\frac{2x - 1}{5} - \frac{3x - 2}{7} = \frac{6}{35}$$

$$35 \cdot \left(\frac{2x - 1}{5} - \frac{3x - 2}{7}\right) = 35 \cdot \frac{6}{35} \qquad \begin{array}{l}\text{Multiply by 35, the l.c.d.}\\ \text{of all the terms.}\end{array}$$

$$7(2x - 1) - 5(3x - 2) = 6 \qquad \text{Simplify.}$$

$$14x - 7 - 15x + 10 = 6 \qquad \text{Distributive property.}$$

$$-x + 3 = 6 \qquad \text{Simplify.}$$

$$-x = 3 \qquad \begin{array}{l}\text{Subtract 3 from both sides}\\ \text{and simplify.}\end{array}$$

$$x = -3 \qquad \text{Multiply by } -1.$$

Some equations with the variable in a denominator are equivalent to linear equations, as in the next example. The first step with such an equation is to multiply both sides by an expression that will remove the variable from the denominator. If we do that, however, then we *must* check the final answer. For equations such as those in Examples 7.9 and 7.10 the only reason to check an answer is to catch an error in calculation; in contrast, for equations such as the following we can get an incorrect answer even without such an error. (See Example 7.8 and the remark that preceded it.) An answer that arises in this way is called an **extraneous solution** (or **extraneous root**); it is *not* a solution of the original equation. Example 7.12 will give another illustration.

Example 7.11

$$\frac{1}{x} - \frac{3}{7x} = \frac{8}{21}$$

$$21x \left(\frac{1}{x} - \frac{3}{7x}\right) = 21x \cdot \frac{8}{21} \qquad \begin{array}{l}\text{Multiply by } 21x, \text{ the}\\ \text{l.c.d. of all the terms.}\end{array}$$

$$21 - 9 = 8x \qquad \text{Simplify.}$$

$$8x = 12 \qquad \begin{array}{l}\text{Simplify and interchange} \\ \text{sides.}\end{array}$$

$$x = \frac{3}{2} \qquad \text{Divide by 8 and simplify.}$$

Check

$$\frac{1}{\dfrac{3}{2}} - \frac{3}{7\left(\dfrac{3}{2}\right)} \overset{?}{=} \frac{8}{21}$$

$$\frac{2}{3} - \frac{2}{7} \overset{?}{=} \frac{8}{21}$$

$$\frac{14 - 6}{21} \overset{?}{=} \frac{8}{21}$$

$$\frac{8}{21} = \frac{8}{21}$$

Thus $\frac{3}{2}$ is a solution of the original equation.

Example 7.12

$$\frac{1}{x} + \frac{1}{x - 1} = \frac{1}{x(x - 1)}$$

$$x(x - 1)\left(\frac{1}{x} + \frac{1}{x - 1}\right) = x(x - 1)\frac{1}{x(x - 1)} \qquad \begin{array}{l}\text{Multiply by } x(x - 1), \text{ the} \\ \text{l.c.d. of all the terms.}\end{array}$$

$$x - 1 + x = 1 \qquad \text{Simplify.}$$

$$x = 1 \qquad \begin{array}{l}\text{Add 1 to both sides, simplify,} \\ \text{and divide by 2.}\end{array}$$

Check The terms $1/(x - 1)$ and $1/[x(x - 1)]$ in the original equation are undefined if $x = 1$, so 1 is not a solution of the original equation; it is an extraneous solution. The original equation has no solution.

Sometimes we must solve an equation for a selected variable in terms of one or more other variables in the equation. To do that we treat the other variables just like real numbers.

Example 7.13 Solve the following equation for z in terms of x and y.

$$\frac{x}{2} + \frac{y}{3} + \frac{z}{4} = 1$$

$$6x + 4y + 3z = 12 \qquad \text{Multiply by 12.}$$

$$3z = 12 - 6x - 4y \qquad \begin{array}{l}\text{Subtract } 6x + 4y \text{ from} \\ \text{both sides.}\end{array}$$

$$z = \frac{1}{3}(12 - 6x - 4y) \qquad \text{Divide by 3 } \left(\text{multiply by } \frac{1}{3}\right).$$

C. APPLICATIONS OF LINEAR EQUATIONS

Before looking at applications of linear equations we consider some general remarks on problem solving.

Our chances of solving a problem are much better if we approach it with a plan. Although no single plan will work for every problem, Steps I–V that follow will at least provide some guidelines. Try to keep these steps in mind, and be patient and persistent. Don't worry if the answer to a problem isn't obvious, or if the problem seems confusing or difficult—the whole reason for using algebra is that it can lead us systematically through problems that we cannot see through at a glance.

Step I. Read the problem carefully. Be sure you know what all of the key words mean. If possible, draw a figure. Also make a note of the units in the problem, if there are any.

Step II. Assign variables to the unknown or unknowns.

Step III. Form all equations that might have something to do with the problem. Make use of what is given in the problem and also of any general formulas that apply. Remember the remarks on units in Section 6D.

Step IV. Use algebra to solve the equations in Step III.

Step V. Check to see if the answers found in Step IV really do solve the original problem. Also be sure to include units (if there are any) in the final answer.

Example 7.14 A tank one-third full has 20 gallons added; then it is one-half full. What is the capacity of the tank?

Solution I and II. The units are gallons. Let C denote the capacity in gallons.

III. One-third of the capacity is $\frac{1}{3}C$. One-half of the capacity is $\frac{1}{2}C$. Therefore, from what is given,

$$\frac{1}{3}C + 20 = \frac{1}{2}C. \tag{7.4}$$

IV. Solve Equation (7.4).

$$6\left(\frac{1}{3}C + 20\right) = 6 \cdot \frac{1}{2}C$$

$$2C + 120 = 3C$$

$$C = 120$$

That is, the capacity is 120 gallons.

V. **Check** $\frac{1}{3}(120) + 20 \overset{?}{=} \frac{1}{2}(120)$

$$40 + 20 = 60. \qquad \blacksquare$$

Example 7.15 The perimeter of a rectangle is 20 centimeters and the length is 3 centimeters more than the width. Determine the length and width.

Solution I. *Perimeter* means *distance around*. The units are centimeters.

II. Let P = perimeter, l = length, and w = width, each measured in centimeters.

III. By the definition of perimeter, $P = 2l + 2w$. Also, from what is given, $P = 20$ and $l = w + 3$ (Figure 7.1).

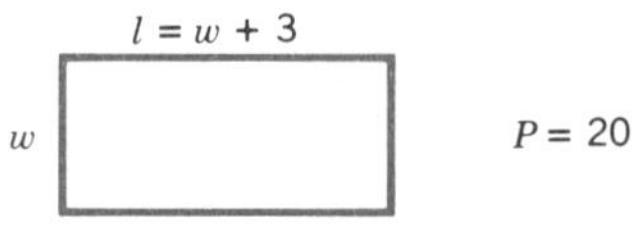

$$l = w + 3$$

w [rectangle] $P = 20$

FIGURE 7.1

IV. From the equations in III,

$$P = 2l + 2w$$
$$20 = 2(w + 3) + 2w \text{ (because } P = 20 \text{ and } l = w + 3)$$
$$20 = 4w + 6$$
$$w = \frac{7}{2} = 3.5.$$

That is, the width is 3.5 centimeters. Thus the length is $3.5 + 3$ or 6.5 centimeters.

V. **Check** $20 \overset{?}{=} 2(6.5) + 2(3.5)$

$20 = 13 + 7.$ ■

Mixture problems such as the following are prime examples of problems that may look difficult at first but become much easier by using equations and moving one step at a time.

Example 7.16 We require 4 liters of a solution that is 15% alcohol. We can draw from two solutions: one that is 12% alcohol and another that is 20% alcohol. How many liters of each should we use?

Solution   I. The units are liters.

II. Let x = number of liters of 12% solution to be used. Then

$$4 - x = \text{number of liters of 20\% solution to be used.}$$

III. The 12% solution will contribute $0.12x$ liters of alcohol to the final solution. The 20% solution will contribute $0.20(4 - x)$ liters of alcohol to the final solution. The total amount of alcohol in the final solution will be $0.15(4) = 0.6$ liters. Therefore, for the volume of alcohol,

$$0.12x + 0.20(4 - x) = 0.6. \tag{7.5}$$

IV. Solve Equation (7.5).

$$0.12x + 0.8 - 0.20x = 0.6$$
$$-0.08x = -0.2$$
$$x = 2.5 \, l.$$

Thus we should use 2.5 liters of 12% solution and $4 - 2.5$ or 1.5 liters of 20% solution.

V. **Check**   Liters of solution $= 2.5 + 1.5 = 4$, as required. Liters of alcohol $= 0.12(2.5) + 0.20(1.5) = 0.6$, as required.   ■

Example 7.17 The airline distance between Chicago and Washington, D.C., is 600 miles. A plane leaves Chicago for Washington at an average rate of 400 miles per hour.

Twenty minutes later, another plane leaves Washington for Chicago at an average rate of 300 miles per hour. How long after the first plane leaves will they meet?

Solution   I. Figure 7.2 summarizes some of the essential information: the planes are labeled A and B, where B leaves 20 minutes, or $\frac{1}{3}$ hour, after A. We need units for distance, time, and (average) rate. The obvious choices for this problem are *miles, hours,* and *miles per hour*. Thus 20 minutes has been converted to $\frac{1}{3}$ hour.

FIGURE 7.2

II. Let T denote time, measured in hours from when the first plane leaves.

III. The basic formula for motion problems such as this is

$$\text{distance} = \text{rate} \times \text{time}$$
$$D = RT.$$

Let D_A, R_A, and T_A denote the values of distance, rate, and time for plane A. Similarly for plane B. Then $T_A = T$, and $T_B = T - \frac{1}{3}$ because plane B leaves $\frac{1}{3}$ hour after plane A. Thus

$$R_A = 400 \text{ mi/hr} \qquad R_B = 300 \text{ mi/hr}$$

$$T_A = T \text{ hr} \qquad T_B = \left(T - \frac{1}{3}\right) \text{ hr}$$

$$D_A = 400T \text{ mi} \qquad D_B = 300\left(T - \frac{1}{3}\right) \text{ mi}$$

The problem requires that we determine when

$$D_A + D_B = 600,$$

that is,

$$400T + 300\left(T - \frac{1}{3}\right) = 600. \tag{7.6}$$

IV. Solve Equation (7.6):

$$400T + 300T - 100 = 600$$
$$700T = 700$$
$$T = 1.$$

Therefore, the planes will meet one hour after the first plane leaves.

V. **Check**   At the end of one hour plane A will have traveled

$$D_A = 400 \cdot 1 = 400 \text{ mi.}$$

At the end of one hour plane B will have traveled

$$D_B = 300\left(1 - \frac{1}{3}\right) = 300 \cdot \frac{2}{3} = 200 \text{ mi.}$$

Thus the two planes will indeed have traveled a total of 600 miles at the end of one hour, and the answer checks. Our calculations also show that the meeting place will be 400 miles from Chicago and 200 miles from Washington.   ■

EXERCISES FOR SECTION 7

Find the domain of each expression.

1. $1/x$

2. $\sqrt{x - 4}$

3. $3x^2 + 2x - 1$

4. $\sqrt{w + 5}$

5. $\dfrac{1}{y + 2}$

6. $\dfrac{\sqrt{z}}{z - 2}$

Find the domain of each equation.

7. $\dfrac{1}{x} = \sqrt{x + 1}$

8. $\dfrac{1}{\sqrt{x - 1}} = \dfrac{1}{x - 3}$

9. $\dfrac{1}{2x - 1} = \dfrac{1}{\sqrt{x}}$

Determine by substitution whether the given number is a solution of the given equation.

10. $4x + 2 = 2(x + 6)$; 5

11. $\dfrac{1}{2}(4 - 6x) = x - 1$; $\dfrac{1}{2}$

12. $2[x - (1 - x)] = 5(x - 1)$; 3

13. $y^3 - y = 7y + 5$; 3

14. $u^4 - 3u^2 = u^2(3u - 1)$; 4

15. $z^3 - 2z = 0.001(20 - z)$; 0.1

16. $\sqrt{x + 5} = 7 - x$; 11

17. $-\sqrt{t + 1} = 5 - t$; 8

18. $\sqrt{2x - 4} = 2 - x$; 2

Solve each equation.

19. $3x - 5 = 2x + 1$

20. $4y = 5y - 7$

21. $6z - 1 = 7z + 1$

22. $10(2u - 3) = 4(5 - 2u)$

23. $5(x - 1) = 2(4x + 1)$

24. $3(2x - 1) = -2(3x + 1)$

25. $6.25(1 - 0.48x) = -72.1(x + 1)$ ⓒ

26. $-115(x + 93) = 312(18 - x)$ ⓒ

27. $0.1772(2x + 45) = 1.649(1 + 3x)$ ⓒ

28. $\dfrac{v}{2} - \dfrac{v}{3} = \dfrac{1}{5}$

29. $\dfrac{2t + 1}{3} - \dfrac{3t + 1}{2} = \dfrac{1}{4}$

30. $\dfrac{1}{2}\left(y + \dfrac{1}{3}\right) = \dfrac{1}{3}\left(y + \dfrac{1}{4}\right)$

31. $\dfrac{1}{x} - \dfrac{2}{3x} = \dfrac{1}{12}$

32. $\dfrac{1}{t} + \dfrac{1}{t^2} = \dfrac{1}{2t}$

33. $\dfrac{2}{z - 2} + \dfrac{3}{z} = \dfrac{4}{z(z - 2)}$

34. $\dfrac{2}{y + 1} - \dfrac{3}{y - 1} = \dfrac{-4}{(y + 1)(y - 1)}$

35. $\dfrac{x}{x - 1} = -3 + \dfrac{1}{x - 1}$

36. $\dfrac{1}{z} - \dfrac{3}{z + 1} = \dfrac{1}{2z}$

Solve each equation twice: (a) for y in terms of x; (b) for x in terms of y.

37. $2x + 3y = 1$

38. $4y = 2(x - 1)$

39. $5(x - 1) = 6(y + 2)$

40. $y + 5(x - 1) = 7(1 - y) - x$

41. $2[x - (y - 1)] = 3$

42. $2x + 1 = 3[y + 2(1 - x)]$

Solve each equation for the variable indicated. Each letter with a subscript, like v_0, r_1, r_2, or r_m, is a separate variable.

43. $I = Prt$ for r.

44. $R^3 = KT^2$ for K.

45. $P = k/V$ for V.

46. $v = gt + v_0$ for t.

47. $P = 2L + 2W$ for L.

48. $T = 2\pi rh + 2\pi r^2$ for h.

49. $B = V - (xV/n)$ for V.

50. $V = (a + b)h/2$ for h.

51. $A = P(1 + rt)$ for r.

52. $\dfrac{1}{r} = \dfrac{1}{r_1} + \dfrac{1}{r_2}$ for r.

53. $R = \dfrac{ab}{a + b}$ for a.

54. $r = r_m\left(\dfrac{K - N}{K}\right)$ for K.

Assign variables as needed, and then translate each statement into an equation or an inequality.

55. The surface area of a sphere is 4π times the square of the radius.

56. The volume of a sphere is four-thirds of π times the cube of the radius (radius raised to the third power).

57. The volume of a pyramid is one-third the area of the base times the height (altitude).

58. The square root of the product of two positive numbers does not exceed one-half of the sum of the two numbers.

59. Less than three-tenths of the surface of the earth is covered by land.

60. The length of the diagonal of a rectangular parallelopiped (the shape of an ordinary box) is the square root of the sum of the squares of the lengths of the length, width, and height.

In each of the following exercises, indicate clearly the method used to obtain the solution.

61. Ten more than four times a number is 16. What is the number?

62. One-half of a number plus one-third of the number is four less than the number. What is the number?

63. One less than the reciprocal of a number is four. What is the number?

64. In the 35 years from 1946 through 1980, Australia won the Davis Cup (in tennis) 3 more times than the United States. All other countries combined won the cup only 4 times. How many times did Australia and the United States each win?

65. An apple contains twice as many calories as a peach. Two apples and one peach contain as many calories as two bananas. If a banana contains 100 calories, how many calories are in a peach?

66. The total world production of wheat in 1983 was approximately 498 million metric tons. The U.S., (mainland) China and the Soviet Union together produced approximately 46% of the total, and China and the Soviet Union each produced 16 million tons more than the U.S. Based on these figures, what was the U.S. production?

67. The perimeter of a basketball court is 288 feet and the length is 6 feet less than twice the width. Determine the length and width.

68. Based on the 1980 census, 169 U.S. cities had population over 100,000. Of these, the number with population less than 250,000 was one more than twice the number with population over 250,000. How many of the 169 had population over 250,000 and how many had population less than 250,000?

69. Assume that the sum of three consecutive integers is 87. What are the integers? (Suggestion: If the smallest integer is n, then the others are $n + 1$ and $n + 2$.)

70. A student has scores of 86, 78, 92, and 87 on the first four exams in a course. What score on the fifth exam will give an average of 85 on the five exams?

71. A multiple-choice exam has 30 questions. Each student's score is computed by adding 30 points to four times the number of correct answers and then subtracting one point for each incorrect answer. A student who answered all but five questions made a score of 95. How many questions were answered correctly?

72. Consider the exam in Exercise 71. Let S and C denote the score and the number of correct answers, respectively. Find a formula giving C in terms of S for students who answer every question.

73. One (interior) angle of a triangle is 15° more than the sum of the other two angles, one of which is twice the other. Find the three angles.

74. A merchant buys an item for 70% of his selling price, and allows 15% of the selling price for overhead costs. If the profit is $12.30 (after deducting purchase and overhead costs), what is the selling price?

75. Suppose that for income over $15,000 but not over $18,200 a tax rate schedule indicates a tax of $2330 plus 27% of the amount over $15,000. What is the income of a person whose tax is $2951?

76. How many liters of solution that is 10% alcohol must be added to 2 liters of solution that is 60% alcohol to produce a solution that is 40% alcohol?

77. Container A holds a solution that is 12% salt and container B holds a solution that is 20% salt. How much of each solution should one use to form 4 liters of solution that is 15% salt?

78. How many gallons of water must be added to 50 gallons of liquid fertilizer that is 30% nitrogen to produce a solution that is 12% nitrogen?

79. The airline distance between Chicago and Washington, D.C., is 600 miles. A plane leaves Chicago for Washington at an average rate of 400 miles per hour. At the same time, another plane leaves Washington for Chicago. If the second plane arrives in Chicago 10 minutes after the first plane arrives in Washington, what was the average rate for the second plane?

80. A plane leaves San Francisco for Cleveland at an average rate of 465 miles per hour. At the same time, another plane leaves Cleveland for San Francisco at an average rate of 430 miles

per hour. The planes meet 2 hours and 5 minutes later. What is the airline distance between the two cities?

81. I step on an escalator to leave a subway. Three seconds later someone else steps on a parallel escalator to enter the subway, and we pass 7 seconds after that. I know the rate of each escalator to be 0.5 meters per second. What is the length of the escalators?

SECTION **8**
Quadratic Equations

A. INTRODUCTION

A **quadratic equation** in a variable x is an equation that can be written in the form

$$ax^2 + bx + c = 0, \tag{8.1}$$

where a, b, and c denote real numbers with $a \neq 0$.

A quadratic equation in any other variable is defined by replacing x by that variable. For an equation to be quadratic in a variable, the equation must contain a second-degree term in that variable after it has been simplified; it may or may not contain a first-degree term or a constant term. We shall see that a quadratic equation may have no real solution, one real solution, or two real solutions.

A quadratic equation with no first-degree term is called a **pure quadratic equation.** A pure quadratic equation with variable x should first be solved for x^2. Then use the fact that, for each nonnegative real number p,

$$x^2 = p \quad \text{iff} \quad x = \pm\sqrt{p}. \tag{8.2}$$

(Remember from Section 3 that $\sqrt{p}$ denotes the principal or nonnegative square root of p.)

Example 8.1 To solve the pure equation $3x^2 - 75 = 0$, write

$$3x^2 = 75$$

$$x^2 = 25$$

$$x = \pm 5.$$

Each of the equations above is equivalent to the one that precedes it, so that $3x^2 - 75 = 0$ has *two* solutions: 5 and -5. ■

Both solutions in Example 8.1 can be checked by direct substitution. However, by the equivalence of the successive equations in the solution process, checking is necessary only to catch mistakes in computation. The same remarks about checking will apply to all of the solutions in this section.

Note well that $p \geq 0$ in Statement (8.2). If $p < 0$, then $x^2 = p$ has no solution in the real numbers since $x^2 \geq 0$ for every real number x. For example, $x^2 = -3$ has no real solution. (Chapter X will introduce the system of complex numbers, which contains the real numbers and has solutions to equations like $x^2 = -3$. If you prefer, the first section of Chapter X can be covered immediately after this section.)

B. SOLUTION BY FACTORING

Any pure quadratic equation can be solved as in the preceding example. The first method to try on any other quadratic equation is *factoring*, which depends on the following fact about real numbers (Section 1B):

$$ab = 0 \quad \text{iff} \quad a = 0 \text{ or } b = 0. \tag{8.3}$$

Example 8.2 Solve $t^2 + t - 6 = 0$.

Solution The statements in the successive lines that follow are equivalent.

$$t^2 + t - 6 = 0$$

$$(t - 2)(t + 3) = 0$$

$$t - 2 = 0 \quad \text{or} \quad t + 3 = 0$$

$$t = 2 \quad \text{or} \quad t = -3$$

Thus the solutions are 2 and -3. ◼

Pure quadratic equations (Subsection A) can also be solved by factoring. For example, if $p \geq 0$, then

$$x^2 = p$$

$$x^2 - p = 0$$

$$(x - \sqrt{p})(x + \sqrt{p}) = 0$$

$$x = \pm\sqrt{p}$$

To use the factoring method we must be able to factor the expression that results after the equation has been written so that only 0 remains on one side of the equality sign. The next example illustrates how to use factoring to solve a quadratic equation with no constant term.

Example 8.3 Solve $x^2 = 3x$.

Solution

$$x^2 = 3x$$

$$x^2 - 3x = 0$$

$$x(x - 3) = 0$$

$$x = 0 \quad \text{or} \quad x = 3$$

That is, the solutions of $x^2 = 3x$ are 0 and 3.

Notice that if we were to divide both sides of $x^2 = 3x$ by x, we would be left with $x = 3$. Thus we would have lost the solution $x = 0$ of the original equation. Remember that a solution may be lost whenever both sides of an equation are divided by either the variable or an expression that involves the variable.

C. SOLUTION BY COMPLETING THE SQUARE

The following example will serve as a bridge between pure quadratic equations and the method to be discussed next.

Example 8.4 Solve $y^2 - 2y + 1 = 9$.

Solution

$$y^2 - 2y + 1 = 9$$

$$(y - 1)^2 = 9 \qquad \text{Because } y^2 - 2y + 1 = (y - 1)^2.$$

$$y - 1 = \pm 3 \qquad \text{Use Statement (8.2) with } x = y - 1 \text{ and } p = 9.$$

$$y = 1 \pm 3 \qquad \text{Add 1 to both sides.}$$

$$y = 4 \quad \text{or} \quad y = -2$$

Both 4 and -2 are solutions of $y^2 - 2y + 1 = 9$ since the statements in the successive lines of the solution process are equivalent.

In Example 8.4 we exploited the fact that $y^2 - 2y + 1$ is a *perfect square,* namely, $(y - 1)^2$. The method of completing the square depends on the fact that if there is not a perfect square, then we can create one. Specifically, for each number d, $x^2 + dx$ becomes a perfect square if we add $(\frac{1}{2}d)^2$:

$$x^2 + dx + \left(\frac{1}{2}d\right)^2 = \left(x + \frac{1}{2}d\right)^2.$$

The method of completing the square will be outlined in the following five steps, and each step will be illustrated with the equation

$$3x^2 + x - 2 = 0.$$

I. Arrange the equation with the constant term on the right of the equality sign and the other terms on the left.

$$3x^2 + x = 2 \qquad \text{Add 2 to both sides of the original equation.}$$

II. Divide both sides by the coefficient of the second-degree term.

$$x^2 + \frac{1}{3}x = \frac{2}{3} \qquad \text{Divide by 3.}$$

III. Add the square of half the coefficient of the first-degree term to both sides.

$$x^2 + \frac{1}{3}x + \frac{1}{36} = \frac{2}{3} + \frac{1}{36} \qquad \text{Add } \left[\frac{1}{2} \cdot \frac{1}{3}\right]^2 = \frac{1}{36} \text{ to both sides.}$$

IV. The previous step made the left side a perfect square; write it as such, and simplify the right side.

$$x^2 + 2\left(\frac{1}{6}\right)x + \left(\frac{1}{6}\right)^2 = \frac{24}{36} + \frac{1}{36}$$

$$\left(x + \frac{1}{6}\right)^2 = \frac{25}{36}$$

V. Use the idea in Statement (8.2), and simplify.

$$x + \frac{1}{6} = \pm\sqrt{\frac{25}{36}}$$

Use Statement (8.2) with x replaced by $x + \frac{1}{6}$ and $p = \frac{25}{36}$.

$$x + \frac{1}{6} = \pm\frac{5}{6}$$

$$x = -\frac{1}{6} \pm \frac{5}{6}$$

$$x = \frac{-1 + 5}{6} \quad \text{or} \quad x = \frac{-1 - 5}{6}$$

$$x = \frac{2}{3} \quad \text{or} \quad x = -1.$$

Example 8.5 Solve $2x^2 = 5x - 1$ by completing the square.

Solution Compare each step with the preceding outline.

$$2x^2 = 5x - 1$$

I. $\qquad\qquad 2x^2 - 5x = -1$ Isolate the constant on the right.

II. $\qquad\qquad x^2 - \frac{5}{2}x = -\frac{1}{2}$ Divide by the coefficient of x^2.

III. $\quad x^2 - \frac{5}{2}x + \left(\frac{-5}{4}\right)^2 = -\frac{1}{2} + \left(\frac{-5}{4}\right)^2$ Complete the square.

IV. $\quad x^2 - 2\left(\frac{5}{4}\right)x + \left(\frac{-5}{4}\right)^2 = -\frac{1}{2}\cdot\frac{8}{8} + \frac{25}{16}$ Simplify.

$$\left(x - \frac{5}{4}\right)^2 = \frac{17}{16}$$

V. $\qquad\qquad x - \frac{5}{4} = \pm\sqrt{\frac{17}{16}}$ Solve for x.

$$x - \frac{5}{4} = \pm\frac{\sqrt{17}}{4}$$

$$x = \frac{5 \pm \sqrt{17}}{4}.$$

The two solutions are $\dfrac{5 + \sqrt{17}}{4}$ and $\dfrac{5 - \sqrt{17}}{4}$.

Check Let's check $(5 + \sqrt{17})/4$ as a reminder of how checking is done. We begin by substituting in the original equation.

Left side	Right side

$$2\left(\frac{5 + \sqrt{17}}{4}\right)^2 \overset{?}{=} 5\left(\frac{5 + \sqrt{17}}{4}\right) - 1$$

$$2\left(\frac{25 + 10\sqrt{17} + 17}{16}\right) \overset{?}{=} \frac{5(5 + \sqrt{17}) - 4}{4}$$

$$\frac{42 + 10\sqrt{17}}{8} \overset{?}{=} \frac{21 + 5\sqrt{17}}{4}$$

$$\frac{21 + 5\sqrt{17}}{4} = \frac{21 + 5\sqrt{17}}{4}$$

D. THE QUADRATIC FORMULA

The solutions of any quadratic equation can be found by completing the square. We can also complete the square once for the most general quadratic equation,

$$ax^2 + bx + c = 0 \; (a \neq 0), \tag{8.4}$$

and then use the formula that results. Here is the calculation, with the steps numbered as before.

I. $\qquad ax^2 + bx = -c \qquad$ *Isolate the constant term on the right.*

II. $\qquad x^2 + \left(\frac{b}{a}\right)x = -\left(\frac{c}{a}\right) \qquad$ *Divide by the coefficient of x^2.*

III. $\qquad x^2 + \left(\frac{b}{a}\right)x + \left(\frac{b}{2a}\right)^2 = -\left(\frac{c}{a}\right) + \left(\frac{b}{2a}\right)^2 \qquad$ *Complete the square.*

IV. $\qquad x^2 + 2\left(\frac{b}{2a}\right)x + \left(\frac{b}{2a}\right)^2 = -\left(\frac{c}{a}\right)\left(\frac{4a}{4a}\right) + \frac{b^2}{4a^2} \qquad$ *Simplify.*

$$\left[x + \left(\frac{b}{2a}\right)\right]^2 = \frac{b^2 - 4ac}{4a^2}$$

V. $\qquad x + \frac{b}{2a} = \pm\sqrt{\frac{b^2 - 4ac}{4a^2}} \qquad$ *Solve for x.*

$$x = -\frac{b}{2a} \pm \frac{\sqrt{b^2 - 4ac}}{2a}$$

Exercise 96 asks you to show by substitution that the $+$ option here does give a solution of Equation (8.4). The proof for the $-$ option is similar. Thus we arrive at the following conclusion, which definitely should be memorized.

Quadratic formula

The solutions of $ax^2 + bx + c = 0$, with $a \neq 0$, are

$$x = \frac{-b \pm \sqrt{b^2 - 4ac}}{2a}.$$

Example 8.6 Solve $6x^2 + x - 2 = 0$ using the quadratic formula.

Solution Use $a = 6$, $b = 1$, and $c = -2$.

$$x = \frac{-1 \pm \sqrt{1^2 - 4(6)(-2)}}{2(6)}$$

$$= \frac{-1 \pm \sqrt{1 + 48}}{12}$$

$$= \frac{-1 \pm \sqrt{49}}{12}$$

$$= \frac{-1 \pm 7}{12}$$

$$= \frac{-1 + 7}{12} \quad \text{or} \quad \frac{-1 - 7}{12}$$

$$= \frac{1}{2} \quad \text{or} \quad -\frac{2}{3}$$

That is, the solutions are $\frac{1}{2}$ and $-\frac{2}{3}$.

The next example emphasizes that the numbers a, b, and c in the quadratic formula are the coefficients after a quadratic equation has been written so that only zero remains on one side of the equality sign.

Example 8.7 Solve $x^2 + 8x = 2$ using the quadratic formula.

Solution First rewrite the equation with only zero on the right of the equality sign:

$$x^2 + 8x - 2 = 0.$$

This gives $a = 1$, $b = 8$, and $c = -2$. Therefore,

$$x = \frac{-8 \pm \sqrt{8^2 - 4(1)(-2)}}{2(1)}$$

$$= \frac{-8 \pm \sqrt{72}}{2}$$

$$= \frac{-8 \pm \sqrt{36 \cdot 2}}{2}$$

$$= \frac{-8 \pm 6\sqrt{2}}{2}$$

$$= -4 \pm 3\sqrt{2}.$$

The solutions are $-4 + 3\sqrt{2}$ and $-4 - 3\sqrt{2}$.

Example 8.8 Solve $4x^2 - 0.4x + 0.01 = 0$ using the quadratic formula.

Solution Here $a = 4$, $b = -0.4$, and $c = 0.01$.

$$x = \frac{0.4 \pm \sqrt{(-0.4)^2 - 4(4)(0.01)}}{2(4)}$$

$$= \frac{0.4 \pm \sqrt{0.16 - 0.16}}{8}$$

$$= \frac{0.4 \pm 0}{8}$$

$$= 0.05$$

Thus there is one solution, 0.05. ∎

Here are suggestions for solving quadratic equations.

- If there is no first-degree term, solve the equation as a pure quadratic equation.
- If the equation is not pure, try to solve it by factoring. (This applies, in particular, if there is no constant term.)
- If the first suggestion does not apply and the second is difficult, use the quadratic formula or complete the square.

E. THE CHARACTER OF THE SOLUTIONS

Example 8.9 If we apply the quadratic formula to

$$x^2 + x + 1 = 0,$$

we obtain

$$x = \frac{-1 \pm \sqrt{1^2 - 4 \cdot 1 \cdot 1}}{2 \cdot 1}$$

$$= \frac{-1 \pm \sqrt{-3}}{2}.$$

There is no real number whose square is -3, so the equation has no real solution. ∎

The following analysis categorizes this and all other quadratic equations.

The expression $b^2 - 4ac$ in the quadratic formula is called the **discriminant** of $ax^2 + bx + c$.

If the discriminant is negative, then the quadratic formula involves the square root of a negative number (as in Example 8.9) so there is no real solution.

If the discriminant is positive, the quadratic formula gives two (unequal) real solutions:

$$\frac{-b + \sqrt{b^2 - 4ac}}{2a} \quad \text{and} \quad \frac{-b - \sqrt{b^2 - 4ac}}{2a}.$$

If the discriminant is zero, then the two numbers given by the quadratic formula are the same:

$$\frac{-b + \sqrt{0}}{2a} = -\frac{b}{2a} \quad \text{and} \quad \frac{-b - \sqrt{0}}{2a} = -\frac{b}{2a}.$$

Table 8.1 summarizes the preceding remarks.

TABLE 8.1

Discriminant	Character of the solutions
$b^2 - 4ac < 0$	No real solution
$b^2 - 4ac = 0$	One real solution
$b^2 - 4ac > 0$	Two real solutions

For examples of the possibilities in Table 8.1, see Examples 8.9, 8.8, and 8.7, respectively.

F. APPLICATION: FALLING OBJECTS

Assume that an object is dropped near the earth's surface, and let s denote the distance it falls in the first t seconds. A law of physics tells us that if we ignore all forces (such as air resistance) except gravity, then

$$s = \frac{1}{2} gt^2. \tag{8.5}$$

Here g denotes a constant whose value depends on the units used to measure s. If s is measured in feet, then $g \approx 32$; if s is measured in meters, then $g \approx 9.8$. Therefore, approximately,

$$s = \begin{cases} 16t^2 \text{ feet} \\ 4.9t^2 \text{ meters}. \end{cases} \tag{8.6}$$

Example 8.10   Figure 8.1 shows the distance, in meters, that an object will have dropped at the end of 1, 2, 3, and 4 seconds. At the end of 3 seconds, for example, Equation (8.6) gives

$$s = 4.9(3)^2 \approx 44 \text{ m.} \qquad \blacksquare$$

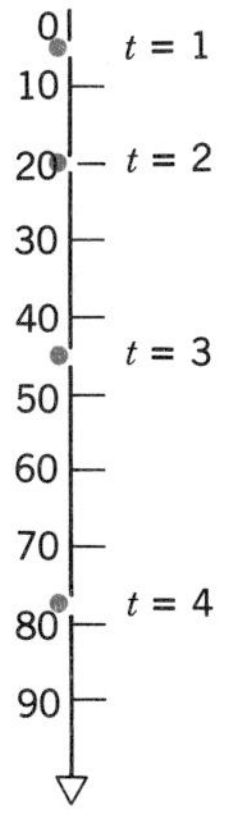

FIGURE 8.1

Example 8.11   A baseball is dropped from the top of the Washington Monument, which is 555 feet tall. Estimate when it will reach the ground.

Solution   We must determine the value of t that will yield $s = 555$ feet in Equation (8.6):

$$555 = 16t^2$$

$$t = \sqrt{555/16} = \sqrt{555}/4 \approx 5.9 \text{ sec.}$$

Only the positive solution has been shown because t represents time, and the negative solution would have no significance. ■

If an object is thrown down near the earth's surface, rather than just dropped, then the distance the object travels in the first t seconds is approximately

$$s = \frac{1}{2} gt^2 + v_0 t, \tag{8.7}$$

where v_0 denotes the *initial velocity* of the object, that is, the velocity when $t = 0$. If we use feet to measure s, feet per second to measure v_0, and $g = 32$, then

$$s = 16t^2 + v_0 t. \tag{8.8}$$

Example 8.12 A rock is thrown directly down from the top of a building 336 feet tall with an initial velocity of 64 ft/sec. When will it reach the ground?

Solution From Equation (8.8),

$$s = 16t^2 + 64t. \tag{8.9}$$

Since the building is 336 feet tall, the object will reach the ground when $s = 336$. Thus we use Equation (8.9) with $s = 336$ and solve for t:

$$336 = 16t^2 + 64t$$

$$16t^2 + 64t - 336 = 0$$

$$t^2 + 4t - 21 = 0$$

$$(t + 7)(t - 3) = 0$$

$$t = -7, 3.$$

The circumstances dictate a positive value for time, so the answer is 3 seconds. ■

EXERCISES FOR SECTION 8

Solve for x, y, *or* t, *or indicate that there is no real solution. Assume A > 0 and B > 0 in Exercises 11–13.*

1. $x^2 = 36$

2. $y^2 = 144$

3. $t^2 - 100 = 0$

4. $16y^2 = 9$

5. $0.8t^2 - 1.8 = 0$

6. $98x^2 - 2 = 0$

7. $5t^2 + 20 = 0$

8. $10x^2 = 0$

9. $9y^2 + 25 = 0$

10. $0.01x^2 - 8 = 0$

11. $A^2 y^2 = 27$

12. $0 = ABt^2$

13. $By^2 = 0$

14. $3t^2 + 12 = 0$

15. $0.49x^2 = 0.04$

Solve by factoring.

16. $x^2 + 4x = 0$

17. $2x^2 - 6x = 0$

18. $5x^2 = -25x$

19. $y^2 - 2y + 1 = 0$

20. $y^2 + 4y = -4$

21. $y^2 = 6y - 9$

22. $2t^2 + 2t = 4$

23. $3t^2 - 9t - 12 = 0$

24. $t^2 - t = 12$

25. $0.1z^2 = 0.3z$

26. $z^2 + 0.2z + 0.01 = 0$

27. $z^2 = z - 0.25$

28. $5x^2 + 4x - 1 = 0$

29. $3x^2 - x - 2 = 0$

30. $5x^2 - 8x + 3 = 0$

Solve with a calculator having a square root key. C

31. $t^2 - 12.31 = 0$ **32.** $509s^2 = 36$

33. $4 \cdot 3.14r^2 = 605$

Solve by completing the square. Check at least one solution (by substitution) in each exercise.

34. $x^2 + 2x = 2$ **35.** $x^2 = 1 - 6x$

36. $x^2 + 4x = 1$ **37.** $3y^2 + 2y = 1$

38. $2y^2 + 9y = 5$ **39.** $5y^2 = 2y + 1$

40. $2x^2 + 2x - 1 = 0$

41. $3x^2 + x - 3 = 0$

42. $4x^2 + 2x - 1 = 0$

Without solving these equations, use the discriminant to determine whether there is no real solution, one real solution, or two real solutions.

43. $2x^2 = 4 - x$

44. $9x^2 = 6x - 1$

45. $4x^2 - 5x + 1 = 0$

46. $2x^2 + 2x + 1 = 0$

47. $x^2 + 2x + 3 = 0$

48. $-x^2 - 6x = 9$

49. $25x^2 + 20x + 4 = 0$

50. $x^2 + 3x - 1 = 0$

51. $x^2 - x + 1 = 0$

Use the quadratic formula to solve for x.

52. $2x^2 - x - 2 = 0$

53. $-3x^2 + x + 1 = 0$

54. $5x^2 - x - 4 = 0$

55. $8x^2 = 3 - 10x$

56. $-3x^2 + 13x = 10$

57. $x^2 = 2x + 4$

58. $qx^2 = qx + r$

59. $x^2 + qx = q$

60. $rx^2 + qx + p = 0$

Solve for x, y, *or* t. *Use any method.*

61. $x^2 - 7x = 0$

62. $y^2 = 5y$

63. $4t^2 + t = 0$

64. $x^2 - 2x = 2$

65. $y^2 + 3y + 1 = 0$

66. $t^2 + t = 1$

67. $0.04x^2 - 1 = 0$

68. $0 = 0.2 - 0.8y^2$

69. $0 = 0.1t^2 - 0.001$

70. $1 = \dfrac{1}{2}(x + x^2)$

71. $2y^2 + 3y = 0$

72. $\dfrac{1}{3}t^2 + \dfrac{5}{3}t + 2 = 0$

73. $3.25x^2 - 1.76x = 2.01$ C

74. $16.1t^2 + 20.0t = 32.5$ C

75. $509y^2 - 792 = 361y$ C

76. Solve $h^2 + (s/2)^2 = s^2$ for h.

77. Solve $V = k(R^2 - r^2)$ for r.

78. Solve $S = 2\pi rh + 2\pi r^2$ for r.

79. If the area of a circle is 8π square inches, what is the radius?

80. If the surface area of a sphere is 20 square centimeters, what is the radius? ($S = 4\pi r^2$)

81. If the surface area of a cube is 150 square inches, what is the length of each edge of the cube?

82. If the square of a number is decreased by 15, the result is equal to the number increased by 15. What is the number? (Find all solutions.)

83. If a principal amount P is invested at a compound rate r for two conversion periods, then the compound amount is $S = P(1 + r)^2$. Solve this equation for r.

84. A water sprinkler that sprays a circular pattern is placed at the center of a rectangular garden whose perimeter is 34 meters (Figure 8.2). A radius of 6.5 meters for the circular pattern is just sufficient to cover the garden. What are the garden's dimensions?

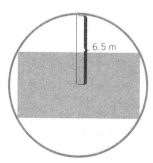

6.5 m

FIGURE 8.2

85. Consider Equation (8.5), with t measured in seconds. What are the units for g: (a) if s is measured in feet? (b) if s is measured in meters? (The units on the two sides of the equation must be the same. See Section 6D.)

86. Consider Equation (8.6) with s measured in feet. Solve for t.

87. If an object is dropped near the earth's surface, how many feet will it fall in the first 5 seconds? How long will it take to fall 225 feet?

88. If an object is dropped near the earth's surface, how many meters will it drop in the first 4 seconds? How long will it take to fall 19.6 meters?

89. A baseball is dropped from the top of the Gateway Arch in St. Louis, which is 630 feet tall. Estimate when it will reach the ground.

90. An object is thrown down near the earth's surface with an initial velocity of 24 feet per second.

(a) Write the equation giving the position after t seconds. [Use Equation (8.8).]

(b) Where is the object 1 second after it is thrown?

(c) How long will it take for the object to travel the first 16 feet?

91. An object is thrown down from the top of a building 400 feet tall, and 4 seconds later it hits the ground. What was the initial velocity? [Use Equation (8.8).]

92. A 25-foot ladder leans against a vertical wall with the foot of the ladder 7 feet from the wall. How far will the foot of the ladder slide if the top slides 2 feet?

93. The larger of two concentric circles subtends a segment of length t centimeters on a tangent drawn to the smaller of the two circles. The radius of the larger circle is R centimeters. What is the radius of the smaller circle?

94. The area between two concentric circles equals twice the area of the smaller circle. The radius of the smaller circle is r. What is the radius of the larger circle?

95. Solve Equation (8.7) for t.

96. Verify by substitution that

$$\frac{-b + \sqrt{b^2 - 4ac}}{2a}$$

is a solution of Equation (8.4).

97. For which values of k does $2x^2 + kx + 1 = 0$ have exactly one real solution?

98. If one solution of $2x^2 + kx + 5 = 0$ is 5, what is the other solution?

Use the quadratic formula to verify each of the following facts about the solutions of $ax^2 + bx + c = 0$ $(a \neq 0)$.

99. Assume $a > 0$ and $c > 0$. Then there is exactly one real solution iff $b = \pm 2\sqrt{ac}$.

100. The sum of the solutions is $-b/a$.

101. The product of the solutions is c/a.

102. The square of the difference of the solutions (in either order—take your choice) is $|b^2 - 4ac|/a^2$.

SECTION 9

Miscellaneous Equations

A. RADICAL EQUATIONS

Frequently, an equation that is not strictly linear (first degree) or quadratic (second degree) can be solved by changing it into an equation that *is* linear or quadratic. The first examples illustrating this will be **radical equations,** so-called because the unknown variable appears under a radical sign (such as a square root sign).

Example 9.1 Solve $\sqrt{2x + 3} = 7$.

Solution To solve this equation we square both sides to remove the radical. If $\sqrt{2x + 3} = 7$, then

$$(\sqrt{2x + 3})^2 = 7^2$$

$$2x + 3 = 49$$

$$2x = 46$$

$$x = 23.$$

Check
$$\sqrt{2 \cdot 23 + 3} \overset{?}{=} 7$$

$$\sqrt{49} = 7.$$

Thus the equation has one solution, 23.

Two points must be made about the method just used. First, if the original equation has solutions, then the method will produce them. This is because if p and q denote expressions such that $p = q$, then $p^n = q^n$ for each positive integer n. Thus, in Example 9.1, $\sqrt{2x + 3} = 7$ implies $(\sqrt{2x + 3})^2 = 7^2$, so that any solution of the first equation will also be a solution of the second; by solving the second equation we find all possible solutions of the first equation.

The second point about the method in Example 9.1 is that as well as producing the solutions of the original equation, the method can also produce answers that are *not* solutions of the original equation. These are called **extraneous solutions,** and they were discussed in Example 7.8 and preceding Example 7.11. Extraneous solutions can arise, in particular, whenever both sides of an equation are squared or whenever both sides of an equation are multiplied by an expression that involves the variable. For this reason we must always check the answers given by the method in Example 9.1. Here is an example to illustrate this point.

Example 9.2 Solve $\sqrt{x + 4} = x - 2$.

Solution If $\sqrt{x + 4} = x - 2$, then
$$(\sqrt{x + 4})^2 = (x - 2)^2$$

$$x + 4 = x^2 - 4x + 4$$

$$x^2 - 5x = 0$$

$$x(x - 5) = 0$$

$$x = 0 \quad \text{or} \quad x - 5 = 0$$

$$x = 0 \quad \text{or} \quad x = 5.$$

Check For $x = 0$:
$$\sqrt{0 + 4} \overset{?}{=} 0 - 2$$

$$2 \neq -2.$$

For $x = 5$:
$$\sqrt{5 + 4} \overset{?}{=} 5 - 2$$

$$3 = 3.$$

Thus 0 is not a solution of the original equation; it is an extraneous solution. But 5 is a solution of the original equation. ∎

In the next example we must square, simplify, and then square again.

Example 9.3 Solve $\sqrt{3x - 3} - \sqrt{x} = 1$.

Solution Before squaring, isolate one of the radicals on one side of the equation; this will make the calculations simpler. If $\sqrt{3x - 3} - \sqrt{x} = 1$, then

$$\sqrt{3x - 3} = 1 + \sqrt{x}$$
$$(\sqrt{3x - 3})^2 = (1 + \sqrt{x})^2$$
$$3x - 3 = 1 + 2\sqrt{x} + (\sqrt{x})^2$$
$$3x - 3 = 1 + 2\sqrt{x} + x$$
$$2x - 4 = 2\sqrt{x}$$
$$x - 2 = \sqrt{x}.$$

Notice that the remaining radical has been isolated on one side of the equation. This was done so there will be no radicals left after we square again. Thus, squaring again,

$$(x - 2)^2 = (\sqrt{x})^2$$
$$x^2 - 4x + 4 = x$$
$$x^2 - 5x + 4 = 0$$
$$(x - 1)(x - 4) = 0$$
$$x = 1 \quad \text{or} \quad x = 4.$$

Check For $x = 1$:

$$\sqrt{3 \cdot 1 - 3} - \sqrt{1} \overset{?}{=} 1$$
$$-1 \neq 1.$$

For $x = 4$:

$$\sqrt{3 \cdot 4 - 3} - \sqrt{4} \overset{?}{=} 1$$
$$\sqrt{9} - \sqrt{4} \overset{?}{=} 1$$
$$1 = 1.$$

Thus 1 is not a solution; it is an extraneous solution. But 4 is a solution. Remember: To check an answer, always substitute in the *original* equation. ∎

Example 9.4 Solve $W^2 = (S/K)^3$ for S.

Solution
$$W^2 = S^3/K^3$$
$$S^3 = K^3 W^2$$
$$S = (K^3 W^2)^{1/3}$$
$$S = KW^{2/3}$$

(This equation gives, approximately, the surface area S of an animal in terms of the weight W of the animal. When S is in square meters and W is in kilograms, the constant K is approximately 0.1 for most animals.) ■

B. THE UNKNOWN IN A DENOMINATOR

To solve an equation with the unknown variable in a denominator, we can first multiply both sides of the equation by a common multiple of the denominators in the equation. We did this in Section 7, where the result was always a linear equation. Here are two examples that produce quadratic equations. Notice that we must check for extraneous solutions in these examples.

Example 9.5 Solve $\dfrac{1}{t + 2} = \dfrac{4}{t} + \dfrac{1}{2}$.

Solution If we multiply both sides of the original equation by the common multiple $2t(t + 2)$ of the denominators, we obtain

$$2t(t + 2)\left(\frac{1}{t + 2}\right) = 2t(t + 2)\left(\frac{4}{t} + \frac{1}{2}\right)$$

$$2t = 8(t + 2) + t(t + 2)$$

$$2t = 8t + 16 + t^2 + 2t$$

$$t^2 + 8t + 16 = 0$$

$$(t + 4)^2 = 0$$

$$t = -4.$$

Check

$$\frac{1}{-4 + 2} \stackrel{?}{=} \frac{4}{-4} + \frac{1}{2}$$

$$\frac{-1}{2} = \frac{-1}{2}.$$

Thus the equation has one solution, -4. ■

Example 9.6 Solve $\dfrac{1}{x} + \dfrac{1}{x + 1} = \dfrac{-1}{x(x - 1)}$.

Solution A common multiple of x, $x + 1$, and $x(x - 1)$ is $x(x + 1)(x - 1)$.

$$x(x + 1)(x - 1)\left(\frac{1}{x} + \frac{1}{x + 1}\right) = x(x + 1)(x - 1)\left(\frac{-1}{x(x - 1)}\right)$$

$$(x + 1)(x - 1) + x(x - 1) = -(x + 1)$$

$$x^2 - 1 + x^2 - x = -x - 1$$

$$2x^2 = 0$$

$$x^2 = 0$$

$$x = 0$$

Check Both sides of the original equation are undefined for $x = 0$; that is, 0 is not in the domain of either side of the equation (Section 7A). Therefore, 0 is extraneous and the original equation has no solution. ∎

C. SUBSTITUTIONS

The equations on the left in the following list are not quadratic equations. If, however, we replace the variable by a new variable through the indicated substitution in each case, then we do get a quadratic equation, as shown on the right. We can solve the equation on the left if we first solve its related quadratic equation and then use each solution to obtain a corresponding value of the original variable. Examples 9.7 and 9.8 will show the details for the first two examples.

Equation	Substitution	New (*quadratic*) equation
$y^4 - 3y^2 - 4 = 0$	$x = y^2$	$x^2 - 3x - 4 = 0$
$x^{-2/3} + 3x^{-1/3} - 10 = 0$	$t = x^{-1/3}$	$t^2 + 3t - 10 = 0$
$\left(y + \dfrac{1}{y}\right)^2 + 5\left(y + \dfrac{1}{y}\right) + 6 = 0$	$x = y + \dfrac{1}{y}$	$x^2 + 5x + 6 = 0$

Example 9.7 Solve $y^4 - 3y^2 - 4 = 0$.

Solution Substitute x for y^2. This gives the quadratic equation

$$x^2 - 3x - 4 = 0.$$

Solve for x:

$$(x + 1)(x - 4) = 0$$

$$x = -1 \quad \text{or} \quad x = 4.$$

Now return to the original variable y by using $x = y^2$:

$$x = -1 \text{ implies } y^2 = -1$$

$$x = 4 \text{ implies } y^2 = 4.$$

There is no real number y such that $y^2 = -1$. But $y^2 = 4$ if $y = 2$ or $y = -2$. You can verify by substitution that both 2 and -2 are solutions of the original equation. ∎

Example 9.8 Solve $x^{-2/3} + 3x^{-1/3} - 10 = 0$.

Solution Substitute t for $x^{-1/3}$. Then the original equation gives

$$t^2 + 3t - 10 = 0$$

$$(t - 2)(t + 5) = 0$$

$$t = 2 \quad \text{or} \quad t = -5.$$

Therefore, since $t = x^{-1/3}$, x will be a solution of the original equation if

$$x^{-1/3} = 2 \quad \text{or} \quad x^{-1/3} = -5,$$

that is, if

$$x^{1/3} = \frac{1}{2} \quad \text{or} \quad x^{1/3} = -\frac{1}{5}$$

$$x = \frac{1}{8} \quad \text{or} \quad x = -\frac{1}{125}.$$

Substitution will show that both $\frac{1}{8}$ and $-\frac{1}{125}$ are indeed solutions of the original equation. ■

D. ABSOLUTE VALUE EQUATIONS

To solve an equation that contains absolute values, we can first convert to one or more equations that do not contain absolute values. To do this we use the fact that, if $c \geq 0$, then $|y| = c$ iff $y = c$ or $y = -c$.

Example 9.9 Solve $|2x + 7| = |1 - x|$.

Solution This equation is equivalent to

$$2x + 7 = 1 - x \quad \text{or} \quad 2x + 7 = -(1 - x)$$
$$3x = -6 \quad \text{or} \quad 2x + 7 = -1 + x$$
$$x = -2 \quad \text{or} \quad x = -8$$

Substitution will verify that both -2 and -8 are solutions of the original equation.

This equation could also be solved by squaring both sides at the first step. This would give $(2x + 7)^2 = (1 - x)^2$, which is a quadratic equation without absolute values. ■

An equation with an absolute value need not have a solution. For example, $|x - 5| = -2$ has no solution since $-2 < 0$ but $|x - 5| \geq 0$ for all x.

EXERCISES FOR SECTION 9

Solve each equation.

1. $\sqrt{3x + 4} = 5$

2. $\sqrt{10 - 4x} = 2$

3. $3 = \sqrt{3x + 11}$

4. $(2x + 2)^{1/2} = x - 3$

5. $(5x - 4)^{1/2} = x$

6. $(4x + 1)^{1/2} = x - 1$

7. $\sqrt{3u + 1} - \sqrt{u + 4} = 1$

8. $\sqrt{3v} - 2\sqrt{v + 4} + 2 = 0$

9. $\sqrt{w} - \sqrt{2w + 7} = -2$

10. $\sqrt{x}\,\sqrt{10 - 9x} = 1$

11. $\sqrt{y}\,\sqrt{4y + 3} = 1$

12. $\sqrt{8z - 2}\,\sqrt{z} = 6$

13. $\sqrt{p - 1} = \dfrac{\sqrt{3}}{\sqrt{p + 1}}$

14. $\sqrt{q} - \dfrac{2}{\sqrt{q}} = 1$

15. $\sqrt{r^2 + 1} - r = 2$

16. $\dfrac{x}{x - 1} = \dfrac{8}{x + 2}$

17. $\dfrac{4}{x} = \dfrac{x}{x-1}$

18. $x + \left(\dfrac{0.01}{x+0.2}\right) = 0$

19. $\dfrac{2}{k^2-1} - \dfrac{k}{k-1} = \dfrac{1}{k+1}$

20. $\dfrac{m}{m+1} + \dfrac{2}{m} = \dfrac{5}{m(m+1)}$

21. $\dfrac{t+2}{3t} + \dfrac{1}{t-3} = \dfrac{3}{t(t-3)}$

22. $x^4 - 9 = 0$

23. $y^{2/3} + 10y^{1/3} + 25 = 0$

24. $x^{-4} - 4x^{-2} = 0$

25. $\left(1 + \dfrac{1}{x}\right)^2 + 5\left(1 + \dfrac{1}{x}\right) + 4 = 0$

26. $0.0016 - y^4 = 0$

27. $\left(z + \dfrac{1}{5}\right)^2 - 4\left(z + \dfrac{1}{5}\right) - 5 = 0$

28. $\left(\dfrac{1}{z+1}\right)^{2/5} + 4\left(\dfrac{1}{z+1}\right)^{1/5} + 4 = 0$

29. $(y^2 - 1)^2 + (y^2 - 1) - 6 = 0$

30. $2z^{1/3} + 3z^{1/6} - 2 = 0$

31. $(x - 1)^{2/3} = 2$

32. $\sqrt{x+1} = \sqrt[4]{x+7}$

33. $2 = \left(1 + \dfrac{x}{4}\right)^4$

34. $|-x + 5| = 3$

35. $\left|2x - \dfrac{1}{2}\right| = \dfrac{1}{3}$

36. $|4 - x| = -1$

37. $|x + 1| = |x - 2|$

38. $|3x + 1| = -x$

39. $2|x - 3| = 3|x - 4|$

40. $\sqrt{0.693 - 0.287x} = 1.54$ C

41. $\dfrac{4.1}{2.7x} - \dfrac{2.3}{6.4x + 1.9} = 0$ C

42. $|1.79x - 0.954| = 2.83$ C

43. Solve $r = \sqrt{A/\pi}$ for A.

44. Solve $t = \sqrt{2s/g}$ for s.

45. Solve $T = 2\pi\sqrt{l/g}$ for g.

46. Solve $d = \sqrt{x^2 + y^2}$ for y if $y > 0$.

47. Solve $S = \pi r\sqrt{r^2 + h^2}$ for h if $h > 0$.

48. Solve $y = \sqrt{(1 - x)/2}$ for x.

49. Solve $\dfrac{1}{R} = \dfrac{1}{R_1} + \dfrac{1}{R_2}$ for R_1.

50. Solve $R^3/T^2 = K$ for T.

51. Solve $r_E = \left(1 + \dfrac{r_N}{n}\right)^n - 1$ for r_N.

52. Express the surface area ($S = 6x^2$) of a cube in terms of the volume ($V = x^3$) of the cube.

53. Express the circumference ($C = 2\pi r$) of a circle in terms of the area ($A = \pi r^2$) of the circle.

54. Express the volume $\left(V = \dfrac{4}{3}\pi r^3\right)$ of a sphere in terms of the surface area ($S = 4\pi r^2$) of the sphere.

55. Suppose a container is filled with water to a height of h centimeters, and suppose there is a small opening of cross-sectional area A square centimeters at the bottom of the container. By *Toricelli's law*, the velocity with which water will pass through the opening is

$$v = 0.6\,A\sqrt{2gh}\ \text{cm}^3/\text{sec}, \qquad (9.1)$$

where $g = 980$ cm/sec^2 (the gravitational constant).

(a) If $A = 2$ cm^2, what is v when $h = 10$ cm?

(b) Solve Equation (9.1) for h in terms of g, A, and v.

(c) A coffee urn is originally filled to a depth of 28 centimeters. What is the depth of the coffee in the urn when, sometime later, coffee is observed to pass through the spigot at the bottom of the urn at one-half of its original rate?   C

56. It has often been claimed that a rectangle of length L and width W will have the most pleasing proportions if L and W satisfy the equation

$$\dfrac{L + W}{L} = \dfrac{L}{W}. \qquad (9.2)$$

(a) Solve Equation (9.2) for L in terms of W.

(b) What is L if $W = 10$ centimeters?

57. A salesman who has always driven the 275 miles between two cities at the same constant rate discovers that he can save 30 minutes on the trip if he drives 5 miles per hour faster. Find the faster rate.

SECTION **10**
Linear and Absolute Value Inequalities

A. ORDER PROPERTIES

Order properties relate to inequality. In Section 1 we looked briefly at inequality in terms of a real line directed to the right:

$$a < b \quad \text{iff} \quad a \text{ is to the left of } b.$$

Also, a number was defined to be *positive* if it is to the right of 0 and *negative* if it is to the left of 0. It is advantageous to restate the condition for $a < b$ in the following equivalent form:

$$a < b \quad \text{iff} \quad b - a \text{ is positive.}$$

Example 10.1 (See Figure 10.1.)

(a) $2 < 7$ because $7 - 2 = 5$ is positive.

(b) $-6 < -4.5$ because $-4.5 - (-6) = -4.5 + 6 = 1.5$ is positive.

(c) $-5 < \dfrac{2}{3}$ because $\dfrac{2}{3} - (-5) = \dfrac{2}{3} + 5 = \dfrac{17}{3}$ is positive. ▪

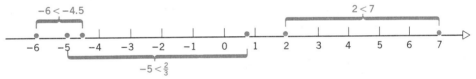

FIGURE 10.1

(Appendix A indicates how to define *positive* without appealing to a real line. When *positive* has been defined independently of a real line, then the equivalence "$a < b$ iff $b - a$ is positive" makes the idea of *inequality* also independent of a real line.)

The notations $a \le b$, $b > a$, and $b \ge a$, as well as the terminology *less than* and *greater than*, are all defined in terms of $a = b$ and $a < b$ as in Section 1C. The basic properties of inequalities are the order properties that follow. Properties (10.1) and (10.2) are actually axioms; Properties (10.3) through (10.6) can be proved using (10.1), (10.2), and other properties of the real numbers. (See Examples 10.2 and 10.3.) *Pay special attention to Properties (10.5) and (10.6).* The former states that if both sides of an inequality are multiplied by the same *positive* number, then the sense (direction) of the inequality remains *unchanged;* the latter states that if both sides are multiplied by the same *negative* number, then the sense is *reversed.*

Order properties

If a is any real number, then exactly one of the following is true:

$$a > 0, a = 0, \text{ or } -a > 0. \tag{10.1}$$

If $a > 0$ and $b > 0$, then $a + b > 0$ and $ab > 0$. $\tag{10.2}$

If $a < b$ and $b < c$, then $a < c$. $\tag{10.3}$

If $a < b$, then

$$a + c < b + c \text{ and } a - c < b - c \tag{10.4}$$

for every c.

If $a < b$ and $c > 0$, then $ac < bc$. $\tag{10.5}$

If $a < b$ and $c < 0$, then $ac > bc$. $\tag{10.6}$

The next two examples give proofs of (10.5) and the first part of (10.4) using "$a < b$ iff $b - a$ is positive" as the definition. We use without hesitation the fact that $b > a$ means the same as $a < b$.

Example 10.2 Prove this part of Property (10.4):

$$\text{If } a < b, \text{ then } a + c < b + c.$$

Proof If $a < b$, then $b - a$ is positive. But $b - a = (b + c) - (a + c)$, so $(b + c) - (a + c)$ is positive and $a + c < b + c$. ■

Example 10.3 Use Property (10.2) to prove Property (10.5).

Proof Assume $a < b$ and $c > 0$. Then $b - a > 0$, and thus $(b - a)c > 0$ by (10.2), since (10.2) states that the product of two positive numbers is positive. Thus $bc - ac > 0$ so $ac < bc$. ■

B. LINEAR INEQUALITIES

The basic terminology about equations with one variable also applies to inequalities with one variable. A real number is a **solution** of such an inequality (or **satisfies** the inequality) if it leads to a true inequality of numbers when it replaces the variable. Thus 4 is a solution of

$$3x < 15$$

because $3 \cdot 4 = 12$ and $12 < 15$.

The set of all solutions of an inequality is called its **solution set.** To **solve** an inequality is to find its solution set. Two inequalities with the same variable are **equivalent** if they have the same solution set.

As with an equation, the most basic plan for solving an inequality is to replace it by simpler and simpler equivalent inequalities until the solutions (if there are any) become obvious. The most frequently used tools for this are Properties (10.4)–(10.6). For the present we concentrate on **linear inequalities**— that is, inequalities that can be written in the form

$$ax + b < 0 \quad \text{or} \quad ax + b \le 0, \text{ with } a \ne 0.$$

Linear inequalities are solved just like linear equations, except that the sense of the inequality sign must be reversed whenever the sides are interchanged or both sides are multiplied (or divided) by a negative number.

Example 10.4

$$9x - 6 < 7x + 4$$

$$9x < 7x + 10 \qquad \text{Add 6 to both sides.}$$

$$2x < 10 \qquad \text{Subtract } 7x \text{ from both sides.}$$

$$x < 5 \qquad \text{Divide both sides by 2.}$$

The original inequality has been transformed into the equivalent inequality $x < 5$. Thus, expressed with set-builder notation (Section 7A), the solution set of the original inequality is $\{x: x < 5\}$. ■

Notice that Example 10.4 has an infinite number of solutions. That is typical of linear inequalities—in contrast with linear equations, which have only one solution each.

The rules for solving inequalities with $\leq$ or $\geq$ are the same as those for $<$ and $>$. Here is an example.

Example 10.5

$$\frac{3u - 2}{5} + 3 \geq \frac{4u - 1}{3}$$

$$3(3u - 2) + 15 \cdot 3 \geq 5(4u - 1) \qquad \text{Multiply by 15.}$$

$$9u - 6 + 45 \geq 20u - 5 \qquad \text{Distributive property.}$$

$$9u \geq 20u - 44 \qquad \text{Subtract 39.}$$

$$-11u \geq -44 \qquad \text{Subtract } 20u.$$

$$u \leq 4 \qquad \begin{array}{l}\text{Divide by } -11 \text{ and} \\ \text{change } \geq \text{ to } \leq.\end{array}$$

Thus the solution set is $\{u: u \leq 4\}$. ■

C. INTERVALS

Assume that a and b are real numbers with $a < b$. The notation $a < x < b$ means that both $a < x$ and $x < b$; that is, x is between a and b. The notations $[a, b]$, $[a, b)$, $(a, b]$, and (a, b) are used to denote the four special sets of real numbers defined in Table 10.1. Each set is called an **interval** with **endpoints** a

TABLE 10.1

Interval	Definition	Figure
$[a, b]$	$\{x: a \leq x \leq b\}$	
$[a, b)$	$\{x: a \leq x < b\}$	
$(a, b]$	$\{x: a < x \leq b\}$	
(a, b)	$\{x: a < x < b\}$	

and b. The keys are that [is used if a is to be *included* in the set, and (is used if a is to be *excluded*. Similarly for] and). [Sometimes a closed circle (●) is used to denote an endpoint that is to be *included*, and an open circle (○) is used to denote an endpoint that is to be *excluded*.]

Table 10.2 shows the five types of **unbounded intervals**—that is, intervals without an endpoint on one or both ends. The symbols $-\infty$ and ∞ *do not* denote real numbers: "$(-\infty$" simply denotes the absence of an endpoint on the left, and "$\infty)$" denotes the absence of an endpoint on the right.

TABLE 10.2

Interval	Definition	Figure
$[a, \infty)$	$\{x: x \geq a\}$	
$(-\infty, b]$	$\{x: x \leq b\}$	
(a, ∞)	$\{x: x > a\}$	
$(-\infty, b)$	$\{x: x < b\}$	
$(-\infty, \infty)$	$\{x: x \text{ is a real number}\}$	

Sometimes it is convenient to use set notation when working with intervals. For this we need the following ideas. (For more examples of these ideas see Appendix B.) Let A and B represent sets (collections of numbers, for example).

$x \in A$ denotes that x is an element of (or is a member of, or belongs to) the set A.

$\{a, b, \ldots\}$ denotes the set whose members are $a, b, \ldots$.

$\emptyset$ denotes the **empty set,** that is, the set that contains no elements.

$A \cup B$ denotes $\{x: x \in A \text{ or } x \in B\}$.

$A \cap B$ denotes $\{x: x \in A \text{ and } x \in B\}$.

The set $A \cup B$ is called the **union** of A and B ("$x \in A$ or $x \in B$" includes the possibility that x belongs to *both* A and B). The set $A \cap B$ is called the **intersection** of A and B. (See Figure 10.2.)

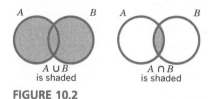

$A \cup B$
is shaded

$A \cap B$
is shaded

FIGURE 10.2

Example 10.6 Figure 10.3 shows the intervals that appear on the left in these equations.

(a) $[-1, 3] \cup [1, 4) = [-1, 4)$

(b) $[-1, 3] \cap [1, 4) = [1, 3]$

(c) $(-\infty, 1] \cap [1, 4) = \{1\}$

(d) $(-\infty, 1] \cap (-1, \infty) = (-1, 1]$

(e) $(-3, -2) \cap [1, 4) = \emptyset$

(f) The set $(-3, -2) \cup [1, 4)$ is not an interval.

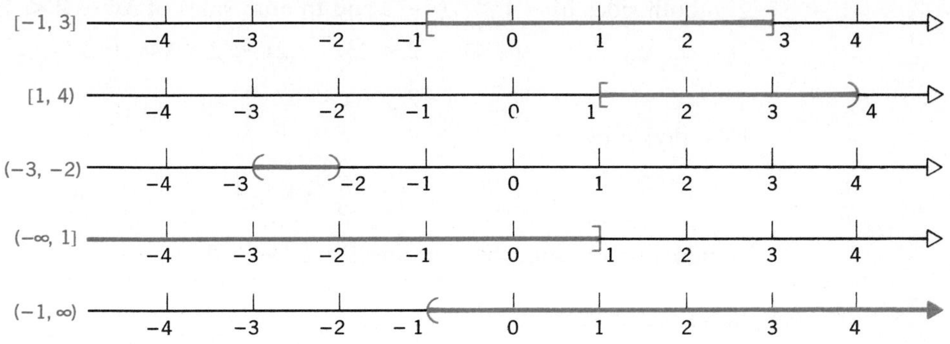

$[-1, 3]$
$[1, 4)$
$(-3, -2)$
$(-\infty, 1]$
$(-1, \infty)$

FIGURE 10.3

D. ABSOLUTE VALUE INEQUALITIES

To solve an inequality that contains absolute values, we can first convert to one or more inequalities that do not contain absolute values. To do this we use the facts that, for $c > 0$,

$$|y| < c \quad \text{iff} \quad y \text{ is } within \ c \text{ units of } 0 \text{ (Figure 10.4)},$$

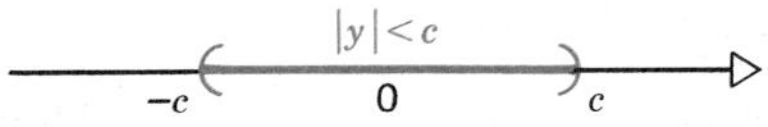

FIGURE 10.4

and

$$|y| > c \quad \text{iff} \quad y \text{ is } more \ than \ c \text{ units from } 0 \text{ (Figure 10.5)}.$$

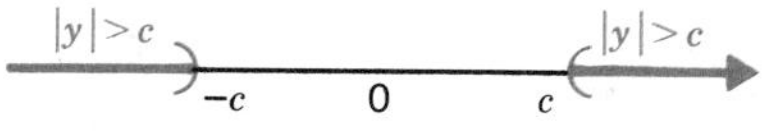

FIGURE 10.5

> For any real numbers y and c such that $c > 0$:
> $$|y| < c \quad \text{iff} \quad -c < y < c \tag{10.7}$$
> $$|y| > c \quad \text{iff} \quad y < -c \text{ or } y > c \tag{10.8}$$

Example 10.7 Solve $|x| < 3$.

Solution By (10.7), with $y = x$ and $c = 3$, the solution set is $\{x: -3 < x < 3\}$. In interval notation, the solution set is $(-3, 3)$. ∎

Example 10.8 Solve $|3x - 2| < 4$.

Solution Using (10.7) with $y = 3x - 2$ and $c = 4$, we see that the given inequality is equivalent to

$$-4 < 3x - 2 < 4.$$

Now proceed as in Subsection B. Add 2 to each part (this is equivalent to adding

2 to both sides of $-4 < 3x - 2$ and to both sides of $3x - 2 < 4$):
$$(-4) + 2 < (3x - 2) + 2 < (4) + 2$$
$$-2 < 3x < 6.$$

Now divide by 3:
$$-\frac{2}{3} < x < 2.$$

In interval notation, the solutions set is $(-\frac{2}{3}, 2)$.

Example 10.9 Solve $|2x| \geq 7$.

Solution Use (10.8) with $y = 2x$ and $c = 7$, and the corresponding statement with $=$ in place of $>$. We see that the given inequality is equivalent to
$$2x \leq -7 \quad \text{or} \quad 2x \geq 7$$
$$x \leq -\frac{7}{2} \quad \text{or} \quad x \geq \frac{7}{2}.$$

Thus the solution set is $(-\infty, -3.5] \cup [3.5, \infty)$.

Example 10.10 Solve $5 \leq |4 - x|$.

Solution This is equivalent to $|4 - x| \geq 5$. Apply (10.8) with $y = 4 - x$ and $c = 5$, and the corresponding statement with $=$ in place of $>$. This leads to
$$4 - x \leq -5 \quad \text{or} \quad 4 - x \geq 5$$
$$-x \leq -9 \quad \text{or} \quad -x \geq 1$$
$$x \geq 9 \quad \text{or} \quad x \leq -1.$$

Thus the solution set is $(-\infty, -1] \cup [9, \infty)$.

E. APPLICATION

Example 10.11 Let C denote temperature in degrees Celsius and F temperature in degrees Fahrenheit. The normal January daily temperature range in Honolulu is 65°F to 79°F. Find the range in degrees Celsius.

Solution The formula for C in terms of F is
$$F = \frac{9}{5}C + 32. \tag{10.9}$$

Therefore, $65 \leq F \leq 79$ implies
$$65 \leq \frac{9}{5}C + 32 \leq 79$$
$$65 - 32 \leq \frac{9}{5}C \leq 79 - 32$$
$$\frac{5}{9}(33) \leq C \leq \frac{5}{9}(47)$$
$$18.\overline{3} \leq C \leq 26.\overline{1}$$

Thus the range is approximately 18°C to 26°C.

EXERCISES FOR SECTION 10

Replace each □ with the appropriate sign, either <
or >.

1. $-1.4 \,\square\, -1.5$

2. $\dfrac{6}{25} \,\square\, \dfrac{7}{25}$

3. $0 \,\square\, -4$

4. $\dfrac{1}{5} \,\square\, \dfrac{1}{6}$

5. $-0.2 \,\square\, -0.19$

6. $1/0.2 \,\square\, 1/0.1$

7. $3 - x^3 \,\square\, 1 - x^3$

8. $-7\pi \,\square\, -5\pi$

9. $-5\sqrt{2} \,\square\, -5\sqrt{3}$

10. $1.2 \times 10^6 \,\square\, 1.2 \times 10^7$

11. $1.2 \times 10^{-13} \,\square\, 1.2 \times 10^{-14}$

12. $-1.2 \times 10^7 \,\square\, -1.2 \times 10^8$

Solve each inequality and express the solution set with
set-builder notation.

13. $6x + 2 > 3x - 4$

14. $5y - 4 < 7y + 2$

15. $1 + 5z < 2 - z$

16. $2(4 - t) \leq 5(1 + 2t)$

17. $\dfrac{1}{3}(u + 1) > \dfrac{1}{4}(u - 1)$

18. $\dfrac{1}{2}(3 - 2x) \geq \dfrac{1}{10}(x + 4)$

19. $0.05[v - 3(1 - v)] \leq 0.4v$

20. $1.46 \geq 0.2[1 + 2.1(r - 3)]$

21. $-y \geq 1.4[2(2 - y) + 0.5]$

Graph each interval on a real line. (See Figure 10.3.)
Then simplify the set (if possible) as in Example 10.6.

22. $(-2, 1] \cup [-1, 2)$

23. $[0, 3] \cap [2, 4]$

24. $[0, 3] \cup [2, 4]$

25. $(1, 5] \cap [-4, 1)$

26. $[1.5, 2.5] \cap [2.5, \infty)$

27. $(-\infty, -1) \cup (2, \infty)$

28. $[0, \infty) \cup (-\infty, 0]$

29. $(-\infty, 0) \cap (0, \infty)$

30. $(-\infty, -1] \cup [1, \infty)$

Solve each inequality and express the solution set with
interval notation.

31. $|x| \leq 4$

32. $|2x| > 14$

33. $|-x| \leq 5$

34. $|6x| > |-3|$

35. $|-6| \geq |-0.1x|$

36. $|-1| < |\tfrac{1}{2}x|$

37. $|2x - 1| < 7$

38. $|3x + 1| \geq 5$

39. $|\tfrac{1}{2}x - 1| \leq \tfrac{3}{2}$

40. $|0.2 - 0.1x| \geq 0.3$

41. $|1 - 5x| < 1$

42. $|3 - 2x| > 1$

43. $4 < |x| < 9$

44. $5 \geq |2x| \geq 1$

45. $2 < |3x - 1| < 5$

46. $|x - 3| < |x|$

47. $|x| \leq |x + 4|$

48. $|x - 5| < |x + 1|$

Find the domain of each expression in Exercises 49–
51.

49. $\sqrt{3x + 4} + \dfrac{1}{\sqrt{2 - x}}$

50. $\dfrac{3}{\sqrt{x}} - \dfrac{4}{\sqrt{3x - 6}}$

51. $\dfrac{1}{\sqrt{x + 5}} + 2\sqrt{-2x - 10}$

In Exercises 52–54: (a) Write an absolute value in-
equality to represent each statement. (b) Solve the
inequality from part (a) and express the answer with
interval notation.

52. The point representing x on a real line is within
2 units of the point representing 5.

53. The point representing x on a real line is at least
6 units from the point representing 3.

54. The point representing x on a real line is more
than twice as far from the point representing -1
as from the point representing 2.

55. Find all values of k for which
$$kx^2 - 3x + 2 = 0$$
has two real solutions.

56. Find all values of p for which
$$-x^2 + 5x + p = 0$$
has no real solution.

57. Prove that $|a + b| \leq |a| + |b|$ for all real num-
bers a and b. This is known as the **triangle in-**
equality. [Suggestion: An equivalent form is
$-|a| - |b| \leq a + b \leq |a| + |b|$. (Why?) Also,
$-|a| \leq a \leq |a|$ and $-|b| \leq b \leq |b|$. (Why?)]

Solve Exercises 58–63 using inequalities.

58. The normal January temperature range in Minneapolis-St. Paul is 3°F to 21°F. Find the range in degrees Celsius.

59. What range of temperatures in degrees Fahrenheit corresponds to a range of 10°C to 20°C?

60. Suppose the pressure P (in pounds) and volume V (in cubic inches) of a confined gas are related by the equation $P = 100/V$. What range for the volume corresponds to a range of 20 pounds per square inch to 40 pounds per square inch

for pressure? (Suggestion: If $a > b > 0$, then $1/b > 1/a$.)

61. If $y = 1/x$ and $|x - 3| < 2$, what can you conclude about y? (Suggestion: If $a > b > 0$, then $1/b > 1/a$.)

62. What can you conclude about the radius of a circle if you know that the circumference exceeds the perimeter of a square whose edges are each 10 centimeters?

63. For which sets of three consecutive positive integers will four times the sum of the first two integers exceed seven times the third of the integers?

SECTION **11**
Nonlinear Inequalities

A. POLYNOMIAL INEQUALITIES

To solve an inequality in which both sides are polynomials, we first rewrite the inequality with only 0 on one side. We next try to factor the other side, and then use the fact that a product of nonzero real numbers is positive iff either all of the factors are positive or an even number of them are negative.

Example 11.1 Solve $x(x - 1) > 6$.

Solution
$$x^2 - x > 6 \qquad \text{Distributive law.}$$
$$x^2 - x - 6 > 0 \qquad \text{Add } -6 \text{ to both sides.}$$
$$(x - 3)(x + 2) > 0 \qquad \text{Factor.}$$

It suffices to solve the latter inequality, since it is equivalent to the original inequality. The factors $x - 3$ and $x + 2$ are 0 at $x = 3$ and $x = -2$, respectively. When these numbers are plotted on a number line, three intervals are formed: $(-\infty, -2)$, $(-2, 3)$, and $(3, \infty)$ (Figure 11.1). The factors $x - 3$ and $x + 2$ will not change sign within these intervals. Figure 11.1 shows the sign of each factor within each interval. To determine the sign of a factor within one of the intervals, it suffices to test with any number in that interval. For example, if $x = 4$ then $x - 3 > 0$, so $x - 3 > 0$ throughout the interval $(3, \infty)$. Since $(x - 3)(x + 2) > 0$ iff both factors are positive or both are negative, the signs of $(x -$

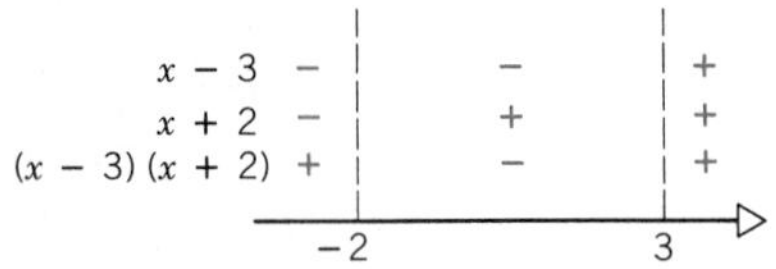

$x - 3$	$-$	$-$	$+$
$x + 2$	$-$	$+$	$+$
$(x - 3)(x + 2)$	$+$	$-$	$+$

FIGURE 11.1

$3)(x + 2)$ are as shown in Figure 11.1. Thus the solution set is

$$(-\infty, -2) \cup (3, \infty).$$

Once the intervals in Figure 11.1 were determined, we could have completed the solution by considering the signs of $x^2 - x - 6$ directly, rather than the signs of the individual factors. In general, choose whichever method seems easier.

Example 11.2 Solve $2x^3 - 15x \le -7x^2$.

Solution
$$2x^3 + 7x^2 - 15x \le 0$$

$$x(2x^2 + 7x - 15) \le 0$$

$$x(2x - 3)(x + 5) \le 0$$

The factors on the left are zero at $x = 0$, $x = \frac{3}{2}$, and $x = -5$, respectively. Four intervals are determined. The signs of the factors and the product within these intervals are shown in Figure 11.2. In this case we want the intervals where the product is negative. Since the values where the product is 0 also provide solutions, we include -5, 0, and $\frac{3}{2}$. Thus the solution set is

$$(-\infty, -5] \cup [0, \tfrac{3}{2}].$$

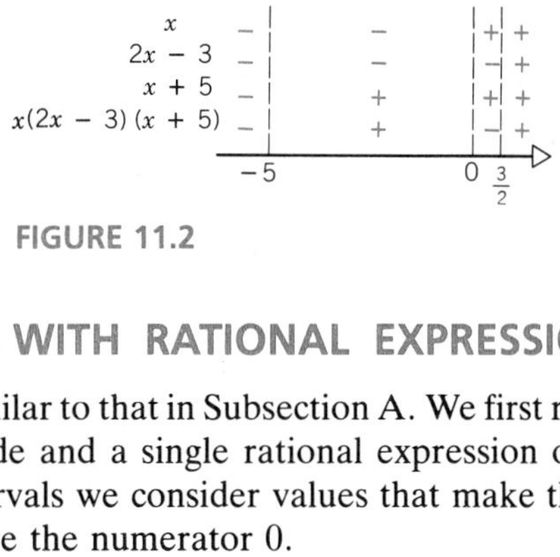

FIGURE 11.2

B. INEQUALITIES WITH RATIONAL EXPRESSIONS

The method here is similar to that in Subsection A. We first rewrite the inequality with only 0 on one side and a single rational expression on the other. Notice that to determine intervals we consider values that make the denominator 0 as well as those that make the numerator 0.

Example 11.3 Solve $\dfrac{x}{x + 3} + 2 \le 0$.

Solution Perform the indicated addition on the left, to obtain a single rational expression.

$$\frac{x}{x + 3} + \frac{2(x + 3)}{x + 3} \le 0 \qquad \text{Obtain common denominators.}$$

$$\frac{3x + 6}{x + 3} \le 0 \qquad \text{Add.}$$

$$\frac{3(x + 2)}{x + 3} \le 0 \qquad \text{Factor the numerator.}$$

$$\frac{x + 2}{x + 3} \le 0 \qquad \text{Divide both sides by 3.}$$

The latter inequality is equivalent to the original inequality. Figure 11.3 shows where $x + 2 = 0$ and $x + 3 = 0$, and also the relevant signs in each resulting interval. The quotient is negative for x the interval $(-3, -2)$. Since $x = -2$ is also a solution of $(x + 2)/(x + 3) \leq 0$, the solution set is $(-3, -2]$. ■

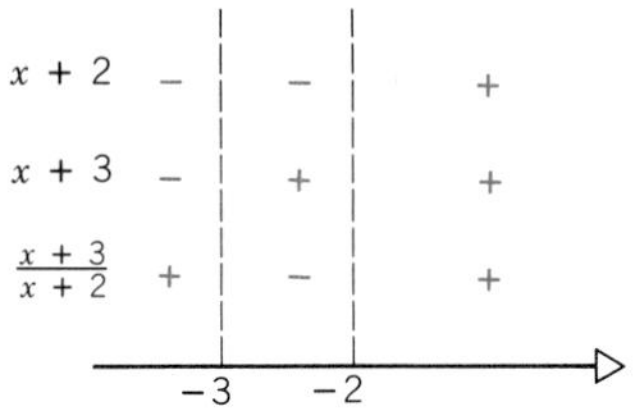

FIGURE 11.3

Example 11.4 Solve $\dfrac{x}{x - 1} \geq \dfrac{5}{x + 5}$.

Solution

$$\frac{x}{x - 1} - \frac{5}{x + 5} \geq 0$$

$$\frac{x(x + 5) - 5(x - 1)}{(x - 1)(x + 5)} \geq 0$$

$$\frac{x^2 + 5}{(x - 1)(x + 5)} \geq 0$$

The numerator is never 0. Figure 11.4 shows where $x - 1 = 0$ and $x + 5 = 0$, and the sign of the quotient in each resulting interval. The solution set is

$$(-\infty, -5) \cup (1, \infty).$$

Since $x^2 + 5 > 0$ for all x, the solution set would not be changed if the sign $\geq$ in the original inequality were changed to $>$. ■

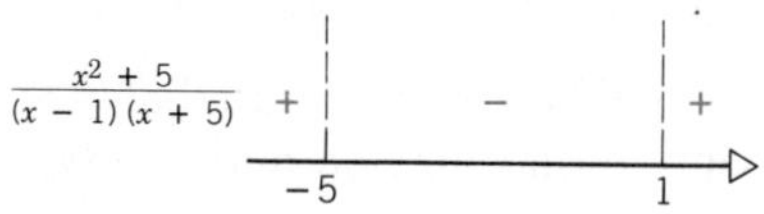

FIGURE 11.4

C. APPLICATION

Example 11.5 If a rocket is fired upward from the earth's surface with an initial velocity of 800 ft/sec, and if s denotes its height t seconds later, then

$$s = -16t^2 + 800t$$

(until the rocket returns to the earth). If the rocket enters the clouds at 8400 feet and returns through the clouds at 8400 feet, during what period will it be above cloud level?

Solution The values of t for which the rocket will be above cloud level are the solutions of

$$-16t^2 + 800t > 8400.$$

This is equivalent to

$$-16t^2 + 800t - 8400 > 0$$

$$t^2 - 50t + 525 < 0 \qquad \text{Divide by } -16 \text{ and change } > \text{ to } <.$$

$$(t - 15)(t - 35) < 0.$$

The solution set of the latter inequality is $\{t: 15 < t < 35\}$. Thus the rocket will be above cloud level from 15 seconds to 35 seconds after it is fired. ∎

EXERCISES FOR SECTION 11

Solve and express the solution set with interval notation.

1. $x^2 - 4x < 5$

2. $10 > x^2 - 3x$

3. $x^2 < 2x + 8$

4. $2(x^2 + x + 2) \geq -7x$

5. $4x^2 + 2x \leq x^2 + x + 2$

6. $(2x + 1)^2 \leq x^2 - x + 3$

7. $y^2 > 1.5y + 1$

8. $y^2 - 2.2y \geq 1.21$

9. $y(y + 0.1) > 0.02$

10. $x^3 \leq x$

11. $(x - 2)x \leq (x - 2)x^2$

12. $x^4 \geq 16$

13. $2x^3 + 3x^2 - 2x \geq 0$

14. $(|x| - 2)(x - 3) < 0$

15. $\dfrac{x - 1}{2x} < 0$

16. $\dfrac{x + 1}{x - 1} > 0$

17. $\dfrac{(x - 1)^2}{x + 4} \leq 0$

18. $\dfrac{t + 1}{t + 2} \leq 3$

19. $\dfrac{1}{t} - t \geq 0$

20. $\dfrac{1}{t + 1} > \dfrac{1}{t - 1}$

21. $\dfrac{2}{x - 2} \geq \dfrac{1}{x + 5}$

22. $\dfrac{4}{x^2 + 2} < \dfrac{3}{x}$

23. $\dfrac{1}{x + 1} + \dfrac{1}{x - 1} < \dfrac{1}{x}$

24. $\dfrac{|x - 2|}{|x| - 2} < 0$

Find the domain of each expression.

25. $\sqrt{1 - x^2}$

26. $\sqrt{x^2 - x - 12}$

27. $\sqrt{0.5 - |2x|}$

28. $\sqrt{\dfrac{x + 5}{|x| - 1}}$

29. $\sqrt{\dfrac{x^2 - 4}{1 - |x|}}$

30. $\sqrt[4]{\dfrac{3x + 1}{1 - x}}$

31. Find all values of k for which $x^2 + kx + 1 = 0$ has two real solutions for x.

32. Find all values of p for which $x^2 + px + p = 0$ has no real solution for x.

33. Prove that if $q \neq 0$, then $x^2 + qx + q^2 = 0$ has no real solution for x.

34. During what period will the rocket in Example 11.5 be above cloud level if the clouds are at 6400 feet rather than 8400 feet?

35. Find all sets of three consecutive integers such that the square of the largest exceeds the sum of the squares of the other two.

36. What can we conclude about the radius of a circle if we know that the area exceeds that of a square whose edges are each 5 centimeters?

37. If resistors of resistances R_1 and R_2 ohms are in parallel in a direct electrical circuit, then the equivalent effective resistance R satisfies

$$\frac{1}{R} = \frac{1}{R_1} + \frac{1}{R_2}.$$

Suppose that $R_1 = 10$ and $6 \leq R_2 \leq 12$. What can we conclude about R?

38. The perimeter of a rectangle is 100 centimeters and the area of the rectangle is known to be between 200 and 400 square centimeters. Let x denote the length of the shorter side of the rectangle. Write two inequalities that x must satisfy and then use them to determine the possible values for x.

39. If the absolute value of a real number exceeds

the reciprocal of the number, what can we conclude about the number?

40. Prove that $\sqrt{ab} \leq \dfrac{a + b}{2}$ for all positive real numbers a and b. [Suggestion: Begin with $(\sqrt{a} - \sqrt{b})^2 \geq 0$.]

41. Prove that $\dfrac{a^2 + b^2}{2} \geq \left(\dfrac{a + b}{2}\right)^2$ for all real numbers a and b. [Suggestion: Show that the inequality is equivalent to $(a - b)^2 \geq 0$.]

42. Prove that if $a < b$ and $(x - a)^2 < (x - b)^2$, then $x < (a + b)/2$.

REVIEW EXERCISES FOR CHAPTER II

Solve each equation for x.

1. $5(x - 2) = 3(1 - 2x)$

2. $\dfrac{x + 1}{2} - \dfrac{2x - 1}{5} = \dfrac{1}{4}$

3. $4(x - 1) = 3y$

4. $y = 4(x + 2) + 1$

5. $\dfrac{1}{x + 2} - \dfrac{2}{x} = 0$

6. $\dfrac{x}{x + 2} + 1 = \dfrac{-2}{x + 2}$

Solve by factoring.

7. $x^2 + 3x - 10 = 0$

8. $2y^2 + 5y = -3$

Solve by completing the square.

9. $2t^2 = 1 - 3t$ **10.** $3u^2 - u = 1$

Solve by any method.

11. $0.09w^2 - 1.21 = 0$

12. $x^2 + px + 1 = 0$

13. $7y^2 - 6y + 1 = 0$

14. $25z^2 = 49$

Use the discriminant to determine whether each equation has no real solution, one real solution, or two real solutions.

15. $r^2 = 4 + r$

16. $0.1s^2 - 0.25s + 0.1 = 0$

17. $\frac{1}{2}t = \frac{1}{4}t^2 + \frac{5}{6}$ **18.** $u^2 = 4u - 4$

Solve each equation.

19. $\sqrt{4x + 6} = 3 + 2x$

20. $\sqrt{x + 7} + x + 1 = 0$

21. $\sqrt{4x + 7} - \sqrt{2 + 4x} = 1$

22. $\sqrt{2x - 3} - \sqrt{x - 2} = 1$

23. $\dfrac{x}{3} + \dfrac{5 - x}{x - 1} = 0$

24. $\dfrac{x - 1}{x - 2} = \dfrac{2x - 3}{x - 2}$

25. Solve $f = (1/2L)\sqrt{T/d}$ for T.

26. Solve $r + 1 = (1 + k)^2$ for k.

Solve each inequality and express the solution set with set-builder notation.

27. $-3(x + 1) < 2(3 - x)$

28. $\frac{1}{2}(y + 1) - \frac{1}{5} \geq \frac{1}{4}(1 - y) + \frac{1}{10}$

Rewrite each set as a single interval.

29. $[-3, 2) \cap [-1, 4)$

30. $(-\pi, \sqrt{2}] \cup [0, \sqrt{3}]$

31. $(-\infty, 7] \cap (-2, \infty)$

32. $[2, 6] \cup (-\infty, 5)$

Solve each inequality and express the solution set with interval notation.

33. $|-x + 1| < 6$ **34.** $|4x + 3| \geq 5$

35. $|5x + 2| \geq 8$ **36.** $|1 - 3x| < 5$

Solve.

37. $x(x^2 - 3) < -2x^2$

38. $\dfrac{2}{y + 1} \geq \dfrac{-1}{y}$

39. If one-half is added to twice the reciprocal of a number, the result is three. What is the number?

40. Ten times the first of three consecutive integers plus nine times the second exceeds seventeen times the third by fifteen. What are the integers?

41. How many liters of solution that is 20% alcohol

must be added to 4 liters of solution that is 60% alcohol to produce a solution that is 50% alcohol?

42. Part of $1200 is invested in an account that pays 8% per year simple interest, and the remainder is invested in an account that pays 10% per year simple interest. The total interest earned from the two accounts in 1 year is $113. How much was invested in each account?

43. If one-half of a positive number is three more than the reciprocal of the number, what is the number?

44. If twice the reciprocal of the square root of a real number equals one-fifth of the square root of the number, what is the number?

45. The top of an equilateral trapezoid has length x, the base has length $2x$, and the height is h. Express s, the length of a slant height, in terms of x and h.

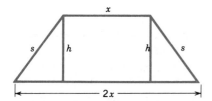

46. The radius of the circle in the illustration is r, and the length of AB is s. Show that the length of CD is $r - \sqrt{r^2 - (s/2)^2}$.

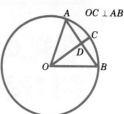

$OC \perp AB$

47. If a basketball is dropped from the rim of a goal, which is 10 feet above the floor, when will it reach the floor?

48. A rock is thrown directly down from the top of a building 73.5 meters tall with an initial velocity of 9.8 m/sec. When will it reach the ground?

49. A circle of radius r is inscribed in a square, and the area of the square exceeds that of the circle by a. Find r (in terms of a).

50. If I stand still on an escalator I can move from one floor of a building to another in 30 seconds. If I walk at my standard pace on a parallel set of stairs the trip will take 25 seconds. How long would the trip take if I walked on the escalator at my standard pace?

CHAPTER III

GRAPHS AND FUNCTIONS

The idea behind the first part of this chapter is so simple that its importance can easily be missed. This idea—to use pairs of numbers to represent points in a plane—is the primary link between algebra and geometry. It allows us to restate geometric problems as problems about numbers, which can often be solved using algebra. Conversely, it allows us to visualize many algebraic problems geometrically. The ability to solve a mathematical problem often rests on little more than the skill to move freely between algebra and geometry in this way.

The chapter also introduces the basic language and notation for functions, which arise whenever one quantity is determined by one or more other quantities.

SECTION **12**

Cartesian Coordinates. Circles

A. COORDINATE SYSTEMS

The basis for Figures 12.1 and 12.2 is two perpendicular real lines: an **x-axis,** directed to the right, and a **y-axis,** directed upward. The point of intersection of the axes, called the **origin,** corresponds to the zero point on each axis. The origin is denoted by the letter O.

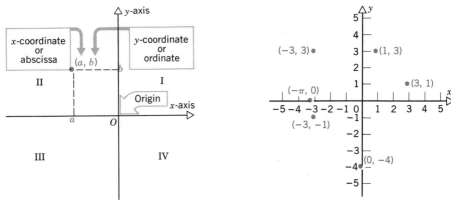

FIGURE 12.1   FIGURE 12.2

Each point in the plane of the axes is represented by an ordered pair* of real numbers, called its **coordinates:**

> The first coordinate is the directed distance of the point from the y-axis. We say *directed* distance because direction is taken into account—the co-ordinate is positive if the point is to the right of the y-axis, and negative if the point is to the left.

> The second coordinate is the directed distance of the point from the x-axis. This coordinate is positive if the point is above the x-axis, and negative if the point is below.

The first coordinate is called the **x-coordinate** (or the **abscissa**). The second coordinate is called the **y-coordinate** (or the **ordinate**).

By representing points by coordinates in this way we achieve a one-to-one correspondence between the set of points in the plane and the set of ordered pairs of real numbers: each point is assigned to precisely one pair, and each pair is assigned to precisely one point. This one-to-one correspondence is called a **Cartesian**† (or **rectangular**) **coordinate system** for the plane. A plane with a

---

*When we say that (a, b) is an **ordered pair,** we mean that it is to be distinguished from (b, a), that is, it is important which symbol appears first (reading from left to right). Thus $(3, 1) \neq (1, 3)$. Coordinate pairs (a, b) and (c, d) are equal iff $a = c$ and $b = d$.

†In honor of René Descartes (1596–1650), who stressed the value of treating algebraic problems geometrically.

Cartesian coordinate system is called a **Cartesian plane,** and the coordinates of its points are called **Cartesian coordinates** (or **rectangular coordinates**). When there is no chance of confusion we often refer to a point P with coordinates (x,y) as "the point (x,y)." The notation $P(x,y)$ will also denote the point P with coordinates (x,y).

The axes divide their plane into four **quadrants,** always labeled I, II, III, and IV as shown in Figure 12.1. The same unit of distance is generally used on the two axes unless the circumstances make some other choice more convenient. Also, the letters x and y are the standard labels for the axes, but circumstances will sometimes lead to other choices.

B. GRAPHS

With each equation involving one or both of the variables x and y we can associate a set of points in the plane, as follows. First, an ordered pair (a,b) of numbers is said to **satisfy** the equation if equality results when x is replaced by a and y is replaced by b.

Definition The **graph** of an equation involving one or both of the variables x and y is the set of all points in the Cartesian plane whose coordinates satisfy the equation.

Example 12.1 The pair $(1,2)$ satisfies $y = 2x$ because $2 = 2 \cdot 1$. The pairs $(-2, -4)$ and $(3,6)$ also satisfy $y = 2x$. These points have been plotted in Figure 12.3, and the graph of $y = 2x$ is the straight line that passes through them. (We'll see later why the complete graph is a straight line.) ■

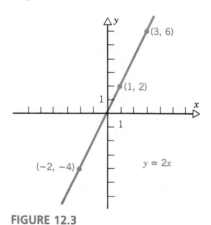

FIGURE 12.3

The next two examples concern equations with only one variable, either x or y.

Example 12.2 Draw the graph (in a Cartesian plane) of the equation $x = 3$.

Solution A point will be on the graph of $x = 3$ if its x-coordinate is 3, regardless of its y-coordinate. Therefore, the graph is the line shown in Figure 12.4. Lines such as this, parallel to the *y*-axis, are called **vertical lines.** If *c* is any real number, the graph of $x = c$ is a vertical line.   ▪

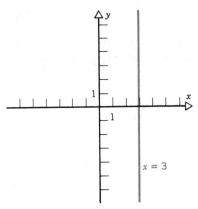

FIGURE 12.4

Example 12.3   Draw the graph of the equation $y = -2$.

Solution   A point will be on the graph of $y = -2$ if its *y*-coordinate is -2, regardless of its *x*-coordinate. Therefore, the graph of $y = -2$ is the line shown in Figure 12.5. Lines such as this, parallel to the *x*-axis, are called **horizontal lines.** If *c* is any real number, the graph of $y = c$ is a horizontal line.   ▪

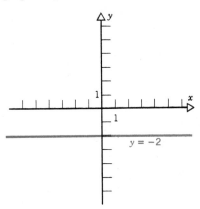

FIGURE 12.5

In the two cases just considered, $x = c$ and $y = c$, we can determine what the graph is just by looking at the equation. In the same way, the distance formula that follows will help us determine all of the equations whose graphs are circles. And in the next section we shall determine all of the equations whose graphs are lines.

C. THE DISTANCE FORMULA

Recall that the distance between points with coordinates *a* and *b* on a real line is $|a - b|$ (Section 1D). We now derive the corresponding formula for points in a Cartesian plane.

Two points $P_1(x_1,y_1)$ and $P_2(x_2,y_2)$ are on the same vertical line iff $x_1 = x_2$ (first coordinates equal). The distance between such points is

$$|y_1 - y_2|. \tag{12.1}$$

Example 12.4   The points $(7,4)$ and $(7,1)$ are on the same vertical line (Figure 12.6). The distance between them is $|4 - 1| = 3$.   ■

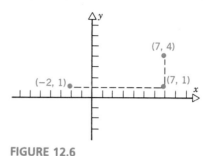

FIGURE 12.6

Two points $P_1(x_1,y_1)$ and $P_2(x_2,y_2)$ are on the same horizontal line iff $y_1 = y_2$ (second coordinates equal). The distance between such points is

$$|x_1 - x_2|. \tag{12.2}$$

Example 12.5   The points $(-2,1)$ and $(7,1)$ are on the same horizontal line (Figure 12.6). The distance between them is $|-2 - 7| = 9$.   ■

The distance d between points $P_1(x_1,y_1)$ and $P_2(x_2,y_2)$ is

$$d = \sqrt{(x_1 - x_2)^2 + (y_1 - y_2)^2}. \tag{12.3}$$

Proof   Figure 12.7 shows typical points P_1 and P_2. The vertical and horizontal lines through these points, as shown, intersect at the point with coordinates (x_1,y_2), which has been labeled Q. Triangle P_1QP_2 is a right triangle with hypotenuse P_1P_2. The legs P_1Q and QP_2 have lengths $|y_1 - y_2|$ and $|x_1 - x_2|$, respectively. Therefore, using the Pythagorean Theorem and the fact that $|a|^2 = a^2$, we have

$$d^2 = |x_1 - x_2|^2 + |y_1 - y_2|^2$$

$$= (x_1 - x_2)^2 + (y_1 - y_2)^2.$$

Equation (12.3) follows by taking the principal square root.   □

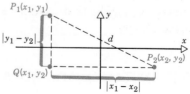

FIGURE 12.7

Notice that it makes no difference which point is labeled P_1 and which is labeled P_2 in using the distance formula, because $(x_1 - x_2)^2 = (x_2 - x_1)^2$ and $(y_1 - y_2)^2 = (y_2 - y_1)^2$.

Example 12.6 Find the distance between $(7,4)$ and $(-2,1)$ (Figure 12.6).

Solution By the distance formula,

$$d = \sqrt{[7 - (-2)]^2 + [4 - 1]^2} = \sqrt{9^2 + 3^2} = \sqrt{90}$$
$$= \sqrt{9}\sqrt{10} = 3\sqrt{10}.$$

D. CIRCLES

Figure 12.8 shows the graph consisting of all points on the circle with center at $P(h,k)$ and radius r. A point $Q(x,y)$ will be on this circle iff its distance from

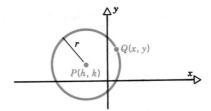

FIGURE 12.8

$P(h,k)$ is r. By the distance formula, this will be true iff

$$\sqrt{(x - h)^2 + (y - k)^2} = r$$

or

$$(x - h)^2 + (y - k)^2 = r^2.$$

This proves the following result.

If r is a positive real number, then the graph of

$$(x - h)^2 + (y - k)^2 = r^2 \qquad\qquad (12.4)$$

is a circle with center at $P(h,k)$ and radius r.

For a circle with center at the origin and radius r, Equation (12.4) becomes $x^2 + y^2 = r^2$.

Example 12.7 Determine an equation of the circle with center at $(3, -2)$ and radius 5.

Solution By Equation (12.4), the answer is

$$[x - 3]^2 + [y - (-2)]^2 = 5^2$$

or

$$(x - 3)^2 + (y + 2)^2 = 25.$$

Example 12.8 Determine an equation of the circle with center at $(-4, 6)$ and passing through $(-1, 2)$.

Solution The circle is shown in Figure 12.9. The radius must be the distance between $(-4, 6)$ and $(-1, 2)$, which is

$$\sqrt{[-4 - (-1)]^2 + [6 - 2]^2} = \sqrt{9 + 16} = 5.$$

Therefore, the answer is

$$(x + 4)^2 + (y - 6)^2 = 25.$$

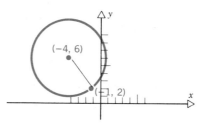

FIGURE 12.9

The equation of the circle in the preceding example can be rewritten as

$$x^2 + 8x + 16 + y^2 - 12y + 36 = 25$$

or

$$x^2 + y^2 + 8x - 12y + 27 = 0.$$

The last equation has the form

$$x^2 + y^2 + ax + by + c = 0, \tag{12.5}$$

where a, b, and c are real numbers. To determine the graph of such an equation we first complete the squares on the terms involving x and the terms involving y, as in the following examples. (Completing the square is discussed in Section 8.) The examples will illustrate the following fact.

> The graph of an equation that can be written in the form (12.5) is either a circle, a single point, or the empty set.

Example 12.9 Characterize the graph of

$$x^2 + y^2 - 10x + 6y + 30 = 0.$$

Solution To complete the square on the x terms add 25 to both sides of the equation. To complete the square on y add 9 to both sides.

$$(x^2 - 10x + 25) + (y^2 + 6y + 9) + 30 = 25 + 9$$

$$(x - 5)^2 + (y + 3)^2 = 4$$

Comparison of this equation with Equation (12.4) shows that the graph is the circle with center at $(5, -3)$ and radius 2.

Example 12.10 Characterize the graph of

$$x^2 + y^2 + 10x - 2y + 26 = 0.$$

Solution If we complete the squares on the x terms and the y terms, the result is

$$(x + 5)^2 + (y - 1)^2 = 0.$$

The only solution of this equation is $(-5, 1)$. Thus the graph consists of a single point. ∎

Example 12.11 Characterize the graph of

$$x^2 - 20x + y^2 + 6y + 113 = 0.$$

Solution If we complete the squares, we obtain

$$(x - 10)^2 + (y + 3)^2 = -4.$$

Since $(x - 10)^2 \geq 0$ for every x and $(y + 3)^2 \geq 0$ for every y, but $-4 < 0$, the equation has no solution. Thus the graph is the empty set (it contains no points). ∎

The method in the last three examples can be used to reduce any equation of the form (12.5) to the form

$$(x - h)^2 + (y - k)^2 = t.$$

The graph will be a circle, a single point, or the empty set, depending on whether the number t on the right is positive, zero, or negative, respectively.

We close this section with an example of a graph of an inequality.

Example 12.12 Describe the set of all points in the Cartesian plane whose coordinates satisfy $x^2 + y^2 < 9$.

Solution The expression $x^2 + y^2$ gives the square of the distance between $P(x,y)$ and the origin. Thus the square of this distance must be less than 9, so the distance itself must be less than 3. Therefore, the answer is the set of all points less than 3 units from the origin, or the set of all points inside (but not on) the circle $x^2 + y^2 = 9$. This set is shaded in Figure 12.10. ∎

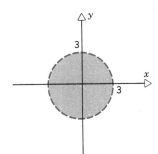

FIGURE 12.10

EXERCISES FOR SECTION 12

For each exercise, draw one set of coordinate axes and plot the given points.

1. $(2, 4), (-2, 4), (0, -3), (3.5, 0), (\sqrt{2}, -\pi), (-1, -2)$

2. $(5, 3), (3, -5), (\pi, 0), (0, -\sqrt{2}), (-1.5, -2.5), (-4, 2)$

3. $(1, 4), (-1, 0), (\sqrt{2}, -1), (0, \pi), (-4, 3), (1.5, 3.5)$

Give the Cartesian coordinates of each point.

4. 3 units above the x-axis and 2 units to the left of the y-axis.

5. 2.5 units to the right of the y-axis and 4 units below the x-axis.

6. 1.7 units below the x-axis and 3 units to the left of the y-axis.

7. On the y-axis and 4 units below the x-axis

8. On the x-axis and 2.5 units to the left of the y-axis.

9. On the y-axis and 3.2 units above the x-axis.

In Exercises 10–15, determine which of the given points are on the graph of the given equation.

10. $y = 3x$; $(2, 6)$, $(0.3, 1)$, $(-\frac{1}{2}, -\frac{3}{2})$

11. $y = 2$; $(0, 2)$, $(2, 0)$, $(4, 2)$

12. $x = -3$; $(0, -3)$, $(-3, 0)$, $(-3, 2)$

13. $y = x^2$; $(1, 1)$, $(2, -4)$, $(-2, 4)$

14. $y^3 = x$; $(1, -1)$, $(-\frac{1}{2}, \frac{1}{8})$, $(-\frac{1}{8}, -\frac{1}{2})$

15. $y^2 + 1 = (x + 1)^2$; $(0, -1)$, $(-1, 0)$, $(-1, -1)$

Draw the graph of each equation.

16. $x = 3$ **17.** $y = -1$

18. $y = 4$ **19.** $y = -1.5$

20. $x = 7/5$ **21.** $x = -2$

22. $x^2 + y^2 = 9$ **23.** $x^2 + y^2 = 16$

24. $x^2 + y^2 = 4$

25. $(x - 1)^2 + y^2 = 10$

26. $(x + 4)^2 + (y - 3)^2 = 15$

27. $(x + 2)^2 + (y + 1)^2 = 12$

Determine equations of the following lines and circles.

28. The line parallel to and 4 units above the x-axis.

29. The line parallel to and 3 units below the x-axis.

30. The line parallel to and 2 units to the left of the y-axis.

31. The line parallel to and 5 units to the right of the y-axis.

32. The line parallel to and 1.5 units to the left of the y-axis.

33. The line parallel to and 5 units above the x-axis.

34. The circle with center at $(1.5, 0)$ and radius $\sqrt{6}$.

35. The circle with center at $(-2, -3)$ and radius $\sqrt{2}$.

36. The circle with center at $(0, 4.5)$ and radius 10.

37. The circle with center at the origin and passing through $(3, -4)$.

38. The circle with center at $(1, 2)$ and passing through the origin.

39. The circle with center at $(-3, 1)$ and tangent to the y-axis.

Find the distance between each pair of points. In 46–48, give the answers to four significant figures.

40. $(2, 5)$, $(-1, 4)$ **41.** $(\frac{3}{2}, -2)$, $(\frac{9}{2}, 3)$

42. $(-1, \frac{8}{3})$, $(7, \frac{2}{3})$ **43.** $(0, \frac{1}{4})$, $(\frac{1}{2}, -\frac{3}{5})$

44. $(-\frac{1}{2}, -2)$, $(0.4, 0)$

45. $(2, 0)$, $(-0.5, -1)$

46. $(\sqrt{2}, 1)$, $(-3, \sqrt{3})$ Ⓒ

47. $(0, \pi)$, $(1, -2)$ Ⓒ

48. $(\sqrt{2}, \sqrt{3})$, $(\sqrt{5}, \sqrt{6})$ Ⓒ

Characterize the graph of each equation.

49. $x^2 + y^2 - 2x + 4y - 4 = 0$

50. $x^2 + y^2 + 12x - 2y - 37 = 0$

51. $x^2 + y^2 - 2x - 8y - 8 = 0$

52. $x^2 + y^2 - 6y = -10$

53. $x^2 + y^2 - 2y = 3$

54. $x^2 - 20x + y^2 = -100$

55. $x^2 + y^2 = 0$

56. $x^2 + y^2 + 9 = 0$

57. $x^2 + y^2 - x + y + 1 = 0$

58. $4x^2 + 4y^2 - 20x + 9 = 0$

59. $9x^2 + 9y^2 - 18x + 12y - 5 = 0$

60. $4x^2 + 4y^2 + 4x - 4y - 22 = 0$

In Exercises 61–69, describe the set of all points in the Cartesian plane whose coordinates satisfy the given condition or conditions. Use a graph where it will help.

61. $x < 0$ **62.** $xy > 0$

63. $x/y < 0$ **64.** $x > y$

65. $x^2 + y^2 > 5$ **66.** $x^2 + y^2 \leq 1$

67. Both $|x| = 2$ and $y > 3$.

68. Both $|x - 3| < 1$ and $y = 4$.

69. Both $|x - 1| \leq 1$ and $|y - 1| \leq 1$.

70. Find all of the points on the x-axis that are five units from $(2, 3)$.

71. Find all of the points where the graph of $(x - 2)^2 + (y + 1)^2 = 9$ intersects a coordinate axis.

72. Find all of the points that are 2 units from the y-axis and on the graph of $(x + 3)^2 + y^2 = 36$.

73. The point $(x, 1)$ is twice as far from $(-1, 0)$ as from $(1, 0)$. What is x? (There are two solutions.)

74. Find an equation that is satisfied by (x, y) iff $P(x, y)$ is on the perpendicular bisector of the segment connecting $P_1(0, 0)$ and $P_2(3, -2)$.

75. Draw the graph of $(x^2 + y^2 - 4)(x^2 + y^2 - 1) \leq 0$.

SECTION **13**
Lines

A. INTRODUCTION

In this section we study lines (straight lines) and the equations whose graphs are lines. We shall see that the equations are precisely those covered by the following definition.

Definition An equation with variables x and y is called a **linear equation** if it can be written in the form

$$Ax + By + C = 0, \qquad (13.1)$$

where A, B, and C are real numbers with A and B not both zero.

Example 13.1 These are linear equations.

$$3x - 2y + 5 = 0 \quad (A = 3, B = -2, C = 5)$$
$$y - 4 = 0 \quad (A = 0, B = 1, C = -4)$$

Also, $2x = -9$ and $y = 2 - x$ are linear equations, since they can be written as

$$2x + 9 = 0 \quad (A = 2, B = 0, C = 9)$$
$$x + y - 2 = 0 \quad (A = 1, B = 1, C = -2).$$

Figure 13.1 shows that the graphs of the linear equations in Example 13.1 are all lines. To move from these examples to more general statements we must first examime some basic facts about lines.

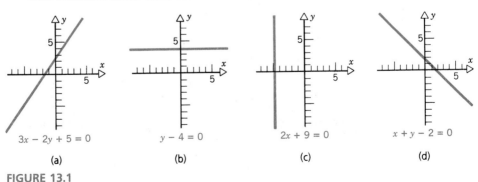

$3x - 2y + 5 = 0$ $y - 4 = 0$ $2x + 9 = 0$ $x + y - 2 = 0$

(a)   (b)   (c)   (d)

FIGURE 13.1

B. THE SLOPE OF A LINE

To study nonvertical lines we begin with the idea of their *slopes*. The slope measures the steepness of a line, and also whether an increase in x causes an increase in y, no change in y, or a decrease in y. In the definition of slope, which follows, we use the capital Greek letter Δ (delta) to denote *change*. In particular, Δx denotes a change in x, and Δy denotes a change in y. In this context Δx and Δy are each to be treated as single symbols: for example, Δx does *not* mean "Δ times x."

Definition The **slope** of the line through $P_1(x_1, y_1)$ and $P_2(x_2, y_2)$, with $x_1 \neq x_2$, is denoted by m and is defined by

$$m = \frac{\Delta y}{\Delta x} = \frac{y_2 - y_1}{x_2 - x_1}. \tag{13.2}$$

From Figure 13.2 we see that the slope is

$$\frac{\text{the change in } y}{\text{the change in } x}$$

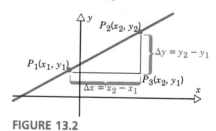

FIGURE 13.2

as we move from one point to another along the line. Notice, in particular, that if $\Delta x = 1$, then $m = \Delta y$. Thus the slope is the number of units that y increases or decreases for each unit of increase in x; if y increases the slope is positive and if y decreases the slope is negative.

Example 13.2 Determine the slope of the line through $(-6, 0)$ and $(2, 4)$.

Solution Use $(x_1, y_1) = (-6, 0)$ and $(x_2, y_2) = (2, 4)$ in Equation (13.2):

$$m = \frac{4 - 0}{2 - (-6)} = \frac{4}{8} = \frac{1}{2}.$$

It makes no difference which point is labeled P_1 and which is labeled P_2 for Equation (13.2), because

$$\frac{y_2 - y_1}{x_2 - x_1} = \frac{y_2 - y_1}{x_2 - x_1} \cdot \frac{(-1)}{(-1)} = \frac{y_1 - y_2}{x_1 - x_2}.$$

Interchanging the order in both the numerator and denominator in Example 13.2, for instance, we get

$$m = \frac{0 - 4}{-6 - 2} = \frac{-4}{-8} = \frac{1}{2},$$

as before. Interchanging the order in *only* the numerator or *only* the denominator would give $-m$ rather than m.

We can choose any other pair of points on the same line and we will get the same number for the slope. For Figure 13.3 this can be seen as follows. The triangles P_1PP_2 and P_3QP_4 are similar (that is, corresponding angles are equal) because the angles labeled α are equal and the angles at P and Q are right

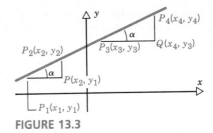

FIGURE 13.3

angles. Therefore, because ratios between corresponding sides of similar triangles are equal,

$$\frac{y_4 - y_3}{x_4 - x_3} = \frac{y_2 - y_1}{x_2 - x_1}. \tag{13.3}$$

If Equation (13.3) were not true, we could not refer to *the* slope of a line, for a line would have more than one slope.

Figure 13.4 shows the lines through the origin having slopes $\pm\frac{1}{2}$, ± 1, ± 2, and ± 4. Also, the x-axis has slope 0. Figure 13.4 illustrates the following facts, which can be proved from Equation (13.2).

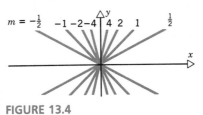

FIGURE 13.4

> $m < 0$ iff y decreases as x increases.
> $m = 0$ iff y is constant as x increases.
> $m > 0$ iff y increases as x increases.
> The larger the number $|m|$, the steeper the line.

Any attempt to compute the slope of a vertical line will lead to division by 0. Therefore, we say that *the slope of a vertical line is undefined.*

C. POINT-SLOPE FORM

There is exactly one line through a given point with a given slope. Its equation is determined as follows.

> An equation for the line through the point $P_1(x_1,y_1)$ with slope m is
>
> $$y - y_1 = m(x - m_1). \tag{13.4}$$
>
> This is called the **point-slope form** for the equation of the line.

Proof Let L denote the line through $P_1(x_1,y_1)$ with slope m. If $P(x, y)$ is any point on L with $x \neq x_1$ (Figure 13.5), then P and P_1 can be used to compute the slope of L, so that

$$\frac{y - y_1}{x - x_1} = m$$

$$y - y_1 = m(x - x_1).$$

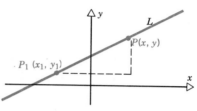

FIGURE 13.5

Since (x_1,y_1) also satisfies this equation, which is the same as Equation (13.4), we see that every point on L satisfies Equation (13.4). Conversely, any point $P(x, y)$ whose coordinates satisfy Equation (13.4) must be on L since there is only one line through $P_1(x_1,y_1)$ with slope m. □

Example 13.3 Find an equation for the line through $(-1, 5)$ with slope 2.

Solution Use Equation (13.4) with $m = 2$ and $(x_1,y_1) = (-1, 5)$:

$$y - 5 = 2[x - (-1)]$$

$$y - 5 = 2x + 2$$

$$2x - y + 7 = 0.$$

If two points on a line are given, the slope can be found from Equation (13.2), and then the equation of the line can be determined by using the point-slope form with either one of the given points.

D. LINES AND LINEAR EQUATIONS

We are now in a position to prove the following statements regarding lines and linear equations.

> Any line is the graph of a linear equation. Conversely, the graph of any linear equation is a line.

Proof The point-slope form shows that a nonvertical line is the graph of an equation of the form

$$y - y_1 = m(x - x_1),$$

which can be rewritten as

$$y - y_1 = mx - mx_1$$

$$mx - y + (y_1 - mx_1) = 0.$$

This has the form

$$Ax + By + C = 0$$

with

$$A = m, B = -1, \text{ and } C = y_1 - mx_1.$$

On the other hand, a vertical line is the graph of a linear equation of the form $x - c = 0$ (Example 12.2). Thus we have verified that *any line is the graph of a linear equation.* Now let's see why, conversely, the graph of any linear equation is a line.

If $B = 0$ in Equation (13.1), $Ax + By + C = 0$, then $A \neq 0$, since we specified that not both A and B are 0. Therefore, the equation can be rewritten as $x = -C/A$, which is the graph of a vertical line.

On the other hand, if $B \neq 0$ in the equation $Ax + By + C = 0$, then the equation can be rewritten as

$$By + C = -Ax$$

$$y + \frac{C}{B} = -\frac{A}{B} x$$

$$y - \left(-\frac{C}{B}\right) = \frac{-A}{B} (x - 0).$$

If we compare this equation with Equation (13.4), we see that this is the equation of the nonvertical line through the point $(0, -C/B)$ with slope $-A/B$. Thus we have shown that *the graph of any linear equation is a line.* □

E. SLOPE-INTERCEPT FORM

Now that we know that the graph of any linear equation is a line, it is easy to draw its graph: plot any two points determined by the equation and then draw the line through those two points. The graph will tend to be more accurate if we choose points that are not too close together. The *intercepts,* whose definitions follow, are usually the easiest points to determine and can be used if they are sufficiently far apart. In any case, it is generally a good idea to plot a third point to serve as a check.

The x-coordinate where a graph intersects the x-axis is called an **x-intercept** of the graph. (To compute the x-intercepts set $y = 0$ and solve for x.)

The y-coordinate where a graph intersects the y-axis is called a **y-intercept** of the graph. (To compute the y-intercepts set $x = 0$ and solve for y.)

Example 13.4 Draw the graph of $3x - 4y + 12 = 0$.

Solution If $y = 0$, then $3x + 12 = 0$ and $x = -4$. Thus -4 is the x-intercept. If $x = 0$, then $-4y + 12 = 0$ and $y = 3$. Thus 3 is the y-intercept. These intercepts have been plotted to determine the graph in Figure 13.6. Also, if $x = 4$, then $12 - 4y + 12 = 0$ and $y = 6$; the point $(4, 6)$ has been plotted as a check. ■

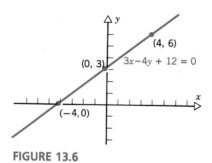

FIGURE 13.6

The following form for the equation of a line is often more useful than the point-slope form.

An equation of the line with slope m and y-intercept b is

$$y = mx + b. \qquad (13.5)$$

This is called the **slope-intercept form** for the equation of the line.

Proof To say that the y-intercept is b is the same as to say that the line passes through the point $(0, b)$. Using the point-slope form, Equation (13.4), with $(0, b)$ for the point and m for the slope, we obtain

$$y - b = m(x - 0)$$
$$y = mx + b. \qquad □$$

Example 13.5 An equation of the line with slope -2 and y-intercept 4 is

$$y = -2x + 4. \qquad ■$$

Example 13.6 Find the slope and y-intercept of the graph of

$$5x - 2y - 4 = 0.$$

Solution Solve the equation for y and compare the answer with Equation (13.5). First,

$$2y = 5x - 4$$

$$y = \frac{5}{2}x - 2.$$

Hence the slope is $\frac{5}{2}$ (the coefficient of x) and the y-intercept is -2 (the constant term). ■

F. PARALLEL AND PERPENDICULAR LINES

Figure 13.7 shows the effect of changing m with b held constant in $y = mx + b$. Figure 13.8 shows the effect of changing b with m held constant. Figure 13.8 illustrates the following fact.

Two lines with slopes m_1 and m_2 are parallel iff $m_1 = m_2$.

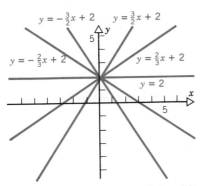

FIGURE 13.7 $y = mx + 2$ for different values of m.

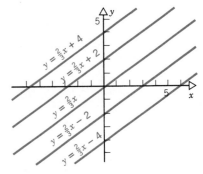

FIGURE 13.8 $y = \frac{2}{3}x + b$ for different values of b.

Proof For the case of parallel lines with positive slopes refer to Figure 13.9. The slope of line L_1 can be computed with points P_1 and Q_1. The slope of line L_2 can be

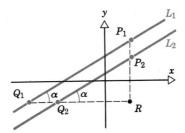

FIGURE 13.9

computed with points P_2 and Q_2. Using, in turn, facts about parallel lines, similar triangles, and slope, we can write

L_1 and L_2 are parallel

iff

the angles labeled by α are equal

iff

triangles P_1Q_1R and P_2Q_2R are similar

iff

$\overline{P_1R}/\overline{Q_1R} = \overline{P_2R}/\overline{Q_2R}$

iff

$m_1 = m_2$.

The proof for lines with negative slopes is similar. □

Example 13.7 The graph of $y = 3x + 5$ is parallel to the graph of $y = 3x - 2$ because each is a line with slope 3, by the slope-intercept form. ■

Example 13.8 Find an equation of the line containing the point $(-1, 3)$ and parallel to the line whose equation is $2x + y - 1 = 0$.

Solution If we rewrite the given equation in the form $y = mx + b$, we have $y = -2x + 1$; thus the slope of the given line, and any parallel line, is -2. Using the point-slope form, Equation (13.4), with the point $(-1, 3)$ and slope -2, we obtain

$$y - 3 = -2(x + 1).$$ ■

Section 34 contains a proof of the following statement.

> Two lines with nonzero slopes m_1 and m_2 are perpendicular iff $m_1 = -1/m_2$.

EXERCISES FOR SECTION 13

Determine the slope of the line through each pair of points.

1. $(3, 1)$ and $(2, 4)$

2. $(1, 3)$ and $(4, 2)$

3. $(-7, 1)$ and $(2, 2)$

4. $(-0.5, 4.3)$ and $(0.1, 1.3)$

5. $\left(-2, \dfrac{7}{3}\right)$ and $\left(2, \dfrac{7}{3}\right)$

6. $(5.2, 3.1)$ and $(7.6, -0.5)$

Determine an equation of the line through the given point with the given slope.

7. $(-3, 1); m = \dfrac{1}{2}$

8. $(-1, -2); m = -0.4$

9. $(4, 1); m = \dfrac{1}{4}$

10. $(6, -1); m = -3$

11. $(3.1, 4); m = 2.5$

12. $(-4, -2); m = -3.1$

Determine an equation of the line through each pair of points. Write the answer in the form $Ax + By + C = 0$.

13. $(1, 2)$ and $(-4, 0)$

14. $(4, 7)$ and $(2, -1)$

15. $(-5, -2)$ and $(2, 5)$

16. $(-3, 5)$ and $(-2, -1)$

17. $(0, -5)$ and $(1, 1)$

18. $(7, 2)$ and $(-1, 0)$

Determine an equation of the line with the given slope (m) and y-intercept (b).

19. $m = -1, b = 2$ **20.** $m = 2, b = -1$

21. $m = 0, b = 2$ **22.** $m = -\dfrac{1}{3}, b = 0$

23. $m = 0, b = 0$ **24.** $m = \dfrac{1}{2}, b = -1$

Determine the slope and y-intercept of the graph of each equation.

25. $2x + y - 3 = 0$

26. $-x + 2y = 5$

27. $5x - 10y - 15 = 0$

28. $y - 5 = 0$

29. $x = y$

30. $2y = -5$

Draw the graph of each equation.

31. $x + y = 0$ **32.** $-2x + y = 0$

33. $x - 2y = 0$ **34.** $x + y + 1 = 0$

35. $x - 2y - 2 = 0$ **36.** $2x + 3y - 6 = 0$

In each of Exercises 37–39, use a single Cartesian plane (coordinate system) to draw the three graphs determined by b = −2, b = 0, *and* b = 3.

37. $y = 2x + b$ **38.** $y = -3x + b$

39. $y = \dfrac{1}{2}x + b$

In each of Exercises 40–42, use a single Cartesian plane (coordinate system) to draw the three graphs determined by m = −2, m = 0, *and* m = 1.

40. $y = mx + 1$

41. $y = mx$

42. $y = mx + 3$

43. Determine an equation of the line through the origin and parallel to the line whose equation is $2x - 3y + 5 = 0$.

44. Determine an equation of the line that has y-intercept 2 and is parallel to the line $x + y = 0$.

45. Determine an equation of the line that has x-intercept -2 and is parallel to the line through $(4, 0)$ and $(0, -4)$.

46. Prove that the line through $(-1, 6)$ and $(2, 4)$ is perpendicular to the line through $(5, 3)$ and $(3, 0)$.

47. Determine an equation of the line that has y-intercept -3 and is perpendicular to the line whose equation is $3x - 4y = 5$.

48. Determine an equation of the line that is perpendicular to the line whose equation is $x + 2y = 0$ and passes through the center of the circle whose equation is $x^2 + y^2 + 6y + 5 = 0$.

49. Prove that an equation of the line through $P_1(x_1, y_1)$ and $P_2(x_2, y_2)$, with $x_1 \neq x_2$, is

$$y - y_1 = \frac{y_2 - y_1}{x_2 - x_1}(x - x_1).$$

(This is called the **two-point form** for the equation of the line.)

50. Prove that an equation of the line with x-inter-cept a and y-intercept b is

$$\frac{x}{a} + \frac{y}{b} = 1 \; (a \neq 0, b \neq 0).$$

(This is called the **intercept form** for the equation of the line.)

51. Prove that the midpoint of the segment joining $P_1(x_1, y_1)$ and $P_2(x_2, y_2)$ is

$$P\left(\frac{x_1 + x_2}{2}, \frac{y_1 + y_2}{2}\right) \quad \text{(midpoint formula)}$$

by showing that (a) P is on the line through P_1 and P_2, and (b) P is equidistant from P_1 and P_2. [Notice that $(x_1 + x_2)/2$ is the average of x_1 and x_2. Similarly for $(y_1 + y_2)/2$.] Where is the midpoint of the segment joining $(-3, 2)$ and $(4, 7)$?

52. Prove that if the x- and y-intercepts of a line are equal, then the slope of the line is -1.

53. Use the slope-intercept form to prove that if a line passes through the point $(1, 1)$, then its slope is one minus its y-intercept.

54. A line cuts the positive x- and y-axes to form an isosceles triangle having area 8 square units. Find its equation. (Exercise 50 may help.)

55. Find an equation of the line that is parallel to the line $y = 2x$ and bisects the circle whose equation is $x^2 - 4x + y^2 + 6y + 9 = 0$.

56. Use the first part of Exercise 51 to help write the equation of the perpendicular bisector of the segment joining $(1, -6)$ and $(4, 3)$.

57. Find the equation of the line tangent to the circle $x^2 + y^2 = 25$ at the point $(3, 4)$.

58. Find an equation of the line that is perpendicular to and has the same x-intercept as $y = mx + b \; (m \neq 0)$.

59. Use the distance formula and the first part of Exercise 51 to prove that the midpoint of the hypotenuse of a right triangle is equidistant from the three vertices of the triangle. [First, explain why the vertices of the triangle can be assumed to be at $P_1(0, 0)$, $P_2(a, 0)$, and $P_3(0, b)$ for some real numbers a and b.]

60. Use the first part of Exercise 51 to prove that the diagonals of every parallelogram bisect each other. [First, explain why the vertices of the parallelogram can be assumed to be at $P_1(0, 0)$, $P_2(a, 0)$, $P_3(a + b, c)$ and $P_4(b, c)$ for some real numbers a, b, and c.]

SECTION **14**
Introduction to Functions

A. DEFINITION

In using mathematics we repeatedly meet problems in which one quantity is determined by one or more other quantities. For example:

The area of a circle is determined by the radius.

The pressure of a confined gas is determined by the volume and temperature of the gas.

The interest on a loan is determined by the principal, rate, and duration of the loan.

The idea of a *function* encompasses all such examples. In most of our examples there will be only two quantities, with one determined by the other.

Example 14.1 If x denotes the length of a side of a square in meters and A denotes the area in square meters, then A is determined by x:

$$A = x^2. \tag{14.1}$$

Example 14.2 Let C denote temperature in degrees Celsius and F temperature in degrees Fahrenheit. Then C is determined by F:

$$C = \frac{5}{9}(F - 32). \tag{14.2}$$

Notice that in each example we can identify one thing as *input* and another thing as *output,* as follows.

Example	Input	Output
14.1	x	$A = x^2$
14.2	F	$C = \frac{5}{9}(F - 32)$

The concept of a function ties together the following observations about these two examples.

• The values that can occur as input form a set (in the first case the possible values for x; in the second case the possible values for F).

• The values that can occur as output also form a set (in the first case the possible values for A; in the second case the possible values for C).

• In each case the output is determined uniquely (unambiguously) by the input.

In the following definitions the set S corresponds to the set of input elements, and the output elements belong to the set T.

Definitions Let S and T denote sets. A **function** from S to T is a relationship (formula, rule, correspondence) that assigns exactly one element of T to each element of S. The set S is called the **domain** of the function. The set of elements of T to which the elements of S are assigned is called the **range** of the function.

Example 14.3 (a) In Example 14.1, x can be any positive real number (length in meters) and A can be any positive real number (area in square meters). Thus the domain and the range are both the set of all positive real numbers.

(b) The domain and range in Example 14.2 are restricted by the laws of nature: F and C both denote real numbers, with $F \geq -459.67$ and $C \geq -273.15$. (These minimum values correspond to absolute zero, the point at which molecules have no heat energy.) In set-builder notation (Section 7),

$$\text{domain} = \{F\colon F \geq -459.67\}$$

and

$$\text{range} = \{C\colon C \geq -273.15\}.$$

Most of the domains and ranges for functions in this book will be sets of real numbers, as in Example 14.3. The next example illustrates a different possibility.

Example 14.4 Let S denote the set of all 50 states and T the set of all cities in the United States. The relationship that assigns a capital to each state is a function: the domain is S and the range is the subset of T consisting of the 50 state capitals.

An alternative but equivalent definition of *function*, in terms of ordered pairs, is given in Appendix B. It will not be used elsewhere in this book.

B. NOTATION

Functions are often denoted by letters, such as f, g, and so on. This allows the following important notational convention.

> If f denotes a function, and x denotes an element in the domain of f, then the element that f assigns to x is denoted by $f(x)$.

That is, if the input is x, then the output is $f(x)$ (Figure 14.1). The notation "$f(x)$" is read "f of x." Notice that in this context $f(x)$ does not mean "f times x;" f represents a function, not a number.

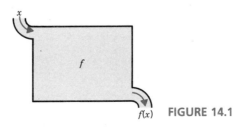

FIGURE 14.1

Example 14.5 Let f denote the function (relationship) associated with $A = x^2$ in Example 14.1. Then

$$f(x) = x^2 \text{ for each positive real number } x.$$

To refer to A or x^2 for a specific value of x, we replace x by that specific value in $f(x)$. For example,

$$f(5) = 5^2 = 25$$
$$f(\tfrac{1}{2}) = (\tfrac{1}{2})^2 = \tfrac{1}{4}$$
$$f(t) = t^2.$$

Replacing x by a number or letter in $f(x)$ is a signal to replace x by that same number or letter in x^2. Furthermore, just as we are using x and $f(x)$ to denote *numbers* in this context, we are using f to denote the *function* (relationship) that assigns x^2 to x. ◼

Example 14.6 If g denotes the function that gives the perimeter P of a square in terms of the length x of a side, then

$$g(x) = 4x \text{ for each positive real number } x.$$

To get the perimeter corresponding to a specific value of x, simply replace x by that value. If, for example, the length of a side is 3 feet, then the perimeter is

$$g(3) = 4 \cdot 3 = 12 \text{ feet.}$$

If the length of a side is 10 centimeters, then the perimeter is

$$g(10) = 4 \cdot 10 = 40 \text{ centimeters.}$$ ◼

Example 14.7 The square root key $\boxed{\vee}$ on a calculator provides another example of a function. Let $h(x) = \sqrt{x}$. If we enter a positive real number x in the calculator and then press $\boxed{\vee}$, the display will exhibit $\sqrt{x}$, that is, $h(x)$. Other examples are given by other function keys. ◼

Suppose f is a function. Sometimes we must replace the x in $f(x)$ by an expression that is more complicated than a single number or letter. The key to doing this is to remember that if x is replaced by any expression representing an element of the domain of f, then x must be replaced by the same expression throughout the formula or rule giving $f(x)$.

Example 14.8 Let $g(x) = x^2 + x$ for each real number x. (That sentence tells us, in particular, that the domain of g is the set of real numbers.) Then

(a) $g(-2) = (-2)^2 + (-2) = 2$
(b) $g(\sqrt{3}) = (\sqrt{3})^2 + \sqrt{3} = 3 + \sqrt{3}$
(c) $g(t) = t^2 + t$
(d) $g(a + 1) = (a + 1)^2 + (a + 1)$
$$= a^2 + 2a + 1 + a + 1$$
$$= a^2 + 3a + 2$$
(e) $g(x/2) = (x/2)^2 + (x/2) = x^2/4 + x/2 = (x^2 + 2x)/4$

(f) $g(\pi^{-2}) = (\pi^{-2})^2 + (\pi^{-2}) = \pi^{-4} + \pi^{-2}$

(g) $g(x^3) = (x^3)^2 + (x^3) = x^6 + x^3$.

Because the domain of g is the set of real numbers, the letters t, a, and x above can also denote any real numbers.

It is important to realize that the function g in example 14.8 could be defined equally well by using any other variable in place of x. That is,

$$g(x) = x^2 + x, \quad g(t) = t^2 + t, \quad \text{and} \quad g(u) = u^2 + u$$

all define the same function—they describe the same relationship between the input (a variable) and the output (the sum of that variable and its square). A similar remark applies to other functions.

Example 14.9 Let $f(x) = 1 + (1/x)$ for each nonzero real number x. Then

(a) $f(-1) = 1 + (1/-1) = 1 - 1 = 0$

(b) $f(\frac{2}{3}) = 1 + (1/\frac{2}{3}) = 1 + \frac{3}{2} = \frac{5}{2}$

(c) $f(a) = 1 + (1/a) \quad (a \ne 0)$

(d) $f(3/(t + 2)) = 1 + \dfrac{1}{3/(t + 2)} = 1 + [(t + 2)/3] = (3 + t + 2)/3 = (t + 5)/3$.

In part (d) we must require $t \ne -2$, for otherwise the input element $3/(t + 2)$ would be undefined.

Expressions that involve functions as well as algebraic operations are simplified by carrying out whatever steps are indicated.

Example 14.10 Let $f(x) = x^2$ and $g(x) = 3x - 2$. Then

(a) $f(x) + g(x) = x^2 + 3x - 2$

(b) $f(2) + g(5) = (2^2) + (3 \cdot 5 - 2) = 17$

(c) $2 \cdot f(1) - 4 \cdot g(-1) = 2(1^2) - 4[3(-1) - 2] = 2(1) - 4(-5) = 22$

(d) $f(t)/g(t) = t^2/(3t - 2)$ for $t \ne \frac{2}{3}$.

Example 14.11 The **difference quotient** of a function f is defined to be

$$\frac{f(x + h) - f(x)}{h}.$$

If $f(x) = 3x^2$, for example, then the difference quotient is

$$\frac{f(x + h) - f(x)}{h} = \frac{3(x + h)^2 - 3x^2}{h}$$

$$= \frac{3x^2 + 6xh + 3h^2 - 3x^2}{h}$$

$$= 6x + 3h \quad (h \ne 0).$$

Difference quotients are important in calculus.

C. MORE ABOUT DOMAINS

The domain of the function f in Example 14.9 is the set of all *nonzero* real numbers: $1/x$ is undefined for $x = 0$, so 0 could not be in the domain of f. If a function f is defined by an algebraic expression, like $f(x) = 1 + (1/x)$, and the domain of f is not specified, then the domain will be assumed to be the domain of the algebraic expression—that is, the set of all real numbers x for which $f(x)$ is also a real number (Section 7A). This domain is called the **natural domain** of the function.

Example 14.12 Determine the domain (that is, the natural domain) of

$$f(x) = \frac{1}{x^2 - 1}.$$

Solution Here $f(x)$ is a real number iff $x^2 - 1 \neq 0$, that is, iff $x \neq 1$ and $x \neq -1$. Thus

$$\text{domain of } f = \{x: x \neq 1 \text{ and } x \neq -1\}. \quad \blacksquare$$

Example 14.13 Determine the domain of

$$g(x) = \frac{\sqrt{x - 2}}{x - 3}.$$

Solution To avoid a square root of a negative number in the numerator we must exclude the real numbers less than 2 from the domain. To avoid zero in the denominator we must exclude $x = 3$. Thus

$$\text{domain of } g = \{x: x \geq 2 \text{ and } x \neq 3\}. \quad \blacksquare$$

If we write $f(x) = x^2$, with no restriction specified, then the domain is understood to be the natural domain—that is, the set of all real numbers. In Examples 14.1 and 14.5 $[A = f(x) = x^2]$, a restriction was dictated by the context: x represented a length in meters so the domain was the set of positive real numbers. In Example 14.3(b) $[C = f(F) = \frac{5}{9}(F - 32)]$ the domain and range were determined by physical laws.

D. COMPOSITION

Example 14.14 Let $f(x) = x + 1$ and $g(x) = x^2$. The variable x in $g(x) = x^2$ can be replaced by any real number or symbol representing a real number; in particular, we can replace it by $f(x)$. The result is

$$g(f(x)) = g(x + 1) = (x + 1)^2.$$

Figure 14.2 shows what we have done. The function f assigns $x + 1$ to x; then, the function g assigns $(x + 1)^2$ to $x + 1$. As a result of the two steps combined, $(x + 1)^2$ is assigned to x. We can think of this result as a single function made up from the two component functions f and g, as in the following definition. $\quad \blacksquare$

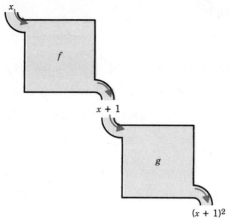

FIGURE 14.2

Definition Assume that f and g are functions. Then the **composition** of f and g, denoted $g \circ f$, is the function defined by

$$(g \circ f)(x) = g(f(x)) \tag{14.3}$$

for each x such that x is in the domain of f and $f(x)$ is in the domain of g.

We read $g \circ f$ as "g composed with f" or "g circle f." It is essential that $g \circ f$ be distinguished from $f \circ g$, as the following example makes clear.

Example 14.15 Let $f(x) = x - 2$ and $g(x) = x^2 + 1$. Determine: (a) $(f \circ g)(x)$; (b) $(g \circ f)(x)$; (c) $(f \circ g)(3)$; (d) $(g \circ f)(3)$.

Solution (a) $(f \circ g)(x) = f(g(x)) = f(x^2 + 1) = (x^2 + 1) - 2 = x^2 - 1$

(b) $(g \circ f)(x) = g(f(x)) = g(x - 2) = (x - 2)^2 + 1$
$= x^2 - 4x + 4 + 1 = x^2 - 4x + 5$

(c) Use the answer to part (a): $(f \circ g)(3) = 3^2 - 1 = 8$.

(d) Use the answer to part (b): $(g \circ f)(3) = 3^2 - 4 \cdot 3 + 5 = 2$. ◼

Composition can be viewed as a way to construct a new function from two given functions. It can also be viewed as a way to break a given function into simpler components. Before illustrating this we need one more definition.

Suppose f and g are the functions, each with the set of real numbers as domain, such that

$$f(x) = (x + 1)^2 \text{ and } g(x) = x^2 + 2x + 1.$$

Because $(x + 1)^2 = x^2 + 2x + 1$ for every real number x, it is reasonable to think of f and g as being equal. This leads us to say that any two functions f and g are **equal** if their domains are equal and if $f(x) = g(x)$ for every x in that common domain; in this case we write $f = g$.

Example 14.16 Let $f(x) = x + 2$ and $g(x) = (x^2 - 4)/(x - 2)$. Then $f \neq g$ even though $g(x) = x + 2$ for $x \neq 2$, because

$$\text{domain } f = \text{the set of all real numbers}$$

and

$$\text{domain } g = \{x: x \neq 2\}.$$ ▪

Example 14.17 Let $f(x) = (2x - 3)^2$. Find functions g and h such that $h \circ g = f$.

Solution Choose $g(x) = 2x - 3$ and $h(x) = x^2$. Then

$$(h \circ g)(x) = h(g(x)) = h(2x - 3) = (2x - 3)^2 = f(x).$$

Therefore, $h \circ g = f$, as required. ▪

Example 14.18 Let $f(x) = \sqrt{x - 2}$. Find functions g and h such that $h \circ g = f$.

Solution The domain of f is $\{x: x \geq 2\}$. Let $g(x) = x - 2$ for $x \geq 2$, and $h(x) = \sqrt{x}$. Then

$$(h \circ g)(x) = h(g(x)) = h(x - 2) = \sqrt{x - 2} = f(x).$$ ▪

EXERCISES FOR SECTION 14

For each function in Exercises 1–6, determine $f(2)$, $f(0)$, $f(-\sqrt{5})$, $f(t)$, $f(a/2)$, $f(b + 1)$, *and* $f(x^2)$.

1. $f(x) = 2x - 3$ **2.** $f(x) = 4x + 1$

3. $f(x) = -x + 5$ **4.** $f(x) = 3x^2$

5. $f(x) = x^2 + 1$ **6.** $f(x) = 4x^2 - 3$

Determine the domain of each function.

7. $f(x) = 2/(x - 1)$

8. $g(x) = -1/(x^2 - 9)$

9. $h(x) = -x/(x^2 + 2x + 1)$

10. $g(x) = \sqrt{2x - 1}$

11. $h(x) = \sqrt{1 - x}$

12. $f(x) = \sqrt{x + 5}$

13. $h(x) = \sqrt{1 - |x|}/x$

14. $f(x) = \sqrt{x}/(x - 1)$

15. $g(x) = \sqrt{4 - x}/(x^2 - 3x - 4)$

In each of Exercises 16–21, determine:
(a) $f(x) + g(x)$ (e) $f(0) + g(1)$
(b) $f(x) - g(x)$ (f) $f(2) - g(-1)$
(c) $f(x) \cdot g(x)$ (g) $f(a) \cdot g(\sqrt{3})$
(d) $f(x)/g(x)$ (h) $f(t)/g(0)$.

16. $f(x) = 2x - 1$ and $g(x) = x^2 + 1$

17. $f(x) = x + 2$ and $g(x) = x^2 - 2$

18. $f(x) = 3x$ and $g(x) = x^2 - 3$

19. $f(x) = x + b$ and $g(x) = x + a$

20. $f(x) = ax$ and $g(x) = ax + b$

21. $f(x) = ax + b$ and $g(x) = x + b$

In each of Exercises 22–30, determine:
(a) $(f \circ g)(x)$ (c) $(f \circ g)(2)$
(b) $(g \circ f)(x)$ (d) $(g \circ f)(-3)$

22. $f(x) = 3x, g(x) = x - 1$

23. $f(x) = x + 2, g(x) = -4x$

24. $f(x) = 1 - x, g(x) = 2x$

25. $f(x) = x^3, g(x) = -3x$

26. $f(x) = 2x - 1, g(x) = x^3$

27. $f(x) = -2x^3, g(x) = x + 1$

28. $f(x) = |x|, g(x) = -x$

29. $f(x) = x^2, g(x) = -|x|$

30. $f(x) = 1/(x - 1), g(x) = 2/x$

In Exercises 31–36, find functions g and h such that $h \circ g = f$.

31. $f(x) = (x - 4)^2$ **32.** $f(x) = \sqrt{x + 3}$

33. $f(x) = \sqrt{2x - 1}$ **34.** $f(x) = x^2 + 1$

35. $f(x) = \sqrt{x} - 1$ **36.** $f(x) = 2\sqrt{x}$

Compute and simplify the difference quotient of each function in Exercises 37–42. (See Example 14.11.)

37. $f(x) = 4x + 2$ **38.** $f(x) = -2x + 1$

39. $f(x) = 1 - x$ **40.** $f(x) = x^2 + 3$

41. $f(x) = 2x^2 - 1$ **42.** $f(x) = 4x^2 + 2$

In Exercises 43–45, find functions g and h such that $f(x) = g(x)h(x)$.

43. $f(x) = 27x^3 - 1000$

44. $f(x) = x^4 + 11x^2 + 30$

45. $f(x) = x^{2/3} - 8^{2/3}$

In Exercises 46–48, find all x such that $f(x) \geq g(x)$.

46. $f(x) = (x - 2)^2$, $g(x) = 3(x - 2)$

47. $f(x) = |x + 2| - 3$, $g(x) = 7$

48. $f(x) = \dfrac{1}{x + 2}$, $g(x) = \dfrac{1}{x - 1}$

For each function f *in Exercises 49–51, use a calculator to find* $f(\sqrt{5})$ *to four significant figures.*

49. $f(x) = (3x - \sqrt{3})^2$ C

50. $f(x) = 1/\sqrt{6 - x}$ C

51. $f(x) = \dfrac{4}{3}\pi x^3$ C

52. Let $g(x) = x^2$ for $x = 0$, ± 1, and ± 2. (Thus the domain of g is $\{0, \pm 1, \pm 2\}$.) What is the range of g?

53. Let $f(x) = x^2 + 1$ for each real number x. What is the range of f?

54. Let $g(x) = 3x - 1$ for each real number x. Show that for each real number y there is a real number x such that $f(x) = y$. (Solve $3x - 1 = y$ for x.) What do you conclude about the range of f?

55. Assume $f(x) = 2x^2 - 1$ and $g(x) = x + 2$. Find every a such that $f(a) - 2[g(a)]^2 = 0$.

56. Assume $f(x) = x^2 - 2$ and $g(x) = x + 2$. Find every a such that $3 \cdot f(a) - g(a) = g(f(a))$.

57. Assume $f(x) = -x + 4$ and $g(x) = x^2$. Find every a such that $3 \cdot f(g(a)) = g(f(a))$.

58. Let A denote the area of a circle of radius r.

  (a) Find $f(r)$ such that $A = f(r)$.

  (b) Find $g(A)$ such that $r = g(A)$.

59. Let C denote temperature in degrees Celsius and F temperature in degrees Fahrenheit. Find $h(C)$ such that $F = h(C)$. What is the domain of h? What is the range of h? [See Example 14.3(b).]

60. Suppose \$200 is invested at 6% per year simple interest for t years. Find $f(t)$ such that $I = f(t)$, where I is the interest earned.

61. Suppose \$2000 is invested at 12% compounded quarterly. Find $g(t)$ such that $S = g(t)$, where S is the compound amount after t years. (See Section 2C.)

62. Let S denote the surface area of a cube and V the volume. Find $f(V)$ such that $S = f(V)$. (Suggestion: If x denotes the length of an edge, then $S = 6x^2$ and $V = x^3$. Solve the second equation for x and substitute in the first equation.)

63. Let S denote the surface area $(S = 4\pi r^2)$ of a sphere and V the volume $(V = \frac{4}{3}\pi r^3)$. Find $f(S)$ such that $V = f(S)$. (The suggestion for Exercise 62 may help.)

SECTION **15**
Graphs. Linear Functions. More Examples

A. THE GRAPH OF A FUNCTION

The domains and ranges of the functions in this section will all be sets of real numbers. One of the best ways to study such functions is to look at their graphs, which are determined as follows. First, choose a plane with a Cartesian coordinate system. If x is in the domain of a function f, then $(x, f(x))$ is a pair of real numbers. Therefore, $(x, f(x))$ determines a point in the coordinate plane— the point with x as first coordinate and $f(x)$ as second coordinate.

Definition The **graph** of a function f is the set of all points having coordinates $(x, f(x))$ for x in the domain of f.

If we use y to represent the second member of a coordinate pair, as usual, then the graph of f is the set of all (x, y) such that $y = f(x)$. Therefore:

> The graph of a *function* f is the same as
> the graph of the *equation* $y = f(x)$.

Example 15.1 Draw the graph of $f(x) = x$. (This function is called the **identity function.**)

Solution The graph is the same as the graph of the equation $y = x$. Typical points on the graph are $(0, 0)$, $(1, 1)$, and $(-5, -5)$. From Section 13 we know that the graph is the line through these points (Figure 15.1). ∎

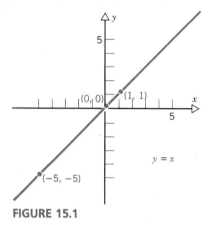

FIGURE 15.1

For each x in the domain of any function f there is only one y such that $y = f(x)$, because each input element determines a unique output element. This means that each vertical line can intersect the graph of a function *at most once:*

If a is in the domain of f, then the vertical line $x = a$ will intersect the graph of f *exactly once*—at the point $(a, f(a))$.

If a is not in the domain of f, then the vertical line $x = a$ will *not* intersect the graph of f.

For example, if the curve C_1 in Figure 15.2 were the graph of a function f, then necessarily $f(a) = b$, and $f(a) = c$, an impossibility because $b \neq c$. On the

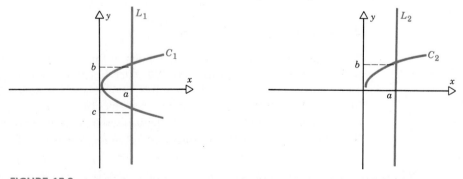

FIGURE 15.2 (a) The curve C_1 is *not* the graph of a function, because the vertical line L_1 intersects the curve more than once. (b) The curve C_2 is the graph of a function, because each vertical line (such as L_2) intersects the curve at most once.

other hand, the curve C_2 in Figure 15.2 *is* the graph of a function; if f denotes this function then, for example, $f(a) = b$. [The curve C_2 is, in fact, the graph of $f(x) = \sqrt{x}$, $x \geq 0$.]

B. LINEAR FUNCTIONS

In this subsection we study the functions whose graphs are lines.

Definition A function f is called a **linear function** if it has the form

$$f(x) = mx + b \qquad (15.1)$$

for some pair of real numbers m and b.

> The graph of a function is a straight line
> iff
> the function is linear.

Proof The graph of a function cannot be a vertical line since the graph of a function cannot contain two or more points with the same x value. On the other hand, the slope-intercept form for the equation of a line (Section 13E) shows that the nonvertical lines are precisely the graphs of the equations of the form $y = mx + b$. ☐

If b is any real number, then the function $f(x) = b$ is called a **constant function.** This is Equation (15.1) with $m = 0$. Thus every constant function is also a linear function. Since the graph of $y = mx + b$ is a horizontal line iff $m = 0$, the graph of a linear function is a horizontal line iff the function is a constant function. Figure 15.3 shows examples.

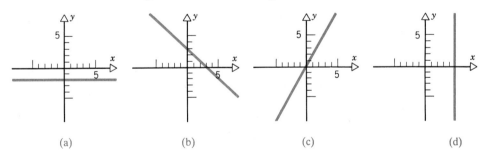

(a)	(b)	(c)	(d)

FIGURE 15.3 (a) Graph of the linear (constant) function $f(x) = -2$. (b) Graph of the linear function $f(x) = -x + 3$. (c) Graph of the linear function $f(x) = 2x$. (d) Graph of $x = 5$, which is not the graph of a function.

Linear functions can be further classified with the following definitions, which will also be useful in studying many other types of functions. Let f denote any function and let S denote any subset of the domain of f.

The function f is **increasing** over S if $x_1 < x_2$ implies $f(x_1) < f(x_2)$ for x_1 and x_2 in S.

The function f is **decreasing** over S if $x_1 < x_2$ implies $f(x_1) > f(x_2)$ for x_1 and x_2 in S.

In terms of a graph, a function is *increasing* if we move *upward* as we move to the right along the graph; and a function is *decreasing* if we move *downward* as we move to the right along the graph. See Figure 15.4.

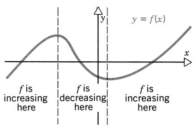

FIGURE 15.4

When we say simply that a function is *increasing,* without specifying a subset, we mean that it is increasing over its entire domain. Similarly for *decreasing.* From our study of slope in Section 13, we know that a linear function $f(x) = mx + b$ is increasing if $m > 0$, and it is decreasing if $m < 0$. For example, the function in Figure 15.3c is increasing, and the function in Figure 15.3b is decreasing.

Because the graph of a linear function is a line, and two distinct points determine a line, it follows that a linear function is completely determined by any two points on its graph. Said differently, a linear function is determined by the output at any two input elements. To apply this observation we use the following form for the equation of a line.

An equation for the line through $P_1(x_1,y_1)$ and $P_2(x_2,y_2)$, with $x_1 \neq x_2$, is

$$y - y_1 = \frac{y_2 - y_1}{x_2 - x_1}(x - x_1). \qquad (15.2)$$

This is called the **two-point form** for the equation of the line.

Proof Equation (15.2) is simply the point-slope form, Equation (13.4), with m replaced by the slope as defined in Equation (13.2). □

Example 15.2 Determine the linear function $f(x) = mx + b$ such that $f(1) = 3$ and $f(2) = -4$.

Solution Since the graph of f is the graph of $y = f(x)$, we need the function whose graph passes through $(1, 3)$ and $(2, -4)$. From Equation (15.2) this is

$$y - 3 = \frac{-4 - 3}{2 - 1}(x - 1)$$

$$y - 3 = -7(x - 1)$$

$$y = -7x + 10.$$

Therefore, the desired function is $f(x) = -7x + 10$.

Check If $f(x) = -7x + 10$, then

$$f(1) = -7 \cdot 1 + 10 = 3 \quad \text{and} \quad f(2) = -7 \cdot 2 + 10 = -4.$$

C. OTHER EXAMPLES

Example 15.3 Draw the graph of $f(x) = |x|$.

Solution Recall (Section 1D) that $|x|$ denotes the *absolute value* of x, which can be defined by

$$|x| = \begin{cases} x & \text{if } x \geq 0 \\ -x & \text{if } x < 0. \end{cases}$$

Therefore, for $x \geq 0$ the graph is the same as the graph of $y = x$. For $x < 0$, however, the graph coincides with the graph of $y = -x$. Typical points are $(1, 1)$, $(-1, 1)$, $(2, 2)$, and $(-2, 2)$. The graph is shown in Figure 15.5.

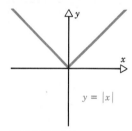

FIGURE 15.5

The function in the next example is defined by one expression for part of its domain and another expression for the remainder of its domain. We say that such functions are defined **piecewise.**

Example 15.4 The graph of

$$f(x) = \begin{cases} -1 & \text{for } x \leq 0 \\ x + 1 & \text{for } x > 0 \end{cases}$$

is shown in Figure 15.6. The symbol ● on the end of a segment means that the endpoint is part of the graph; the symbol ○ means that the endpoint is not part of the graph.

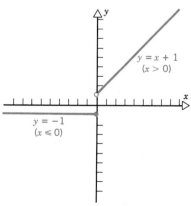

FIGURE 15.6

Example 15.5 The cost of a phone call from one city to another, during a given time period, is usually determined by the duration of the call. For instance, the cost might be $1.00 for the first 3 minutes (or portion thereof), and then $0.25 for each additional minute (or portion thereof). The graph of the corresponding function is shown in Figure 15.7, with T (for *time* in minutes) in place of x, and C (for *cost*) in place of y. Both this function and the one in the next example are illustrations of **step functions.**

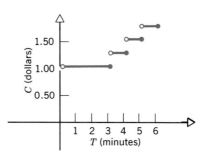

FIGURE 15.7

Example 15.6 If x is any real number, then $[x]$ is often used to denote the greatest integer not exceeding x. In other words, if x is an integer then $[x] = x$; otherwise, x "rounded down" to the next integer is $[x]$. For example,

$$[5] = 5, \quad [1.2] = 1, \quad [\pi] = 3, \quad \text{and} \quad [-17.2] = -18.$$

Figure 15.8 shows the graph of $f(x) = [x]$. This function is called the **greatest integer function.**

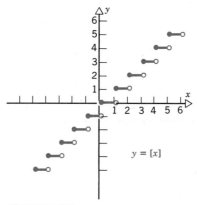

FIGURE 15.8

D. FUNCTIONS OF SEVERAL VARIABLES

The next two examples concern functions of *several variables,* in which one quantity depends on more than one other quantity.

Example 15.7 The formula for simple interest was given in Equation (1.1):

$$I = Prt,$$

where I, P, r, and t denote interest, principal, rate, and time, respectively. The variable I is a function of *three* variables: P, r, and t. To handle this we extend the notation $f(x)$ for functions of a single variable by writing

$$I = f(P, r, t) = Prt.$$

If $P = \$100$, $r = 0.05$, and $t = 2$ years, for example, then

$$I = f(100, 0.05, 2) = 100 \times 0.05 \times 2 = \$10.$$ ▪

Example 15.8 The formula for the volume V of a right circular cylinder of radius r and height h is

$$V = \pi r^2 h. \tag{15.3}$$

(a) Determine a function f such that $h = f(V, r)$.

(b) Determine a function g such that $r = g(V, h)$.

Solution (a) From (15.3), $h = V/\pi r^2$. Thus the function is

$$f(V, r) = \frac{V}{\pi r^2}.$$

If, for example, the volume is 30 cubic meters and the radius is 2 meters, then the height is

$$h = f(30, 2) = \frac{30}{\pi 2^2} \approx 2.39 \text{ meters.}$$ Ⓒ

(b) From (15.3), $r^2 = V/\pi h$, so $r = \sqrt{V/\pi h}$. Therefore,

$$g(V, h) = \sqrt{\frac{V}{\pi h}}.$$ ▪

E. APPLICATION: VARIATION

If n is an integer and

$$y = kx^n \quad (k \neq 0), \tag{15.4}$$

then we say that y **varies as the nth power of** x, or that y is **proportional to the nth power of** x. The number k is called the **constant of proportionality.** In functional notation, Equation (15.4) is represented by

$$y = f(x) \quad \text{where} \quad f(x) = kx^n.$$

Here are some special cases.

$$
\begin{array}{lll}
n = 1 & y = kx & y \text{ varies directly as } x \\
n = -1 & y = k/x & y \text{ varies inversely as } x \\
n = 2 & y = kx^2 & y \text{ varies as the square of } x
\end{array}
$$

Other variation possibilities include the case $y = k\sqrt{x}$, where y **varies as the square root of** x.

The following example illustrates that if a single pair of corresponding values of x and y is known (for a given k), then any other pair can be computed.

Example 15.9 *Boyle's law* states that if the temperature is constant, then the pressure P of a confined gas varies inversely as the volume V of the gas:

$$P = k/V. \qquad (15.5)$$

Problem Suppose the volume of a confined gas is 100 cubic inches when it is subjected to a pressure of 40 pounds per square inch. Determine the volume when the pressure is increased to 45 pounds per square inch.

Solution First we determine k, using Equation (15.5) with $P = 40$ and $V = 100$:

$$40 = k/100$$

$$k = 4000$$

Thus

$$P = 4000/V.$$

Multiplication of both sides by V/P yields

$$V = 4000/P.$$

When $P = 45$, this gives

$$V = 4000/45 \approx 89 \text{ cubic inches.} \qquad \boxed{C} \quad \blacksquare$$

EXERCISES FOR SECTION 15

Draw the graph of each function. Also indicate whether the function is constant, increasing, or decreasing.

1. $f(x) = 3x$

2. $f(x) = -2x$

3. $f(x) = \dfrac{1}{2}x$

4. $g(x) = -x + 2$

5. $g(x) = 6$

6. $g(x) = -3x - 2$

7. $h(x) = -\dfrac{1}{2}$

8. $h(x) = 4x - 1$

9. $h(x) = -3.5$

In 10–18, decide whether each graph is the graph of a function. Give a reason for each answer. Also, if the graph does represent a function, give the set of values of x over which it is increasing and the set over which it is decreasing.

10.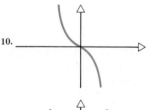

11.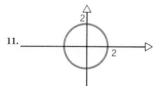

12.

13.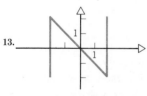

14.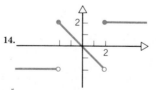

15.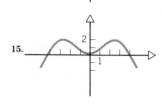

16.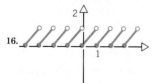

17.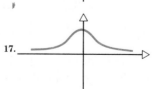

18.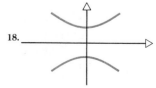

Evaluate. (See Example 15.6 for the meaning of [ ].)

19. $[1.99]$

20. $[-3.2]$

21. $[\pi/2]$

22. $\dfrac{1}{2}[5]$

23. $\left[\dfrac{1}{2}\cdot 5\right]$

24. $5\left[\dfrac{1}{2}\right]$

25. $|[-3.1]|$

26. $[|-3.1|]$

27. $|-[3.1]|$

Draw the graph of each function.

28. $f(x) = |x| + 1$

29. $f(x) = |x| - 1$

30. $f(x) = |2x|$

31. $f(x) = [x] - 1$

32. $f(x) = [2x]$

33. $f(x) = [x] + 1$

34. $f(x) = \begin{cases} -2 & \text{for } x \le 1 \\ x & \text{for } x > 1 \end{cases}$

35. $f(x) = \begin{cases} x & \text{for } x < 0 \\ 2x & \text{for } x \ge 0 \end{cases}$

36. $f(x) = \begin{cases} x + 1 & \text{for } x \le 0 \\ -x + 1 & \text{for } x > 0 \end{cases}$

Determine the linear function f satisfying the two given conditions.

37. $f(1) = 5$ and $f(3) = 9$

38. $f(0) = -1$ and $f(2) = 3$

39. $f(-2) = 1$ and $f(1) = 4$

40. $f(-3) = -1$ and $f(0) = -3$

41. $f(2) = 12$ and $f(5) = 12$

42. $f(0) = 5$ and $f(4) = -1$

43. Suppose $f(x) = -\sqrt{2}$ and $g(x) = x^2 - 1$. Determine $(f \circ g)(x)$ and $(g \circ f)(x)$.

44. Suppose $f(x) = -|x|$ and $g(x) = -7$. Determine $(f \circ g)(x)$ and $(g \circ f)(x)$.

45. Suppose $f(x) = [x]$ and $g(x) = \pi$. Determine $(f \circ g)(x)$ and $(g \circ f)(x)$.

46. The formula for the volume V of a right circular cone having height h and base of radius r is

$$V = \frac{1}{3}\pi r^2 h.$$

(a) Determine a function f such that $h = f(V, r)$.

(b) Determine a function g such that $r = g(V, h)$.

47. The formula for the weight W of a sphere having

radius r and constant density d is $W = \dfrac{4}{3}\pi dr^3$.

(a) Determine a function f such that $d = f(W, r)$.

(b) Determine a function g such that $r = g(W, d)$.

48. Determine a function f such that $S = f(P, r, n)$, where S is the compound amount from a principal amount P invested for one year at an annual rate r (in decimal form) compounded n times annually. [See Equation (2.11).] Compute $f(100, 0.08, 4)$.

49. Suppose that y varies directly as x and that $y = 5$ when $x = 20$. (a) If $y = f(x)$, what is $f(x)$? (b) What is y when $x = 13$? (c) If $x = g(y)$, what is $g(y)$?

50. Suppose that y varies as the square of x and that $y = 70$ when $x = 20$. (a) If $y = f(x)$, what is $f(x)$? (b) What is x when $y = 28$? (c) If $x = g(y)$, what is $g(y)$?

51. Suppose that y varies as the square of x and that $y = 10$ when $x = 5$. (a) If $y = f(x)$, what is $f(x)$? (b) What is y when $x = 7$? (c) If $x = h(y)$, what is $h(y)$?

52. Suppose the volume of a confined gas is 50 cubic inches when it is subjected to a pressure of 20 pounds per square inch. Determine the volume when the pressure is decreased to 18 pounds per square inch. (See Example 15.9.)

53. The intensity I of a sound wave is inversely proportional to the square of the distance r from the source: $I = k/r^2$. (This assumes ideal conditions, including the absence of nearby reflecting or absorbing surfaces.) If r is measured in meters, then I is measured in watts/m^2. Suppose the intensity of a sound wave is 0.05 watts/m^2 when measured 30 meters from the source. What is the intensity 20 meters from the source?

54. The stopping distance d for a car, after the brakes have been applied, varies directly as the square of the velocity v.

(a) If the stopping distance d for a particular car is 125 feet at 50 miles per hour, what is the stopping distance at 60 miles per hour?

(b) In general, how will doubling the velocity affect the stopping distance?

(c) In general, by what factor must the velocity increase in order to double the stopping distance?

55. The period T of a simple pendulum varies as the square root of its length L. Specifically, $T =$

$2\pi\sqrt{L/g}$, where $g \approx 980$ when L is in centimeters and T is in seconds.

(a) What is the period of a 20 centimeter pendulum? Ⓒ

(b) What increase in the length of a pendulum will double the period?

56. If d denotes the distance in miles to the horizon from a point h feet above the earth's surface, then d varies as the square root of h (for reasonably small d). From the top of the Sears Tower in Chicago, which is 1434 feet tall, the distance to the horizon is approximately 46.7 miles. Find the distance to the horizon from the top of the CN Tower in Toronto, which is 1820 feet tall. Ⓒ

57. Newton's *law of universal gravitation* states that any two objects are attracted by a force that is directly proportional to the product of their masses and inversely proportional to the square of the distance between them:

$$F = G\,\frac{m_1 m_2}{r^2},$$

where G, the constant of proportionality, is the *universal constant of gravitation*. If the masses of two objects are doubled and the distance between them is tripled, how will the force between them change?

58. If T denotes the period for a planet (in Earth years), and R denotes the average distance between the planet and the sun (in millions of miles), then $R^3 = KT^2$. Show that T varies as the $\frac{3}{2}$ power of R, that is, $T = cR^{3/2}$ for an appropriate constant c. The average distance between the Earth and sun is approximately 93 million miles. Use this (with $T = 1$) to determine the numerical value of c. Ⓒ

59. The strength S of a horizontal beam of rectangular cross section, supported at its ends, varies jointly as its breadth B and the square of its depth D, and inversely as its length L:

$$S = k\,\frac{BD^2}{L}$$

If a 10-foot long 2 inch by 4 inch beam can support a weight of 800 pounds when resting on its 4-inch side, what weight can it support when resting on its 2-inch side? (The value of k is determined by the material of the beam.)

60. Consider a beam as described in Exercise 59. If the depth of the beam is reduced by 25%, by what percentage must the breadth be increased to restore the beam's original strength? (Assume that the length remains constant.)

SECTION **16**
Quadratic Functions

A. INTRODUCTION. INTERCEPTS

Definition A function f is called a **quadratic function** if it has the form

$$f(x) = ax^2 + bx + c, \qquad (16.1)$$

where a, b, and c are real numbers with $a \neq 0$.

Example 16.1 Draw the graph of the quadratic function $f(x) = x^2$.

Solution Table 16.1 shows some typical coordinate pairs. The graph is shown in Figure 16.1. ▪

The graph of any quadratic function is called a **parabola.** Each figure in this section gives an example. If the parabola opens upward (like Figure 16.1),

TABLE 16.1

x	x^2
0	0
$\pm\frac{1}{2}$	$\frac{1}{4}$
± 1	1
± 2	4
± 3	9

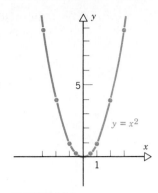

FIGURE 16.1

then the **vertex** is the minimum or lowest point on the graph; if the parabola opens downward (like Figure 16.2*a*), then the **vertex** is the maximum or highest point on the graph. In either case we call the vertex the **extreme point** of the graph. (Parabolas are also discussed in Section 65, but this section is independent of that.)

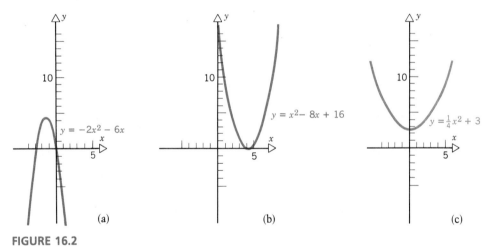

FIGURE 16.2

To analyze a quadratic function we concentrate on three questions, which can be stated in terms of the graph as follows:

Question I. What are the *x*-intercepts, if any?

Question II. Does the graph open upward? or downward?

Question III. Where is the extreme point on the graph?

We can draw the graph with very little effort once these questions have been answered. Moreover, these questions get to the heart of many other problems besides graphing, as Subsection C will show. We treat Question I here and Questions II and III in Subsection B.

Question I is just that of finding the real numbers that are solutions of

$$ax^2 + bx + c = 0. \tag{16.2}$$

This was discussed in Section 8: if possible, solve Equation (16.2) by factoring;

otherwise use the quadratic formula. For convenience, here is the quadratic formula again.

The solutions of $ax^2 + bx + c = 0$ are

$$x = \frac{-b \pm \sqrt{b^2 - 4ac}}{2a}.$$ (16.3)

If the discriminant, $b^2 - 4ac$, is negative, then Equation (16.2) has no real solution; therefore, the graph of (16.1) has no x-intercept.

If the discriminant is 0, then (16.2) has one real solution and the graph of (16.1) has one x-intercept.

If the discriminant is positive, then (16.2) has two unequal real solutions and the graph of (16.1) has two x-intercepts.

Example 16.2 Determine the x-intercepts of the graph of each function.

(a) $f(x) = -2x^2 - 6x$

(b) $f(x) = x^2 - 8x + 16$

(c) $f(x) = \frac{1}{4}x^2 + 3$

Solution (a) Since $-2x^2 - 6x = -2x(x + 3) = 0$ iff $x = 0$ or $x = -3$, there are two x-intercepts, 0 and -3. See Figure 16.2a.

(b) Since $x^2 - 8x + 16 = (x - 4)^2 = 0$ iff $x = 4$, there is one x-intercept, 4. See Figure 16.2b.

(c) In this case, $b^2 - 4ac = 0^2 - 4(\frac{1}{4})(3) = -3 < 0$. Thus there is no x-intercept. Notice that $f(x) > 0$ for every x because $x^2 \geq 0$ and hence $\frac{1}{4}x^2 + 3 > 0$. See Figure 16.2c. ∎

B. MAXIMUM AND MINIMUM VALUES

The vertex or extreme point of the graph of

$$f(x) = ax^2 + bx + c \quad (a \neq 0)$$

occurs at the point $(-b/2a, f(-b/2a))$.

If $a > 0$, the graph opens upward and this point is a minimum.

If $a < 0$, the graph opens downward and this point is a maximum.

(Here is a memory tip: Think of the parabola as a container. If a is *positive*, the container *will* hold water—it opens upward. If a is *negative*, the container *will not* hold water—it opens downward.)

Proof To begin, we rewrite $ax^2 + bx + c$ by completing the square (Section 8).

$$f(x) = ax^2 + bx + c$$

$$= a\left(x^2 + \frac{b}{a}x\right) + c$$

$$= a\left(x^2 + \frac{b}{a}x + \frac{b^2}{4a^2}\right) - \frac{b^2}{4a} + c$$

$$= a\left(x + \frac{b}{2a}\right)^2 + \frac{4ac - b^2}{4a}. \tag{16.4}$$

Since the square of every real number is nonnegative,

$$\left(x + \frac{b}{2a}\right)^2 \geq 0 \text{ for all real numbers } x.$$

Thus

$$a\left(x + \frac{b}{2a}\right)^2 \geq 0 \text{ for all } x \text{ if } a > 0,$$

and

$$a\left(x + \frac{b}{2a}\right)^2 \leq 0 \text{ for all } x \text{ if } a < 0.$$

Therefore, from Equation (16.4),

$$f(x) \geq \frac{4ac - b^2}{4a} \text{ for all } x \text{ if } a > 0, \tag{16.5}$$

and

$$f(x) \leq \frac{4ac - b^2}{4a} \text{ for all } x \text{ if } a < 0. \tag{16.6}$$

Also, (16.4) makes it easy to see that

$$f(x) = \frac{4ac - b^2}{4a} \text{ iff } x = -\frac{b}{2a}. \tag{16.7}$$

By (16.5) and (16.7), if $a > 0$ there is a minimum where $x = -b/2a$. By (16.6) and (16.7), if $a < 0$ there is a maximum where $x = -b/2a$. ☐

Example 16.3 Determine the extreme value of $f(x) = -2x^2 + 8x + 3$ and state whether it is a minimum or a maximum value.

Solution Here $a = -2$ and $b = 8$, so that

$$-\frac{b}{2a} = -\frac{8}{2(-2)} = 2.$$

Thus the extreme value is $f(2) = -2 \cdot 2^2 + 8 \cdot 2 + 3 = 11$. Since $a = -2 < 0$, this is a maximum value. ■

The vertical line $x = -b/2a$, which passes through the vertex, is called the **axis** of the parabola (Figure 16.3). The graph of the parabola is symmetric about

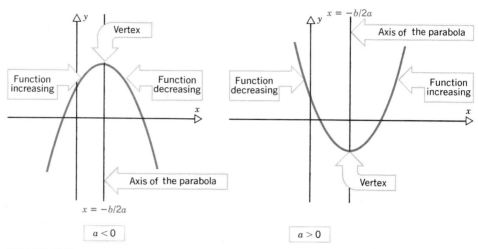

FIGURE 16.3

its axis; that is, for each point of the graph on one side of the axis there is a corresponding point on the other side with the same y-coordinate and the same distance from the axis [compare $(2, 4)$ and $(-2, 4)$ in Figure 16.1]. Figure 16.3 indicates the increasing and decreasing characteristics of quadratic functions.

Example 16.4 Draw the graph of $f(x) = 2x^2 + 3x - 2$. Give the coordinates of the vertex, the equation of the axis, the x-intercepts (if there are any), and where the function is increasing and decreasing.

Solution Here $a = 2$, $b = 3$, and $c = -2$. Since $a > 0$, the parabola opens upward. The minimum is where $x = -b/2a = -3/(2 \cdot 2) = -\frac{3}{4}$. The corresponding y-value is

$$f(-\tfrac{3}{4}) = 2(-\tfrac{3}{4})^2 + 3(-\tfrac{3}{4}) - 2 = -\tfrac{25}{8}.$$

Thus the vertex is at $(-\frac{3}{4}, -\frac{25}{8})$. The axis is the line $x = -\frac{3}{4}$.

Because $2x^2 + 3x - 2 = (2x - 1)(x + 2)$, $f(x) = 0$ iff $x = \frac{1}{2}$ or $x = -2$. Thus the x-intercepts are at $x = \frac{1}{2}$ and $x = -2$.

The function is decreasing for $x < -\frac{3}{4}$ (to the left of the vertex) and increasing for $x > -\frac{3}{4}$ (to the right of the vertex).

To draw the graph, first plot the vertex and the x-intercepts. Then plot several other points to determine the shape more closely. Some sample points in this case are $(-3, 7)$, $(-1, -3)$, $(0, -2)$, and $(1, 3)$. Remember also that a parabola must be symmetric about its axis. Figure 16.4 shows the graph. ■

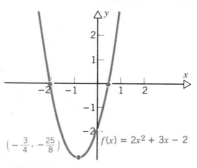

FIGURE 16.4

C. APPLICATIONS

Example 16.5 Suppose that a small rocket is fired upward near the earth's surface. The distance s of the rocket above the earth's surface after t seconds is given by the quadratic function

$$f(t) = -\frac{1}{2}gt^2 + v_0t + s_0 \quad (t \geq 0), \tag{16.8}$$

where

$g \approx 32$ if s is measured in feet,

$g \approx 9.8$ if s is measured in meters,

$v_0 = $ velocity when $t = 0$, and

$s_0 = $ height when $t = 0$.

Remark This type of problem was also considered in Section 8. If you compare Equation (8.7) with Equation (16.8), you will see that $\frac{1}{2}gt^2$ has been replaced by $-\frac{1}{2}gt^2$ and a new term s_0 has been added. The sign on $\frac{1}{2}gt^2$ has been changed because the positive direction is upward here and it was downward in Section 8. The term s_0 has been added because distance is measured from the ground here, rather than from the initial position, as in Section 8. The number v_0 in Equation (16.8) will be negative if the rocket is fired downward rather than upward. Finally, Equation (16.8) will describe the motion of any other object as well as a rocket, at least under the restricted conditions that we consider here.

Problem A small rocket is fired upward from the top of an 80-foot tall building with an initial velocity of 64 feet per second. Determine its maximum altitude and when it will reach the ground.

Solution The function describing the position s is given by Equation (16.8) with $g = 32$, $v_0 = 64$, and $s_0 = 80$:

$$s = f(t) = -16t^2 + 64t + 80.$$

This has the form of Equation (16.1) with $a = -16$, $b = 64$, $c = 80$, and t in place of x. See Figure 16.5 for the graph.

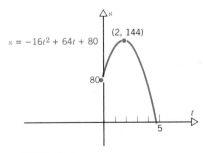

FIGURE 16.5

The rocket will reach its maximum altitude when $t = -b/2a = -64/2(-16) = 2$ seconds. The maximum altitude will be

$$f(2) = -16(2)^2 + 64(2) + 80 = 144 \text{ feet}$$

above the earth's surface.

The rocket will reach the ground when $s = f(t) = 0$:

$$-16t^2 + 64t + 80 = 0$$

$$-16(t^2 - 4t - 5) = 0$$

$$(t + 1)(t - 5) = 0$$

$$t = -1 \text{ or } t = 5.$$

The time $t = -1$ has no physical significance in this problem (remember $t \geq 0$). Therefore, the rocket will reach the ground $t = 5$ seconds after it is fired.

Example 16.6 A rectangular field is to be formed with one side along a straight river bank. If the side along the river bank requires no fencing, and if 400 yards of fencing are available for the other three sides, what is the largest possible area for the field?

Solution Let x denote the length of the sides perpendicular to the bank. Then the other length must be $400 - 2x$, as shown in Figure 16.6. Therefore, the total area is

$$A = f(x) = x(400 - 2x) = -2x^2 + 400x.$$

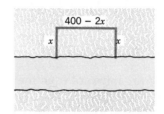

FIGURE 16.6

This is a quadratic function with $a = -2$, $b = 400$, and $c = 0$. Therefore, the x-value that maximizes the function is

$$-b/2a = -400/[2(-2)] = 100 \text{ yards.}$$

The maximum possible area is

$$f(100) = -2(100)^2 + 400(100)$$

$$= 20,000 \text{ square yards.}$$

The graph of f is shown in Figure 16.7.

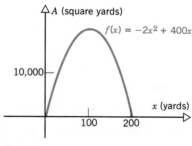

FIGURE 16.7

It is generally harder to determine maximum or minimum values for more complicated functions (such as those involving x^3). That is one of the problems that is treated in calculus.

Other applications of parabolas are shown in Figures 16.8 and 16.9.

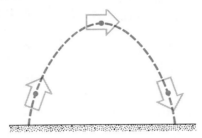

FIGURE 16.8   If air resistance is ignored, then the path followed by the center of gravity of a leaping leopard or other animal is a parabola. So is the path of a projectile fired from a cannon.

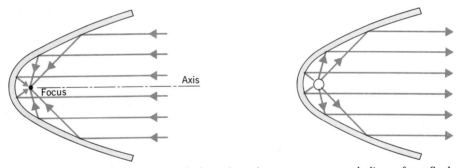

FIGURE 16.9   A parabola rotated about its axis generates a *parabolic surface*. Such surfaces are used for reflecting telescopes because they concentrate the light from distant objects at a single point (called the *focus*), giving sharp images of stars or other sources of light. Conversely, parabolic surfaces are used for such things as headlights because a light placed at the focus will be reflected outward in rays parallel to the axis.

EXERCISES FOR SECTION 16

1. On a single Cartesian plane (coordinate system) draw the graphs of $y = x^2$, $y = 2x^2$, and $y = -\frac{1}{2}x^2$. How does changing a affect the graph of $y = ax^2$?

2. On a single Cartesian plane (coordinate system) draw the graphs of $y = x^2$, $y = -3x^2$, and $y = \frac{1}{3}x^2$. How does changing a affect the graph of $y = ax^2$?

3. On a single Cartesian plane (coordinate system) draw the graphs of $y = x^2$, $y = 4x^2$, and $y = -\frac{1}{4}x^2$. How does changing a affect the graph of $y = ax^2$?

4. On a single Cartesian plane (coordinate system) draw the graphs of $y = x^2$, $y = x^2 + 1$, and $y = x^2 - 2$. How does changing c affect the graph of $y = x^2 + c$?

5. On a single Cartesian plane (coordinate system) draw the graphs of $y = x^2$, $y = x^2 + 2$, and $y = x^2 - 1$. How does changing c affect the graph of $y = x^2 + c$?

6. On a single Cartesian plane (coordinate system) draw the graphs of $y = x^2$, $y = x^2 + 3$, and $y = x^2 - 3$. How does changing c affect the graph of $y = x^2 + c$?

Determine the extreme value of each function and state whether it is a minimum or a maximum value.

7. $f(x) = 2x^2 + x + 1$

8. $g(x) = x^2 - 2$

9. $h(x) = x^2 - 6x$

10. $f(x) = -x^2 - 5$

11. $g(x) = -3x^2 + x + 1$

12. $h(x) = 2x^2 + 4$

Draw the graph of each function. Give the coordinates of the vertex and the equation of the axis. Also determine the x-intercepts if there are any.

13. $f(x) = -2x^2 - 1$

14. $f(x) = -3x^2 + 12$

15. $f(x) = 3x^2 + 3$

16. $g(x) = x^2 - 2x + 1$

17. $g(x) = -x^2 - x$

18. $g(x) = -2x^2 + 12x - 18$

19. $h(x) = (x + 1)^2 - 1$

20. $h(x) = -x^2 + 5x$

21. $h(x) = x^2 + x + 1$

22. $f(x) = 2x^2 + 3x$

23. $f(x) = (3 + x)^2 + 3$

24. $f(x) = (x - 2)^2 + 1$

By completing the square, $ax^2 + bx + c$ can be written in the form $a(x - h)^2 + k$. (This is the idea behind Subsection B.) The graph of $y = a(x - h)^2 + k$ is a parabola with vertex at (h, k), opening upward if $a > 0$ and downward if $a < 0$.

In Exercises 25–30, determine the vertex of the graph by completing the square rather than by using the formula derived in Subsection B. [The derivation of Equation (16.4) illustrates the method.]

25. $f(x) = 2x^2 - 4x + 5$

26. $f(x) = 3x^2 + 12x + 11$

27. $f(x) = 4x^2 + 8x + 5$

28. $f(x) = -4x^2 - 4x - 2$

29. $f(x) = 3x^2 - 18x + 25$

30. $f(x) = -4x^2 - 12x + 1$

By solving the inequality $f(x) > 0$ (as in Section 11), determine where the graph of f is above the x-axis.

31. $f(x) = 2x^2 + 7x - 4$

32. $f(x) = 3x^2 + 22x + 7$

33. $f(x) = x^2 - 3$

34. $f(x) = -x^2 + 8x - 15$

35. $f(x) = -x^2 + 2x - 3$

36. $f(x) = 16x^2 + 24x + 9$

37. If the graph of $f(x) = ax^2 + 2x + 3$ contains the point $(1, -2)$, what is a?

38. If the graph of $f(x) = ax^2 + bx + c$ contains the origin, what is c?

39. If the graph of $f(x) = x^2 + bx + 1$ has an x-intercept at $x = -2$, what is b?

40. Find c so that the lowest point on the graph of $f(x) = x^2 + x + c$ is on the x-axis.

41. For which values of b will the graph of $f(x) = x^2 + bx + 1$ not intersect the x-axis?

42. If the graph of $f(x) = ax^2 + x + 1$ is symmetric about the line $x = 3$, what is a?

43. An object is thrown upward from the top of a 64-foot tall building with an initial velocity of 48 feet per second. Determine its maximum altitude and when it will reach the ground.

44. An object is thrown upward from the top of a 29.4-meter tall building with an initial velocity of 24.5 meters per second. Determine its maximum altitude and when it will reach the ground. (Remember to use $g = 9.8$.) [C]

45. The altitude in feet after t seconds for a rocket fired from the ground is given by $f(t) = -16t^2 + v_0t$, where v_0 is the initial velocity [Equation (16.8) with $s_0 = 0$].

(a) Find the maximum altitude in terms of v_0.

(b) What happens to the maximum altitude if v_0 is doubled?

46. Repeat Example 16.6 assuming that 1000 meters of fence are available.

47. Find two numbers whose sum is 7 and whose product is as large as possible. [Suggestion: If x is one number, then the other is $7 - x$. Maximize $x(7 - x)$.]

48. Find two numbers whose sum is 17 and whose product is as large as possible. (Exercise 47 gives a suggestion.)

49. Prove that if x and y are numbers whose sum is s, then the product xy will be maximum when $x = y = s/2$.

50. Find the largest possible shaded area for a rectangle inscribed in a triangle as shown in Figure 16.10. [Suggestion: First, use similar triangles to verify that $y/4 = (3 - x)/3$. Then explain why the area of the rectangle is $4x(3 - x)/3$. Finally, find the maximum value for this area.]

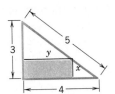

FIGURE 16.10

51. The arches on many bridges have the shape of a parabola. Suppose such an arch has span l feet and rise h feet at the center (Figure 16.11). Prove that the rise x feet from an end is given by

$$f(x) = 4h \left(1 - \frac{x}{l}\right)\frac{x}{l}, \ 0 \le x \le l.$$

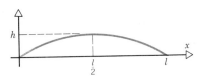

FIGURE 16.11

52. Use Exercise 51 to solve the following problems about parabolic arches.

(a) Show that one-fourth of the way from one end toward the other end (measured along the horizontal), the rise of an arch is three-fourths of the rise at the center.

(b) How far from one end toward the other is the rise one-half of the rise at the center?

53. A farmer plans to use 500 meters of fencing to enclose a rectangular area and then divide it into

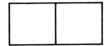

FIGURE 16.12

two equal rectangular plots (Figure 16.12). What dimensions will yield the maximum area? What is the maximum area?

54. Find the dimensions of the rectangle of maximum area having two vertices on the x-axis and two vertices above the x-axis and on the graph of $y = 4 - |3x|$.

55. A window consists of a rectangle surmounted by a semicircle (Figure 16.13). What dimensions (values for r and h) will maximize the total area if the perimeter is to be 10 meters?

FIGURE 16.13

56. Prove that if a parabola is the graph of a quadratic function and has two (unequal) x-intercepts, then the axis of the parabola bisects the segment connecting the x-intercepts. [Suggestion: Consider the average of the x-values in Equation (16.3).]

57. The graph of an equation $x = ay^2 + by + c$ will be a parabola with a horizontal axis. Use analogy with the results of this section to answer the following questions.

(a) Under what conditions will the graph have no y-intercept? one y-intercept? two y-intercepts?

(b) What is the x-intercept?

(c) Under what conditions will the parabola open to the left? to the right?

(d) What are the coordinates of the vertex?

SECTION **17**
Inverse Functions

A. INTRODUCTION. ONE-TO-ONE FUNCTIONS

Consider the following statements, where x represents a real number in each case.

If we choose x, add 5, and then subtract 5 from the result, we get back x.

If we choose x, multiply by 3, and then divide the result by 3, we get back x.

If we choose x positive, square it, and then take the principal square root of the result, we get back x.

In this section we shall see that each of the three statements represents a very special case of a general idea involving functions—the idea of the *inverse* of a function. Before considering inverse functions, however, we need the following facts about *one-to-one* functions.

If f is any function, then for each x in the domain of f there is only *one* y such that $y = f(x)$. In other words, unequal *output* elements of f cannot correspond to the same *input* element. If we also require of f that unequal *input* elements not correspond to the same *output* element, then we get the important class of functions covered by the following definition.

Definition A function f is **one-to-one** iff

$$x_1 \neq x_2 \text{ implies } f(x_1) \neq f(x_2) \tag{17.1}$$

for all x_1 and x_2 in the domain of f.

Condition (17.1) is equivalent to

$$f(x_1) = f(x_2) \text{ implies } x_1 = x_2 \tag{17.2}$$

for all x_1 and x_2 in the domain of f.

Example 17.1 Verify that the function $f(x) = 2x + 5$ is one-to-one.

First Solution Use (17.1): $x_1 \neq x_2$ implies $2x_1 \neq 2x_2$, which implies $2x_1 + 5 \neq 2x_2 + 5$; that is, $f(x_1) \neq f(x_2)$.

Second Solution Use (17.2): $f(x_1) = f(x_2)$ means $2x_1 + 5 = 2x_2 + 5$, which implies $2x_1 = 2x_2$, which in turn implies $x_1 = x_2$. ■

Example 17.2 The function $f(x) = x^2$ is not one-to-one, because (for example) $3 \neq -3$ but $(3)^2 = (-3)^2$. Notice that to show that a function is *not* one-to-one, it suffices to find *one* pair of numbers x_1 and x_2 such that $x_1 \neq x_2$ but $f(x_1) = f(x_2)$. ■

The definition of one-to-one function can be interpreted geometrically as follows. First, recall that each *vertical* line intersects the graph of a function at most once (Figure 15.2). If a function is one-to-one, then we can also say that each *horizontal* line intersects the graph at most once. The graph on the left in Figure 17.1 represents a function that is not one-to-one: the horizontal line L_1 intersects the graph in points such that $x_1 \neq x_2$ but $f(x_1) = f(x_2)$, violating (17.1).

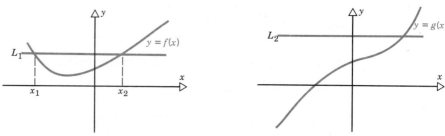

FIGURE 17.1 (a) The function f is not one-to-one, because the horizontal line L_1 intersects the graph more than once: $x_1 \neq x_2$ but $f(x_1) = f(x_2)$. (b) The function g is one-to-one, because each horizontal line (such as L_2) intersects the graph at most once.

If a function is linear and nonconstant, then its graph is a nonhorizontal line. Thus every horizontal line intersects the graph only once, and the function is one-to-one. The same argument applies to any other function that is either increasing (over its entire domain) or decreasing (over its entire domain). This gives us the following statements.

Every increasing function is one-to-one.
Every decreasing function is one-to-one.
Every nonconstant linear function is one-to-one.

B. THE INVERSE OF A ONE-TO-ONE FUNCTION

Assume that f is a one-to-one function and that y is in the range of f. Then there is exactly one element x in the domain of f such that $y = f(x)$. Denote this element x by $g(y)$. Thus

$$x = g(y) \text{ iff } y = f(x). \tag{17.3}$$

In this way we obtain a new function g. The domain of g is the range of f, and the range of g is the domain of f.

The relationship in (17.3) can be represented schematically as follows.

$$\underset{x \quad \overset{g}{\longleftarrow} \quad y}{\overset{f}{\longrightarrow}} \qquad \begin{array}{l} y = f(x) \text{ so } f \text{ applied to } x \text{ gives } y \\[4pt] x = g(y) \text{ so } g \text{ applied to } y \text{ gives } x \end{array}$$

Viewed in this way, it is easy to see what happens if we apply f and then g, or if we apply g and then f; we write the results using composition (Section 14):

$$(g \circ f)(x) = g(f(x)) = g(y) = x \text{ for each } x \text{ in the domain of } f$$

and

$$(f \circ g)(y) = f(g(y)) = f(x) = y \text{ for each } y \text{ in the domain of } g.$$

The conclusion of the last line—that $(f \circ g)(y) = y$ for each y in the domain of g—will remain the same if we replace y throughout by any other variable. In particular, we can use x, which gives "$(f \circ g)(x) = x$ for each x in the domain of g."

The preceding remarks show that g is an *inverse* of f, in the sense of the following definition.

Definition A function g is an **inverse** of a function f if

$$(g \circ f)(x) = x \text{ for each } x \text{ in the domain of } f$$

and

$$(f \circ g)(x) = x \text{ for each } x \text{ in the domain of } g.$$

Example 17.3 Verify that if $f(x) = 2x + 5$ and $g(x) = \frac{1}{2}(x - 5)$, then g is an inverse of f.

Solution
$$(g \circ f)(x) = g(f(x)) = g(2x + 5) = \frac{1}{2}[(2x + 5) - 5] = x$$
$$(f \circ g)(x) = f(g(x)) = f(\tfrac{1}{2}(x - 5)) = 2[\tfrac{1}{2}(x - 5)] + 5 = x \quad \blacksquare$$

Using the terminology of Example 15.1, if g is an inverse of f then $g \circ f$ is the identity function on the domain of f and $f \circ g$ is the identity function on the domain of g.

From the symmetric roles of f and g in the definition of *inverse*, it follows that g is an inverse of f iff f is an inverse of g. It can be proved that f *has an inverse iff f is one-to-one*. Moreover, if a function does have an inverse, then it has only one. The unique inverse of f, if it exists, is denoted by f^{-1} (read "f inverse.")

Note: $f^{-1}(x)$ *does not, in general, mean* $\dfrac{1}{f(x)}$. This use of -1 as a superscript is not the same as its use as an exponent.

---

f^{-1} denotes the inverse of f.
$$y = f(x) \text{ iff } f^{-1}(y) = x.$$

It follows from the definition of an inverse function that the domain of f^{-1} is the range of f, and the range of f^{-1} is the domain of f.

The three examples at the beginning of this section can be described in the language of functions as follows:

If $f(x) = x + 5$, then $f^{-1}(x) = x - 5$.
If $f(x) = 3x$, then $f^{-1}(x) = x/3$.
If $f(x) = x^2$ for $x \geq 0$, then $f^{-1}(x) = \sqrt{x}$ for $x \geq 0$.

In the last example, we require $x \geq 0$ because otherwise f would not have an inverse; the function defined by $f(x) = x^2$, with no restriction on x, is not one-to-one.

C. FINDING INVERSE FUNCTIONS

If the equation $y = f(x)$ can be solved uniquely for x, the result will be $f^{-1}(y)$. If we want an expression for f^{-1} in terms of x, then we simply replace y by x throughout the equation giving $f^{-1}(y)$. Summarizing:

To find $f^{-1}(x)$, given $f(x)$.

I. Let $y = f(x)$.
II. Solve for x to get $x = f^{-1}(y)$.
III. Replace y by x to get $f^{-1}(x)$.

Example 17.4 Find $f^{-1}(x)$ if $f(x) = 4x - 3$.

Solution Let $y = f(x)$ and solve for x.

$$y = 4x - 3$$
$$y + 3 = 4x$$
$$x = \frac{1}{4}(y + 3)$$

Therefore

$$f^{-1}(y) = \frac{1}{4}(y + 3).$$

Replace y by x to get $f^{-1}(x)$.

$$f^{-1}(x) = \frac{1}{4}(x + 3)$$

You can check the answer by direct computation, as in Example 17.3.　■

Example 17.5 Find $f^{-1}(x)$ if $f(x) = x^2 + 5$ for $x \geq 0$. (We require $x \geq 0$ so that f will be one-to-one.)

Solution Let $y = f(x)$ and solve for x:

$$y = x^2 + 5$$
$$y - 5 = x^2$$
$$x = \sqrt{y - 5}$$

(The positive square root was chosen in the last step because $x \geq 0$ for x in the domain of f. Also $y \geq 5$ since $y = x^2 + 5$, so $\sqrt{y - 5}$ will be a real number.) Thus

$$f^{-1}(y) = \sqrt{y - 5} \text{ for } y \geq 5,$$

and

$$f^{-1}(x) = \sqrt{x - 5} \text{ for } x \geq 5.$$

In interval notation (Section 10):

$$\text{domain of } f = \text{range of } f^{-1} = [0,\infty)$$
$$\text{range of } f = \text{domain of } f^{-1} = [5,\infty).$$　■

D. GRAPHS

Let f denote a function with an inverse. Then:

> (a, b) is on the graph of $y = f(x)$
> iff
> (b, a) is on the graph of $y = f^{-1}(x)$.

Proof (a, b) is on the graph of $y = f(x)$

iff

$$b = f(a)$$

iff

$$a = f^{-1}(b)$$

iff

(b, a) is on the graph of $y = f^{-1}(x)$. ☐

Figure 17.2 shows several pairs of points whose coordinates have this relationship of (a, b) to (b, a). Each pair is located symmetrically about the line $y = x$. That will always be true, and thus we have the following conclusion.

The graphs of $y = f(x)$ and $y = f^{-1}(x)$ are located symmetrically about the line $y = x$.

It follows that if we graph $y = f(x)$ and $y = f^{-1}(x)$ in the same coordinate system, and then turn the figure so that we are looking along the line $y = x$, the figure will be "left-right symmetric."

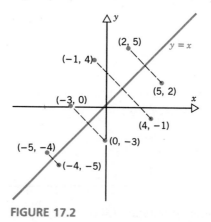

FIGURE 17.2

Example 17.6 If $f(x) = \sqrt{x}$ for $x \geq 0$, then $f^{-1}(x) = x^2$ for $x \geq 0$. Figure 17.3 shows the symmetrical locations of the graphs of f and f^{-1}. ■

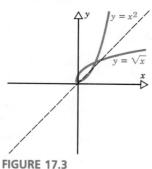

FIGURE 17.3

Example 17.7 In Example 17.4 we saw that if $f(x) = 4x - 3$, then $f^{-1}(x) = \frac{1}{4}(x + 3)$. Figure 17.4 shows the symmetrical locations of the graphs.

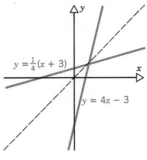

$y = \frac{1}{4}(x + 3)$

$y = 4x - 3$

FIGURE 17.4

EXERCISES FOR SECTION 17

Decide in each case whether the function is one-to-one. Justify your answer. (See Examples 17.1 and 17.2, and the statements at the end of Subsection A.)

1. $f(x) = 10x$ **2.** $f(x) = -7$

3. $f(x) = 7x + 1$ **4.** $g(x) = 0$

5. $g(x) = -5x + 7$

6. $g(x) = 2 - x^2$

7. $h(x) = 3x^2$ for $x \geq 0$

8. $h(x) = |x|$

9. $h(x) = 1 + \sqrt{x}$ for $x \geq 0$

For each function f in Exercises 10–24:
(a) Determine $f^{-1}(x)$.
(b) Verify that $(f^{-1} \circ f)(x) = x$ for each x in the domain of f and $(f \circ f^{-1})(x) = x$ for each x in the domain of f^{-1}.

(c) Draw the graphs of f and f^{-1} in the same Cartesian plane.
(Assume $x \geq 0$ in Exercises 19–24.)

10. $f(x) = 2x$ **11.** $f(x) = 4x$

12. $f(x) = -3x$ **13.** $f(x) = x + 1$

14. $f(x) = 5 - x$ **15.** $f(x) = x - 3$

16. $f(x) = 3 - 2x$ **17.** $f(x) = 3x + 4$

18. $f(x) = 6 - x$ **19.** $f(x) = \sqrt{x} + 1$

20. $f(x) = \sqrt{2x}$ **21.** $f(x) = 1 - 2\sqrt{x}$

22. $f(x) = 2x^2$ **23.** $f(x) = x^2 - 1$

24. $f(x) = 1 - x^2$

In 25–33, decide whether each graph is the graph of a function. If it is, decide whether the function is one-to-one. If the graph is the graph of a one-to-one function, draw the graph of the inverse function.

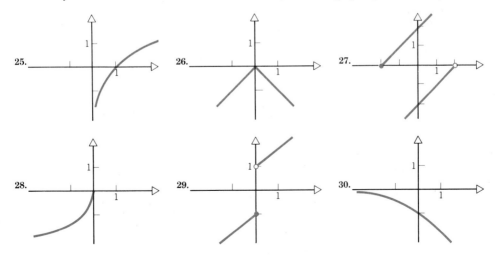

25.

26.

27.

28.

29.

30.

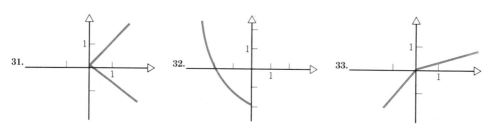

31.   32.   33.

34. Compute $f^{-1}(10.92)$ if $f(x) = 4.123 - 6.708x$. [C]

35. Compute $f^{-1}(0.213)$ if $f(x) = 0.447 \sqrt{x}$ $(x \geq 0)$. [C]

36. Compute $f^{-1}(\sqrt{2})$ if $f(x) = 9.434x + 5.916$. [C]

37. Suppose $f(x) = \sqrt{1 - x^2}$ for $0 \leq x \leq 1$.

(a) Verify that f is its own inverse.

(b) Draw the graph of f and observe its symmetry with respect to the line $y = x$. (The graph is part of the graph of $x^2 + y^2 = 1$.)

38. Suppose f is defined by $f(x) = x + 1$ for $-1 \leq x < 0$, and $f(x) = x - 1$ for $0 \leq x \leq 1$. (Thus the domain of f is the interval $[-1, 1]$.)

(a) Verify that f is its own inverse.

(b) Draw the graph of f and observe its symmetry with respect to the line $y = x$.

39. Suppose $f(x) = -x^2 + 4$ for $-2 \leq x \leq 0$.

(a) Draw the graph of f and then explain how you know that f has an inverse.

(b) Find f^{-1} and draw its graph.

(c) Determine the domain and range of both f and f^{-1}.

40. Find the inverse of f if $f(x) = x^2$ for $x \leq 0$. Also determine the domain and range of f^{-1}.

41. Find the inverse of f if $f(x) = \dfrac{2x - 1}{x + 3}$ for $x \neq -3$. Also determine the domain and range of f^{-1}.

42. Find the inverse of f if $f(x) = -|x - 1|$ for $x \leq 1$. Also determine the domain and range of f^{-1}.

43. Find the inverse of the general linear function $f(x) = ax + b$ $(a \neq 0)$.

44. Find the inverse of the quadratic function $f(x) = ax^2 + bx + c$, $a \neq 0$, $x \geq -b/2a$. (The condition $x \geq -b/2a$ guarantees that f has an inverse. Why?)

45. Prove that if f and g are functions having inverses, then $f \circ g$ also has an inverse and $(f \circ g)^{-1} = g^{-1} \circ f^{-1}$. [Begin with $((g^{-1} \circ f^{-1}) \circ (f \circ g))(x)$. Along the way, you will need to use $f^{-1}(f(g(x))) = g(x)$. Why is the last equation true?]

REVIEW EXERCISES FOR CHAPTER III

Determine the distance between each pair of points.

1. $(-3, 4)$ and $(1, 2)$

2. $(6, 0)$ and $(1, -2)$

3. Find an equation for the circle with center at $(-5, 2)$ and radius 4.

4. Find an equation for the circle with center at $(2, -4)$ and passing through $(-2, -1)$.

5. Characterize the graph of $x^2 + y^2 - 12x + 4y + 39 = 0$.

6. Characterize the graph of $x^2 + y^2 - 2x + 4y + 6 = 0$.

Determine an equation of the line through each pair of points. Write the answer in the form $Ax + By + C = 0$.

7. $(4, -5)$ and $(-1, 2)$

8. $(0, 3)$ and $(-3, 1)$

9. Find an equation of the line with slope -4 and y-intercept 2.

10. Find an equation of the line through $(-2, 1)$ with slope $-\frac{4}{3}$.

11. Find an equation of the line containing the point $(2, -1)$ and parallel to the line whose equation is $x - y + 5 = 0$.

12. Find an equation of the line containing the point $(1, 1)$ and perpendicular to the line whose equation is $4x + 2y - 1 = 0$.

Draw the graph of each equation.

13. $x^2 + y^2 - 5 = 0$

14. $x^2 + 2y = 2$ **15.** $3x + y = 1$

16. $x^2 + y^2 - 12 = 0$ **17.** $3x^2 + y = 1$

18. $3x - 2 = 0$ **19.** $2y - 5 = 0$

20. $4x - y = 3$ **21.** $x^2 + y^2 + 6x = 0$

22. $x^2 + y^2 - 2y - 8 = 0$

23. $x^2 - 2y = 4$ **24.** $-3x^2 + y = 1$

25. $x^2 + y^2 + 10x - 4y + 13 = 0$

26. $4x^2 + 4y^2 - 16x + 4y + 1 = 0$

Determine the domain of each function.

27. $f(x) = \sqrt{x + 1}/(x^2 - 4)$

28. $f(x) = \sqrt{2x - 1}/(x + 1)$

In Exercises 29 and 30, determine:

(a) $f(x) + g(x)$ (b) $f(x)g(x)$

(c) $f(0) - g(2)$ (d) $g(1)/f(-1)$

(e) $(f \circ g)(x)$ (f) $(g \circ f)(3)$

29. $f(x) = 3x + 5$ and $g(x) = x^2 - 1$

30. $f(x) = x^2 + 1$ and $g(x) = -2x + 3$

31. Let $f(x) = \sqrt{4x + 1}$. Find functions g and h such that $h \circ g = f$.

32. Let $f(x) = x^3 - 1$. Find functions g and h such that $h \circ g = f$.

Determine the difference quotient of each function.

33. $f(x) = -3x + 4$ **34.** $g(x) = x^2 + 2$

Draw the graph of each function.

35. $f(x) = -x^2 + 2x + 2$

36. $f(x) = 2x^2 - 6x + 1$

37. $f(x) = \begin{cases} x + 1 & \text{for } x \le 0 \\ x^2 & \text{for } x > 0 \end{cases}$

38. $f(x) = \begin{cases} -x^2 + 1 & \text{for } x < 1 \\ x & \text{for } x \ge 1 \end{cases}$

By solving the inequality f(x) < 0, *determine where the graph of* f *is below the x-axis.*

39. $f(x) = 3x^2 + 13x - 10$

40. $f(x) = |2x - 1| - 3$

41. Determine the maximum value:
$f(x) = x(1 - x)$.

42. Determine the minimum value:
$f(x) = (x + 1)(x + 2)$.

43. Find two real numbers such that the first minus the second is 10 and the product of the two numbers is as small as possible.

44. Find two real numbers such that the first plus twice the second is 10 and the product of the two numbers is as large as possible.

45. If the lowest point on the graph of $f(x) = 2x^2 + bx + c$ is $(-1, 5)$, what are b and c?

46. If $f(3) = 10$, $f(5) = 7$, and f is linear, what is $f(-2)$?

Determine in each case whether the function is one-to-one. Justify your answer.

47. $f(x) = 2|x| + 3$ **48.** $g(x) = x^2 - 4$

Determine f⁻¹(x) *and draw the graphs of* f *and* f⁻¹ *in the same Cartesian plane.*

49. $f(x) = 5 - 3x$

50. $f(x) = x^2 + 2$ $(x \ge 0)$

51. Determine a function f such that $y = f(x)$, if y varies directly as the square of x, and $y = 5$ when $x = 2$.

52. Determine a function g such that $u = g(v)$, if u varies inversely as v, and $u = 10$ when $v = 3$.

53. Suppose that the total production costs of a company per day consist of a fixed cost of \$1000 (independent of the number of items produced) and a variable cost of \$50 per item produced. Determine the function f such that $T = f(n)$, where T denotes the total cost per day for the production of n items. Your answer should be a linear function of n. What is the significance of the variable cost per item in terms of the graph of f?

54. Assume that a state's income tax is computed at the rate of 3% for the first \$2000 of taxable income, 4% for the next \$2000 of taxable income, and 5% for any additional taxable income beyond \$4000. Determine a function f such that $T = f(I)$ for $0 \le I \le 10,000$, where I denotes taxable income and T denotes the amount of tax, both in dollars. (The definition will have three parts.)

55. Newton's law of universal gravitation is given

by

$$F = G \frac{m_1 m_2}{r^2} \quad \text{(Exercise 57}$$
$$\text{of Section 15).}$$

(a) Determine a function f such that $G = f(m_1, m_2, r, F)$.

(b) Determine a function h such that $r = h(m_1, m_2, G, F)$.

56. The formula $\dfrac{1}{f} = \dfrac{1}{d_o} + \dfrac{1}{d_i}$ relates the focal length (f), object distance (d_o), and image distance (d_i) of a simple lens. (Notice that f represents a number, not a function, in this exercise.)

(a) Determine a function g such that $f = g(d_o, d_i)$.

(b) Determine a function h such that $d_o = h(f, d_i)$.

57. Suppose $f(x) = 2x + 1$, $g(x) = 5 - x$, and $h = f \circ g$. Find f^{-1}, g^{-1}, and h^{-1}.

58. The formula for the surface area S of a right circular cone having height h and base of radius r is $S = \pi r \sqrt{r^2 + h^2}$. Determine a function f

such that $h = f(S, r)$. (Suggestion: Begin by squaring both sides of the equation giving S.)

59. Let $f(x) = x^3 + 0.125$.

(a) Find functions g and h such that $f = g \circ h$.

(b) Find functions r and s such that $f(x) = r(x)s(x)$.

60. Find the domain of $f(x) = \sqrt{|2x| - 3}/(x + 2)$.

61. Suppose the graph of a function f is the upper half of the circle whose equation is $x^2 + (y - 3)^2 = 25$. What is $f(x)$?

62. Find all of the points in the Cartesian plane that are five units from both $(3, 4)$ and $(2, -3)$.

63. Find the values of x for which the graph of $f(x) = 10x^2 + 3x - 1$ is above the graph of $g(x) = 3x^2 - 10x + 1$ [that is, for which $f(x) > g(x)$].

64. Find $f^{-1}(x)$ if $f(x) = \dfrac{a^2}{\pi^2}(x - b)^2$ for $x \geq b$. (Assume $a > 0$ and $b > 0$.)

65. If $f(x) = x^2 - 3x - 4$, what is the smallest value of k such that f is one-to-one when the domain is restricted to the interval $[k, \infty)$?

CHAPTER IV

POLYNOMIAL AND RATIONAL FUNCTIONS

In Chapter III we analyzed the polynomial functions of degrees one and two—the linear and quadratic functions. In the first four sections of this chapter we consider polynomial functions of degree greater than two. In the last two sections we consider quotients of polynomials, which are called *rational functions*. Most of the chapter will draw heavily on the ideas about division that are covered in the first section of the chapter. In particular, the first section introduces synthetic division, which provides a very efficient method for evaluating a polynomial at a number.

SECTION **18**
Division of Polynomials

A. THE DIVISION ALGORITHM

If we divide any integer (the *dividend*) by a positive integer (the *divisor*), we obtain a *quotient* and a *remainder*. For example,

dividend ⟶   ⟵ remainder

$$\frac{38}{5} = 7 + \frac{3}{5} \longleftarrow \text{divisor} \qquad (18.1)$$

divisor ⟶   ⟵ quotient

This can be rewritten as

$$38 = 5 \cdot 7 + 3.$$

The remainder will always be nonnegative and less than the divisor (in the example, $0 \le 3 < 5$). The general form of Equation (18.1) is

$$\frac{a}{b} = q + \frac{r}{b}.$$

This can be rewritten as

dividend ⟶   ⟵ remainder

$$a = bq + r, \quad \text{with} \quad 0 \le r < b. \qquad (18.2)$$

divisor ⟶   ⟵ quotient

Equation (18.2) summarizes a fact about integers known as the *Division Algorithm*. Our concern now is with the corresponding fact about polynomials. In the statement of this fact, which follows, $b(x) \ne 0$ means that the polynomial $b(x)$ has at least one nonzero coefficient. And $r(x) = 0$ means that $r(x)$ is identically zero; that is, $r(x)$ has *no* nonzero coefficient.

Division Algorithm

If $a(x)$ and $b(x)$ are polynomials such that $b(x) \ne 0$, then there are unique polynomials $q(x)$ and $r(x)$ such that

$$a(x) = b(x)q(x) + r(x)$$

with either $r(x) = 0$ or deg $r(x) <$ deg $b(x)$. $\qquad (18.3)$

The polynomials $q(x)$ and $r(x)$ are called the **quotient** and the **remainder,** respectively.

Here is an illustration of ordinary long division of integers. The examples that follow the illustration will show that division of polynomials is similar; they will also show how to compute $q(x)$ and $r(x)$ in (18.3).

$$
\begin{array}{r}
345 \longleftarrow \text{quotient} \\
\text{divisor} \longrightarrow 12\ \overline{\smash{\big)}\,4151} \longleftarrow \text{dividend} \\
\text{subtract}\quad \dfrac{36}{55} \\
\text{subtract}\quad \dfrac{48}{71} \\
\text{subtract}\quad \dfrac{60}{11} \longleftarrow \text{remainder}
\end{array}
$$

The division process actually consists of subtracting multiples of the divisor from the dividend, until we arrive at a remainder less than the dividend. If you look carefully at the preceding calculation, for example, you will see that we successively subtracted 300×12, 40×12, and 5×12. The same idea applies to division of polynomials.

Example 18.1 Determine the quotient and remainder when $6x^4 + 5x^3 - 3x^2 + x + 3$ (the dividend) is divided by $2x^2 - x + 1$ (the divisor).

Solution Write the dividend and divisor as shown, and concentrate first on the terms that are circled.

$$
\boxed{2x^2} - x + 1\ \overline{\smash{\big)}\ \boxed{6x^4} + 5x^3 - 3x^2 + x + 3}
$$

Divide $2x^2$ into $6x^4$. Write the result, $3x^2$, above $6x^4$. Then multiply $3x^2$ times $2x^2 - x + 1$, and write the result, $6x^4 - 3x^3 + 3x^2$, below the dividend.

$$
\begin{array}{r}
3x^2 \qquad\qquad\qquad\qquad \\
2x^2 - x + 1\ \overline{\smash{\big)}\ 6x^4 + 5x^3 - 3x^2 + x + 3} \\
6x^4 - 3x^3 + 3x^2 \qquad\qquad = 3x^2(2x^2 - x + 1)
\end{array}
$$

Now subtract, as in long division of integers.

$$
\begin{array}{r}
3x^2 \qquad\qquad\qquad\qquad \\
2x^2 - x + 1\ \overline{\smash{\big)}\ 6x^4 + 5x^3 - 3x^2 + x + 3} \\
\text{subtract}\ \underline{6x^4 - 3x^3 + 3x^2} \qquad\qquad \\
\boxed{8x^3} - 6x^2 \qquad\qquad
\end{array}
$$

Next, divide $2x^2$ into $8x^3$ and proceed as before, bringing down the term x from the dividend for convenience.

$$
\begin{array}{r}
3x^2 + 4x \qquad\qquad\qquad \\
2x^2 - x + 1\ \overline{\smash{\big)}\ 6x^4 + 5x^3 - 3x^2 + x + 3} \\
\underline{6x^4 - 3x^3 + 3x^2}\quad \downarrow \qquad \\
8x^3 - 6x^2 + x \qquad \\
\text{subtract}\ \underline{8x^3 - 4x^2 + 4x} = 4x(2x^2 - x + 1)
\end{array}
$$

Subtract again, and repeat the process until the result of the subtraction is either 0 or of degree less than the degree of the divisor.

$$
\begin{array}{r}
3x^2 + 4x - 1 \qquad \longleftarrow \text{quotient} \\
2x^2 - x + 1 \overline{\smash{\big)}\, 6x^4 + 5x^3 - 3x^2 + x + 3} \\
\underline{6x^4 - 3x^3 + 3x^2} \\
8x^3 - 6x^2 + x \\
\underline{8x^3 - 4x^2 + 4x} \\
-2x^2 - 3x + 3 \\
\underline{-2x^2 + x - 1} \\
-4x + 4 \longleftarrow \text{remainder}
\end{array}
$$

Then

$$
\frac{6x^4 + 5x^3 - 3x^2 + x + 3}{2x^2 - x + 1} = 3x^2 + 4x - 1 + \frac{-4x + 4}{2x^2 - x + 1}
$$

and

$$
6x^4 + 5x^3 - 3x^2 + x + 3 = (2x^2 - x + 1)(3x^2 + 4x - 1) + (-4x + 4).
$$

For the division process the terms of the divisor and dividend should be arranged so that the powers of x decrease from left to right. Also, in writing the dividend, space should be left for any missing powers of the variable; in the following example space has been left for an x^3 term.

Example 18.2 Determine the quotient and remainder when $4 + 5x - 6x^2 + x^4$ is divided by $x - 2$.

Solution

$$
\begin{array}{r}
x^3 + 2x^2 - 2x + 1 \qquad \longleftarrow \text{quotient} \\
x - 2 \overline{\smash{\big)}\, x^4 \qquad\quad - 6x^2 + 5x + 4} \\
\underline{x^4 - 2x^3} \\
2x^3 - 6x^2 \\
\underline{2x^3 - 4x^2} \\
-2x^2 + 5x \\
\underline{-2x^2 + 4x} \\
x + 4 \\
\underline{x - 2} \\
6 \longleftarrow \text{remainder}
\end{array}
$$

B. SYNTHETIC DIVISION

Synthetic division is a process that shortens the work in any division problem in which the divisor has the form $x - c$, as in Example 18.2. Example 18.3 will summarize the rules that describe synthetic division. But first we run through Example 18.2 again, showing how the process of synthetic division is derived.

Synthetic division depends on two facts about division: First, there is no need to write the powers of the variable as long as the numbers in the various columns are kept carefully aligned. Second, many coefficients occur more than once—we can save time by not writing the duplications.

Here is Example 18.2 with the powers of x deleted and with the duplications in each column circled or connected by an arrow.

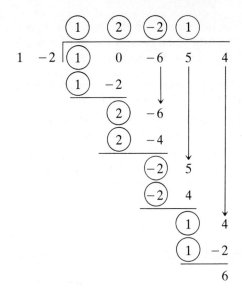

We can rewrite this keeping just one of each of the various duplicated numbers.

$$
\begin{array}{c|cccccc}
1 & -2 & 1 & 0 & -6 & 5 & 4 \\
& & & -2 & & & \\
\hline
& & & 2 & & & \\
& & & & -4 & & \\
\hline
& & & & -2 & & \\
& & & & & 4 & \\
\hline
& & & & & 1 & \\
& & & & & & -2 \\
\hline
& & & & & & 6
\end{array}
$$

Now we can compress what remains. Also, we can delete the 1 at the left in the first line, since it accounts only for the x in $x - c$ and we use this process only when the divisor has the form $x - c$. For convenience in what comes later, we repeat in the third row the first number under the division symbol, as indicated by the arrow.

$$
\begin{array}{c|ccccc}
-2 & 1 & 0 & -6 & 5 & 4 \\
& \downarrow & -2 & -4 & 4 & -2 \quad \text{subtract} \\
\hline
& 1 & 2 & -2 & 1 & 6
\end{array}
$$

Each number in the second row (such as -4) arose originally from multiplying the -2 in $x - 2$ by the number in the third row and one column to the left of the specified number in the second row (thus $-4 = -2 \cdot 2$). Because addition is easier than subtraction, we now change the sign on the leading -2; this will change the sign on each number in the second row, and so we compensate by replacing subtraction by addition, as desired. When we remove the bar on top, we get this abbreviated form.

$$
\begin{array}{c|ccccc}
2 & 1 & 0 & -6 & 5 & 4 \\
& & 2 & 4 & -4 & 2 \quad \text{add} \\
\hline
& 1 & 2 & -2 & 1 & 6
\end{array}
$$

The last number in the third row, 6, is the remainder (compare Example 18.2). The other numbers in the third row are the coefficients in the quotient, $x^3 + 2x^2 - 2x + 1$, which has degree one less than the degree of the dividend.

Example 18.3 Use synthetic division to determine the quotient and remainder when $2x^5 - 10x^3 + 20x^2 - 9x + 5$ is divided by $x + 3$.

Solution For synthetic division the divisor has the form $x - c$. In this case, $x + 3 = x - (-3)$, so that $c = -3$. Thus -3 will be the leading number in the first row. The other numbers in the first row are the coefficients of the dividend when the terms are written with exponents decreasing from left to right, and when zeros are supplied for missing powers of the variable.

$$-3 \;\big|\; 2 \quad 0 \quad -10 \quad 20 \quad -9 \quad 5$$

Bring down 2, multiply it by -3, and write the result, -6, as shown.

$$
\begin{array}{r|rrrrrr}
-3 & 2 & 0 & -10 & 20 & -9 & 5 \\
   & \downarrow & -6 & & & & \quad\text{add} \\
\hline
   & 2 & & & & &
\end{array}
$$

Add 0 and -6, multiply the result by -3, and write the result, 18 as shown.

$$
\begin{array}{r|rrrrrr}
-3 & 2 & 0 & -10 & 20 & -9 & 5 \\
   & \downarrow & -6 & 18 & & & \quad\text{add} \\
\hline
   & 2 & -6 & & & &
\end{array}
$$

Now repeat; alternately adding and multiplying until the end.

$$
\begin{array}{r|rrrrrr}
-3 & 2 & 0 & -10 & 20 & -9 & 5 \\
   & \downarrow & -6 & 18 & -24 & 12 & -9 \quad\text{add} \\
\hline
   & 2 & -6 & 8 & -4 & 3 & -4
\end{array}
$$

The last number is the remainder. The other numbers in the third row are the coefficients of the quotient.

The quotient is $2x^4 - 6x^3 + 8x^2 - 4x + 3$.

The remainder is -4.

Example 18.4 Determine the quotient and remainder when $x^4 + x^2 + 1$ is divided by $x - \frac{1}{2}$.

Solution

$$
\begin{array}{r|rrrrr}
\frac{1}{2} & 1 & 0 & 1 & 0 & 1 \\
            &   & \frac{1}{2} & \frac{1}{4} & \frac{5}{8} & \frac{5}{16} \\
\hline
            & 1 & \frac{1}{2} & \frac{5}{4} & \frac{5}{8} & \frac{21}{16}
\end{array}
$$

The quotient is $x^3 + \frac{1}{2}x^2 + \frac{5}{4}x + \frac{5}{8}$.

The remainder is $\frac{21}{16}$.

C. REMAINDER THEOREM

The following theorem provides an indirect but convenient way to compute $f(c)$ for each polynomial $f(x)$ and each real number c.

Remainder Theorem

If c is a real number and a polynomial $f(x)$ is divided by $x - c$, then the remainder is $f(c)$.

Proof By the Division Algorithm, when $f(x)$ is divided by $x - c$ (which has degree one), the remainder is either 0 or a polynomial of degree 0 (that is, a constant). Thus, in any case, the remainder is a real number, which we can denote by r. If $q(x)$ denotes the quotient, then

$$f(x) = (x - c)q(x) + r.$$

Therefore,

$$f(c) = (c - c)q(c) + r$$

$$= 0 \cdot q(c) + r$$

$$= r. \qquad \square$$

Example 18.5 Use the Remainder Theorem and synthetic division to compute $f(3)$ if $f(x) = x^4 - 2x^3 + x - 1$.

Solution We want the remainder when $f(x)$ is divided by $x - 3$. Here is the calculation using synthetic division.

$$\left.\begin{array}{l} \text{To compute} \\ f(3), \text{ use} \\ 3 \text{ here.} \end{array}\right\} \nearrow 3 \quad \begin{array}{|rrrrr} 1 & -2 & 0 & 1 & -1 \\ & 3 & 3 & 9 & 30 \\ \hline 1 & 1 & 3 & 10 & 29 = f(3) \end{array}$$

The remainder is 29, so $f(3) = 29$, as indicated.

Here is the calculation by direct substitution:

$$f(3) = 3^4 - 2 \cdot 3^3 + 3 - 1 = 81 - 54 + 3 - 1 = 29.$$

Generally, the total number of operations (additions, subtractions, and multiplications) needed to compute $f(c)$ by synthetic division will be less than the number needed by direct substitution.

Example 18.6 Use the Remainder Theorem and synthetic division to show that $-\frac{2}{3}$ is a solution of $3x^3 + 2x^2 + 3x + 2 = 0$.

Solution Remember that c is a solution of $f(x) = 0$ iff $f(c) = 0$. Therefore, we let $f(x) = 3x^3 + 2x^2 + 3x + 2$ and then use synthetic division to show that $f(-\frac{2}{3}) = 0$.

$$-\tfrac{2}{3} \quad \begin{array}{|rrrr} 3 & 2 & 3 & 2 \\ & -2 & 0 & -2 \\ \hline 3 & 0 & 3 & 0 \end{array} \quad = f(-\tfrac{2}{3})$$

Synthetic division is especially easy with a calculator. Each step involves either addition to or multiplication by the number remaining in the calculator's display after the previous step. The intermediate steps do not even have to be written down—although you should write them when you do the exercises. Rounding, if necessary, can be done on the final answer.

Example 18.7 Use the Remainder Theorem and synthetic division to compute $f(1.62)$ if $f(x) = -0.14x^3 + 4.2x^2 + 3x - 1.8$.

Solution

$$
\begin{array}{r|rrrr}
1.62 & -0.14 & 4.2 & 3 & -1.8 \\
& & -0.2268 & 6.436584 & 15.28726608 \\
\hline
& -0.14 & 3.9732 & 9.436584 & 13.48726608
\end{array}
$$

Rounding gives $f(1.62) = 13.5$. ⬜ ■

Remark The calculations used to evaluate $f(c)$ by synthetic division and the Remainder Theorem, as in Example 18.7, are the same as those used when $f(x)$ has been written using **nested multiplication.** The right sides of the following equations show the form for nested multiplication for polynomials of degrees two, three, and four. The idea extends in an obvious way to polynomials of higher degree.

$$a_2x^2 + a_1x + a_0 = (a_2x + a_1)x + a_0$$

$$a_3x^3 + a_2x^2 + a_1x + a_0 = ((a_3x + a_2)x + a_1)x + a_0$$

$$a_4x^4 + a_3x^3 + a_2x^2 + a_1x + a_0 = (((a_4x + a_3)x + a_2)x + a_1)x + a_0$$

For $f(x)$ as in Example 18.7, the nested form is

$$f(x) = ((-0.14x + 4.2)x + 3)x - 1.8,$$

from which

$$f(1.62) = ((-0.14(1.62) + 4.2)1.62 + 3)1.62 - 1.8.$$

Comparison will reveal that the computational steps indicated here, starting from the left, are the same as those used in Example 18.7.

EXERCISES FOR SECTION 18

Determine the quotient and remainder when f(x) is divided by g(x).

1. $f(x) = 3x^3 - 2x^2 + 5x + 7,\ g(x) = 3x + 1$

2. $f(x) = 4x^3 + x - 3,\ g(x) = 2x - 1$

3. $f(x) = 2x^3 - 14x - 8,\ g(x) = 2x + 4$

4. $f(x) = 3x^4 - 10x^2 - 2x + 2,\ g(x) = x^2 + 2x + 1$

5. $f(x) = 3x^3 - 8x - 4,\ g(x) = x^2 + 2x + 1$

6. $f(x) = x^5 + 2x^3 - 1,\ g(x) = x^3 - 1$

7. $f(x) = 2x^3 + 3x^2 + x,\ g(x) = x^3 + 1$

8. $f(x) = x^3 - x^2 - 2,\ g(x) = 2x^2 - 1$

9. $f(x) = x^4 + x^3 + x + 1,\ g(x) = 5x^2 - x$

10. $f(x) = x^5 - x^3 + x + 1,\ g(x) = 4x^3 + 2$

11. $f(x) = x^3 + x^2 - 4,\ g(x) = x^4 + 1$

12. $f(x) = x^3 + x^2 + x + 1,\ g(x) = x^3 + 1$

Use synthetic division to determine the quotient and remainder when f(x) is divided by g(x).

13. $f(x) = 2x^3 - x^2 + 5x + 1,\ g(x) = x - 1$

14. $f(x) = x^3 - x + 4,\ g(x) = x - 3$

15. $f(x) = x^3 + 2x^2 - 1,\ g(x) = x - 2$

16. $f(x) = x^4 + x^2 + 1,\ g(x) = x + 2$

17. $f(x) = x^3 - x,\ g(x) = x + 5$

18. $f(x) = 5x^3 - 5x + 2,\ g(x) = x + 5$

19. $f(x) = x^4 + x^3,\ g(x) = x + \frac{1}{2}$

20. $f(x) = x^2 + x + 1,\ g(x) = x + 0.1$

21. $f(x) = x^3 + x,\ g(x) = x + 0.5$

22. $f(x) = x^4 - 1,\ g(x) = x + 0.2$

23. $f(x) = x^5 + x^4 + x^3 + x^2 + x + 1,\ g(x) = x - \frac{1}{2}$

24. $f(x) = 3x^4 - x^2 + 1,\ g(x) = x - \frac{1}{3}$

Use the Remainder Theorem and synthetic division to compute the indicated values $f(c)$.

25. $f(x) = x^3 - x + 1$; $f(2)$, $f(-5)$

26. $f(x) = 2x^4 - x^2 + x$; $f(3)$, $f(-4)$

27. $f(x) = x^5 + 2x^3 - x + 1$; $f(1)$, $f(-2)$

28. $f(t) = t^3 + t^2 + t + 1$; $f(0.2)$, $f(-\frac{1}{2})$

29. $f(t) = 5t^4 - 2t^2 - 1$; $f(\frac{1}{3})$, $f(-0.1)$

30. $f(t) = 2t^3 + t^2 + t + 3$; $f(\frac{1}{2})$, $f(-0.1)$

31. $f(x) = 2.01x^3 - 1.47x + 0.16$; $f(0.4)$, $f(-0.72)$ $\boxed{C}$

32. $f(x) = 12.5x^3 - 28.7x^2 + 32.5$; $f(3)$, $f(-4.2)$ $\boxed{C}$

33. $f(x) = -0.02x^4 + 0.11x^2 - 0.05$; $f(1.4)$, $f(-0.95)$ $\boxed{C}$

Use the Remainder Theorem and synthetic division to show that the given number is a solution of the given equation.

34. 2; $x^3 - 3x^2 + 3x - 2 = 0$

35. -1; $x^4 + x^3 + 2x + 2 = 0$

36. $-\frac{1}{2}$; $2x^3 - x^2 + x + 1 = 0$

37. 0.1; $10x^3 - x^2 + 10x - 1 = 0$

38. -0.4; $5x^3 + 2x^2 + 5x + 2 = 0$

39. -0.3; $10x^4 + 3x^3 + 20x + 6 = 0$

40. 6.91; $3x^4 - 22.73x^3 + 4.82x^2 + 68.19x - 41.46 = 0$ $\boxed{C}$

41. -1.73; $2x^3 + 8.46x^2 + 11.65x + 5.19 = 0$ $\boxed{C}$

42. 9.24; $0.2x^4 - 1.848x^3 + 5.1x^2 - 47.824x + 6.468 = 0$ $\boxed{C}$

SECTION **19**
Factors and Real Zeros of Polynomials

A. FACTOR THEOREM

Recall that a polynomial $g(x)$ is a *factor* of a polynomial $f(x)$ if $f(x) = q(x)g(x)$ for some polynomial $q(x)$ (Section 4D). The following theorem provides a quick test to determine whether a polynomial of the form $x - c$ is a factor of another polynomial.

Factor Theorem

If $f(x)$ is a polynomial and c is a real number, then

$$x - c \text{ is a factor of } f(x) \text{ iff } f(c) = 0.$$

Proof Divide $f(x)$ by $x - c$. By the Division Algorithm and the Remainder Theorem (Section 18),

$$f(x) = q(x) \cdot (x - c) + f(c).$$

This equation shows that if $f(c) = 0$, then $f(x) = q(x)(x - c)$ so that $x - c$ is a factor of $f(x)$. Conversely, since $f(c)$ is the *unique* remainder when $f(x)$ is divided by $x - c$, the equation shows that if $x - c$ is a factor of $f(x)$, then $f(c) = 0$. $\square$

We can compute $f(c)$ by synthetic division, and by the Factor Theorem $f(c)$ will tell us whether $x - c$ is a factor of $f(x)$. But synthetic division does more: it furnishes $q(x)$ such that $f(x) = q(x)(x - c)$ whenever $x - c$ is a factor of $f(x)$. Here is an illustration.

Example 19.1 Use synthetic division and the Factor Theorem to show that $x - 4$ is a factor of

$$f(x) = 6x^3 - 23x^2 - 6x + 8.$$

Also determine $q(x)$ such that $f(x) = (x - 4)q(x)$.

Solution

$$
\begin{array}{r|rrrr}
4 & 6 & -23 & -6 & 8 \\
  &   & 24 & 4 & -8 \\
\hline
  & 6 & 1   & -2 & 0 = f(4)
\end{array}
$$

This shows that $f(4) = 0$. Thus $x - 4$ is a factor of $f(x)$ by the Factor Theorem. Moreover, the numbers 6 1 -2 in the third row show that the quotient is $q(x) = 6x^2 + x - 2$. Thus

$$6x^3 - 23x^2 - 6x + 8 = (x - 4)(6x^2 + x - 2).$$ ◼

Example 19.2 Is $x + 0.1$ a factor of

$$f(x) = x^3 + 0.1x^2 + 0.1x - 0.01?$$

Solution Because $x + 0.1 = x - (-0.1)$, we use -0.1 (not 0.1) in testing by synthetic division.

$$
\begin{array}{r|rrrr}
-0.1 & 1 & 0.1 & 0.1 & -0.01 \\
     &   & -0.1 & 0 & -0.01 \\
\hline
     & 1 & 0   & 0.1 & -0.02
\end{array}
$$

The remainder, $f(-0.1) = -0.02$, is not zero. Thus $x + 0.1$ is not a factor of $f(x)$. ◼

Example 19.3 Use the Factor Theorem to prove that $x - a$ is a factor of $x^n - a^n$ for each real number a and each positive integer n.

Solution Let $f(x) = x^n - a^n$. Then $f(a) = a^n - a^n = 0$. Therefore, by the Factor Theorem, $x - a$ is a factor of $f(x)$. ◼

B. FACTORS AND ZEROS

A number c is called a **zero** of a polynomial $f(x)$ if $f(c) = 0$. Thus c is a zero of a polynomial $f(x)$ iff c is a solution of the equation $f(x) = 0$. For example, 2 is a zero of $f(x) = x^2 - 4$ because $f(2) = 2^2 - 4 = 0$.

The conclusion of the Factor Theorem can be restated like this:

$x - c$ if a factor of $f(x)$

iff

c is a zero of $f(x)$.

This means that every factor of the form $x - c$ determines a zero, and, conversely, every zero determines a factor $x - c$.

Example 19.4 Find the three zeros of $f(x) = x^3 - x^2 - 2x$ by factoring.

Solution Because

$$f(x) = x(x^2 - x - 2) = x(x - 2)(x + 1),$$

the Factor Theorem tells us that 0, 2, and -1 are zeros. ◾

Example 19.5 Form a polynomial having -3, 2, and 5 as zeros.

Solution By the Factor Theorem, it suffices to form a polynomial having $x + 3$, $x - 2$, and $x - 5$ as factors. Thus

$$\begin{aligned} f(x) &= (x + 3)(x - 2)(x - 5) \\ &= (x + 3)(x^2 - 7x + 10) \\ &= x^3 - 4x^2 - 11x + 30 \end{aligned}$$

has the required zeros. ◾

C. COMPLETE FACTORIZATION

Assume that c is a zero of the polynomial $f(x)$. Then $x - c$ is a factor of $f(x)$, and it may even be true that $(x - c)^2$ or $(x - c)^3$ or some higher power of $x - c$ is a factor of $f(x)$. If $(x - c)^m$ is the highest power of $x - c$ that is a factor of $f(x)$, then c is said to be a zero of **multiplicity** m.

Example 19.6 Determine the multiplicity of each zero of $x^3(x + 4)(x - 5)^2$.

Solution   0 is a zero of multiplicity 3   [from $x^3 = (x - 0)^3$]

-4 is a zero of multiplicity 1   [from $x + 4 = x - (-4)$]

5 is a zero of multiplicity 2   [from $(x - 5)^2$]. ◾

Example 19.7 Form a polynomial having 0, -4, and 1 as zeros of multiplicities 1, 2, and 3, respectively.

Solution The answer must have x^1, $(x + 4)^2$, and $(x - 1)^3$ as factors. The simplest such polynomial is

$$x(x + 4)^2(x - 1)^3.$$ ◾

It will be instructive to expand the answer to the last example:

$$\begin{aligned} x(x + 4)^2(x - 1)^3 &= x(x^2 + 8x + 16)(x^3 - 3x^2 + 3x - 1) \\ &= (x^3 + 8x^2 + 16x)(x^3 - 3x^2 + 3x - 1) \\ &= x^6 + 5x^5 - 5x^4 - 25x^3 + 40x^2 - 16x. \end{aligned}$$

The degree of this polynomial is 6, which is the sum of the multiplicities of the zeros ($6 = 1 + 2 + 3$). We can see that, in general, a zero of multiplicity m will contribute m to the degree of a polynomial, because the degree of $(x - c)^m$ is m. This leads to the following conclusion.

> The degree of a polynomial must be at least as great as the sum of the multiplicities of its real zeros.   (19.1)

The next example shows that we cannot replace "at least as great as" by "equal to" in Statement (19.1). We will be able to make such a replacement in Chapter X, however, after we extend the number system beyond the real numbers to the complex numbers and replace "real zeros" by "complex zeros." All of the polynomials in this chapter have real numbers as coefficients, and the zeros considered are real zeros.

Example 19.8 Determine the zeros of $x^3 + x$, along with their multiplicities.

Solution
$$x^3 + x = 0 \text{ iff } x(x^2 + 1) = 0$$
$$\text{iff } x = 0 \text{ or } x^2 + 1 = 0.$$

The factor $x^2 + 1$ is never 0 for x a real number (because $x^2 \geq 0$ for all real x). Thus the only real zero of $x^3 + x$ is 0, and its multiplicity is 1.

The degree of $x^3 + x$ is 3, and that is greater than the sum of the degrees of the multiplicities of the real zeros (which is 1). ∎

If a zero of multiplicity m is counted as a zero m times, then we can rephrase Statement (19.1) as follows.

Each polynomial of degree n has at most n real zeros.

(19.2)

Very often we are interested in finding *all* of the zeros of a polynomial. The higher the degree of the polynomial the more difficult this problem tends to be. The next example illustrates that once we have a zero we can use synthetic division and factorization to reduce the degree of our problem by one: in this example we are given a zero of a polynomial of degree three and we use it to reduce the problem to that of a polynomial of degree two.

Example 19.9 Given that -1 is a zero of
$$f(x) = x^3 + 4x^2 - 7x - 10,$$
determine all of the zeros along with their multiplicities.

Solution If -1 is a zero, then $x + 1$ must be a factor. We can use synthetic division to find $q(x)$ such that $f(x) = (x + 1)q(x)$.

$$
\begin{array}{r|rrrr}
-1 & 1 & 4 & -7 & -10 \\
   &   & -1 & -3 & 10 \\
\hline
   & 1 & 3 & -10 & 0
\end{array}
$$

The coefficients 1 3 -10 from the last row show that
$$x^3 + 4x^2 - 7x - 10 = (x + 1)(x^2 + 3x - 10).$$

The factor $x^2 + 3x - 10$ contributes two zeros:
$$x^2 + 3x - 10 = 0$$
$$(x + 5)(x - 2) = 0$$
$$x = -5, 2.$$

The zeros of $f(x)$ are, therefore, -1, -5, and 2, each of multiplicity 1. That is,

$$f(x) = (x + 1)(x + 5)(x - 2).$$ ■

Recall from Section 5D that a polynomial has been *factored completely* when it has been written as a product of irreducible polynomials, where a polynomial is irreducible if it cannot be written as a product of two other polynomials that are both of positive degree. Since each linear polynomial is irreducible, the factorization

$$x^3 + 4x^2 - 7x - 10 = (x + 1)(x + 5)(x - 2)$$

in Example 19.9 is complete. As stated in Section 5D, some second-degree polynomials are irreducible and some are not; a second-degree polynomial is reducible iff it has a real zero. It can be shown that each polynomial of degree greater than two is reducible.

Example 19.10 Given that 3 is a zero of

$$f(x) = x^3 - 5x^2 + 8x - 6,$$

determine all of the real zeros along with their multiplicities. Also, factor $f(x)$ completely.

Solution

$$
\begin{array}{r|rrrr}
3 & 1 & -5 & 8 & -6 \\
  &   & 3 & -6 & 6 \\
\hline
  & 1 & -2 & 2 & 0 \\
\end{array}
$$

This shows that $f(x) = (x - 3)(x^2 - 2x + 2)$. The quadratic factor $x^2 - 2x + 2$ has no real zero, because its discriminant is negative: $b^2 - 4ac = (-2)^2 - 4(1)(2) = -4 < 0$. Therefore, 3 is the only real zero of $f(x)$; its multiplicity is 1. The quadratic factor is irreducible since it has no real zero. The complete factorization is

$$x^3 - 5x^2 + 8x - 6 = (x - 3)(x^2 - 2x + 2).$$ ■

EXERCISES FOR SECTION 19

Use synthetic division and the Factor Theorem to determine whether $x - c$ *is a factor of* $f(x)$. *If* $x - c$ *is a factor, determine* $q(x)$ *such that* $f(x) = (x - c)q(x)$.

1. $x - 2$; $f(x) = 6x^3 - 11x^2 - 4x + 4$

2. $x - 3$; $f(x) = 5x^3 - 8x^2 - 13x - 20$

3. $x - 1$; $f(x) = x^4 - 3x^3 + 2x^2 + x - 1$

4. $x + 1$; $f(x) = 2x^3 - x^2 + 10$

5. $x + 5$; $f(x) = x^4 + 5x^3 - x^2 - x + 20$

6. $x + 4$; $f(x) = x^3 + 4x^2 + 5x + 20$

7. $x - 0.1$; $f(x) = 10x^4 - x^3 - 2x$

8. $x - 0.2$; $f(x) = 5x^6 - x^5 + 10x^2 - 2x$

9. $x + 0.4$; $f(x) = 5x^3 + x + 1$

Determine the multiplicity of each zero of each polynomial.

10. $(x - 1)^3(x + 2)^5$

11. $x(x + 1)(x - 5)^2$

12. $(x - 10)^3(x + 2)^2$

13. $x^3(x - 10)^2(x + 0.1)^5$

14. $(x - 0.5)^2(x + 3)^5$

15. $x(x + 2)(x + 0.4)^2$

Form a polynomial having the specified zeros with the specified multiplicities. In 19-21, multiply the factors and simplify.

16. 0, 1, and -3 of multiplicities 2, 2, and 1, respectively.

17. 0, $\frac{1}{2}$, and -5 of multiplicities 1, 2, and 3, respectively.

18. $\sqrt{2}$, -1, and 3 of multiplicities 2, 1, and 2, respectively.

19. 0, -41, and 13 of multiplicities 4, 2, and 1, respectively. ⓒ

20. 14, $-\sqrt{3}$, and 8 of multiplicities 1, 1, and 3, respectively. ⓒ

21. -2.7 and 1.2, both of multiplicity 2. ⓒ

Determine the real zeros of each polynomial, along with their multiplicities; one zero is given in each case. Also, factor each polynomial completely.

22. $x^3 + x^2 - 5x - 5$; -1 is a zero.

23. $x^3 - x^2 - 4x + 4$; 1 is a zero.

24. $x^3 - 5x^2 - 4x + 20$; 5 is a zero.

25. $2x^3 - 5x^2 + 4x - 1$; $\frac{1}{2}$ is a zero.

26. $5x^3 - x^2 - 10x + 2$; 0.2 is a zero.

27. $5x^3 + 12x^2 + 9x + 2$; -0.4 is a zero.

28. $x^3 + 3x^2 - 2x - 6$; -3 is a zero.

29. $x^3 - 5x^2 - 5x + 25$; 5 is a zero.

30. $2x^3 - 9x^2 + 11x - 2$; 2 is a zero.

31. $x^3 + 2x^2 + x + 2$; -2 is a zero.

32. $2x^3 - x^2 + 8x - 4$; $\frac{1}{2}$ is a zero.

33. $3x^3 - 8x^2 + 6x + 3$; $-\frac{1}{3}$ is a zero.

34. By Example 19.3, $x - 2$ is a factor of $x^6 - 64$.

Use synthetic division to find $q(x)$ such that $x^6 - 64 = (x - 2)q(x)$.

35. Use the Factor Theorem to show that $x - 2$ is not a factor of $x^6 + 64$. (Compare Exercise 34.)

36. (a) Use the Factor Theorem and direct substitution (as in Example 19.3) to prove that $x + 2$ is a factor of $x^5 + 32$. (b) Use synthetic division to find $q(x)$ such that $x^5 + 32 = (x + 2)q(x)$.

37. For which positive integers n (if any) is $x - a$ a factor of $x^n + a^n$? Assume $a \neq 0$.

38. For which positive integers n (if any) is $x + a$ a factor of $x^n - a^n$? Assume $a \neq 0$.

39. For which positive integers n (if any) is $x + a$ a factor of $x^n + a^n$? Assume $a \neq 0$.

40. Determine c if $x - 2$ is a factor of $x^3 - 5x^2 + cx - 2$.

41. Determine a if $x - 2$ is a factor of $x^4 + x^3 - 3x^2 + ax + a$.

42. Find a polynomial $f(x)$ of degree 3 such that $f(-1) = 0$, $f(2) = 0$, $f(5) = 0$, and $f(0) = 2$. [Suggestion: For each real number $k \neq 0$, the zeros of $k(x - a)(x - b)(x - c)$ are a, b, and c.]

43. Find a polynomial $f(x)$ such that $f(-3) = 0$, $f(2) = 0$, $f(4) = 0$, and $f(x) \geq 0$ for every real number x. [The degree of $f(x)$ will be greater than three.]

44. Prove that the graph of a polynomial function of degree n can intersect the x-axis at most n times.

45. Prove that if $f(x)$ and $g(x)$ are polynomials of degrees m and n, respectively, and if $m \leq n$, then the graphs of $f(x)$ and $g(x)$ can intersect at most n times.

SECTION **20**

Graphs of Polynomial Functions. Some General Principles

A. POWER FUNCTIONS

We know that the graph of every linear (first-degree) function is a straight line and the graph of every quadratic (second-degree) function is a parabola. In this section we study the graphs of polynomial functions of degree greater than two.

We also consider some general principles that apply to other functions as well as to polynomial functions.

A function f is called a **power function** if it has the form

$$f(x) = kx^n, \tag{20.1}$$

where k and n are nonzero real numbers.* Thus a power function of x is simply a constant times a power of x. If n is a nonnegative integer, then kx^n is a polynomial. These polynomials have the simplest graphs and so they will be considered first. Figures 20.1–20.4 show examples. Some coordinate pairs for the graphs of $y = x^3$ and $y = x^4$ are given in Tables 20.1 and 20.2.

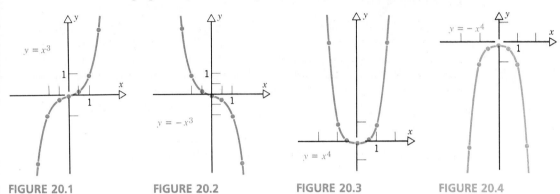

FIGURE 20.1   FIGURE 20.2   FIGURE 20.3   FIGURE 20.4

TABLE 20.1

x	0	0.5	-0.5	1	-1	1.5	-1.5
x^3	0	0.125	-0.125	1	-1	3.375	-3.375

TABLE 20.2

x	0	0.5	-0.5	1	-1	1.5	-1.5
x^4	0	0.0625	0.0625	1	1	5.0625	5.0625

The graphs of $y = x^5$, $y = x^7$, . . . (odd powers) all have the same general shape as that of $y = x^3$; the graphs of $y = x^4$, $y = x^6$, . . . (even powers) all have the same general shape as that of $y = x^2$. The differences are that as n increases, the graph of $y = x^n$ becomes flatter near the origin and steeper away from the origin. (See Exercises 1–3.)

Changing the sign of k in kx^n will reflect the graph through the x-axis— that is, turn it upside down. Compare the graph of $y = x^3$ with the graph of $y = -x^3$ (Figures 20.1 and 20.2). Also, compare the graph of $y = x^4$ with the graph of $y = -x^4$ (Figures 20.3 and 20.4).

Changing k but not its sign in kx^n will change the details of the graph but not its general characteristics. If $k > 1$, then the graph of $y = kx^n$ is the graph of $y = x^n$ *stretched vertically* by a factor of k. For example, each point on the graph of $y = x^n$ corresponds to a point that is on the graph of $y = 2x^n$ and twice as far from the x-axis. If $0 < k < 1$, then the graph of $y = kx^n$ is the graph of $y = x^n$ *compressed vertically*. Exercises 4–6 ask you to verify the

*For the present n will be an integer. Other cases will be considered in Section 24.

statements in this and the preceding paragraph for some particular values of k and n. The statements are special cases of the following general principles.

If f is a function and k is a positive constant, then the graph of $y = k \cdot f(x)$ is the graph of $y = f(x)$ stretched vertically if $k > 1$, and compressed vertically if $k < 1$. The graph of $y = -f(x)$ is the graph of $y = f(x)$ reflected through the x-axis.

B. SYMMETRY

A graph is *symmetric about a line L* if for each point A on the graph there is a corresponding point B on the graph such that L is the perpendicular bisector of AB (Figure 20.5). Figure 20.3 shows that the graph of $f(x) = x^4$ is symmetric about the y-axis. If the graph of a function f is symmetric about the y-axis and a is in the domain of f, then the point corresponding to $(a, f(a))$ must be $(-a, f(-a))$; thus necessarily $f(-a) = f(a)$ (Figure 20.5 again). In the case of $f(x) = x^4$, the symmetry is a result of the even exponent, 4, which makes a negative sign inconsequential: for example, $(-2)^4 = 16 = 2^4$. These observations about symmetry and even powers lead to the following definition and remark.

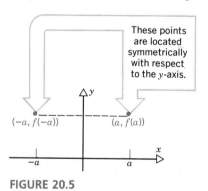

These points are located symmetrically with respect to the y-axis.

$(-a, f(-a))$ $(a, f(a))$

FIGURE 20.5

A function f is **even** if $f(-x) = f(x)$ for all x.
The graph of f is symmetric about the y-axis
iff
f is even.

A graph is *symmetric about a point P* if for each point A on the graph there is a corresponding point B on the graph such that P is the midpoint of AB (Figure 20.6). Figure 20.1 shows that the graph of $f(x) = x^3$ is symmetric about the origin. If the graph of a function f is symmetric about the origin and a is in the domain of f, then the point corresponding to $(a, f(a))$ must be $(-a, f(-a))$, so that necessarily $f(-a) = -f(a)$ (Figure 20.6 again). In the case of $f(x) = x^3$, the symmetry is a result of the odd exponent, 3. This leads to the next definition and remark.

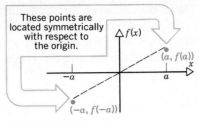

FIGURE 20.6

A function f is **odd** if $f(-x) = -f(x)$ for all x.
The graph of f is symmetric about the origin
iff
f is odd.

The next statements are illustrated by the examples already given and by the examples that follow.

A polynomial function is even
iff
it contains only even powers of its variable.†

A polynomial function is odd
iff
it contains only odd powers of its variable.

Example 20.1   The function $f(x) = -x^4$ is even and its graph is symmetric about the y-axis (Figure 20.4). The function $f(x) = -x^3$ is odd and its graph is symmetric about the origin (Figure 20.2).

Example 20.2   The function $f(x) = x^4 - 6x^2 + 3$ is even because it contains only even powers of x. Here is a direct verification that $f(-x) = f(x)$:

$$f(-x) = (-x)^4 - 6(-x)^2 + 3 = x^4 - 6x^2 + 3 = f(x).$$

Example 20.3   The function $g(x) = x^3 - 2x$ is odd because it contains only odd powers of x. Here is a direct verification that $g(-x) = -g(x)$:

$$g(-x) = (-x)^3 - 2(-x) = -x^3 + 2x = -(x^3 - 2x) = -g(x).$$

Example 20.4   The function $h(x) = 2x^3 + x^2 + 2$ is neither even nor odd, because it contains both odd and even powers of x.

†Nonzero constant terms have degree zero, which is even.

There are also even and odd functions that are not polynomials; we shall meet some of those in Section 22. Knowing that a function is even or odd essentially cuts in half the work of graphing the function—once we determine what happens to the right of the y-axis we automatically know what happens to the left.

C. TRANSLATIONS

Figure 20.7 shows the graphs of $y = x^3$ and $y = (x - 2)^3$. It shows that the graph of $y = (x - 2)^3$ is the graph of $y = x^3$ *translated* (shifted) 2 units to the right. For example, the point $(2, 0)$ on the graph of $y = (x - 2)^3$ corresponds

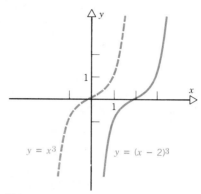

$y = x^3$

$y = (x - 2)^3$

FIGURE 20.7

to the point $(0, 0)$ on the graph of $y = x^3$. The relationship between these two graphs is a special case of the following principle.

> If f is a function and c is a constant, then the graph of $y = f(x - c)$ is the graph of $y = f(x)$ translated to the right by $|c|$ units if $c > 0$, and to the left by $|c|$ units if $c < 0$.　　(20.2)

Proof First, the point (x_1, y_1) is on the graph of $y = f(x - c)$ iff $y_1 = f(x_1 - c)$ iff $(x_1 - c, y_1)$ is on the graph of $y = f(x)$. Now simply observe that (x_1, y_1) is $|c|$ units to the right of $(x_1 - c, y_1)$ if $c > 0$, and (x_1, y_1) is $|c|$ units to the left of $(x_1 - c, y_1)$ if $c < 0$.　　□

Example 20.5 Draw the graph of $f(x) = -(x + 5)^2$.

Solution The graph of $y = -x^2$ is a parabola through the origin and opening downward, as shown in Figure 20.8. Since $-(x + 5)^2 = -[x - (-5)]^2$, we have a case with $c = -5 < 0$. Thus the graph of $y = -(x + 5)^2$ is obtained by translating the graph of $y = -x^2$ to the left by 5 units. Figure 20.8 shows the result.　■

> If f is a function and d is a constant, then the graph of $y = f(x) + d$ is the graph of $y = f(x)$ translated upward by $|d|$ units if $d > 0$, and translated downward by $|d|$ units if $d < 0$.　　(20.3)

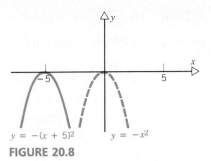

$y = -(x + 5)^2$ $y = -x^2$

FIGURE 20.8

Proof The point (x_1, y_1) is on the graph of $y = f(x) + d$ iff $y_1 = f(x_1) + d$ iff $y_1 - d = f(x_1)$ iff $(x_1, y_1 - d)$ is on the graph of $y = f(x)$. But (x_1, y_1) is $|d|$ units above $(x_1, y_1 - d)$ if $d > 0$, and $|d|$ below $(x_1, y_1 - d)$ if $d < 0$. $\square$

Example 20.6 Draw the graph of $f(x) = 2x^2 - 12x + 23$.

Solution Since this is a quadratic function, we could graph it as in Section 16. To illustrate translation, however, we proceed as follows. First, factor 2 from each term involving x and then complete the square.

$$2x^2 - 12x + 23 = 2(x^2 - 6x) + 23$$
$$= 2(x^2 - 6x + 9) + 23 - 18$$
$$= 2(x - 3)^2 + 5$$

To get the graph of f, we first graph $y = 2x^2$, then translate it 3 units to the right [because f involves $(x - 3)^2$ rather than x^2], and, finally, translate the result 5 units upward. The final result and the intermediate steps are shown in Figure 20.9. ◼

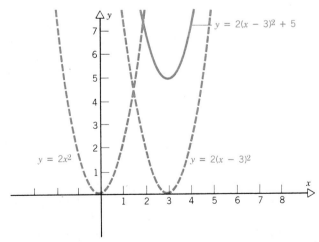

$y = 2(x - 3)^2 + 5$

$y = 2x^2$ $y = 2(x - 3)^2$

FIGURE 20.9

The principles in this section often allow us to determine the graph of a function, or at least its general characteristics, from the graph of a related function. But it is still advisable to compute some coordinate pairs directly from the given function, both to provide details and to serve as a check.

D. MORE ON POLYNOMIAL FUNCTIONS

The properties in this subsection apply to the graph of a polynomial function of degree n,

$$f(x) = a_n x^n + a_{n-1} x^{n-1} + \cdots + a_1 x + a_0, \qquad (20.4)$$

where $a_n \neq 0$. It can be proved that the graph of any such function has no breaks or jumps, in contrast with the graphs of the step functions in Examples 15.5 and 15.6, for example. This means that we can trace the graph between any two of its points without lifting our pencil from the paper. It can also be proved that the graph of any polynomial function is smooth; it will not have a sharp point like the graph of $y = |x|$ (Figure 15.5), for example.

The next property is true because a polynomial of degree n has at most n real zeros [Statement (19.2)].

The graph of a polynomial function of degree n has at most n x-intercepts.

For example, the graph of a third-degree function has at most three x-intercepts. The graph of $f(x) = x^3$ (Figure 20.1) shows that the number of x-intercepts may, in fact, be less than the degree of the polynomial.

We say that a function f has a **local** (or **relative**) **maximum** at $x = c$ if $f(x) \leq f(c)$ for all x sufficiently near c. The graph in Figure 20.10 has local maxima at c_1 and c_3. If $f(x) \leq f(c)$ for *all* x in the domain of f, then f has an **absolute maximum** at c. The definitions of **local minimum** and **absolute minimum** are similar. The graph in Figure 20.10 has a local minimum at c_2, but the graph has no absolute minimum.

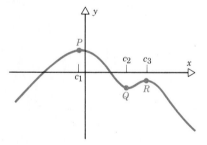

FIGURE 20.10

If there is a local maximum at $x = c$, with $f(x) < f(c)$ immediately to the left and right of c, then we say that the graph has a **turning point** at c. We also say there is a **turning point** at c if $f(x) > f(c)$ immediately to the left and right of c. Figure 20.10 gives illustrations. The following property can be proved with calculus. Also, the turning points can be located with calculus.

The graph of a polynomial function of degree n has at most $n - 1$ turning points.

For example, the graph of a fourth-degree polynomial has at most three turning points. It may have less, as is shown by $f(x) = x^4$ (Figure 20.3).

For each polynomial function there is a real number c such that the function is either increasing for all $x > c$ or decreasing for all $x > c$. Also, there is a real number d such that the function is either increasing for all $x < d$ or decreasing for all $x < d$. If Figure 20.10 gives the graph of a fourth-degree polynomial, then there can be only the three turning points that are shown; thus the function must be increasing for all $x < c_1$ and decreasing for all $x > c_3$. The increasing or decreasing behavior of a polynomial for input elements of large absolute value will be the same as the behavior of its term of highest degree. The following notation and example should help clarify these remarks.

We use the notation $x \to \infty$, which is read "x approaches infinity," to mean that x increases without bound—that is, beyond any real number, no matter how large. In the same way, we write $x \to -\infty$, read "x approaches minus infinity," to mean that x decreases beyond any negative real number, no matter how large its absolute value. As was stressed in the discussion of unbounded intervals (Section 10), ∞ is not a real number.

Example 20.7 (a) $x^3 \to \infty$ as $x \to \infty$, so $x^3 - 3x + 2 \to \infty$ as $x \to \infty$.

(b) $x^3 \to -\infty$ as $x \to -\infty$, so $x^3 - 3x + 2 \to -\infty$ as $x \to -\infty$.

(c) $-8x^3 \to -\infty$ as $x \to \infty$, so $-8x^3 + 2x^2 \to -\infty$ as $x \to \infty$.

(d) $3x^4 \to \infty$ as $x \to -\infty$, so $3x^4 - 15x^3 + 1 \to \infty$ as $x \to -\infty$. ◼

Here is a brief outline to consider when graphing a polynomial function.

Step I. *Check for symmetry* by determining if the function is either even or odd. If the function is either even or odd, we can concentrate on non-negative values of x in most of the remaining steps; what happens for negative x will follow automatically.

Step II. *Locate intercepts.*

(a) The y-intercept is $f(0)$.

(b) The x-intercepts are the zeros of $f(x)$. Use the ideas in Section 19 to help find these. (More information on zeros is given in the next section. In the exercises for this section, if no zero is given then you should be able to find one by factoring or the quadratic formula.)

Step III. *Compute and plot points.* Remember that the computations are especially easy with synthetic division (Section 18).

Step IV. *Determine the behavior for values of x with large absolute value.* The function will be increasing or decreasing for such values depending on whether its term of highest degree is increasing or decreasing, respectively.

Step V. *Draw the graph* as carefully as possible, remembering that it should be unbroken and smooth.

Example 20.8 Draw the graph of $f(x) = x^3 - 3x^2 + 2$.

Solution I. The function is neither even nor odd.

II. (a) The y-intercept is $f(0) = 2$.

(b) Inspection shows that $f(1) = 0$. Thus $x - 1$ is a factor of $f(x)$ and synthetic division produces a complementary factor.

$$
\begin{array}{r|rrr}
1 & 1 & -3 & 0 & 2 \\
  &   & 1 & -2 & -2 \\
\hline
  & 1 & -2 & -2 & 0
\end{array}
$$

$x^3 - 3x^2 + 2 = (x - 1)(x^2 - 2x - 2)$

From $x^2 - 2x - 2 = 0$ the quadratic formula gives $x = 1 \pm \sqrt{3}$. Thus the x-intercepts are at 1, $1 + \sqrt{3} \approx 2.7$, and $1 - \sqrt{3} \approx -0.7$.

III. The first row of Table 20.3 lists the coefficients of $f(x)$. Each remaining

TABLE 20.3

	1	-3	0	2
1	1	-2	-2	0
2	1	-1	-2	-2
3	1	0	0	2
-1	1	-4	4	-2
0.5	1	-2.5	-1.25	1.375
-0.5	1	-3.5	1.75	1.125
1.5	1	-1.5	-2.25	-1.375
2.5	1	-0.5	-1.25	-1.125

row gives the third row from the synthetic division using the number at the left end of that row; if c is at the left end of a row, then the number at the right end if $f(c)$. The entries in such a table can often be calculated mentally.

IV. $x^3 \to \infty$ as $x \to \infty$, so $f(x) \to \infty$ as $x \to \infty$.

$x^3 \to -\infty$ as $x \to -\infty$, so $f(x) \to -\infty$ as $x \to -\infty$.

V. Figure 20.11 shows the graph. Notice that there are three x-intercepts and two turning points. ■

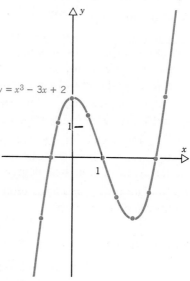

$y = x^3 - 3x + 2$

FIGURE 20.11

EXERCISES FOR SECTION 20

1. On a single Cartesian plane draw the graphs of $y = x^2$, $y = x^3$, and $y = x^4$, for $-1 \le x \le 1$. Make the unit of length as long as possible. Ⓒ

2. On a single Cartesian plane draw the graphs of $y = x^2$ and $y = x^4$, for $-2 \le x \le 2$. Ⓒ

3. On a single Cartesian plane draw the graphs of $y = x^3$ and $y = x^5$, for $-1.5 \le x \le 1.5$. Ⓒ

4. On a single Cartesian plane draw the graphs of $y = x^3$, $y = 2x^3$, and $y = \frac{1}{2}x^3$. How does changing k affect the graph of $y = kx^3$ for $k > 0$?

5. On a single Cartesian plane draw the graphs of $y = x^4$, $y = 2x^4$, and $y = \frac{1}{2}x^4$. How does changing k affect the graph of $y = kx^4$ for $k > 0$?

6. On a single Cartesian plane draw the graphs of $y = -x^3$, $y = -2x^3$, and $y = -\frac{1}{2}x^3$. How does changing k affect the graph of $y = kx^3$ for $k < 0$?

In each exercise, state whether the function is even, odd, or neither. Also state what your answer indicates about the symmetry of the graph. In Exercises 13–18 you will need to compute $f(-x)$.

7. $f(x) = 2x^3 - 2x$

8. $f(x) = x^3 + x^2 + x + 1$

9. $f(x) = x^4 + 1$

10. $f(x) = x^3 - 1$

11. $f(x) = 4x^5 + 2x$

12. $f(x) = 3x^4 + 5x$

13. $f(x) = |x|$

14. $f(x) = |2x| + 1$

15. $f(x) = |x + 1|$

16. $f(x) = 1/x$

17. $f(x) = x + (1/x^2)$

18. $f(x) = \sqrt{x^2 + 1}$

19. On a single Cartesian plane draw the graphs of $y = |x|$, $y = |x + 2|$, $y = -|x + 2|$, and $y = -|x + 2| - 3$.

20. On a single Cartesian plane draw the graphs of $y = |x|$, $y = 2|x|$, $y = -2|x|$, and $y = -2|x| + 1$.

21. On a single Cartesian plane draw the graphs of $y = x^3$, $y = (x + 4)^3$, and $y = (x + 4)^3 - 2$.

22. On a single Cartesian plane draw the graphs of $y = -x^4$, $y = -(x - 2)^4$, and $y = -(x - 2)^4 + 3$.

23. Figure 20.12 shows the graph of a function f with domain $[-4, 4]$. Draw the graphs of $y = f(x - 3)$, $y = -2f(x - 3)$, and $y = -2f(x - 3) + 1$.

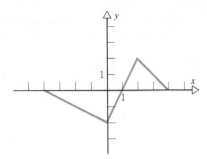

FIGURE 20.12

24. Figure 20.13 shows the graph of a function g with domain $[-2, 4]$. Draw the graphs of $y = g(x + 2)$, $y = -3g(x + 2)$, and $y = -3g(x + 2) - 1$.

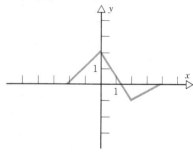

FIGURE 20.13

Draw the graph of each function by first completing the square, as in Example 20.6.

25. $f(x) = 3x^2 + 6x - 1$

26. $f(x) = -2x^2 + 8x - 5$

27. $f(x) = \frac{1}{2}x^2 - x + \frac{5}{2}$

Draw the graph of each function.

28. $f(x) = x^3 - 3x^2$

29. $f(x) = -x^3 + x$

30. $f(x) = x^4 - x^2$

31. $f(x) = 2x^3 - 2x$

32. $f(x) = x^3 + x^2 - 7x - 7$

33. $f(x) = -x^3 - x^2 + 9x + 9$

34. $f(x) = (x^2 - 4)(x^2 + 1)$

35. $f(x) = -(4x^2 - 1)(x^2 - 4)$
36. $f(x) = -(x + 1)^2(x - 4)$

Exercises 37–39 are based on the fact that for any real numbers k, a, b, *and* c (k $\neq$ 0), *the x-intercepts of the graph of* f(x) = k(x − a)(x − b)(x − c) *are* a, b, *and* c *(Factor Theorem).*

37. Find a cubic function f with leading coefficient 1 whose graph has x-intercepts -2, 3, and 4.

38. Find a cubic function f whose graph has x-intercepts -1, 2, and 3 and y-intercept 12. [The last condition will determine k in $k(x - a)(x - b)(x - c)$.]

39. Find a cubic function f whose graph has x-in-

tercepts -1, 0, and 2, and also passes through $(1, -6)$.

40. If f is an even function, g is an odd function, $f(a) = 3$, and $g(b) = 4$, what is $f(-a) - g(-b)$?

41. Assume that f is an even function and g is an odd function. Prove that the composite function $f \circ g$ is even. (Do not assume that f and g are polynomial functions.)

42. The product fg of functions f and g is defined by $(fg)(x) = f(x)g(x)$ for all x that are in the domain of both f and g. Assume that f is an even function and g is an odd function. Prove that the product fg is an odd function. (Do not assume that f and g are polynomial functions.)

SECTION **21**

More About Real Zeros of Polynomials (Optional)

A. RATIONAL ZEROS

The general form for a polynomial $f(x)$ of degree n is

$$f(x) = a_nx^n + a_{n-1}x^{n-1} + \cdots + a_1x + a_0, \qquad (21.1)$$

where $a_n, a_{n-1}, \ldots, a_1$, and a_0 denote the coefficients and $a_n \neq 0$. We know how to find the zeros of polynomials of degrees one and two; the ideas in this section will help with polynomials of higher degree. The following theorem puts a narrow restriction on the possible *rational zeros* of polynomials with *integral coefficients*.

Rational Zero Theorem

Assume that the coefficients of $f(x)$ in Equation (21.1) are integers. If a rational number r/s (reduced to lowest terms) is a zero of $f(x)$, then r is a factor of a_0 (the constant term) and s is a factor of a_n (the leading coefficient).

Proof If r/s is a zero of $f(x)$, then

$$a_n \left(\frac{r}{s}\right)^n + a_{n-1} \left(\frac{r}{s}\right)^{n-1} + \cdots + a_1 \left(\frac{r}{s}\right) + a_0 = 0.$$

Multiply both sides of this equation by s^n, and then subtract a_0s^n from both sides. The result is

$$a_nr^n + a_{n-1}r^{n-1}s + \cdots + a_1rs^{n-1} = -a_0s^n.$$

Since r is a factor of all of the terms on the left side of this equation, r must also be a factor of $-a_0 s^n$. Therefore, since r and s have no common factor (recall that r/s is reduced to lowest terms), r must be a factor of a_0.* In the same way, from

$$a_{n-1} r^{n-1} s + \cdots + a_1 r s^{n-1} + a_0 s^n = -a_n r^n$$

we can deduce that s must be a factor of a_n. $\qquad\qquad\square$

Example 21.1 Find the rational zeros of

$$x^3 - 3x^2 + 2x - 6.$$

Solution   Possible numerators (factors of -6): $\pm 1, \pm 2, \pm 3, \pm 6$

Possible denominators (factors of 1): ± 1

Possible rational zeros: $\pm 1, \pm 2, \pm 3, \pm 6$

We can test these possible zeros by synthetic division. If we choose the order $1, -1, 2, -2, \ldots$, then the first zero we find will be 3.

$$
\begin{array}{r|rrrr}
3 & 1 & -3 & 2 & -6 \\
  &   & 3 & 0 & 6 \\
\hline
  & 1 & 0 & 2 & 0
\end{array}
$$

We could continue through the list of possible rational zeros, trying $-3, 6$, and -6, but it is more efficient to use the Factor Theorem and reduce the degree by one. Thus $x^3 - 3x^2 + 2x - 6 = (x - 3)(x^2 + 2)$(recall that the factor $x^2 + 2$ comes from the numbers 1 0 2 in the third row of the synthetic division). The factor $x^2 + 2$ has no real zero (because its discriminant is negative, or because $x^2 + 2 > 0$ for every real number x). Therefore, the only real zero (rational or irrational) of $x^3 - 3x^2 + 2x - 6$ is 3. $\qquad\blacksquare$

The preceding example, in which the only possible denominators for zeros were ± 1, leads to the following observation.

> If the coefficients in Equation (21.1) are integers and $a_n = 1$, then any rational zero of $f(x)$ must be an integer.

Example 21.2 Find the rational solutions of

$$12x^3 - 8x^2 = 3x - 2.$$

Solution Remember that the zeros of $f(x)$ are the same as the solutions of $f(x) = 0$. Thus we can rewrite the given equation as $12x^3 - 8x^2 - 3x + 2 = 0$ and apply the Rational Zero Theorem.

Possible numerators (factors of 2): $\pm 1, \pm 2$

Possible denominators (factors of 12): $\pm 1, \pm 2, \pm 3, \pm 4, \pm 6, \pm 12$

Possible rational zeros: $\pm 1, \pm 2, \pm \frac{1}{2}, \pm \frac{1}{3}, \pm \frac{2}{3}, \pm \frac{1}{4}, \pm \frac{1}{6}, \pm \frac{1}{12}$

---

*Here we have used the following fact about integers: If b, c, and d denote nonzero integers such that d is a factor of bc, and b and d have no common factor except ± 1, then d must be a factor of c. Elementary number theory books give a proof.

If we choose the order $1, -1, 2, -2, \ldots$, then the first zero we find will be $\frac{1}{2}$.

$$
\begin{array}{r|rrrr}
\frac{1}{2} & 12 & -8 & -3 & 2 \\
 & & 6 & -1 & -2 \\
\hline
 & 12 & -2 & -4 & 0
\end{array}
$$

Thus $12x^3 - 8x^2 - 3x + 2 = (x - \frac{1}{2})(12x^2 - 2x - 4)$. The factor $12x^2 - 2x - 4$ gives

$$12x^2 - 2x - 4 = 0$$

$$6x^2 - x - 2 = 0$$

$$(2x + 1)(3x - 2) = 0$$

$$x = -\tfrac{1}{2} \quad \text{or} \quad x = \tfrac{2}{3}.$$

Therefore, the rational zeros are $\frac{1}{2}$, $-\frac{1}{2}$, and $\frac{2}{3}$. ◼

Sometimes we can use the Rational Zero Theorem to find rational zeros and, at the same time, reduce the polynomial by factoring so that we also get irrational zeros.

Example 21.3 Find the real zeros of

$$f(x) = 3x^4 + 5x^3 - 8x^2 - 10x + 4.$$

Solution   Possible numerators: $\pm 1, \pm 2, \pm 4$

Possible denominators: $\pm 1, \pm 3$

Possible rational zeros: $\pm 1, \pm 2, \pm 4, \pm\frac{1}{3}, \pm\frac{2}{3}, \pm\frac{4}{3}$

If we choose the order $1, -1, 2, -2, \ldots$, then the first zero we find will be -2.

$$
\begin{array}{r|rrrrr}
-2 & 3 & 5 & -8 & -10 & 4 \\
 & & -6 & 2 & 12 & -4 \\
\hline
 & 3 & -1 & -6 & 2 & 0
\end{array}
$$

Thus $f(x) = (x + 2)(3x^3 - x^2 - 6x + 2)$. Now we can restrict our attention to possible rational zeros of

$$g(x) = 3x^3 - x^2 - 6x + 2.$$

These are $\pm 1, \pm 2, \pm\frac{1}{3}$, and $\pm\frac{2}{3}$. Since $1, -1$, and 2 are not zeros of $f(x)$, they will not be zeros of $g(x)$. Therefore, we need to try only $-2, \pm\frac{1}{3}$, and $\pm\frac{2}{3}$.

$$
\begin{array}{r|rrrr}
\frac{1}{3} & 3 & -1 & -6 & 2 \\
 & & 1 & 0 & -2 \\
\hline
 & 3 & 0 & -6 & 0
\end{array}
$$

Thus

$$f(x) = (x + 2)g(x)$$

$$= (x + 2)(x - \tfrac{1}{3})\,(3x^2 - 6).$$

The factor $3x^2 - 6$ gives

$$3x^2 - 6 = 0$$
$$x^2 = 2$$
$$x = \sqrt{2} \quad \text{or} \quad x = -\sqrt{2}.$$

Therefore, the zeros of $f(x)$ are $-2, \frac{1}{3}, \sqrt{2}$, and $-\sqrt{2}$.

B. DESCARTES' RULE OF SIGNS

For the next theorem we need the idea of the *variation in signs* of a polynomial. For this purpose, we assume that the terms of the polynomial $f(x)$ have been written in order by decreasing powers of x. Also, any term with coefficient zero is omitted. Then a **variation in sign** occurs whenever two adjacent terms have opposite signs. For example,

$$f(x) = 3x^5 - 2x^4 + 5x^2 + x - 1 \tag{21.2}$$

has three variations in sign: one from $3x^5$ to $-2x^4$, one from $-2x^4$ to $5x^2$, and one from x to -1.

Descartes' Rule of Signs

(a) The number of positive zeros of a polynomial $f(x)$ is either equal to the number of variations in sign of $f(x)$ or less than that by an even integer.

(b) The number of negative zeros of a polynomial $f(x)$ is either equal to the number of variations in sign of $f(-x)$ or less than that by an even integer.

The proof of Descartes' Rule will be omitted.

Example 21.4 Apply Descartes' Rule of Signs to discuss the zeros of $f(x)$ given by Equation (21.2).

Solution (a) Since $f(x)$ has three variations in sign, the number of positive zeros is either three or one.

(b) Since $f(-x) = -3x^5 - 2x^4 + 5x^2 - x - 1$, there are two variations in sign for $f(-x)$. Therefore, $f(x)$ has either two negative zeros or no negative zero.

In applying Descartes' Rule, a zero of multiplicity m must be counted m times. For example, $x^2 - 10x + 25$, which equals $(x - 5)^2$, has 5 as a zero of multiplicity two. That is consistent with Descartes' Rule, since $x^2 - 10x + 25$ has two variations in sign.

C. BOUNDS ON REAL ZEROS

A number c is an **upper bound** for the zeros of a polynomial if no zero exceeds c. A number c is a **lower bound** for the zeros of a polynomial if no zero is less than c. Knowing such bounds can shorten the work of finding zeros, by eliminating possible rational zeros, for example. The following theorem can be used to find upper and lower bounds.

Theorem on Bounds

Let $f(x)$ be a polynomial whose leading coefficient is positive, and divide $f(x)$ synthetically by $x - c$.

(a) If $c > 0$ and the numbers in the third row of the synthetic division are all nonnegative, then c is an upper bound for the zeros of $f(x)$.

(b) If $c < 0$ and the numbers in the third row of the synthetic division alternate in sign (with 0 thought of as positive or negative, whichever will produce a variation in sign), then c is a lower bound for the zeros of $f(x)$.

The proof of the Theorem on Bounds will be omitted, but here is the idea. Suppose $c > 0$ and the numbers in the third row of the synthetic division are all nonnegative. If c is replaced by any larger number d, then the numbers in the third row will all increase, so that the last number in the third row cannot be zero. Thus d cannot be a zero of $f(x)$. (A specific example will make the idea clear. Try $c = 4$ and then $d = 5$ with the polynomial in Example 21.1, for example.) Similar remarks apply if $c < 0$.

Example 21.5 Apply the Theorem on Bounds to find integral bounds for the zeros of

$$f(x) = 2x^4 - 3x^3 - 8x^2 + 9x + 6.$$

Solution We try 1, 2, 3, . . . to search for an upper bound, and then $-1, -2, -3, . . .$ to search for a lower bound.

The first row of Table 21.1 lists the coefficients of $f(x)$. Each remaining row gives the third row from the synthetic division using the number at the left end of that row; thus if c is at the left end of a row, the number at the right end is $f(c)$.

TABLE 21.1

	2	-3	-8	9	6	Information about zeros
1	2	-1	-9	0	6	No information.
2	2	1	-6	-3	0	2 is a zero.
3	2	3	1	12	42	3 is an upper bound because these numbers are nonnegative.
-1	2	-5	-3	12	-6	No information.
-2	2	-7	6	-3	12	-2 is a lower bound because the signs alternate.

The information from the table shows that -2 is a lower bound for the zeros and 3 is an upper bound. (The zeros of $f(x)$ are, in fact, $-\sqrt{3} \approx -1.7$, -0.5, $\sqrt{3} \approx 1.7$, and 2.)

Example 21.6 The following computations show that -1 is a lower bound for the zeros of $x^3 + x^2 + 3x - 1$.

$$
\begin{array}{r|rrrr}
-1 & 1 & 1 & 3 & -1 \\
   &   & -1 & 0 & -3 \\
\hline
   & 1 & 0 & 3 & -4
\end{array}
\;\longleftarrow \left\{ \begin{array}{l} \text{Think of 0 as negative to} \\ \text{give alternating signs.} \end{array} \right.
$$

D. ISOLATING ZEROS

We can use the next theorem to isolate the real zeros of a polynomial even if we cannot find the zeros exactly.

Intermediate-Value Theorem

If $f(x)$ is a polynomial and a and b are real numbers such that $f(a)$ and $f(b)$ have opposite signs, then $f(x)$ has at least one zero between $x = a$ and $x = b$.

The Intermediate-Value Theorem is proved in books on calculus. The theorem is usually stated in a more general form than that used here.

Intuitively, the theorem is true because the graph of a polynomial is an unbroken curve, so that we can trace the graph between any two points without lifting our pencil from the paper. Thus, if we move from one point where the graph of $f(x)$ is above the x-axis to another point where the graph is below the x-axis, then we must cross the axis at least once. Any crossing-point (such as at c in Figure 21.1) gives a zero of $f(x)$.

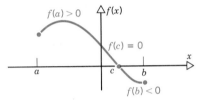

FIGURE 21.1

Example 21.7 The polynomial $2x^3 + 7x^2 - 10x - 35$ has three real zeros, none of which is an integer. Isolate the zeros between consecutive integral integers.

Solution Either synthetic division or substitution will give values of $f(x)$ for consecutive values of x.

x	-4	-3	-2	-1	0	1	2	3
$f(x)$	-11	4	-3	-20	-35	-36	-11	52

The sign changes in $f(x)$ are indicated. By the Intermediate-Value Theorem there is at least one zero in each of the intervals $(-4, -3)$, $(-3, -2)$, and $(2, 3)$. In fact, since a polynomial of degree three has at most three zeros, there is exactly one zero in each interval. (You can verify that the zeros are -3.5, $-\sqrt{5} \approx -2.24$, and $\sqrt{5} \approx 2.24$.)

E. APPROXIMATING ZEROS

The Intermediate-Value Theorem can be used to approximate the real zeros of a polynomial to any desired degree of accuracy. Suppose, for example, that $f(x)$ is a polynomial and that $f(2) > 0$ and $f(3) < 0$. Then there is at least one zero between 2 and 3. Compute $f(2.5)$. If $f(2.5) = 0$, then 2.5 is a zero. Otherwise, either $f(2.5) < 0$ or $f(2.5) > 0$. If $f(2.5) < 0$, then there is at least one zero between 2 and 2.5; if $f(2.5) > 0$, then there is at least one zero between 2.5 and 3. If there is a zero between 2 and 2.5, try 2.2 (or any other number between

2 and 2.5); if there is a zero between 2.5 and 3, try 2.7, for example. Continuing in this way we will either locate a zero exactly or generate a sequence of intervals of decreasing lengths, each of which must contain a zero. This method is especially practical when used with a computer or calculator.

Example 21.8 The polynomial

$$f(x) = x^4 - 4x^3 - x^2 + 10x + 4$$

has one irrational zero between 2 and 3. Find it to the nearest 0.5.

Solution We use the Intermediate-Value Theorem and synthetic division, writing only the third row of the synthetic division in each case.

	1	-4	-1	10	4	
2	1	-2	-5	0	4	$= f(2) > 0$
3	1	-1	-4	-2	-2	$= f(3) < 0$
2.5	1	-1.5	-4.75	-1.875	-0.6875	$= f(2.5) < 0$
2.25	1	-1.75	-4.9375	-1.109375	1.503906	$= f(2.25) > 0$

Because $f(2.25) > 0$ and $f(2.5) < 0$, the zero is between 2.25 and 2.5. Therefore, the zero is closer to 2.5 than to 2, and to the nearest 0.5 the zero is 2.5. (The zeros of $f(x)$ are, in fact, $1 \pm \sqrt{2}$ and $1 \pm \sqrt{5}$. The zero between 2 and 3 is $1 + \sqrt{2} \approx 2.414$.)

F. SUMMARY AND EXAMPLE

Here is an outline for searching for the real zeros of a polynomial $f(x)$ of degree greater than two. The computations are to be done by synthetic division. An example follows the outline.

* Apply Descartes' Rule of Signs to determine possibilities for the number of positive and number of negative zeros. Use this information as a guide in the remaining steps.

* Factor the highest possible power of x from the polynomial. If this highest power is x^k ($k \geq 1$), then 0 will be a zero of multiplicity k. Now work with the factor that remains.

* Test nonnegative integers in increasing order beginning with 0. (See Remark 1 that follows this outline.)

* If any positive integer is a zero, use the factor (quotient) produced by synthetic division to reduce the degree of the problem by one. Otherwise, continue testing positive integers until an upper bound for the zeros has been reached.

* Test negative integers in decreasing order beginning with -1. (See Remark 1 that follows this outline.)

* Continue as with the positive integers. Stop when a lower bound for the zeros has been reached.

* Use the Rational Zero Theorem and the Intermediate-Value Theorem to help determine which rational possibilities to test next (all of the possible integers between the bounds will have been tested already).

* Always factor the polynomial when a zero is found. If, at any step, only a

second-degree factor remains, apply the methods for treating quadratics (Section 8).

* If the previous techniques do not yield all of the real zeros, test any remaining rational possibilities or use the Intermediate-Value Theorem to compute approximations.

Remark 1   Instead of testing every positive integer until an upper bound for the zeros has been reached, you may want to test only those that are possible zeros as dictated by the Rational Zero Theorem. The same comment applies to negative integers.

Remark 2   There may be a zero between a and b even if $f(a)$ and $f(b)$ have the same sign. For example, if there are two unequal zeros between a and b, then the graph of $f(x)$ will intersect the x-axis twice in that interval. In many cases locating the zeros is not easy.

Remark 3   Remember that a zero may have multiplicity greater than 1. Therefore, if you find that c is a zero of $f(x)$ and that $f(x) = (x - c)g(x)$, then you should also test c as a possible zero of $g(x)$ [if, by the Rational Zero Theorem, it is a possible zero of $g(x)$].

Example 21.9   Find the real zeros of

$$f(x) = 2x^5 - 3x^4 - 11x^3 + 9x^2 + 15x.$$

Solution   We work through the steps in the preceding outline. Tables 21.2 and 21.3 summarize the computations.

TABLE 21.2 Synthetic Division for $g(x)$

	2	−3	−11	9	15
0	2	−3	−11	9	15
1	2	−1	−12	−3	12
2	2	1	−9	−9	−3
3	2	3	−2	3	24
4	2	5	9	45	195
−1	2	−5	−6	15	0

TABLE 21.3 Synthetic division for $h(x)$.

	2	−5	−6	15
−2	2	−9	12	−9
$\frac{3}{2}$	2	−2	−9	$\frac{3}{2}$
$\frac{5}{2}$	2	0	−6	0

There are two variations of sign in $f(x)$, so by Descartes' Rule of Signs $f(x)$ has either two positive zeros or no positive zero. There are also two variations of sign in $f(-x) = -2x^5 - 3x^4 + 11x^3 + 9x^2 - 15x$, so $f(x)$ has either two negative zeros or no negative zero.

The number 0 is a zero of multiplicity one. Write

$$f(x) = x(2x^4 - 3x^3 - 11x^2 + 9x + 15);$$

call the second factor $g(x)$ and work only with it hereafter.

$g(0) = 15 > 0$
$g(1) = 12 > 0$

$g(2) = -3 < 0$	There is a zero between 1 and 2.
$g(3) = 24 > 0$	There is a zero between 2 and 3. We now know that there are two positive zeros less than 3. Therefore, by our application of Descartes' Rule of Signs, there are no zeros greater than 3. We next compute $g(4)$ anyway, to give a separate proof that 4 is an upper bound for the zeros.
$g(4) = 195 > 0$	The signs are nonnegative in this row of Table 21.2, so 4 is an upper bound for the zeros of $g(x)$ and $f(x)$. Now test negative integers.
$g(-1) = 0$	Thus -1 is a zero. Also, $g(x) = (x + 1)h(x)$, where $h(x) = 2x^3 - 5x^2 - 6x + 15$. Consider $h(x)$ (Table 21.3).
$h(-2) = -9 < 0$	Because the signs alternate in this row of Table 21.3, -2 is a lower bound for the zeros of $h(x)$, $g(x)$, and $f(x)$. Now test $\frac{3}{2}$, the rational possibility between 1 and 2.
$h(\frac{3}{2}) = \frac{3}{2} > 0$	Not a zero. Test $\frac{5}{2}$, the rational possibility between 2 and 3.
$h(\frac{5}{2}) = 0$	Thus $\frac{5}{2}$ is a zero. Also, $h(x) = (x - \frac{5}{2})(2x^2 - 6)$.

Solve $2x^2 - 6 = 0$: $x^2 = 3$ and $x = \pm\sqrt{3}$. Thus the polynomial $f(x)$ has five real zeros, each of multiplicity one. In increasing order they are $-\sqrt{3}$, -1, 0, $\sqrt{3}$, and $\frac{5}{2}$. ■

EXERCISES FOR SECTION 21

Use the Rational Zero Theorem to construct a list of possible rational zeros for each polynomial.

1. $5x^3 - x^2 + 30x - 6$

2. $4x^3 - 2x^2 + 20x - 5$

3. $14x^3 + x^2 - 28x - 2$

Apply Descartes' Rule of Signs to discuss the zeros of each polynomial.

4. $3x^4 - 5x^3 + x^2 - 5x - 2$

5. $-x^5 + 3x^4 + 7x^3 - 2x - 1$

6. $x^3 - 2x - 1$

7. $-x^5 + 2x^4 - x^2 - 1$

8. $5x^4 - x^3 - 2x^2 + x - 1$

9. $-x^4 + 2x^2 - x + 7$

Use the Theorem on Bounds to find integral bounds for the real zeros of each polynomial.

10. $3x^3 + x^2 - 21x - 7$

11. $2x^3 - 5x^2 - 10x + 25$

12. $x^3 + 5x^2 - 6x - 30$

Each polynomial in Exercises 13–15 has three real zeros, none of which is an integer. Use the Inter-mediate-Value Theorem to isolate the zeros between consecutive integers.

13. $2x^3 - x^2 - 14x + 7$

14. $2x^3 - 11x^2 + 14x - 3$

15. $3x^3 - x^2 - 9x - 3$

Find the real zeros of each polynomial.

16. $x^3 - x^2 - 5x + 5$

17. $x^3 + x^2 - 3x - 3$

18. $x^3 + x^2 - 7x - 7$

19. $4x^3 - 7x - 3$

20. $2 - 11x + 17x^2 - 6x^3$

21. $2x^3 - 7x^2 + 4x + 4$

22. $-2 + 3x - 2x^2 + 3x^3$

23. $2x^3 - x^2 + 8x - 4$

24. $-1 + 5x - x^2 + 5x^3$

25. $x^3 - 2x^2 + 2x$

26. $x^3 - 2x^2 - 3x$

27. $x^3 - 10x^2 + 22x$

28. $4x^4 - 3x^3 + 3x^2 - 3x - 1$

29. $4x^4 - 4x^3 - 15x^2 + 16x - 4$

30. $2x^4 - 7x^3 - 7x^2 + 35x - 15$

31. $x^4 - 5x^3 + 4x^2 - 20x$

32. $3x^4 + 2x^3 - 6x^2 - 4x$

33. $3x^4 + 4x^3 - 9x^2 - 12x$

34. The polynomial $f(x) = x^4 - 6x^3 + 8x^2 - 6x + 7$ has one irrational zero between 1 and 2. Use the Intermediate-Value Theorem to find it to the nearest 0.5. $\boxed{C}$

35. The polynomial $f(x) = x^4 - 2x^3 - 5x^2 - 2x - 6$ has one irrational zero between 3 and 4. Use the Intermediate-Value Theorem to find it to the nearest 0.5. $\boxed{C}$

36. The polynomial $f(x) = x^4 - 2x^3 - 2x - 1$ has one irrational zero between -1 and 0. Use the Intermediate-Value Theorem to find it to the nearest 0.5. $\boxed{C}$

37. The fourth power of an integer is 75 more than the integer plus its square. What is the integer?

38. The square of a rational number plus 6 times the reciprocal of the number is 2 more than 3 times the number. What is the number?

39. The cube of an integer plus 3 times the square of the integer is 12 more than 20 times the integer. What is the integer?

40. The shortest edge of a box is half as long as another edge and one inch shorter than the third edge. If the volume of the box is 504 cubic inches, how long is each edge?

41. The height of a pyramid with a square base is one meter longer than each edge of the base. If the volume of the pyramid is $1\frac{7}{8}$ cubic meters, what is the height? (Use $V = \frac{1}{3}hb^2$, where h is the height and b is the length of each edge of the base.)

42. A box with volume 72 cubic centimeters and an open top has been formed from a square piece of cardboard 10 centimeters on a side by cutting equal squares from the corners and turning up the sides (Figure 21.2).

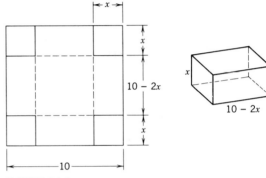

FIGURE 21.2

(a) Show that if x denotes the length of each edge of the removed squares, then

$$x^3 - 10x^2 + 25x - 18 = 0.$$

(b) Solve the equation in part (a), and use the answer to determine the dimensions of the box.

SECTION **22**

Graphs of Rational Functions

A. POWER FUNCTIONS

> **Definition** A function of a variable x is called a **rational function** if it can be written as a fraction in which both the numerator and denominator are polynomials in x.

The simplest examples of rational functions, other than polynomials, are the power functions that have the form

$$f(x) = kx^{-n} = \frac{k}{x^n},\tag{22.1}$$

where n is a positive integer. Figures 22.1 and 22.2 show the graphs of the power functions $f(x) = 1/x$ and $f(x) = 1/x^2$. These cases will be analyzed in detail because they illustrate the two significant ways in which graphs of rational functions can differ from graphs of polynomials: First, a polynomial function of x is

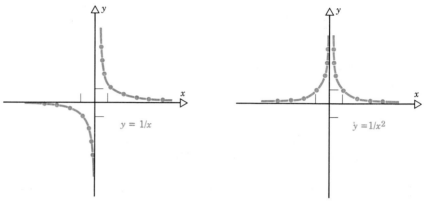

FIGURE 22.1   FIGURE 22.2

defined for every value of x, but a rational function will be undefined for any value of x that is a zero of the denominator; special care is required in graphing the function near such a value of x. Second, the graph of a polynomial function will either increase or decrease beyond all bounds as we move to the left or right along the graph, but the graph of a rational function can approach a horizontal line as we move to the left or right.

Example 22.1 Draw the graph of $f(x) = \dfrac{1}{x}$.

Solution First, notice that the function f is odd, because

$$f(-x) = \frac{1}{-x} = -\frac{1}{x} = -f(x).$$

Therefore, the graph is symmetric about the origin, so we can concentrate on positive values of x and then use symmetry to take care of negative values of x. Table 22.1 gives some coordinate pairs for points on the graph. These points have been plotted in Figure 22.1.

TABLE 22.1

x	$\frac{1}{4}$	$\frac{1}{3}$	$\frac{1}{2}$	1	2	3	4
$1/x$	4	3	2	1	$\frac{1}{2}$	$\frac{1}{3}$	$\frac{1}{4}$

As we compute $f(1)$, $f(2)$, . . . , we quickly realize that $f(x)$ approaches 0 as x gets larger. Using the notation introduced in Section 20, we abbreviate this by

$$f(x) \longrightarrow 0 \text{ as } x \longrightarrow \infty.$$

When a graph approaches a horizontal line, as in the case of $1/x \to 0$ as $x \to \infty$, we say that the line is a **horizontal asymptote** for the graph. This means that the distance between the line (asymptote) and the points on the graph approaches 0 as $x \to \infty$ or as $x \to -\infty$.

Now consider $f(\frac{1}{2})$, $f(\frac{1}{3})$, ... Here we see that $f(x) \to \infty$ as $x \to 0^{+}$. The notation $x \to 0^{+}$ is used to indicate that x is approaching 0 from the right. (We use $x \to 0^{-}$ when we approach 0 from the left.) As we move along the x-axis from the right toward the origin, the corresponding points on the graph rise without bound.

When a graph approaches a vertical line, as in the case of $1/x \to \infty$ as $x \to 0^{+}$, we say that the line is a **vertical asymptote** for the graph. Thus $x = 0$ is a vertical asymptote for $f(x) = 1/x$. In general, a line $x = c$ is a vertical asymptote for the graph of $f(x)$ if $f(x) \to \infty$ or $f(x) \to -\infty$ as $x \to c^{+}$ or as $x \to c^{-}$.

Making use of the symmetry about the y-axis, we obtain the portion of Figure 22.1 that lies in the third quadrant. Again, $y = 0$ is a horizontal asymptote:

$$\frac{1}{x} \longrightarrow 0 \text{ as } x \longrightarrow -\infty.$$

And $x = 0$ is a vertical asymptote:

$$\frac{1}{x} \longrightarrow -\infty \text{ as } x \longrightarrow 0^{-}.$$

Example 22.2 Draw the graph of $f(x) = \dfrac{1}{x^2}$.

Solution This function is even, because

$$f(-x) = \frac{1}{(-x)^2} = \frac{1}{x^2} = f(x).$$

Thus the graph is symmetric about the y-axis.

TABLE 22.2

x	$\frac{1}{4}$	$\frac{1}{3}$	$\frac{1}{2}$	1	2	3	4
$1/x^2$	16	9	4	1	$\frac{1}{4}$	$\frac{1}{9}$	$\frac{1}{16}$

Table 22.2 gives some coordinate pairs for points on the graph. These points have been plotted in Figure 22.2. Also, symmetry has been used to plot the corresponding points in the second quadrant. Observe the following similarities with Figure 22.2:

$f(x) \to 0$ as $x \to \infty$, so $y = 0$ is a horizontal asymptote.

$f(x) \to \infty$ as $x \to 0^{+}$, so $x = 0$ is a vertical asymptote.

$f(x) \to 0$ as $x \to -\infty$, so $y = 0$ is a horizontal asymptote.

$f(x) \to \infty$ as $x \to 0^{-}$, so $x = 0$ is a vertical asymptote.

The graph of any function of the form

$$f(x) = \frac{k}{(x - a)^n} \quad (n > 0) \tag{22.2}$$

will be similar to the graph $f(x) = 1/x$ if n is odd, and similar to the graph of $f(x) = 1/x^2$ if n is even. Figure 22.3 and 22.4 give two typical examples. By statement (20.2), the effect of replacing x by $x - a$ is to shift the graph to the right if $a > 0$ and to the left if $a < 0$. Thus $x = a$ becomes an asymptote.

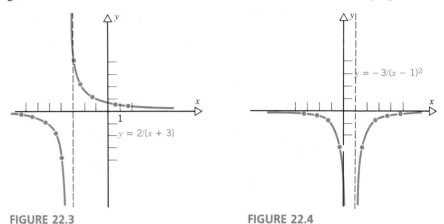

FIGURE 22.3   FIGURE 22.4

The graph of

$$f(x) = \frac{k}{(x - a)^n} \quad (n > 0)$$

has horizontal asymptote $y = 0$ and vertical asymptote $x = a$.

Changing k in $k/(x - a)^n$ will change the details of the graph but not its asymptotes. In particular, changing the sign of k will reflect the graph through the x-axis—that is, turn it upside down.

Example 22.3   Draw the graph of $f(x) = \dfrac{2}{x + 3}$.

Solution   The graph is shown in Figure 22.3. Here are some of its features.

Horizontal asymptote: $y = 0$.
Vertical asymptote: $x = -3$ (shown as a dashed line).
Symmetry about the point $(-3, 0)$.

x	0	1	2	-1	-2	$-\frac{5}{2}$	$-\frac{7}{2}$	-4	-5	-6	-7	-8
$2/(x + 3)$	$\frac{2}{3}$	$\frac{1}{2}$	$\frac{2}{5}$	1	2	4	-4	-2	-1	$-\frac{2}{3}$	$-\frac{1}{2}$	$-\frac{2}{5}$

Example 22.4   Draw the graph of $f(x) = \dfrac{-3}{(x - 1)^2}$.

Solution   The graph is shown in Figure 22.4. Here are some of its features.

Horizontal asymptote: $y = 0$.
Vertical asymptote: $x = 1$ (shown as a dashed line).
Symmetry about the line $x = 1$.

x	$\frac{3}{2}$	2	3	4	5	$\frac{1}{2}$	0	-1	-2	-3
$-3/(x-1)^2$	-12	-3	$-\frac{3}{4}$	$-\frac{1}{3}$	$-\frac{3}{16}$	-12	-3	$-\frac{3}{4}$	$-\frac{1}{3}$	$-\frac{3}{16}$

B. A GENERAL PLAN

The following plan for graphing rational functions is based on what we learned from the preceding examples. It is modified slightly to account for a wider range of functions. The examples in Subsection C will follow the plan step-by-step. *If you have trouble with Steps I–V, move ahead to Step VI and plot some points; this may help you see through I–V.*

Step I. *Check for symmetry* by determining if the function is either even or odd.

Step II. *Check for horizontal asymptotes.* Assume that $a(x)$ and $b(x)$ are polynomials, and that $f(x) = a(x)/b(x)$ has been reduced to lowest terms.

(a) If degree $a(x) <$ degree $b(x)$, then $y = 0$ is a horizontal asymptote. (See the examples in Subsection A.)

(b) If degree $a(x) =$ degree $b(x)$ and a_n and b_n are the leading coefficients of $a(x)$ and $b(x)$, respectively, then $y = a_n/b_n$ is a horizontal asymptote. [For instance, in Example 22.5 we'll see that $f(x) = 3x/(x + 1)$ has $y = 3$ as a horizontal asymptote; in this case $a_n = 3$ and $b_n = 1$. Also see Exercise 32.]

(c) If degree $a(x) >$ degree $b(x)$, then there is no horizontal asymptote.

Step III. *Locate the intercepts.*

(a) The y-intercept is at $(0, f(0))$, provided $f(0)$ is defined.

(b) The x-intercepts occur at the zeros of $f(x)$. These are the zeros of the numerator that are not also zeros of the denominator. (For what happens when a number is a zero of both the numerator and the denominator, see Exercises 19–24.)

Step IV. *Check for vertical asymptotes.* These will occur at any value $x = c$ where the denominator is zero but the numerator is not zero.

Step V. *Determine the behavior of the function near the vertical asymptotes.* Consider $x \to c^+$ (x approaching c from the right) and $x \to c^-$ (x approaching c from the left). For example, determine whether $f(x) \to \infty$ or $f(x) \to -\infty$ as $x \to c^+$.

Step VI. *Plot points.* The number and location of required points will depend on the particular function. We will at least need points near and on both sides of any vertical asymptote.

Step VII. *Draw the graph* as carefully as possible. Breaks can occur only at x-values for which the denominator is zero.

C. MORE EXAMPLES

Example 22.5 Draw the graph of $f(x) = \dfrac{3x}{x + 1}$.

Solution Follow the steps in Subsection B.

I. $f(-x) = 3(-x)/[(-x) + 1] = -3x/(-x + 1)$. Thus $f(-x) \neq f(x)$ and $f(-x) \neq -f(x)$ so that f is neither even nor odd.

II. Because the degrees of the numerator and denominator are equal, there is a horizontal asymptote. To see why this is at $y = 3$, divide both the numerator and denominator by the highest power of x in each [this amounts to multiplying by $(1/x)/(1/x)$ with $x \neq 0$, and thus leaves $f(x)$ unchanged for $x \neq 0$]:

$$f(x) = \frac{3x}{x + 1} \cdot \frac{1/x}{1/x} = \frac{3}{1 + (1/x)}.$$

Because $1/x \to 0$ as $x \to \infty$, it is clear that

$$f(x) \longrightarrow 3 \text{ as } x \longrightarrow \infty.$$

Similarly, $f(x) \to 3$ as $x \to -\infty$. The asymptote $y = 3$ is shown as a dashed line in Figure 22.5.

III. (a) There is a y-intercept at $(0, 0)$ because $f(0) = 0$. (b) Because $3x = 0$ iff $x = 0$, the (only) x-intercept is also at $(0, 0)$.

IV. $x + 1 = 0$ iff $x = -1$, so $x = -1$ is the (only) vertical asymptote.

V. For $x > -1$ but x close to -1, $3x < 0$ but $x + 1 > 0$. Thus $f(x) = 3x/(x + 1) < 0$ and $f(x) \to -\infty$ as $x \to -1^+$. For $x < -1$ but x close to -1, $3x < 0$ and $x + 1 < 0$. Thus $f(x) = 3x/(x + 1) > 0$ and $f(x) \to \infty$ as $x \to -1^-$.

VI. Here is a table of coordinate pairs.

x	0	1	2	4	5	6	7	$-\frac{1}{2}$	$-\frac{3}{4}$	$-\frac{5}{4}$	$-\frac{3}{2}$	-2	-3	-4	-5
$3x/(x + 1)$	0	$\frac{3}{2}$	2	$\frac{12}{5}$	$\frac{5}{2}$	$\frac{18}{7}$	$\frac{21}{8}$	-3	-9	15	9	6	$\frac{9}{2}$	4	$\frac{15}{4}$

VII. The graph is shown in Figure 22.5. ∎

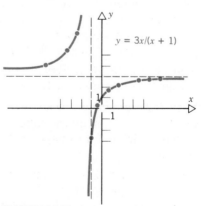

$y = 3x/(x + 1)$

FIGURE 22.5

Example 22.6   Draw the graph of $f(x) = \dfrac{x}{x^2 - 1}$.

Solution   I. $f(-x) = \dfrac{-x}{(-x)^2 - 1} = -\dfrac{x}{x^2 - 1} = -f(x)$. Therefore, f is odd and the graph is symmetric about the origin.

II. The degree of the numerator is less than the degree of the denominator. Thus $y = 0$ is a horizontal asymptote.

III. (a) The y-intercept is at $(0, 0)$ because $f(0) = 0$. (b) The x-intercept is also at $(0, 0)$.

IV. Since $f(x) = \dfrac{x}{(x + 1)(x - 1)}$, there are vertical asymptotes at $x = -1$ and $x = 1$.

V. For $x > 1$ but x close to 1, $f(x) > 0$, and $f(x) \to \infty$ as $x \to 1^+$. For $x < 1$ but x close to 1, $f(x) < 0$, and $f(x) \to -\infty$ as $x \to 1^-$. (The behavior near $x = -1$ will follow by symmetry.)

VI. Here is a table of coordinate pairs.

x	0	$\frac{1}{2}$	$\frac{3}{4}$	$\frac{5}{4}$	$\frac{3}{2}$	2	3
$x/(x^2 - 1)$	0	$-\frac{2}{3}$	$-\frac{12}{7}$	$\frac{20}{9}$	$\frac{6}{5}$	$\frac{2}{3}$	$\frac{3}{8}$

VII. The graph is shown in Figure 22.6. The points in Step VI have been plotted and symmetry has been used. ∎

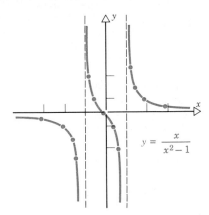

$$y = \frac{x}{x^2 - 1}$$

FIGURE 22.6

EXERCISES FOR SECTION 22

In Exercises 1–6, find all of the horizontal and vertical asymptotes.

Draw the graph of each function.   Ⓒ

1. $f(x) = \dfrac{2x}{x + 5}$

2. $f(x) = \dfrac{x^2 - 1}{3x^2 + 2}$

3. $f(x) = \dfrac{7}{x^2 - 9}$

4. $f(x) = \dfrac{-x + 10}{(x - 10)^3}$

5. $f(x) = \dfrac{x^3}{8x^3 - 1}$

6. $f(x) = \dfrac{5x^2}{x^2 + 1}$

7. $f(x) = \dfrac{-1}{x}$

8. $f(x) = \dfrac{-1}{x^2}$

9. $f(x) = \dfrac{-1}{x^3}$

10. $f(x) = \dfrac{1}{(x - 1)^3}$

11. $f(x) = \dfrac{3}{x - 3}$

12. $f(x) = \dfrac{2}{(x + 2)^2}$

13. $f(x) = \dfrac{2x}{x - 2}$

14. $f(x) = \dfrac{-x}{x + 1}$

15. $f(x) = \dfrac{-x}{x - 1}$

16. $f(x) = \dfrac{x}{(x - 1)(x + 2)}$

17. $f(x) = \dfrac{-x}{x^2 - 4}$

18. $f(x) = \dfrac{-x}{(x + 1)(x - 3)}$

Suppose that f(x) = a(x)/b(x) *and that c is a zero of both* a(x) *and* b(x). *Then* f(x) *is undefined at* x = c, *but the line* x = c *need not be a vertical asymptote as it is when c is zero of* b(x) *but not* a(x). *Exercises 19–24 provide illustrations. In Exercise 19, for example,* f(0) *is undefined, but* f(x) = x + 1 *for* x ≠ 0; *thus the graph is a straight line with one point removed. Draw the graph of each function.*   Ⓒ

19. $f(x) = \dfrac{x^2 + x}{x}$ **20.** $f(x) = \dfrac{x^2 - 1}{x + 1}$

21. $f(x) = \dfrac{2x^2 - 3x}{x}$ **22.** $f(x) = \dfrac{x - 1}{x^2 - x}$

23. $f(x) = \dfrac{x}{x^2 - 2x}$ **24.** $f(x) = \dfrac{x^4 - x^2}{x^2 - 1}$

Suppose that f(x) = a(x)/b(x), *that* a(x) *and* b(x) *have no common zero, and that the degree of* a(x) *is one more than the degree of* b(x). *Then division will yield* f(x) = cx + d + [r(x)/b(x)] *where c and d are real numbers,* c ≠ 0, *and deg* r(x) < *deg* b(x). *In this case* r(x)/b(x) → 0 *as* x → ∞ *or as* x → −∞, *so that* f(x) → cx + d. *The graph of* y = cx + d, *which is a straight line, is called an* **oblique asymptote** *for the graph of* f(x). *Exercises 25–30 provide illustrations. In Exercise 25, for example,* f(x) = (x² + 1)/x = x + (1/x), *so that* y = x *is an oblique asymptote. Draw the graph of each function.*   Ⓒ

25. $f(x) = \dfrac{x^2 + 1}{x}$ **26.** $f(x) = \dfrac{-x^2 + 1}{x}$

27. $f(x) = \dfrac{x^2 - 1}{2x}$ **28.** $f(x) = \dfrac{x^3}{x^2 + 1}$

29. $f(x) = \dfrac{x^3}{x^2 - 1}$ **30.** $f(x) = \dfrac{2x^2 + 1}{x + 1}$

31. Assume $f(x) = a(x)/b(x)$.

    (a) Prove that if $a(x)$ and $b(x)$ are both even then $f(x)$ is even.

    (b) Prove that if $a(x)$ and $b(x)$ are both odd then $f(x)$ is even.

    (c) Prove that if one of $a(x)$ and $b(x)$ is even and the other is odd then $f(x)$ is odd.

32. Assume $f(x) = a(x)/b(x)$, $a(x) = a_n x^n + \cdots + a_1 x + a_0$, and $b(x) = b_n x^n + \cdots + b_1 x + b_0$, with $a_n \neq 0$ and $b_n \neq 0$. Prove that $y = a_n/b_n$ is a horizontal asymptote for the graph of $f(x)$. [Suggestion: Divide both the numerator and denominator of $f(x)$ by x^n. Then consider $x \to \infty$ and $x \to -\infty$.]

33. Boyle's law states that if the temperature is constant, then the pressure P of a confined gas varies inversely as the volume of the gas: $P = k/V$ (Example 15.9). Graph P as a function of V in the special case $k = 4000$, with V measured in cubic inches and P in pounds per square inch. You will need to choose the unit distance on each axis so that the graph is manageable.)

34. The intensity I of a sound is inversely proportional to the square of the distance r from the source: $I = k/r^2$. Graph I as a function of r in the special case $k = 45$, with r measured in meters and I in watts per square meter.

35. The time t required to fill a tank from a water hose is inversely proportional to the square of the diameter of the hose: $t = k/d^2$. Graph t as a function of d in the special case $k = 2$, with d measured in inches and t measured in minutes.

36. The weight $w(x)$ of an object at altitude x miles above sea level is given by

$$w(x) = \frac{w_0 R^2}{(x + R)^2} \quad (x \geq 0), \quad (22.3)$$

where w_0 is the weight at sea level and R is the radius of the earth in miles. Determine the w-intercept and a horizontal asymptote for the graph of Equation (22.3). Draw a graph of w as a function of x. (Without specific values for w_0 and R, only the general shape can be shown, of course.) What does the graph reveal about the weight as the altitude is increased?

SECTION **23**
Partial Fractions (Optional)

A. INTRODUCTION

To add rational functions we convert to common denominators, add, and then simplify. For example,

$$\frac{3}{x + 1} - \frac{x}{x^2 - x + 2} = \frac{3(x^2 - x + 2) - x(x + 1)}{(x + 1)(x^2 - x + 2)}$$

$$= \frac{2x^2 - 4x + 6}{x^3 + x + 2}. \tag{23.1}$$

In this section we consider how to reverse that process—how to begin with an expression like that on the right in (23.1) and end with an expression like that on the left. This reversed process is needed in calculus, for example.

The method we use rests on the following fact, whose proof is omitted: With real coefficients, every polynomial can be written as a product of linear and irreducible quadratic factors. That fact, together with the division algorithm for polynomials, can be used to justify the following statement.

> Every rational function can be written uniquely as a sum of a polynomial (which may be identically zero) and fractions of the form
>
> $$\frac{A}{(ax + b)^m} \quad \text{and} \quad \frac{Bx + C}{(ax^2 + bx + c)^n}$$
>
> where $ax^2 + bx + c$ is irreducible.

The resulting form is called the **partial fraction decomposition** of the original rational function. The linear $(ax + b)$ and quadratic $(ax^2 + bx + c)$ factors in the decomposition are the irreducible factors of the original denominator. The first step is always to factor the original denominator into powers of irreducible factors. (Remember that a quadratic factor is irreducible iff its discriminant is negative.)

B. DISTINCT LINEAR FACTORS

We call an irreducible factor of the denominator *distinct* (as opposed to *repeated*) if it occurs only with exponent one when we factor the denominator.

> Each distinct linear factor $ax + b$ contributes a term of the form
>
> $$\frac{A}{ax + b}$$
>
> to the decomposition.

Example 23.1 Decompose $\dfrac{x + 12}{(x - 2)(x + 5)}$ into partial fractions.

Solution The factors $x - 2$ and $x + 5$ will contribute terms

$$\frac{A}{x - 2} \quad \text{and} \quad \frac{B}{x + 5}$$

respectively. If

$$\frac{x + 12}{(x - 2)(x + 5)} = \frac{A}{x - 2} + \frac{B}{x + 5} \tag{23.2}$$

then on multiplying by $(x - 2)(x + 5)$ to clear the denominators, we obtain

$$x + 12 = A(x + 5) + B(x - 2). \tag{23.3}$$

The last equation must be true for all values of x, so in particular it must be true for $x = -5$ and $x = 2$, the zeros of the two linear factors. By substituting these in turn we can determine A and B.

Use $x = -5$ in (23.3):

$$-5 + 12 = A(0) + B(-7)$$

$$B = -1.$$

Use $x = 2$ in (23.3):

$$2 + 12 = A(7) + B(0)$$

$$A = 2.$$

With $A = 2$ and $B = -1$, Equation (23.2) gives the answer:

$$\frac{x + 12}{(x - 2)(x + 5)} = \frac{2}{x - 2} - \frac{1}{x + 5}.$$

To check the answer perform the subtraction indicated on the right and show that the result simplifies to the expression on the left. ∎

C. REPEATED LINEAR FACTORS

Each repeated linear factor $(ax + b)^m$ contributes a sum of the form

$$\frac{A_1}{ax + b} + \frac{A_2}{(ax + b)^2} + \cdots + \frac{A_m}{(ax + b)^m}$$

to the decomposition.

Example 23.2 Decompose $\dfrac{x^2 - 5x - 2}{x^3 + 2x^2 + x}$ into partial fractions.

Solution First, factor the denominator completely.

$$x^3 + 2x^2 + x = x(x^2 + 2x + 1) = x(x + 1)^2$$

The distinct factor x contributes

$$\frac{A}{x}.$$

The repeated linear factor $(x + 1)^2$ contributes

$$\frac{B}{x + 1} + \frac{C}{(x + 1)^2}.$$

If

$$\frac{x^2 - 5x - 2}{x(x + 1)^2} = \frac{A}{x} + \frac{B}{x + 1} + \frac{C}{(x + 1)^2} \tag{23.4}$$

then

$$x^2 - 5x - 2 = A(x + 1)^2 + Bx(x + 1) + Cx. \tag{23.5}$$

With $x = 0$ in (23.5) we will get $A = -2$. With $x = -1$ in (23.5) we will get $C = -4$. Now we use any other value of x in (23.5), along with $A = -2$ and $C = -4$, to get B. With $x = 1$ we will get $B = 3$.

If we use $A = -2$, $B = 3$, and $C = -4$ in Equation (23.5), and revert to the original form of the denominator on the left, we get

$$\frac{x^2 - 5x - 2}{x^3 + 2x^2 + x} = -\frac{2}{x} + \frac{3}{x + 1} - \frac{4}{(x + 1)^2}. \qquad ■$$

D. DISTINCT QUADRATIC FACTORS

> Each distinct quadratic factor $ax^2 + bx + c$ contributes a term of the form
>
> $$\frac{Ax + B}{ax^2 + bx + c}$$
>
> to the decomposition.

Example 23.3 Decompose $\dfrac{3x^2 + 2x}{(x + 1)(x^2 + x + 1)}$ into partial fractions.

Solution The factor $x^2 + x + 1$ is irreducible. The factors $x + 1$ and $x^2 + x + 1$ contribute terms

$$\frac{A}{x + 1} \quad \text{and} \quad \frac{Bx + C}{x^2 + x + 1}$$

respectively. If

$$\frac{3x^2 + 2x}{(x + 1)(x^2 + x + 1)} = \frac{A}{x + 1} + \frac{Bx + C}{x^2 + x + 1} \tag{23.6}$$

then

$$3x^2 + 2x = A(x^2 + x + 1) + (Bx + C)(x + 1). \tag{23.7}$$

With $x = -1$ in (23.7) we will get $A = 1$. With $x = 0$ and $A = 1$ in (23.7) we will get $C = -1$. With $x = 1$, $A = 1$, and $C = -1$ in (23.7) we will get $B = 2$. Thus

$$\frac{3x^2 + 2x}{(x + 1)(x^2 + x + 1)} = \frac{1}{x + 1} + \frac{2x - 1}{x^2 + x + 1}.$$

E. REPEATED QUADRATIC FACTORS

Each repeated quadratic factor $(ax^2 + bx + c)^n$ contributes a sum of the form

$$\frac{A_1x + B_1}{ax^2 + bx + c} + \frac{A_2x + B_2}{(ax^2 + bx + c)^2} + \cdots + \frac{A_nx + B_n}{(ax^2 + bx + c)^n}$$

to the decomposition.

In the previous examples the coefficients in the partial fraction decompositions have been determined by substituting appropriate numbers for x. The next example uses a method based on the fact that a polynomial in x is identically zero (that is, zero for every value of x) iff each of its coefficients is zero. (This will be proved in Section 48.)

Example 23.4 Decompose $\dfrac{1}{x(x^2 + 2)^2}$ into partial fractions.

Solution The factors x and $(x^2 + 2)^2$ contribute

$$\frac{A}{x} \quad \text{and} \quad \frac{Bx + C}{x^2 + 2} + \frac{Dx + E}{(x^2 + 2)^2}$$

respectively. If

$$\frac{1}{x(x^2 + 2)^2} = \frac{A}{x} + \frac{Bx + C}{x^2 + 2} + \frac{Dx + E}{(x^2 + 2)^2} \tag{23.8}$$

then

$$1 = A(x^2 + 2)^2 + (Bx + C)x(x^2 + 2) + (Dx + E)x$$

$$1 = A(x^4 + 4x^2 + 4) + B(x^4 + 2x^2) + C(x^3 + 2x) + Dx^2 + Ex$$

$$(A + B)x^4 + Cx^3 + (4A + 2B + D)x^2 + (2C + E)x + 4A - 1 = 0. \tag{23.9}$$

Since Equation (23.9) is to be satisfied for every value of x, each of the coefficients must be zero. (See the remark preceding this example.)

$$x^4: \quad A + B = 0 \qquad\qquad x^3: \quad C = 0$$

$$x^2: \quad 4A + 2B + D = 0 \qquad x: \quad 2C + E = 0$$

$$x^0: \quad 4A - 1 = 0$$

The equation for x^0 yields $A = \frac{1}{4}$. The equation for x^4 then yields $B = -\frac{1}{4}$. The equation for x^3 yields $C = 0$. The equation for x then yields $E = 0$. With $A = \frac{1}{4}$ and $B = -\frac{1}{4}$, the equation for x^2 yields $D = -\frac{1}{2}$. With these values for A, B, C, D, and E, Equation (23.8) becomes

$$\frac{1}{x(x^2 + 2)^2} = \frac{1}{4x} - \frac{x}{4(x^2 + 2)} - \frac{x}{2(x^2 + 2)^2}.$$

F. IMPROPER FRACTIONS

A rational function is called a **proper fraction** if the degree of the numerator is less than the degree of the denominator; otherwise it is called an **improper fraction.** By division, an improper fraction can be written as a sum of a polynomial and a proper fraction. The preceding methods can then be applied to decompose the proper fraction into partial fractions.

Example 23.5 Decompose

$$f(x) = \frac{x^4 + 3x^3 - 11x^2 - 2x + 22}{x^2 + 3x - 10}$$

into partial fractions.

Solution Division yields

$$f(x) = x^2 - 1 + \frac{x + 12}{x^2 + 3x - 10}.$$

The proper fraction on the right is the same as the fraction in Example 23.1. Therefore, using the solution of that example, we have

$$f(x) = x^2 - 1 + \frac{2}{x - 2} - \frac{1}{x + 5}.$$

EXERCISES FOR SECTION 23

Decompose into partial fractions.

1. $\dfrac{x - 8}{(x + 1)(x - 2)}$

2. $\dfrac{3x + 3}{x(x - 3)}$

3. $\dfrac{6x - 18}{(x + 2)(x - 4)}$

4. $\dfrac{2}{3x^2 + 4x}$

5. $\dfrac{-1}{2x^2 + x - 1}$

6. $\dfrac{8x + 3}{6x^2 + 2x}$

7. $\dfrac{3x - 4}{(x - 1)^2}$

8. $\dfrac{4x + 5}{(2x + 1)^2}$

9. $\dfrac{-4x + 11}{(2x - 3)^2}$

10. $\dfrac{-x^2 + 2x + 4}{x^3 + x^2}$

11. $\dfrac{2x^2 - 2x + 3}{x^3 - 2x^2 + x}$

12. $\dfrac{7x^2 - 9x}{(x + 1)(x - 1)^2}$

13. $\dfrac{4x^2 + x + 4}{x^3 + x^2 + x}$

14. $\dfrac{2x^2 + 5x + 2}{x^3 + x}$

15. $\dfrac{-3x^2 + 4x - 6}{x^3 + 2x}$

16. $\dfrac{7x^2 + 13}{(x^2 + 1)(x^2 + 2)}$

17. $\dfrac{x^2 - 2}{x^4 + x^2}$

18. $\dfrac{-x^2}{(x^2 + 1)(x^2 + 3)}$

19. $\dfrac{x^3 - x + 1}{(x^2 + 1)^2}$

20. $\dfrac{3x^2 + 4x + 5}{(x^2 + 2)^2}$

21. $\dfrac{2x^3 + x^2 + 4x - 1}{(x^2 + x + 1)^2}$

22. $\dfrac{-2x^2 + 2x + 3}{x^4 + x^3}$

23. $\dfrac{4x^2 - 11x + 8}{(x - 1)^3}$

24. $\dfrac{-2x^3 + x^2 + 2x + 4}{(x^2 + 1)^3}$

25. $\dfrac{3x^3 + 2x^2 - 2x + 1}{x^2 + x}$

26. $\dfrac{x^4 - 2x^3 - x^2 + 6x - 5}{(x - 1)^2}$

27. $\dfrac{2x^5 - x^4 + 5x^3 - 3x^2 + x - 2}{(x^2 + 1)^2}$

REVIEW EXERCISES FOR CHAPTER IV

Determine the quotient and remainder when f(x) *is divided by* g(x).

1. $f(x) = 2x^3 - 6x^2 + 5,\ g(x) = 2x^2 - 1$

2. $f(x) = x^4 - x^3 + 3x^2 - x + 2,\ g(x) = x^2 + 2$

Use synthetic division to determine the quotient and remainder when f(x) *is divided by* g(x).

3. $f(x) = x^4 - 3x^2 + x + 4,\ g(x) = x - 2$

4. $f(x) = 3x^3 - 5x^2 + 2,\ g(x) = x + 1$

Use the Remainder Theorem and synthetic division to compute the indicated values of f(c).

5. $f(x) = 2x^3 + x^2 - 5;\ f(3),\ f(-1)$

6. $f(t) = t^4 + 3t^2 - t;\ f(2),\ f(-2)$

Use the Remainder Theorem and synthetic division to show that the given value of c *is a zero of the given polynomial. Also determine* q(x) *such that* f(x) = (x − c)q(x).

7. $c = 4;\ f(x) = x^3 - 3x^2 - 2x - 8$

8. $c = -\dfrac{1}{2};\ f(x) = 4x^4 + 2x^3 - 2x - 1$

Determine the real zeros of each polynomial along with their multiplicities. Also, factor each polynomial completely. For Exercises 11 and 12 use the Rational Zero Theorem from Section 21.

9. $x^3 - 4x^2 + x + 6$

10. $x^3 - 2x^2 - 3x + 6$

11. $2x^4 - 3x^3 - 3x^2 + 6x - 2$

12. $3x^4 + x^3 - 11x^2 - 3x + 6$

Each polynomial in Exercises 13 and 14 has three real zeros, none of which is an integer. Use the Intermediate-Value Theorem to isolate the zeros between consecutive integers.

13. $2x^3 + 7x^2 - 10x - 35$

14. $3x^3 + 5x^2 - 30x - 50$

Apply Descartes' Rule of Signs to discuss the zeros of each polynomial.

15. $x^6 - x^5 + 3x^4 - x^3 + 5$

16. $-2x^5 + 4x^4 - x^2 + x$

In each exercise, state whether the function is even, odd, or neither. Also state what your answer indicates about the symmetry of the graph of the function.

17. $f(x) = x^3 - x + 1$

18. $f(x) = 3x^6 + x^4 - 2x^2 - 1$

19. $f(x) = \dfrac{2x}{x^3 - x}$

20. $f(x) = \dfrac{x^2 + 1}{x^3 - 5x}$

Draw the graph of each function.

21. $f(x) = |x + 1| - 2$

22. $f(x) = -(x - 3)^3 + 1$

23. $f(x) = x^3 - 4x$

24. $f(x) = -2x^3 + 3x^2$

25. $f(x) = -x^4 + x^3 + x^2 - x$

26. $f(x) = 2x^4 + x^3 - 6x^2$

27. $f(x) = \dfrac{3}{(x - 1)^2}$

28. $f(x) = \dfrac{-2x}{x + 2}$

29. $f(x) = \dfrac{x^2}{x^2 + 1}$

30. $f(x) = \dfrac{2x}{x^2 - x - 2}$

Decompose into partial fractions.

31. $\dfrac{3x + 2}{(x + 2)^2}$

32. $\dfrac{4x^2 - 3x + 4}{x^3 + x}$

33. $\dfrac{x^3 + x^2 + 2x + 5}{x^2 + x - 2}$

34. $\dfrac{x^5 - x^4 + 3x^2 - 4x^2 + 7x - 9}{(x - 1)(x^2 + 2)}$

35. If 3 times the cube of a rational number minus

9 times the number is 6 less than 2 times the square of the number, what is the number?

36. The shortest edge of a box is one-third as long as another edge and two centimeters shorter than the third edge. If the volume of the box is 1323 cubic centimeters, how long is each edge?

CHAPTER V

EXPONENTIAL AND LOGARITHMIC FUNCTIONS

A *power function* of x has the form x^b where the exponent b is constant (Section 20). In this chapter we interchange the roles of x and b to obtain *exponential functions*, which have the form b^x where b is constant. We also consider the inverses of exponential functions, which are called *logarithmic functions*. We study the general properties of exponential and logarithmic functions and also look at some typical applications. This chapter will draw freely on ideas from Section 3 (Rational Exponents and Radicals) and Section 17 (Inverse Functions).

SECTION **24**
Exponential Functions

A. EXAMPLE

Consider the function $f(x) = 2^x$. To make the domain the set of all real numbers, we must give a meaning to 2^x for *every* real number x. From Section 3, 2^x already has a meaning if x is rational. For example,

$$2^3 = 2 \cdot 2 \cdot 2 = 8,$$

$$2^{-4} = \frac{1}{2^4} = \frac{1}{16},$$

$$2^{5/3} = (2^{1/3})^5 = (\sqrt[3]{2})^5$$

and, in general,

$$2^{m/n} = (2^{1/n})^m = (\sqrt[n]{2})^m. \tag{24.1}$$

By looking at these powers geometrically, we shall see how to go further.
Figure 24.1 indicates the pattern of the points

$$(x, 2^x) \text{ for } x = 0, \pm 1, \pm 2, \ldots.$$

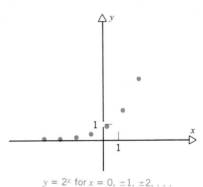

$y = 2^x$ for $x = 0, \pm 1, \pm 2, \ldots$

FIGURE 24.1

Some coordinate pairs are shown in Table 24.1.

TABLE 24.1

x	$\cdots$	-3	-2	-1	0	1	2	3	$\cdots$
2^x	$\cdots$	$\frac{1}{8}$	$\frac{1}{4}$	$\frac{1}{2}$	1	2	4	8	$\cdots$

We can add to Figure 24.1 any point $(x, 2^x)$ for x a rational number by using Equation (24.1). Table 24.2 shows some specific calculations, and the points determined from Tables 24.1 and 24.2 are all shown in Figure 24.2.

TABLE 24.2

x	2^x
0.5	$2^{1/2} = \sqrt{2} \approx 1.414$
1.5	$2^{3/2} = \sqrt{2^3} = \sqrt{2^2}\sqrt{2} = 2\sqrt{2} \approx 2.828$
2.5	$2^{5/2} = \sqrt{2^5} = \sqrt{2^4}\sqrt{2} = 4\sqrt{2} \approx 5.656$
$\vdots$	$\vdots$
-0.5	$2^{-1/2} = 1/\sqrt{2} = \sqrt{2}/2 \approx 0.707$
-1.5	$2^{-3/2} = 1/2\sqrt{2} = \sqrt{2}/4 \approx 0.354$
-2.5	$2^{-5/2} = 1/4\sqrt{2} = \sqrt{2}/8 \approx 0.177$
$\vdots$	$\vdots$

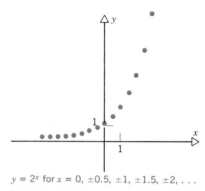

$y = 2^x$ for $x = 0, \pm 0.5, \pm 1, \pm 1.5, \pm 2, \ldots$

FIGURE 24.2

If we continue in this way with other pairs $(x, 2^x)$ for x rational, the points will continue to fall into the pattern suggested by Figures 24.1 and 24.2. This leads to a geometric interpretation of 2^x for every real number x. First, draw a smooth curve through the points $(x, 2^x)$ for rational values of x. This curve is shown in Figure 24.3. Then, given any real number x (rational or irrational), use the corresponding y-value on the curve as an approximation of 2^x.

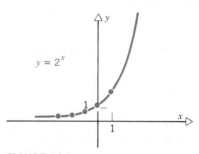

FIGURE 24.3

Here is the idea in nongeometric terms for the special case 2^π. The following sequence of rational numbers gives successively more accurate approximations of π.

3

3.1

3.14

3.141

3.1415

3.14159

$\vdots$

The equation $2^{m/n} = (2^{1/n})^m$ permits us to interpret 2^x for each number x in this sequence. For instance,

$$2^{3.14} = 2^{314/100} = (2^{1/100})^{314}.$$

TABLE 24.3

x	2^x
3	8
3.1	$8.57418 \cdots$
3.14	$8.81524 \cdots$
3.141	$8.82135 \cdots$
3.1415	$8.82441 \cdots$
3.14159	$8.82496 \cdots$
$\cdots$	$\cdots$

This idea leads to the sequence of pairs in Table 24.3. (The values in Table 24.3 can be verified with a calculator having a $\boxed{y^x}$ key.) It can be proved that the sequence of numbers on the right in Table 24.3 approaches a unique real number, and this unique number is taken as the definition of 2^π. To seven significant figures $2^\pi = 8.824978$. If we were to plot the pairs from Table 24.3 on a graph, they would approach the point $(\pi, 2^\pi)$ as x approaches π.

Because every real number has a decimal representation, we can apply the idea just used for 2^π to define 2^x for every real number x. This yields the function $f(x) = 2^x$ whose domain is the set of all real numbers. The graph of this function is the smooth curve in Figure 24.3.

B. DEFINITION

The idea in Subsection A can be carried out with any other positive real number in place of 2. This leads to the class of functions in the following definition.

Definition If b is any positive real number except 1, then the function f defined by

$$f(x) = b^x \text{ for every real number } x \qquad (24.2)$$

is called the **exponential function** with **base** b.

[The reason for excluding 1 in this definition is that later we shall want to consider inverses of such functions, and $b = 1$ would lead to the constant function $f(x) = 1$, which does not have an inverse.]

Example 24.1 The graph of $f(x) = 4^x$ is shown in Figure 24.4. Some coordinate pairs for points on the graph are $(-2, \frac{1}{16})$, $(-1, \frac{1}{4})$, $(0, 1)$, $(1, 4)$, and $(2, 16)$.

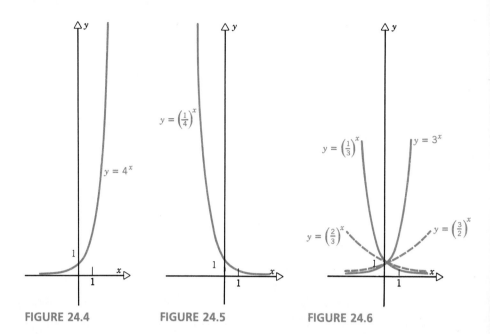

FIGURE 24.4   FIGURE 24.5   FIGURE 24.6

Example 24.2 The graph of $f(x) = (\frac{1}{4})^x = 4^{-x}$ is shown in Figure 24.5. Some coordinate pairs for points on the graph are $(-2, 16), (-1, 4), (0, 1), (1, \frac{1}{4})$, and $(2, \frac{1}{16})$. ▨

Example 24.3 Figure 24.6 shows the graphs of the four exponential functions

$$f(x) = 3^x, \ f(x) = (\tfrac{1}{3})^x, \ f(x) = (\tfrac{3}{2})^x, \text{ and } f(x) = (\tfrac{2}{3})^x.$$ ▨

Compare the graphs of the two functions in each of the following pairs:

$$f(x) = 4^x \quad \text{and} \quad f(x) = (\tfrac{1}{4})^x = 4^{-x}$$
$$f(x) = 3^x \quad \text{and} \quad f(x) = (\tfrac{1}{3})^x = 3^{-x}$$
$$f(x) = (\tfrac{3}{2})^x \quad \text{and} \quad f(x) = (\tfrac{2}{3})^x = (\tfrac{3}{2})^{-x}.$$

These comparisons illustrate that the graph of $y = (1/b)^x = b^{-x}$ is the reflection through the y-axis of the graph of $y = b^x$. To prove that this relationship between graphs is true, let $f(x) = b^x$ and $g(x) = b^{-x}$. Then

(x, y) is on the graph of $y = f(x)$ iff $(-x, y)$ is on the graph of $y = g(x)$,

because $y = f(x)$ iff $y = b^x$ iff $y = b^{-(-x)}$ iff $y = g(-x)$.

[Here we have used the laws for real exponents to be given in (24.3)–(24.7).]
    Recall (from Section 15) that a function f is

increasing if $x_1 < x_2$ implies $f(x_1) < f(x_2)$,

and

decreasing if $x_1 < x_2$ implies $f(x_1) > f(x_2)$.

The increasing and decreasing properties of exponential functions are included in the following list. The properties in this list are suggested and illustrated by the examples in Figures 24.3–24.6. There is no need to memorize these properties; instead, just remember the representative graphs in Figure 24.7.

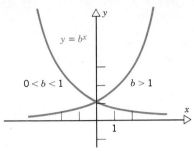

FIGURE 24.7

Let $f(x) = b^x$ be an exponential function. The *domain* of f is the set of all real numbers. The *range* of f is the set of all positive real numbers. If $b > 1$, then

> f is increasing,
> $f(x) \to \infty$ as $x \to \infty$, and
> $f(x) \to 0$ as $x \to -\infty$.

If $0 < b < 1$, then

> f is decreasing,
> $f(x) \to 0$ as $x \to \infty$, and
> $f(x) \to \infty$ as $x \to -\infty$.

It can be proved that the laws for rational exponents (Section 3) carry over to any real exponents. Here is a summary.

If a, b, r, and s denote real numbers with $a > 0$ and $b > 0$, then

$$a^r a^s = a^{r+s} \tag{24.3}$$

$$(a^r)^s = a^{rs} \tag{24.4}$$

$$(ab)^r = a^r b^r \tag{24.5}$$

$$\frac{a^r}{b^r} = \left(\frac{a}{b}\right)^r \tag{24.6}$$

$$\frac{a^r}{a^s} = a^{r-s} \tag{24.7}$$

Applications of exponential functions will come later in the chapter.

EXERCISES FOR SECTION 24

For each function in Exercises 1–6, compute decimal values or approximations of f(x) *for* x = 0, $\pm\frac{1}{2}$, ± 1, $\pm\frac{3}{2}$, ± 2, $\pm\frac{5}{2}$, *and* ± 3. *Then draw the graph of each function.* (*If you do not have a calculator with a* $\boxed{y^x}$ *key, use the given approximate square root in each case, along with the ideas in Tables 24.1 and 24.2.*) $\boxed{C}$

1. $f(x) = 5^x$ (Use $\sqrt{5} \approx 2.24$.)

2. $f(x) = 2.5^x$ (Use $\sqrt{2.5} \approx 1.58$.)

3. $f(x) = 3.5^x$ (Use $\sqrt{3.5} \approx 1.87$.)

4. $f(x) = (\frac{1}{5})^x$ (Use $\sqrt{1/5} \approx 0.45$.)

5. $f(x) = 0.4^x$ (Use $\sqrt{0.4} \approx 0.63$.)

6. $f(x) = (\frac{2}{7})^x$ (Use $\sqrt{2/7} \approx 0.53$.)

In Exercises 7–9, assume f(x) = 10,000^x *and compute each functional value.* (*Do not use a calculator.*)

7. $f(3/4)$ **8.** $f(-1/4)$

9. $f(1.5)$

In Exercises 10–18, assume

$$g(x) = 2^x, \qquad h(x) = 3^x, \qquad k(x) = 6^x,$$

rewrite the given expression without using the letters g, h, *and* k, *and simplify.*

10. $h(-3)$ **11.** $k(2) - g(2)$

12. $-h(-2)$ **13.** $(h \circ k)(0)$

14. $(h \circ g)(1)$ **15.** $(g \circ h)(0)$

16. $g(x)h(x)k(-x)$

17. $h(x - y)g(y - x)k(x - y)$

18. $[k(x) - g(x)]/[h(x) - 1]$

19. Draw the graph of $y = 2^{|x|}$.

20. Draw the graph of $y = 2^{-|x|}$.

21. Draw the graph of $y = \frac{1}{2}(2^x + 2^{-x})$.

22. Use Statement (20.2) to find a relationship between the graph of $y = 2^{x+2}$ and the graph of $y = 2^x$. Then find another relationship between the two graphs by using $2^{x+2} = 4 \cdot 2^x$ and the statement at the end of Section 20A.

23. Show that, for all x and all $b > 0$,
$$(b^x + b^{-x})^2 - (b^x - b^{-x})^2 = 4.$$

24. If $f(x) = a^x$, then Equation (24.3) can be written as $f(r)f(s) = f(r + s)$. Rewrite Equations (24.4) and (24.7) using the same idea.

In Exercises 25–27 assume that f(x, y) = xb^y, *where* b > 0 *and* b ≠ 1.

25. Prove that $f(x_1 + x_2, y) = f(x_1, y) + f(x_2, y)$ for all x_1, x_2, and y.

26. Prove that $f(x, y_1 + y_2) = f(f(x, y_1), y_2)$ for all x, y_1, and y_2.

27. Prove that
$$f(x_1 x_2, y_1 + y_2) = f(x_1, y_1) f(x_2, y_2)$$
for all x_1, x_2, y_1, and y_2.

In Exercises 28–30 use a calculator with a $\boxed{y^x}$ *key.* $\boxed{C}$

28. Compute 2^x for $x = 1.4, 1.41, 1.414, 1.4142$, and 1.41421. Is the sequence of answers approaching the calculator's value for $2^{\sqrt{2}}$?

29. Compute x^x for $x = 3.1, 3.14, 3.141, 3.1415$, and 3.14159. Is the sequence of answers approaching the calculator's value for π^π?

30. Verify that $1.5^{1.5} < 2$ and $1.6^{1.6} > 2$. Is $1.55^{1.55} < 2$ or is $1.55^{1.55} > 2$? By repeatedly adjusting the estimate from step to step, find a solution of $x^x = 2$ accurate to the nearest 0.0001.

SECTION **25**

Logarithmic Functions

A. DEFINITION

Let b denote any positive number except 1. We know from Section 24 that the graph of the exponential function $f(x) = b^x$ is either increasing (if $b > 1$) or decreasing (if $b < 1$). Therefore, in either case the function f has an *inverse function* (Section 17).

Definition The inverse of the exponential function $f(x) = b^x$ $(b \neq 1)$ is called the **logarithmic function** with **base** b. If we denote this inverse function by g, then

$$g(y) = x \quad \text{iff} \quad y = f(x). \tag{25.1}$$

The output of this inverse function g for any input element y is denoted by $\log_b y$ (which is read "log to the base b of y"). With this notation, Relation (25.1) becomes

$$\boxed{\log_b y = x \quad \text{iff} \quad y = b^x.} \tag{25.2}$$

Example 25.1 (a) $\log_2 8 = 3$ because $2^3 = 8$.

(b) $\log_5 25 = 2$ because $5^2 = 25$.

(c) $\log_{10} 10000 = 4$ because $10^4 = 10000$.

(d) $\log_5 1 = 0$ because $5^0 = 1$.

(e) $\log_3 (\frac{1}{9}) = -2$ because $3^{-2} = \frac{1}{9}$.

(f) $\log_4 8 = \frac{3}{2}$ because $4^{3/2} = (\sqrt{4})^3 = 8$.

(g) $\log_{0.2} 25 = -2$ because $(0.2)^{-2} = (\frac{1}{5})^{-2} = 5^2 = 25$. ◼

Relation (25.2) allows us to write many equations in either an exponential form or a logarithmic form. Here are some illustrations.

Example 25.2 Solve each equation for x by using the logarithmic form.

(a) $6^x = 3$ (b) $b = a^{2x}$ (c) $\sqrt{2} = 3 \cdot 5^x$

Solution (a) The logarithmic form gives the solution:
$$x = \log_6 3.$$

(b) The logarithmic form of $b = a^{2x}$ is $\log_a b = 2x$. Therefore, $x = \frac{1}{2} \log_a b$.

(c) To use Relation (25.2) we first write the given equation as $\sqrt{2}/3 = 5^x$. This gives $x = \log_5(\sqrt{2}/3)$. ◼

Example 25.3 Solve each equation for x by using the exponential form.

(a) $\log_2 x = \sqrt{2}$ (b) $u = \log_b(x/2)$ (c) $4 \log_3 x = 5$

Solution (a) The exponential form gives the solution:
$$x = 2^{\sqrt{2}}.$$

(b) The exponential form of the equation is $x/2 = b^u$. Therefore, $x = 2b^u$.

(c) To use Relation (25.2) we first rewrite the given equation as $\log_3 x = \frac{5}{4}$. This gives
$$x = 3^{5/4} = 3^{1.25}.$$ ◼

B. GRAPHS

We can describe the graph of any logarithmic function by using the information about the graphs of exponential functions from Section 24 along with what we know about the graphs of inverse functions from Section 17. Specifically, we know that if g is the inverse of a function f, then the graph of g can be obtained from the graph of f by reflection through the line $y = x$. Figure 25.1 shows the result for a typical case with $b > 1$.

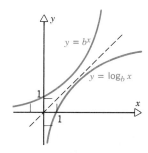

FIGURE 25.1

Example 25.4 Draw the graphs of $y = 4^x$ and $y = \log_4 x$ on the same Cartesian plane.

Solution Figure 25.2 shows the solution. The graph of $y = 4^x$ has been taken from Figure 24.4. The graph of $y = \log_4 x$ results from the graph of $y = 4^x$ by reflection through the line $y = x$. ◾

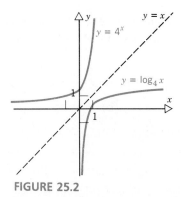

FIGURE 25.2

The graphs of the logarithmic functions for different values of b will furnish convincing evidence of the following properties. For the case $b > 1$, see Figure 25.1.

Let $f(x) = \log_b x$ be a logarithmic function. The *domain* of f is the set of all positive real numbers. The *range* of f is the set of all real numbers. If $b > 1$, then

f is increasing,

$f(x) \to \infty$ as $x \to \infty$, and

$f(x) \to -\infty$ as $x \to 0$.

If $0 < b < 1$, then

> f is decreasing,
> $f(x) \to -\infty$ as $x \to \infty$, and
> $f(x) \to \infty$ as $x \to 0$.

Other properties of logarithmic functions will be considered in Section 26.

EXERCISES FOR SECTION 25

Determine each of the following logarithms.

1. $\log_2 2$

2. $\log_3 9$

3. $\log_4 4$

4. $\log_2 16$

5. $\log_3 3$

6. $\log_5 125$

7. $\log_5 \left(\frac{1}{25}\right)$

8. $\log_7 \left(\frac{1}{49}\right)$

9. $\log_8 2$

10. $\log_{100} 10$

11. $\log_6 1$

12. $\log_3 \left(\frac{1}{27}\right)$

13. $\log_{10} 1$

14. $\log_{0.1} 100$

15. $\log_{0.5} 16$

16. $\log_9 27$

17. $\log_{25} 125$

18. $\log_4 1$

19. $\log_{0.2} 125$

20. $\log_{25} 5$

21. $\log_{16} 8$

Solve each equation for x by using the logarithmic form.

22. $2^x = 3$

23. $u^x = v$

24. $5^x = 10$

25. $4^{3x} = 2$

26. $3 = \pi^{2x}$

27. $y^{4x} = z$

28. $a = b \cdot c^x$

29. $2 \cdot 5^x = 4$

30. $6 = 2 \cdot \pi^x$

Solve each equation for x by using the exponential form.

31. $\log_3 x = \pi$

32. $\log_5 x = -2$

33. $\log_{10} x = \sqrt{2}$

34. $y = \log_4 3x$

35. $2y = \log_3 2x$

36. $v = \log_b cx$

37. $5 \log_a x = 20$

38. $4 = 3 \log_b x$

39. $y = 3 \log_2 x$

40. Draw the graphs of $y = 2^x$ and $y = \log_2 x$ on the same Cartesian plane.

41. Draw the graphs of $y = \left(\frac{1}{3}\right)^x$ and $y = \log_{1/3} x$ on the same Cartesian plane.

42. Draw the graphs of $y = 5^x$ and $y = \log_5 x$ on the same Cartesian plane.

43. Draw the graph of $y = \log_4 |x|$, for x both positive and negative.

44. Draw the graph of $y = |\log_4 x|$.

45. Draw the graphs of $y = \log_2 x$, $y = \log_2(x + 1)$, and $y = (\log_2 x) + 1$ on the same Cartesian plane. (Use the largest possible domain in each case.)

Determine the domain of each function.

46. $f(x) = \log_2(2x - 3)$

47. $f(x) = \log_3 |x - 1|$

48. $f(x) = \log_{10}(x^2 - 1)$

49. Assume that $0 < b < 1$ and that $f(x) = b^x$ has a graph as shown in Figure 24.7. Draw the graph of $y = \log_b x$.

50. Assume that $0 < b < 1$.

  (a) For which values of x is $\log_b x > 0$?

  (b) For which values of x is $\log_b x < 0$?

51. Assume that $b > 1$.

  (a) For which values of x is $\log_b x > 0$?

  (b) For which values of x is $\log_b x < 0$?

For Exercises 52–54, remember that $\log_b$ x is an increasing function of x if b > 1, and a decreasing function of x if 0 < b < 1.

52. Solve $3 < \log_2 x < 5$.

53. Solve $-\frac{1}{2} < \log_9 x < \frac{1}{2}$.

54. Solve $\frac{1}{2} < \log_{1/3} x < 2$.

SECTION **26**
Properties of Logarithmic Functions

A. BASIC PROPERTIES

Laws of logarithms

Each of the following equations is valid for all real numbers b, x, y, and r for which both sides are defined. Thus the only restrictions are $b > 0$, $b \neq 1$, $x > 0$, and $y > 0$.

$$\log_b b = 1 \tag{26.1}$$

$$\log_b 1 = 0 \tag{26.2}$$

$$\log_b xy = \log_b x + \log_b y \tag{26.3}$$

$$\log_b \left(\frac{x}{y}\right) = \log_b x - \log_b y \tag{26.4}$$

$$\log_b x^r = r \cdot \log_b x \tag{26.5}$$

Proof Each law can be proved by using an appropriate law of exponents together with the defining condition

$$\log_b x = y \quad \text{iff} \quad b^y = x. \tag{26.6}$$

Following are proofs of Laws (26.1), (26.2), and (26.3). The proofs of (26.4) and (26.5) will be left as exercises. In (26.5) $\log_b x^r$ means $\log_b (x^r)$, *not* $(\log_b x)^r$.

Law (26.1) is simply the logarithmic form of $b^1 = b$. Law (26.2) is the logarithmic form of $b^0 = 1$.

To prove Law (26.3), let

$$u = \log_b x \quad \text{and} \quad v = \log_b y.$$

Then

$$b^u = x \quad \text{and} \quad b^v = y.$$

Therefore, by a law of exponents [Equation (24.3)],

$$xy = b^u b^v = b^{u+v}.$$

In logarithmic form, this is

$$\log_b xy = u + v = \log_b x + \log_b y,$$

as required. □

The next two laws merely restate that the logarithmic function with base b

is the inverse of the exponential function with base b. (See the proof that follows the laws.)

$$\log_b b^x = x \qquad (26.7)$$

$$b^{\log_b x} = x \qquad (26.8)$$

Proof Let $f(x) = b^x$ and $g(x) = \log_b x$. Then

$$g(f(x)) = x,$$

so

$$g(b^x) = x \quad \text{and} \quad \log_b b^x = x,$$

which is Equation (26.7). Also,

$$f(g(x)) = x,$$

so

$$f(\log_b x) = x \quad \text{and} \quad b^{\log_b x} = x,$$

which is Equation (26.8). □

B. EXAMPLES

Example 26.1 Express each logarithm in terms of $\log_b x$, $\log_b y$, and $\log_b z$.

(a) $\log_b(1/x)$ (b) $\log_b \sqrt{x}$
(c) $\log_b(x^2 y/z)$ (d) $\log_b \sqrt[3]{yz^4/x}$

Solution (a) $\log_b(1/x) = \log_b x^{-1}$
$$= -\log_b x \qquad \text{by (26.5) with } r = -1$$
(b) $\log_b \sqrt{x} = \log_b x^{1/2}$
$$= (\tfrac{1}{2}) \log_b x \qquad \text{by (26.5) with } r = \tfrac{1}{2}$$
(c) $\log_b(x^2 y/z) = \log_b x^2 y - \log_b z \qquad \text{by (26.4)}$
$$= \log_b x^2 + \log_b y - \log_b z \qquad \text{by (26.3)}$$
$$= 2 \cdot \log_b x + \log_b y - \log_b z \qquad \text{by (26.5)}$$
(d) $\log_b \sqrt[3]{yz^4/x} = \log_b(yz^4/x)^{1/3}$
$$= \tfrac{1}{3} \log_b(yz^4/x) \qquad \text{by (26.5) with } r = \tfrac{1}{3}$$
$$= \tfrac{1}{3}(\log_b y + 4 \cdot \log_b z - \log_b x) \qquad \text{by (26.3), (26.4), and} \\ \text{(26.5)} \quad ■$$

Example 26.2 Rewrite each expression as a single logarithm.

(a) $\log_b u - \log_b v$ (b) $\log_b u + 3 \cdot \log_b v$
(c) $\tfrac{1}{2}(\log_b u^2 + 6 \cdot \log_b v - 4 \cdot \log_b u)$ (d) $\log_b 4 + \log_b 3 - \log_b 6$

Solution (a) $\log_b u - \log_b v = \log_b(u/v) \qquad \text{by (26.4)}$

(b) $\log_b u + 3 \cdot \log_b v = \log_b u + \log_b v^3$ by (26.5)
$$= \log_b uv^3 \quad \text{by (26.3)}$$

(c) $\frac{1}{2} (\log_b u^2 + 6 \cdot \log_b v - 4 \cdot \log_b u)$

$\quad = \frac{1}{2} (\log_b u^2 + \log_b v^6 - \log_b u^4)$ by (26.5)

$\quad = \frac{1}{2} \log_b(u^2 v^6/u^4)$ by (26.3) and (26.4)

$\quad = \frac{1}{2} \log_b(v^6/u^2)$ by a law of exponents

$\quad = \log_b(v^6/u^2)^{1/2}$ by (26.5)

$\quad = \log_b(v^3/u)$ by laws of exponents

(d) $\log_b 4 + \log_b 3 - \log_b 6 = \log_b \left(4 \cdot \dfrac{3}{6} \right)$ by (26.3) and (26.4)

$$= \log_b 2$$

Example 26.3 Determine each logarithm that follows, given that $\log_7 3 = 0.5646$ and $\log_7 5 = 0.8271$.

    (a) $\log_7 15$ (b) $\log_7 25$ (c) $\log_7 \sqrt{3}$ (d) $\log_7 21$

Solution   (a) $\log_7 15 = \log_7 3 \cdot 5 = \log_7 3 + \log_7 5 = 0.5646 + 0.8271 = 1.3917$

    (b) $\log_7 25 = \log_7 5^2 = 2 \log_7 5 = 2(0.8271) = 1.6542$

    (c) $\log_7 \sqrt{3} = \log_7 3^{1/2} = \frac{1}{2} \log_7 3 = \frac{1}{2}(0.5646) = 0.2823$

    (d) $\log_7 21 = \log_7 3 \cdot 7 = \log_7 3 + \log_7 7 = 0.5646 + 1 = 1.5646$

Example 26.4 Simplify.

    (a) $\log_2 2^5 = 5$ by (26.7)

    (b) $\log_{10} 10^{-\pi} = -\pi$ by (26.7)

    (c) $3^{\log_3 7} = 7$ by (26.8)

    (d) $10^{\log_{10} \sqrt{2}} = \sqrt{2}$ by (26.8)

C. CHANGE OF BASE

The next two equations allow us to change from one logarithmic base to another.

$$\log_a c = (\log_a b)(\log_b c) \tag{26.9}$$

$$\log_a b = \frac{1}{\log_b a} \tag{26.10}$$

To verify Equation (26.9), let

$$u = \log_a b \quad \text{and} \quad v = \log_b c.$$

Then

$$a^u = b \quad \text{and} \quad b^v = c.$$

Therefore,

$$c = b^v = (a^u)^v = a^{uv},$$

so

$$\log_a c = uv = (\log_a b)(\log_b c).$$

Equation (26.10) is the special case of (26.9) with $c = a$ (Exercise 54).

Equation (26.9) shows that there is a close relationship between any two logarithmic functions. If

$$f(x) = \log_a x \quad \text{and} \quad g(x) = \log_b x,$$

for example, then by Equation (26.9) (with x in place of c)

$$f(x) = (\log_a b)g(x) \text{ for all } x$$

or

$$\log_a x = (\log_a b) \log_b x \text{ for all } x > 0. \qquad (26.11)$$

The number $\log_a b$ is a constant that depends only on a and b.

Example 26.5 Determine a constant k such that

$$\log_2 x = k \cdot \log_8 x \text{ for all } x.$$

Solution By Equation (26.11) the proper value of k is $\log_2 8 = 3$. That is,

$$\log_2 x = 3 \cdot \log_8 x \text{ for all } x.$$

EXERCISES FOR SECTION 26

Express each logarithm in terms of $\log_b$ x, $\log_b$ y, and $\log_b$ z.

1. $\log_b(xy/z)$

2. $\log_b(x/yz)$

3. $\log_b(1/xy)$

4. $\log_b \sqrt[3]{xy}$

5. $\log_b x\sqrt{z}$

6. $\log_b \sqrt{xyz}$

7. $\log_b(\sqrt{x/yz})^3$

8. $\log_b \sqrt{xy}/z$

9. $\log_b \sqrt[3]{x} \sqrt[4]{y} \sqrt[5]{z}$

10. $\log_b \sqrt[4]{x^2y^4}/ \sqrt{x}$

11. $\log_b \sqrt{x/(yz)^3}$

12. $\log_b x\sqrt[3]{1/(x^6y)}$

Rewrite each expression as a single logarithm and simplify.

13. $\log_b u^2 + \log_b v^2$

14. $\log_b v - 2 \cdot \log_b u$

15. $2 \cdot \log_b u - \log_b u^2v$

16. $\frac{1}{2} (\log_b u^4 - \log_b v^2)$

17. $\frac{1}{3} (9 \cdot \log_b v + 3 \cdot \log_b u)$

18. $\frac{1}{2} (\log_b u + 2 \cdot \log_b v + 3 \cdot \log_b u)$

19. $\log_b 5 + \log_b 12 - \log_b 15$

20. $\log_b 1 + \log_b 6 - \log_b 12$

21. $\log_b 0.2 + \log_b 5$

22. $2 \cdot \log_b b - \log_b b^3 - \log_b(1/b)$

23. $\log_b \sqrt{b} + 3 \cdot \log_b \sqrt{b}$

24. $\log_b \sqrt[3]{b} + \frac{1}{3} \log_b b^2$

Determine the logarithms in Exercises 25–33 given that $\log_7 2 = 0.3562$, $\log_7 3 = 0.5646$, and $\log_7 5 = 0.8271$.

25. $\log_7 6$

26. $\log_7 10$

27. $\log_7 9$

28. $\log_7 \sqrt[3]{2}$

29. $\log_7 \sqrt{2}$

30. $\log_7 \sqrt[3]{5}$

31. $\log_7 \dfrac{1}{8}$

32. $\log_7 \dfrac{1}{50}$

33. $\log_7 \dfrac{5}{6}$

Simplify.

34. $\log_5 5^{-2}$

35. $2^{\log_2 3}$

36. $\log_{10} 10^{12}$

37. $7^{\log_7 \pi}$

38. $\log_3 3^{-0.5}$

39. $10^{\log_{10} 1.73}$

40. $a^{-\log_a \pi}$

41. $\log_a a^{-x}$

42. $a^{2\log_a x}$

43. $\log_a x + \log_a(1/x)$

44. $b^{(\log_b a)(\log_a x)}$

45. $\log_a(x + \sqrt{x^2 - 1}) + \log_a(x - \sqrt{x^2 - 1})$

46. Determine a constant k such that $\log_{100} x = k \cdot \log_{10} x$ for all x.

47. Determine a constant k such that $\log_{0.1} x = k \cdot \log_{10} x$ for all x.

48. Determine a constant k such that $\log_8 x = k \cdot \log_{16} x$ for all x.

49. Solve for x if $\log_{10}(\log_5(\log_2 x)) = 0$.

50. Solve for x if $\log_2(\log_x 3) = 1$.

51. Prove that if both a and b are positive and different from 1, then $\log_a(1/b) = 1/\log_b(1/a)$.

52. Prove Law (26.4).

53. Prove Law (26.5).

54. Use Equation (26.9) to verify that Equation (26.10) is true.

SECTION 27

Common Logarithms. Exponential and Logarithmic Equations

A. DEFINITION AND EXAMPLES

Although any positive number except 1 can be used as the base for a logarithmic function, two particular bases are of paramount importance. One is the number 10, which will be used in this section. The other is the number e, which will be defined and used in the section that follows. Base 10 logarithms are important because our number system is based on 10. The importance of base e logarithms lies deeper; the number e has the same kind of intrinsic significance in mathematics as the number π.

A logarithm with base 10 is called a **common logarithm.** These logarithms will be written with the base omitted. That is:

> In this book log x will mean $\log_{10} x$.

(This is a widely used convention. On the other hand, in higher mathematics the notation log x is often reserved for $\log_e x$.) From the definition of logarithm, it follows that

> $\log x = y$ iff $10^y = x$.

Therefore, log x is the power to which 10 must be raised to produce x.

Example 27.1   (a) $\log 10 = 1$ because $10^1 = 10$.

(b) $\log 100{,}000 = 5$ because $10^5 = 100{,}000$.

(c) $\log 0.001 = -3$ because $10^{-3} = 0.001$.

(d) $\log 10^n = n$ because $10^n = 10^n$.

If x is not a readily recognizable power of 10, then more advanced ideas are needed to compute log x. Fortunately, these ideas have been programmed into calculators and used to construct logarithm tables. The easiest way to determine log x is with a calculator having a [log] key: simply enter x, press [log], and read log x from the display. For example, a calculator with an eight-digit display will give log 497.2 = 2.6965311. Values of logarithms from either calculators or tables are nearly always approximate. Thus we should write log 497.2 ≈ 2.6965311, but we follow the more customary practice of using an equality sign.

B. COMMON LOGARITHMS FROM TABLES

This subsection explains how to determine values of log x using a table. *If you will be determining common logarithms with a calculator, rather than with a table, then you may skip to Subsection C.*

Table 27.1 is a condensed version of Table II at the back of the book. The values in these tables have been rounded off to four significant figures.

TABLE 27.1   Common logarithms to four significant figures (also see Table II)

x	log x	x	log x
1.0	0.0000	5.5	0.7404
1.5	0.1761	6.0	0.7782
2.0	0.3010	6.5	0.8129
2.5	0.3979	7.0	0.8451
3.0	0.4771	7.5	0.8751
3.5	0.5441	8.0	0.9031
4.0	0.6021	8.5	0.9294
4.5	0.6532	9.0	0.9542
5.0	0.6990	9.5	0.9777

With the equation log 10^n = n [Example 27.1(d)] and the property log xy = log x + log y, we can determine the common logarithm of any positive number if we know the common logarithms of the numbers between 1 and 10. Here is how: First write x in scientific notation (Section 2B).

$$x = a \times 10^n, \text{ with } 1 \le a < 10.$$

Then Law (26.3) implies that

$$\log x = \log a + \log 10^n,$$

so

$$\log x = n + \log a.$$

That is:

If $x = a \times 10^n$, with $1 \le a < 10$, then

$$\log x = n + \log a. \tag{27.1}$$

The number n in Equation (27.1) is called the **characteristic** of log x, and log a is called the **mantissa.** By the definition of scientific notation, n is the number of places the decimal point in x must be moved to bring it just to the right of the leftmost nonzero digit, counting movement to the left as positive and to the right as negative. For example, $4735 = 4.735 \times 10^3$, so the characteristic of log 4735 is 3.

Example 27.2 Use Table 27.1 to determine each logarithm.

(a) log 450   (b) log 0.0007

Solution (a) Because $450 = 4.5 \times 10^2$, the characteristic is 2. From Table 27.1, the mantissa is 0.6532. Therefore,

$$\log 450 = 2 + \log 4.5 = 2 + 0.6532 = 2.6532.$$

(b) The characteristic of log 0.0007 is -4 and (from Table 27.1) the mantissa is 0.8451. Therefore,

$$\log 0.0007 = -4 + \log 7 = -4 + 0.8451 = -3.1549.$$ ∎

Example 27.3 Use Table II to determine each logarithm.

(a) log 57100   (b) log 0.0369

Solution (a) The characteristic is 4 and (from Table II) the mantissa is 0.7566. Therefore,

$$\log 57100 = 4.7566.$$

(b) The characteristic is -2 and (from Table II) the mantissa is 0.5670. Therefore,

$$\log 0.0369 = -2 + 0.5670 = -1.4330.$$

It is important to realize that -1 is not the characteristic and 0.4330 is not the mantissa. ∎

To determine a logarithm on a calculator with a ⌐log⌐ key, the characteristic and mantissa do not have to be treated separately. Simply enter the given number (such as 0.0369), press ⌐log⌐, and then read the logarithm in the display. For log 0.0369 a calculator with an eight-digit display will show -1.4329736. Ⓒ

C. EXPONENTIAL EQUATIONS

An **exponential equation** is an equation with an unknown variable in an exponent, such as $2^x = 5$. Many exponential equations can be solved using logarithms and the property

$$\log y^r = r \cdot \log y. \tag{27.2}$$

[This is Law (26.5) with $b = 10$ and y in place of x.]

Example 27.4 Solve the equation $2^x = 5$.

Solution If $2^x = 5$, then log 2^x = log 5. Therefore, since log $2^x = x \log 2$,

$$x \log 2 = \log 5$$

$$x = \frac{\log 5}{\log 2}.$$

For an approximate solution we can use a calculator or the values for log 5 and log 2 from Table 27.1:

$$x = \frac{\log 5}{\log 2} = \frac{0.6990}{0.3010} \approx 2.32.$$

$\boxed{C}$

Alternate The logarithmic form of $2^x = 5$ is $\log_2 5 = x$. Thus $\log_2 5$ is a solution. To find
Solution an approximate decimal solution, we use Equation (26.9) in the form

$$\log_b c = \frac{\log_a c}{\log_a b},$$

with $a = 10$, $b = 2$, and $c = 5$. This gives

$$\log_2 5 = \frac{\log_{10} 5}{\log_{10} 2} = \frac{\log 5}{\log 2} \approx 2.32,$$

as in the first solution. ■

The main idea in the next example is the same as in the first solution of Example 27.4: take the logarithm of both sides of the equation, and then use Law (27.2) to remove x as an exponent and obtain a linear equation in x. Other properties of logarithms are then used to reduce the equation to the form $ax = b$.

Example 27.5 Solve the equation

$$5^{2x-1} = 3^{1-x}.$$

Solution If $5^{2x-1} = 3^{1-x}$, then

$$\log 5^{2x-1} = \log 3^{1-x}$$

$$(2x - 1) \log 5 = (1 - x) \log 3 \qquad \text{by (27.2)}$$

$$2x \cdot \log 5 - \log 5 = \log 3 - x \cdot \log 3$$

$$(2 \cdot \log 5 + \log 3)x = \log 3 + \log 5$$

$$(\log 5^2 + \log 3)x = \log (3 \cdot 5) \qquad \text{by (27.2) and (26.3)}$$

$$x \cdot \log (5^2 \cdot 3) = \log 15 \qquad \text{by (26.3)}$$

$$x = \frac{\log 15}{\log 75}.$$

The solution $(\log 15)/(\log 75)$ is exact. For an approximate solution use Table 27.1, Table II, or a calculator:

$$x = \frac{\log 15}{\log 75} = \frac{1.1761}{1.8751} \approx 0.627.$$

$\boxed{C}$ ■

D. LOGARITHMIC EQUATIONS

A **logarithmic equation** is an equation with an unknown variable in a logarithm. If a logarithmic equation has the form $\log_b f(x) = c$, try to solve it by using the exponential form $f(x) = b^c$.

Example 27.6 Solve $\log(2x - 1) = 3$.

Solution Changing to the exponential form, we have

$$2x - 1 = 10^3$$

$$2x = 1001$$

$$x = 1001/2 = 500.5.$$

The answer can be checked by substitution. ◼

If the equation is not given in the form $\log_b f(x) = c$, we may be able to convert it to that form using the laws of logarithms, as in the next example.

Example 27.7 Solve $\log_2 (5 - x) + \log_2 (5 + x) = 4$.

Solution
$$\log_2 [(5 - x)(5 + x)] = 4 \qquad \text{by (26.3)}$$

$$\log_2 (25 - x^2) = 4$$

$$25 - x^2 = 2^4$$

$$x^2 = 9$$

$$x = \pm 3$$

Both answers can be checked by substitution. (Other logarithmic equations may have extraneous solutions, so the answers should always be checked. See Exercise 21, for example.) ◼

Some logarithmic equations cannot be solved explicitly for x. For example, the equation $\log x = -x$ is equivalent to $10^{-x} = x$, which can be solved only by approximation. (See Exercise 45.)

E. APPLICATION: COMPOUND INTEREST

In Section 2C we derived the following equation for the compound amount S from a principal amount P invested at an annual interest rate r compounded n times per year for t years:

$$S = P \left(1 + \frac{r}{n}\right)^{nt}. \tag{27.3}$$

Example 27.8 Solve Equation (27.3) for t.

Solution
$$\frac{S}{P} = \left(1 + \frac{r}{n}\right)^{nt}$$

$$\log \left(\frac{S}{P}\right) = \log \left(1 + \frac{r}{n}\right)^{nt}$$

$$\log \left(\frac{S}{P}\right) = nt \log \left(1 + \frac{r}{n}\right)$$

$$t = \frac{\log \left(\frac{S}{P}\right)}{n \log \left(1 + \frac{r}{n}\right)} \tag{27.4}$$

◼

Example 27.9 Determine the minimum number of whole years that $1000 must be invested at 6% compounded annually so that the compound amount will be $2000.

Solution Use $S = \$2000$, $P = \$1000$, $r = 0.06$, and $n = 1$ in Equation (27.4):

$$t = \frac{\log(2000/1000)}{\log 1.06} = \frac{\log 2}{\log 1.06} = \frac{0.3010}{0.0253} \approx 11.9.$$

Therefore, the minimum number of whole years is 12.

Notice that in Equation (27.4) only the *ratio* of S to P matters, not the specific values of S and P. For instance, the answer to Example 27.9 would be the same for any amounts S and P such that $S/P = 2$. In other words, Example 27.9 shows that it takes money approximately 12 years to double at 6% compounded annually. This is a special case of the following rule of thumb.

Rule of 72

> If a quantity is increasing at a constant rate of $p\%$ per year, then the quantity will double in approximately $72/p$ years.

Here p has been used in place of r to denote rate in order to stress that p is not to be changed to decimal form, in contrast to r in Equations (27.3) and (27.4). For example, if the rate is 6% then we use $p = 6$, *not* $p = 0.06$.

Although the value $72/p$ is only approximate, it is reasonably close for the rates that arise in most interest rate problems. For example, if $5 \le p \le 12$, then the Rule of 72 will give the doubling time with at most a 2% error. Thus money invested at 12% compounded annually will double in approximately $72/12 = 6$ years, and 6 years is off by at most 2%, that is, by at most $0.02 \times 6 = 0.12$ years.

For the rates of increase that occur in most population problems (which we shall consider shortly) a closer approximation is given by using 70 in place of 72; the result is called the **Rule of 70.** The Rule of 70 will give the doubling time with at most a 2% error if $0.1 \le p \le 5$.

In Example 27.8 we proved that Equation (27.4) is correct. In contrast, the Rule of 72 will not be proved. However, in the next example and the exercises the reliability of the rule will be verified in special cases.

Example 27.10 Suppose that money is invested in an account that pays 9% interest compounded annually.

(a) Compute the time required for doubling by using Equation (27.4).

(b) Compute the time required for doubling by using the Rule of 72.

(c) Compute the percentage error from using the Rule of 72.

Solution (a) Use Equation (27.4) with $S/P = 2$, $r = 0.09$, and $n = 1$:

$$t = \frac{\log 2}{\log 1.09} = \frac{0.3010}{0.0374} \approx 8.04 \text{ years.}$$

(b) The Rule of 72 gives doubling in $72/9 = 8$ years.

(c) The *percentage error* is $100 \cdot \left| \dfrac{\text{error}}{\text{correct value}} \right|$. In this case the correct value is 8.04 [from part (a)], and the error is $8.04 - 8 = 0.04$. Thus the percentage error is

$$100 \cdot \left| \frac{0.04}{8.04} \right| \approx 0.50\%,$$

or approximately one-half of one percent.

F. APPLICATION: POPULATION GROWTH

Compound interest on money is just one example of **exponential growth** or **decay,** which occurs whenever a quantity is increasing or decreasing at a constant percentage rate. The basic ideas and formulas are similar in all examples of exponential growth or decay—the notation and the specific numbers and units are all that change from one example to another. We illustrate this now with population growth.

Suppose that P represents the population of a country at a certain time, and that the population is increasing at a constant rate r per period of time (such as a year). Then the population at the end of the first period, which we denote by P_1, will be

$$P_1 = P + Pr = P(1 + r).$$

The population at the end of the second period will be

$$P_2 = P_1 + P_1 r = P(1 + r) + P(1 + r)r$$

$$= P(1 + r)(1 + r) = P(1 + r)^2.$$

If you compare this with the derivation of Equation (2.11) [which is the same as Equation (27.3)], you will see that after t periods the population will be

$$\boxed{P_t = P(1 + r)^t.}$$

(27.5)

Example 27.11 The 1970 population of America (both continents and the adjacent areas) was 509 million. Between 1970 and 1980, the population increased at an average rate of 1.9% per year. *If* this trend remained constant, what would the population be in the year 2000?

Solution Use Equation (27.5) with $P = 509$, $r = 0.019$, and $t = 30$:

$$P_{30} = 509(1.019)^{30} \approx 895.$$

That is, in the year 2000 the population would be approximately 895 million.

Example 27.12 The population of the U.S. was 76.0 million in 1900 and 92.0 million in 1910. If the population had continued to grow at the same percentage rate for each 10-year period, what would the population have been in 1980?

Solution The increase from 1900 to 1910 was $92.0 - 76.0 = 16.0$ million. Since $16.0/76.0 \approx 0.21$, this means that the increase between 1900 and 1910 was approximately 21%.

Now use Equation (27.5) with $P = 76.0$, $r = 0.21$, and $t = 8$ (each time period is 10 years). The answer will be in millions:

$$P_8 = 76.0(1.21)^8 \approx 349 \text{ million.}$$ [C]

In contrast, the actual population in 1980 was only 227 million. (See Figure 27.1, where the projections are based on the percentage increase from 1900 to 1910, which was used in this example.)

The U.S. population increased by at least 25% over each decade from 1800 to 1890, but the increase has been less than 20% for each decade since 1910. This brings out the extremely important point that a past rate of growth will not necessarily continue into the future. ■

Problems in which t must be determined from Equation (27.5) can be solved just like Examples 27.8 and 27.9.

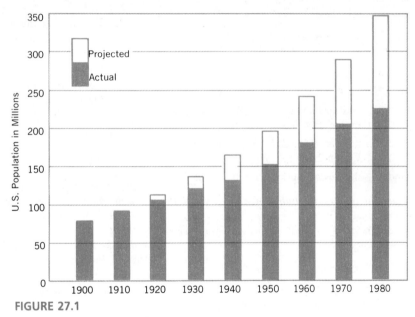

FIGURE 27.1

EXERCISES FOR SECTION 27

Use Table II to determine the common logarithm of each number.

1. (a) 8.51   (b) 851   (c) 0.0851

2. (a) 6.7   (b) 0.00067   (c) 67

3. (a) 3.4   (b) 3400   (c) 0.34

4. (a) 570   (b) 5700   (c) 0.57

5. (a) 16200   (b) 1.620   (c) 0.0162

6. (a) 0.000222   (b) 222   (c) 22200

7. Use a calculator to determine the common log-arithms of the numbers in part (a) in each of Exercises 1–6. [C]

8. Use a calculator to determine the common logarithms of the numbers in part (b) in each of Exercises 1–6. [C]

9. Use a calculator to determine the common logarithms of the numbers in part (c) in each of Exercises 1–6. [C]

Solve each equation for x. Give both an exact answer (involving logarithms) and an approximate decimal answer. (*Table 27.1 will provide the logarithms you need.*)

10. $3^x = 2$

11. $5^x = 7$

12. $2^x = 10$

13. $4^{3x-1} = 5^x$

14. $9^{x-1} = 2^{x+1}$

15. $7^{2x} = 5^{x-3}$

16. $5 = 2 \cdot 3^{2x}$

17. $11 = 2 \cdot 5^{-x}$

18. $3 = 2 \cdot 4^{5x}$

Solve for x.

19. $\log(x^2 + 19) = 2$

20. $\log_3(x^2 - 9) - \log_3(x + 3) = 2$

21. $\log_5 x + \log_5(x - 24) = 2$

22. $\log_7(x^3 - 1) = 1 + \log_7(x - 1)$

23. $\log|x| = 2$

24. $\log(x^2 - x + 1) + \log(x + 1) = 1$

25. $\log_{1/2} \sqrt{x} - \log_{1/2} x = 4$

26. $\log(\log x) = 1$

27. $\log(x^{\log x}) = 4$

28. Find all solutions of $2^x < 3$. (Suggestion: If $2^x < 3$, then $\log 2^x < \log 3$. Why?)

29. Find all solutions of $\log 2x < 3$. (Suggestion: If $\log 2x < 3$, then $2x < 10^3$. Why?)

30. Find all solutions of $\log(\log x) < 0$.

Determine the minimum number of whole years that the given principal P *must be invested at the given rate* r *compounded annually to produce the given compound amount* S.

31. $P = \$1000, r = 6\%, S = \$1500$

32. $P = \$1000, r = 8\%, S = \$1800$

33. $P = \$2000, r = 10\%, S = \$4500$

34. $P = \$500, r = 6.5\%, S = \$950$

35. $P = \$1500, r = 7\%, S = \$2500$

36. $P = \$1000, r = 5\%, S = \$3000$

In each of Exercises 37–42, assume that money is invested in an account that pays the given rate of interest compounded annually. Then do each of the following.

(a) Compute the time required for doubling by using Equation (27.4).

(b) Compute the time required for doubling by using the Rule of 72.

(c) Compute the percentage error from using the Rule of 72.

37. 6%

38. 8%

39. 12%

40. 1%

41. 24%

42. 18%   C

43. Find the minimum number of whole years that an amount must be invested at 6% compounded annually so that it will triple in value.

44. Formulate a statement (like the Rule of 72) that will give, approximately, the number of years required for an amount to increase to eight times its original value if it is increasing at a constant rate of $p\%$ per year. (Assume $5 \le p \le 12$; see the paragraph following the Rule of 72.)

45. The equation $\log x = -x$ has exactly one solution, which is the x-coordinate of the point of intersection of the graphs of $y = \log x$ and $y = -x$.

(a) By drawing the graphs of $y = \log x$ and $y = -x$ on the same Cartesian plane, show that the solution of $\log x = -x$ is between 0 and 1.

(b) Verify that $\log 0.40 > -0.40$ and that $\log 0.39 < -0.39$. What does this imply about the solution of $\log x = -x$?

(c) Find the solution of $\log x = -x$ to the nearest 0.001 by refining the results in part (b).   C

Exercises 46–57 are based on Table 27.2, which gives population figures for the six major areas of the world used in U.N. statistical summaries. Notice that the rates of increase are annual rates. In each computation, take 1980 as year 0 and assume that the annual rate of increase shown in the table will continue

TABLE 27.2 Estimates of mid-year population in millions for major areas of the world

Area	1980	1985	Annual rate of increase 1980–1985
Africa	491	566	2.9
America	616	674	1.8
Asia	2593	2831	1.8
Europe	484	492	.3
Oceania	23	24	1.5
USSR	266	278	.9
World[a]	4474	4865	1.7

[a]The slight discrepancy in the totals is due to rounding.
Source: 1986 Statistical Abstract of the United States.

throughout the period being considered. (*In most cases this will probably give a population estimate that is slightly high.*)   C

46. Estimate the population of Africa in the year 1995.

47. Estimate the population of America in the year 1995.

48. Estimate the population of Europe in the year 2000.

49. When will the population of Asia reach 3500 million?

50. When will the population of Oceania reach 26 million?

51. When will the population of the USSR reach 320 million?

52. By the Rule of 70, in how many years will the population of Africa double?

53. By the Rule of 70, in how many years will the population of America double?

54. By the Rule of 70, in how many years will the population of the USSR double?

55. Estimate the world population in the year 2000.

56. When will the world population reach 5.5 billion?

57. (a) In how many years will the world population double according to Equation (27.4)?
(b) In how many years will the world population double according to the Rule of 70?
(c) What do you get if you use the Rule of 72 rather than the Rule of 70 in part (b)?

SECTION **28**
Applications Involving *e*

A. THE NUMBER *e*

The small letter *e* is used to denote a special irrational number that is approximately equal to 2.71828. This number, like π, arises naturally throughout mathematics and its applications. The following example contains a more precise definition of *e*.

Example 28.1 Suppose that $1 is invested at 100% interest compounded *n* times per year for one year. The amount accumulated is given by either Equation (2.11) or Equation (27.3): the interest rate (in decimal form) per conversion period is $1/n$, so the compound amount, in dollars, is

$$S = 1\left(1 + \frac{1}{n}\right)^n = \left(1 + \frac{1}{n}\right)^n.$$

Table 28.1 gives some specific examples. The table shows, for instance, that

TABLE 28.1

Compounding interval	n	$[1 + (1/n)]^n$ (to five decimal places)
annually	1	2
semiannually	2	2.25
quarterly	4	2.44141
monthly	12	2.61304
daily	365	2.71457
hourly	8,760	2.71813
each minute	525,600	2.71828
each second	31,536,000	2.71828

quarterly compounding would yield $2.44. Compounding either hourly or each minute or each second would yield $2.72 (rounded to the nearest cent). (A calculator with a $\boxed{y^x}$ key can be used to verify most of the entries in the table.)

It can be proved that the values of $[1 + (1/n)]^n$ approach an irrational number as n increases without bound. This number is called the *limit of* $[1 + (1/n)]^n$ *as n approaches infinity,* and it is the number e. Concisely,

$$e = \lim_{n \to \infty} \left(1 + \frac{1}{n} \right)^n. \tag{28.1}$$

To 20 decimal places,

$$e = 2.71828\ 18284\ 59045\ 23536.$$

Besides giving a definition of e, this example illustrates a general point about compound interest: For a fixed principal and rate of interest, an increase in the frequency of compounding will increase the compound amount; but beyond a certain point this increase in compound amount will be negligible. In **continuous compounding** the compound amount is the limit of the compound amounts as $n \to \infty$, where n is the number of compounding periods per year. Continuous compounding might make an investor feel better, but it will not make him or her significantly richer than daily compounding. The next subsection will give a more general formula for continuous compounding.

Any precise definition of e must involve some kind of limit, as in Equation (28.1). Such limits are studied in more detail in calculus.

A logarithm with base e is called a **natural logarithm.** We use the widely adopted notation $\ln x$ for these logarithms. That is:

> In this book $\ln x$ will mean $\log_e x$.

It follows that

> $\ln x = y$ iff $e^y = x$.

To determine a natural logarithm on a calculator with an $\boxed{\ln}$ key, simply enter the given number, press $\boxed{\ln}$, and then read the logarithm in the display. For $\ln 0.064$ a calculator with an eight-digit display will show -2.7488722. Subsection D explains how to find natural logarithms with tables.

B. MORE ABOUT CONTINUOUS COMPOUNDING

Example 28.1 shows that if $1 is invested at 100% interest compounded continuously for one year, then the compound amount will be e (rounded off). We now generalize this example to derive a formula for the compound amount from continuous compounding for any amount invested at any annual rate for any number of years.

First, from either Equation (2.11) or Equation (27.3), we know that if a principal P is invested at an interest rate r compounded n times per year for t

years, then the compound amount is

$$S = P\left(1 + \frac{r}{n}\right)^{nt}. \tag{28.2}$$

We can analyze the last equation more easily if we let $x = n/r$. Then $r/n = 1/x$. Also, $n = xr$ so that $nt = xrt$. Thus Equation (28.2) can be written as

$$S = P\left(1 + \frac{1}{x}\right)^{xrt}$$

$$S = P\left[\left(1 + \frac{1}{x}\right)^x\right]^{rt}. \tag{28.3}$$

For continuous compounding we let the number n of conversion periods per year increase without bound. This will not change P, r, or t. However, $x \to \infty$ as $n \to \infty$, and by Equation (28.1) (with x in place of n)

$$e = \lim_{x \to \infty}\left(1 + \frac{1}{x}\right)^x.$$

Thus the right-hand side of Equation (28.3) approaches Pe^{rt} as $x \to \infty$. This leads to the following conclusion.

> If a principal amount P is invested for t years at an annual rate r compounded continuously, then the compound amount will be
>
> $$S = Pe^{rt}. \tag{28.4}$$

As usual, the rate r must be in decimal form in this equation. The special case of Equation (28.4) with $P = 1$, $t = 1$, and $r = 1$ is Example 28.1.

Powers of e needed to apply Equation (28.4) can be obtained from Table III or from a calculator having an $\boxed{e^x}$ key, a $\boxed{y^x}$ key, or $\boxed{\text{inv}}$ and $\boxed{\ln}$ keys.

Example 28.2 What is the compound amount if \$500 is invested for 2 years at 12% compounded continuously?

Solution Use Equation (28.4) with $P = \$500$, $r = 0.12$, and $t = 2$. A calculator with an $\boxed{e^x}$ key will give

$$S = 500e^{0.24}$$

$$S = \$635.62$$

Alternatively, Table III will give $e^{0.24} = 1.2712$, from which $S = 500 \times 1.2712 = \$635.60$. In Example 2.11 we solved this problem with quarterly compounding in place of continuous compounding. The compound amount under quarterly compounding was \$633.39. ◼

C. APPLICATION: CONTINUOUS GROWTH AND DECAY

In Section 27 we used population growth to illustrate that the basic ideas and formulas for compound interest carry over to any example of exponential growth or decay, that is, to any example where a quantity is increasing or decreasing at a constant percentage rate. In the same way, the ideas and formulas for *continuous* compounding carry over to other examples; this produces what is called **continuous growth** or **decay.** The basic formulas for continuous growth and decay are

$$Q = Q_0 e^{kt} \text{ (growth)} \tag{28.5}$$

and

$$Q = Q_0 e^{-kt} \text{ (decay)}. \tag{28.6}$$

In these equations Q represents a quantity that changes with time, k is a positive constant that depends on the particular problem, t denotes time expressed in the appropriate units, and Q_0 is the **initial value** of Q, that is, the value of Q when $t = 0$.

Example 28.3 If a principal amount P of money is invested for t years at an annual rate r compounded continuously, then the compound amount will be

$$S = Pe^{rt}.$$

This equation is just Equation (28.4). It is the special case of Equation (28.5) with S in place of Q, r in place of k, and P in place of Q_0. ■

The decay of any radioactive substance is described by Equation (28.6) with the constant k depending on the particular substance. An especially useful number connected with a radioactive substance is the **half-life** of the substance, which is the time required for an amount of the substance to be reduced by half. The following example will show how the half-life of a substance is related to the constant k in Equation (28.6).

Example 28.4 Suppose that the amount Q of a radioactive substance remaining after t units of time is determined by

$$Q = Q_0 e^{-kt}.$$

Let T denote the half-life of the substance. Then $Q = \frac{1}{2} Q_0$ when $t = T$. Therefore,

$$\frac{1}{2} Q_0 = Q_0 e^{-kT}$$

$$\frac{1}{2} = e^{-kT}.$$

The logarithmic form of the last equation is

$$\ln \frac{1}{2} = -kT.$$

Since $\ln \frac{1}{2} = \ln 2^{-1} = -\ln 2$, we see that $-\ln 2 = -kT$, so that

$$T = \frac{\ln 2}{k}. \tag{28.7}$$

Also, $k = (\ln 2)/T$, so the equation describing radioactive decay can be written as

$$Q = Q_0 e^{-[(\ln 2)/T]t}, \tag{28.8}$$

where T denotes the half-life of the substance. Since $e^{\ln 2} = 2$ [Equation (26.8)], (28.8) can be rewritten as

$$Q = Q_0(e^{\ln 2})^{-t/T} \qquad (28.9)$$
$$Q = Q_0\, 2^{-t/T}. \qquad \blacksquare$$

Example 28.5 The half-life of carbon-14 is 5730 years. Answer each question based on an initial amount of 1 gram of the substance.

    (a) How much will remain after two half-lives?

    (b) How much will remain after three half-lives?

    (c) How much will remain after 1000 years?

    (d) When will 0.9 grams remain?

Solution (a) After one half-life $\frac{1}{2}$ gram will remain. After another half-life $\frac{1}{2}$ of $\frac{1}{2}$, or $\frac{1}{4}$ gram, will remain.

    (b) Continuing as in part (a), we see that after three half-lives $\frac{1}{2}$ of $\frac{1}{4}$, or $\frac{1}{8}$ gram, will remain. (Notice that three half-lives is more than 17 thousand years.)

    (c) The amount Q in this example is given by Equation (28.8) with $T = 5730$ years and $Q_0 = 1$ gram:

$$Q = e^{-[(\ln 2)/5730]t} \qquad (28.10)$$

We are asked for Q when $t = 1000$. A calculator with an $\boxed{e^x}$ key will give

$$Q = e^{-[(\ln 2)/5730]1000}$$
$$\approx 0.886 \text{ grams.}$$

    (d) To determine when 0.9 grams will remain, let $Q = 0.9$ in Equation (28.10)

$$0.9 = e^{-[(\ln 2)/5730]t}.$$

We must solve this equation for t. The logarithmic form of the equation is

$$-[(\ln 2)/5730]t = \ln 0.9.$$

Thus

$$t = -5730(\ln 0.9)/(\ln 2).$$

With a calculator you can verify that this gives

$$t \approx 871 \text{ years.} \qquad \boxed{C}$$

(The decaying property of carbon-14 is critical in carbon-dating, which is used to determine the age of once-living objects found at archaeological sites.) $\blacksquare$

    In Section 27 we used the equation

$$P_t = P(1 + r)^t$$

to study population growth [Equation (27.5)]. This equation is essentially the same as the formula for interest compounded over fixed time intervals. We can also study population growth by using Equation (28.5), the equation for continuous growth,

$$Q = Q_0 e^{kt}.$$

Here is an illustration.

Example 28.6 The world population was estimated to be 4.1 billion in 1975 and 4.5 billion in 1980. Assume that after 1975 world population has grown and will continue to grow according to Equation (28.5). Compute an estimate for the world population in the year 2000.

Solution Use the equation

$$Q = Q_0 e^{kt}$$

with $t = 0$ in 1975 and Q measured in billions and t in years. Then $Q_0 = 4.1$, and $Q = 4.5$ when $t = 5$ (that is, in 1980). Therefore,

$$4.5 = 4.1 e^{5k}$$

$$4.5/4.1 = e^{5k}.$$

The logarithmic form of this equation is

$$5k = \ln(4.5/4.1),$$

so that

$$k = \frac{1}{5} \ln (4.5/4.1).$$

This gives

$$k \approx 0.0186. \qquad \boxed{C}$$

Therefore, our equation for estimating population becomes

$$Q = 4.1 e^{0.0186t}, \qquad (28.11)$$

where t denotes years measured from 1975. For the year 2000 we use $t = 25$. The result is

$$Q = 4.1 e^{25(0.0186)}$$

$$Q \approx 6.5 \text{ billion (using Table III).}$$

Remember that this estimate is based on the assumptions that the world population will continue to increase between 1980 and 2000 according to Equation (28.5) and at the same rate that it increased between 1975 and 1980. Such assumptions are safe only for short-range projections. The difficulty of making projections about population growth is illustrated by the fact that the U.S. population growth rate dropped from an average of 1.46 percent per year over 1960–65 to an average of 1.06 percent per year over 1975–80. ∎

D. MORE ON NATURAL LOGARITHMS

We now consider how to compute ln x using tables. One way is to use a table of common logarithms and the equation for changing from one logarithmic base to another. Specifically, use $a = 10$ and $b = e$ in

$$\log_b x = \frac{\log_a x}{\log_a b}$$

[from Equation (26.11)]. The result is

$$\log_e x = \frac{\log_{10} x}{\log_{10} e}$$

or

$$\ln x = \frac{\log x}{\log e}.$$

It can be shown that $\log e = 0.4343$ and $1/\log e = 2.3026$ (both to four decimal places). Therefore

$$\boxed{\ln x = 2.3026 \log x.} \tag{28.12}$$

Example 28.7 Use Table II and Equation (28.12) to compute $\ln 5130$.

Solution Using Table II, we find that $\log 5130 = 3.7101$. Therefore, by Equation (28.12),

$$\ln 5130 = 2.3026 \log 5130$$
$$= 2.3026 \times 3.7101$$
$$= 8.5429. \qquad \boxed{C} \ \blacksquare$$

A second way to compute $\ln x$ is to use Table IV, which gives the natural logarithms of numbers between 1 and 10. First, write x in scientific notation,

$$x = a \times 10^n, \ 1 \leq a < 10.$$

Then

$$\ln x = \ln a + \ln 10^n$$
$$\ln x = n \ln 10 + \ln a.$$

Since

$$\ln 10 = \log_e 10 = \frac{\log_{10} 10}{\log_{10} e} = \frac{1}{\log_{10} e} = 2.3026,$$

we have the following conclusion.

$$\boxed{\begin{array}{c} \text{If } x = a \times 10^n, \text{ with } 1 \leq a < 10, \text{ then} \\ \ln x = 2.3026n + \ln a. \end{array}} \tag{28.13}$$

Notice that this is similar to Equation (27.1), which states that $\log x = n + \log a$, except that in the present case $2.3026n$ appears in place of n.

Example 28.8 Use Table IV and Equation (28.13) to compute $\ln 0.064$.

Solution First, write $0.064 = 6.4 \times 10^{-2}$. From Table IV, $\ln 6.4 = 1.8563$. Therefore,

$$\ln 0.064 = 2.3026(-2) + 1.8563$$
$$\ln 0.064 = -2.7489.$$

EXERCISES FOR SECTION 28

For Exercises 1–3, use Example 28.1, including Table 28.1.

1. What is the compound amount if $10 is invested for 1 year at 100% compounded quarterly? monthly? daily?

2. What is the compound amount if $100 is invested for 1 year at 100% compounded semiannually? hourly? each second?

3. What is the compound amount if $1000 is invested for 1 year at 100% compounded monthly? daily? each minute?

Determine the compound amount for the given principal invested for the given time period at the given annual rate compounded continuously.

4. $1000 for 3 years at 6%.

5. $5000 for 5 years at 10%.

6. $4000 for 2 years at 8%.

7. $200 for 6 months at 10%.

8. $3000 for 18 months at 7%.

9. $5000 for 30 months at 12%.

Compute each logarithm in Exercises 10–18 by three methods:
(a) using Table II and Equation (28.12),
(b) using Table IV and Equation (28.13), and
(c) using a calculator with an $\boxed{\ln}$ *key.* $\boxed{C}$

10. $\ln 470$
11. $\ln 92$
12. $\ln 4500$
13. $\ln 0.04$
14. $\ln 0.61$
15. $\ln 0.00035$
16. $\ln(34 \times 10^5)$
17. $\ln(0.07 \times 10^{-2})$
18. $\ln(550 \times 10^{-3})$

19. The half-life of krypton-85 is 11 years. Write an equation that gives the amount of the substance after t years if the initial amount is 0.1 grams.

20. The half-life of radon-222 is 3.8 days. Write an equation that gives the amount of the substance after t days if the initial amount is 10^{-5} grams.

21. The half-life of plutonium-239 is 2.4×10^4 years.

Write an equation that gives the amount of the substance after t years if the initial amount is 0.001 grams.

22. The half-life of radium-226 is 1600 years. Answer each question based on an initial amount of 2 grams.

    (a) How much will remain after three half-lives?

    (b) How much will remain after 500 years?

    (c) When will 1.5 grams remain?

23. The half-life of strontium-90 is 29 years. Answer each question based on an initial amount of 1 milligram.

    (a) How much will remain after four half-lives?

    (b) How much will remain after 12 years?

    (c) When will 0.7 milligrams remain?

24. The half-life of polonium-210 is 139 days. Answer each question based on an initial amount of 1 gram.

    (a) How much will remain after two half-lives?

    (b) How much will remain after 200 days?

    (c) When will 0.2 grams remain?

Use the growth equation (28.11) in Example 28.6 to estimate the world population in each of the following years.

25. 1990
26. 2005
27. 2010

28. Estimate the population in 20 years for a country whose population is currently 10 million and growing 1.5% annually. (Use a continuous growth equation.)

29. Estimate the population in 15 years for a city whose population is currently 500,000 and growing 3% annually. (Use a continuous growth equation.)

30. At its present rate of growth the population of a city will double in the next 20 years. Estimate the population 8 years from now if the popu-

lation is currently 200,000. [Begin by determining the growth equation (28.5) based on the data given.]

31. Solve Equation (28.5) for t.

32. Prove that if the growth of a quantity is given by Equation (28.5) and Q has the values Q_1 and Q_2 at times t_1 and t_2, respectively, then

$$k = \frac{1}{t_1 - t_2} \ln \left(\frac{Q_1}{Q_2} \right).$$

33. Prove that

if $I = \dfrac{E}{R} (1 - e^{-Rt/L})$,

then $t = -\dfrac{L}{R} \ln \left(1 - \dfrac{IR}{E} \right)$.

(The first equation expresses the electric current I in a series circuit in terms of the voltage E, the resistance R, the inductance L, and the time t, when all are expressed in standard units.)

34. Prove that if a quantity grows according to Equation (28.5), then the time required for it to double is $(\ln 2)/k$.

35. The graph of an equation of the form

$$y = \frac{a}{2} (e^{x/a} + e^{-x/a}), \qquad (28.14)$$

where $a > 0$, is called a **catenary.** Draw the graph of Equation (28.14) for $a = 1$ and $-3 \le x \le 3$. (This graph has the shape of a homogeneous cable suspended by its ends and subject only to the force of its own weight; the value of a is determined by the position and physical properties of the cable.) $\boxed{C}$

36. The Gateway Arch in St. Louis, designed by Eero Saarinen, is 630 feet high and has a span that is also 630 feet. With x and y both measured in feet, its shape is given by the equation

$$y = 757.7 - \frac{127.7}{2} (e^{x/127.7} + e^{-x/127.7}),$$

$$-315 \le x \le 315.$$

Draw the graph. (The Gateway Arch is sometimes described as being parabolic, but it is actually an inverted catenary. See Exercise 35 for the definition of a catenary.) $\boxed{C}$

REVIEW EXERCISES FOR CHAPTER V

Determine each of the following logarithms without using a calculator or tables.

1. $\log_3 27$

2. $\log_6 \dfrac{1}{36}$

3. $\log_{100} 1000$

4. $\log_{0.4} 2.5$

Solve for x.

5. $\log_7 x = 2$

6. $\log_{0.2} x = -1$

7. $\log_{15} 3x = 2$

8. $a \log_b cx = d$

Express in terms of $\log_b$ x, $\log_b$ y, and $\log_b$ z.

9. $\log_b(xz/y)$

10. $\log_b \sqrt[5]{yz}$

Rewrite as a single logarithm and simplify.

11. $2 \log_b uv - \log_b(u^3/v)$

12. $\log_b 25 - \log_b \dfrac{125}{3} + \dfrac{1}{2} \log_b \dfrac{125}{3}$

Solve each equation for x. Give both an exact answer (involving logarithms) and an approximate decimal answer.

13. $5^x = 4$

14. $3^{2x} = 0.5$

15. $5^x = 3^{2+x}$

16. $4 \cdot 7^{-x} = 5$

Solve each equation for x.

17. $\log(x^2 + x + 44) = 2$

18. $\log(\ln x) = 1$

Compute each logarithm by three methods:
(a) using Table II and Equation (28.12),
(b) using Table IV and Equation (28.13), and
(c) using a calculator with an $\boxed{\ln}$ key. $\boxed{C}$

19. $\ln 31$

20. $\ln 360$

21. Show that if $\frac{1}{2}Q_0 = Q_0 e^{-kT}$, then $T = (\ln 2)/k$.

22. Simplify $(2^{\log_2 3})^3/(3^{\log_3 2})^{-2}$.

23. Suppose $f(x) = 2 \ln x$ and $g(x) = e^{3x+1}$. Find and simplify each of the following.
(a) $(f \circ g)(x)$ (b) $g^{-1}(x)$

24. Find all of the points of intersection of the graphs of $y = 4$ and $y = e^{|x|}$.

25. If the y-intercept of $3x - 2y - \ln k = 0$ is greater than 2 but less than 5, what are the possible values of k?

26. Suppose $f(x) = 2 - |\log(|x| - 1)|$.

(a) Find the domain of f.

(b) Find the x-intercepts of the graph of f.

27. On a single Cartesian plane draw the graphs of $y = \log x$, $y = \log(x - 1)$, and $y = \log[1/(x - 1)]$. (Notice that $\log[1/(x - 1)] = -\log(x - 1)$.)

28. Find all of the real zeros of
$$(x - 1)e^{x-2} \log(x - 3).$$

29. Suppose $f(x) = (e^x + e^{-x})/(e^x - e^{-x})$ and $g(x) = 2/(e^x - e^{-x})$. Show that $[f(x)]^2 - [g(x)]^2 = 1$.

30. The world population in 1985 was approximately 4.87 billion. Write an expression that will give the approximate world population in 1995 if the population were to increase 1.8% per year from 1985 through 1995. If you have a calculator, compute a decimal approximation. Use a continuous growth equation.

31. Assume that money is invested in an account that pays 10% interest compounded annually.

(a) Compute the time required for doubling by using Equation (27.4).

(b) Compute the time required for doubling by using the Rule of 72.

(c) Compute the percentage error from using the Rule of 72.

32. Determine the compound amount if $1000 is invested for 4 years at an annual rate of 6% compounded continuously.

33. Show that if
$$x = \frac{e^y - e^{-y}}{2},$$
then
$$y = \ln(x + \sqrt{x^2 + 1}).$$

[Begin by multiplying both sides of the first equation by $2e^y$. The result can be viewed as a quadratic equation with e^y as the variable; solve for e^y and then convert to logarithmic form to get y.]

CHAPTER VI

INTRODUCTION TO TRIGONOMETRY

At the heart of trigonometry are six special functions called *trigonometric functions*. These functions first arose from questions about relationships between the sides and angles of triangles. From this viewpoint, the trigonometric functions have applications from surveying to engineering to astronomy. The functions have many other uses that have no apparent connection with triangles, however, such as applications to the study of electricity, sound waves, and mechanical vibrations.

In this chapter—after a preliminary section on angles—we introduce the trigonometric functions in terms of angles and apply them to the study of right triangles. In the following chapters we study the functions in terms suitable for other types of applications.

SECTION 29
Angles

A. STANDARD POSITION. DEGREES

A **ray** consists of a point P on a line together with all of the points on one side of P on the line. The point P is called the **endpoint** of the ray [Figure 29.1(*a*)]. In geometry, an **angle** is defined as a union of two rays with a common endpoint. The rays are called the **sides** of the angle, and the common endpoint is called the **vertex** [Figure 29.1(*b*)].

P
Ray with endpoint P

(a)

P Side Side
Angle with vertex P

(b)

FIGURE 29.1

In trigonometry, we often distinguish between the sides of an angle by calling one the **initial side** and the other the **terminal side.** We think of the angle as the result of rotating a ray from the initial side to the terminal side with the endpoint of the ray left fixed (Figure 29.2). An angle with initial and terminal sides is said to be in **standard position** relative to a Cartesian coordinate system if the vertex is at the origin and the initial side is along the positive x-axis. If the rotation from the initial to the terminal side is counterclockwise, then the angle is said to be **positive.** If the rotation is clockwise, then the angle is said to be **negative** (Figure 29.3).

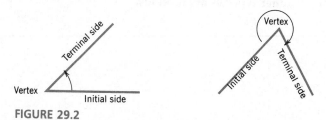

Terminal side
Vertex
Initial side

FIGURE 29.2

Vertex
Initial side
Terminal side

The units most commonly used to measure angles are *degrees* (defined here) and *radians* (defined in the next subsection). A **degree** is the measure of $\frac{1}{360}$th of the angle formed when a ray is rotated through one complete revolution. For angles in standard position, the degree measure is positive or negative depending on whether the angle is positive or negative. Thus one complete counterclockwise revolution produces an angle of $360°$ (360 degrees). One-fourth of a complete clockwise revolution produces an angle of $-90°$. An angle of $0°$ results from no rotation. There is no restriction on the sizes of degree measures for angles: for example, three complete clockwise revolutions produce an angle of $-1080°$.

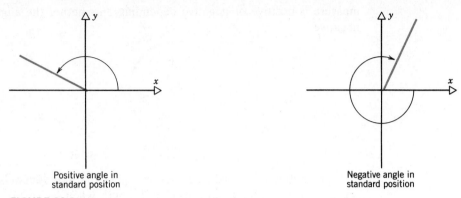

Positive angle in
standard position

Negative angle in
standard position

FIGURE 29.3

Two or more angles in standard position are said to be **coterminal** if their terminal sides coincide. Figure 29.4 shows three pairs of coterminal angles. Two angles in standard position will be coterminal iff their measures differ by an integral multiple of 360°. For example, angles of $-30°$ and 690° are coterminal because $690° - (-30°) = 720° = 2 \cdot 360°$. It is important to realize that coterminal angles are not *equal* unless their measures are equal: for example, $0° \neq 360°$.

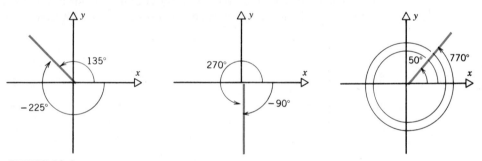

FIGURE 29.4

An angle in standard position is called a **first quadrant angle** if its terminal side lies in the first quadrant. Similarly for the other quadrants. If the terminal side lies along a coordinate axis, the angle is called a **quadrantal angle.**

One-sixtieth ($\frac{1}{60}$th) of one degree is one **minute** (1'). One-sixtieth ($\frac{1}{60}$th) of one minute is one **second** (1"). Parts of degrees can be expressed with decimals as well as with minutes and seconds. For example, a calculator will use 23.5° rather than 23°30'. In this book the use of minutes and seconds will be minimal.

B. RADIANS

To define the measure of an angle in *radians,* place the vertex of the angle at the center of a circle of radius 1 unit. As a ray rotates from one side of the angle to the other, the point of intersection of the ray with the circle moves some distance θ (theta) along the circle. This distance θ is defined to be the **radian measure** of the angle (Figure 29.5). For angles in standard position, the radian measure is positive or negative depending on whether the angle is positive or negative.

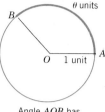

Angle AOB has
radian measure θ.   **FIGURE 29.5**

The definition of radian measure implies that **one radian** is the measure of an angle that subtends an arc of unit length when its vertex is placed at the center of a circle with unit radius. And, since the circumference of a circle with unit radius is 2π units ($C = 2\pi r$ with $r = 1$), we see that one complete revolution produces an angle with radian measure 2π. It follows that 2π radians $= 360°$, so that

$$\pi \text{ radians } = 180°. \tag{29.1}$$

For reasons that will become clear in the next subsection, we generally do not indicate the units when writing the radian measure of an angle. If θ denotes the measure of an angle, then θ (without units) means θ radians. For θ degrees we write $\theta°$. Figure 29.6 shows examples of angles with their radian measures.

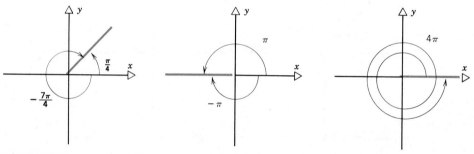

FIGURE 29.6

Equation (29.1) implies that to convert from radians to degrees we must multiply by $180°/\pi$. And to convert from degrees to radians we must multiply by $\pi/180°$. In particular,

$$1 \text{ radian } = \left(\frac{180}{\pi}\right)° \approx 57.3° \quad \text{and} \quad 1° = \left(\frac{\pi}{180}\right) \approx 0.0175 \text{ radian.} \quad \boxed{\text{C}}$$

Example 29.1 Convert from radians to degrees.

(a) $\dfrac{\pi}{4} = \dfrac{\pi}{4} \cdot \dfrac{180°}{\pi} = 45°$

(b) $-\dfrac{3\pi}{2} = -\dfrac{3\pi}{2} \cdot \dfrac{180°}{\pi} = -270°$

(c) $3 = 3 \cdot \dfrac{180°}{\pi} = \left(\dfrac{540}{\pi}\right)° \approx 171.9°$, or $3 \approx 3(57.3)° = 171.9°$

Example 29.2 Convert from degrees to radians.

(a) $75° = 75° \cdot \dfrac{\pi}{180°} = 5\pi/12$

(b) $-540° = -540° \cdot \dfrac{\pi}{180°} = -3\pi$

Remark Many calculators provide three modes for working with angular measure: degrees, radians, and grads (see Exercise 44). *You must pay careful attention to which mode the calculator is in when working with angles and trigonometry.* Read the calculator's instruction booklet for advice.

C. APPLICATION: ARC LENGTH

If the vertex of an angle of measure θ radians is placed at the center of a circle of radius r, and if s denotes the length of the subtended arc, then

$$s = r\theta. \qquad (29.2)$$

Proof Figure 29.7 shows an angle and a circle satisfying the given conditions. The figure shows another circle with the same center and with radius 1. (The figure assumes

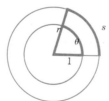

FIGURE 29.7

$r > 1$. The proof is the same if $r \le 1$.) Because ratios between corresponding parts of similar figures are equal, we have

$$\frac{s}{r} = \frac{\theta}{1},$$

which yields (29.2).

Notice that Equation (29.2) implies that radians are unitless, because s and r both denote lengths so that the units balance in the equation without units on θ. (Section 6E discusses units in equations.) That is why we generally do not indicate units when writing the radian measure of an angle. In a sense, radians provide a "natural" measure for angles—there is no arbitrary choice such as that of 360 in the case of degrees.

Equation (29.2) can be used to compute any one of s, r, and θ once the other two are known. In using the equation, remember that θ *must be expressed in radians.* Also, the units must be the same for both s and r.

Example 29.3 An angle subtends an arc of length 6 centimeters when its vertex is at the center of a circle of radius 4 centimeters. Find the measure of the angle in both radians and degrees.

Solution With $s = 6$ and $r = 4$, we have $\theta = s/r = \frac{6}{4} = 1.5$ radians. In degrees, the measure of the angle is $1.5(180/\pi) = 270/\pi \approx 85.9°$. $\boxed{C}$ ■

Example 29.4 The equator crosses the *west* coast of South America near Quito, Ecuador, approximately 80° west of the prime meridian. The equator crosses the *east* coast of South America at the mouth of the Amazon River in Brazil, approximately 50° west of the prime meridian. What is the distance across South America at the equator?

Solution Figure 29.8 represents a cross section of the earth at the equator, viewed from the direction of the North Pole. (The prime meridian connects the North and South Poles and passes through Greenwich, which is a borough of London. Also see Figure 29.9.) The equator is a circle of radius 3960 miles (the radius of the

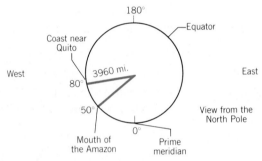

FIGURE 29.8

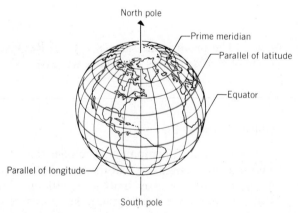

FIGURE 29.9

earth). And we are concerned with a central angle of 30°, the difference between 50° and 80°. To apply $s = r\theta$, first convert 30° to radians: $30(\pi/180) = \pi/6$. This leads to the answer:

$$s = 3960 \times \frac{\pi}{6} = 660\pi \approx 2070 \text{ miles.}$$ $\boxed{C}$ ■

EXERCISES FOR SECTION 29

In each of Exercises 1–15, draw the angle with the given measure in standard position. Also show two other angles, one positive and one negative, that are coterminal with the given angle, and give the measure of each of the two angles.

1. 0°   **2.** 45°

3. 90°   **4.** 225°

5. −180°   **6.** 30°

7. −120°   **8.** 450°

9. −630°   **10.** π/2

11. 3π/4   **12.** −π/2

13. −π/4   **14.** −π/6

15. 2π/3

Convert to degrees, without using minutes in the answer. [For example, 8° 15′ = 8 + ($\frac{15}{60}$)° = 8.25°.]

16. 2° 30′   **17.** −16° 20′

18. 145° 45′

Convert to degrees and minutes. [For example, −20.3° = −20° − (0.3)(60′) = −20°18′.]

19. 14.75°   **20.** 300.7°

21. −82.15°

Convert from radians to degrees.

22. −π/3   **23.** 5π

24. −7π/4   **25.** −1

26. 6   **27.** 11π/6

Convert from degrees to radians.

28. 225°   **29.** 135°

30. 30°   **31.** −54°

32. 1000°   **33.** −(100/π)°

34. Convert −82° 13′ 10″ to degrees, without using minutes and seconds in the final answer. (See Exercises 16–18.) [C]

35. Convert 47.1375° to degrees, minutes, and seconds. (See Exercises 19–21.) [C]

36. Give the degree measure of every positive quadrantal angle less than 750°.

37. Give the radian measure of every negative quadrantal angle greater than −4π.

38. An angle subtends an arc of length 3 inches when its vertex is at the center of a circle of radius 2 inches. Find the measure of the angle in both radians and degrees.

39. The vertex of a 50° angle is placed at the center of a circle of radius 5 centimeters. How long is the subtended arc?

40. A 135° angle subtends a 10 inch arc when the vertex of the angle is at the center of a circle. What is the circle's radius?

41. Rome and Chicago are both approximately 42° north of the equator (that is, at approximately 42° north latitude). How far north is that in miles? [C]

42. The Tropic of Cancer and the Tropic of Capricorn are 23° 27′ north and south of the equator, respectively. How far apart are they in miles? [C]

43. Cape Horn is approximately 2400 miles from the South Pole. What is Cape Horn's latitude? (That is, how many degrees south of the equator is it? The South Pole is 90° south latitude.) [C]

44. A **grad** is a unit of angular measure for an angle whose subtended arc is $\frac{1}{400}$th of a complete circle.

(a) Convert 80 grads to radians.

(b) Convert 10 degrees to grads.

45. A **mil** is a unit of angular measure for an angle whose subtended arc is $\frac{1}{6400}$th of a complete circle.

(a) Convert 1200 mils to degrees.

(b) Convert π/6 radians to mils.

46. Mils (Exercise 45) are used to measure angles in artillery. Here is an application. Suppose that two objects are at the same range (distance from an observer), and that the angle between them, as measured by the observer, is relatively small. Then the distance between the objects is given approximately by the formula

$$W = RM, \qquad (29.3)$$

where W denotes the distance (width) between the objects in meters, R denotes the distance (range) from the observer to the objects in thousands of meters, and M denotes the angle between the objects, in mils, as measured by the observer (Figure 29.10). For example, if two

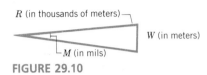

FIGURE 29.10

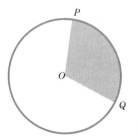

FIGURE 29.11

objects 2000 meters away subtend an angle of 150 mils at an observer, then the objects are approximately 2 · 150 or 300 meters apart. Explain why the approximation given by (29.3) is a consequence of Equation (29.2). [Remember that W is small. Equation (29.3), in the form W/R = M, is called the WORM formula, from Width Over Range equals Mils.]

47. A **nautical mile** is the length of an arc subtended on the earth's surface by an angle of 1′ (one minute) placed at the center of the earth.

   (a) Using 3960 statute (ordinary) miles for the earth's radius, compute the number of statute miles in a nautical mile. $\boxed{\text{C}}$

   (b) The length of a nautical mile used officially in the U.S. is 6076.115 feet. To what value for the earth's radius does this correspond? (Express the answer in statute miles, with 1 statute mile = 5280 feet.) $\boxed{\text{C}}$

48. The shaded area in Figure 29.11 is called a **sector** of a circle. Prove that if the central angle POQ has radian measure θ and the radius of the circle is r, then the area of the sector is $\frac{1}{2}r^2\theta$. (Suggestion: The area of the sector is to the area of the circle as θ is to 2π.)

Exercises 49–51 concern objects that are moving at a constant rate around a circle of radius r. *If* t *denotes time, then* s = rθ *implies* s/t = r(θ/t). *Therefore, if* v *denotes* s/t (*the change in arc length per unit time*) *and* ω *denotes* θ/t (*the change in the central angle, expressed in radians, per unit time*), *then* v = rω. *We call* v *the* **linear speed** *and* ω *the* **angular speed.**

49. A blade of a ceiling fan has diameter 54 inches and is rotating at 50 revolutions per minute. How fast is a point on the end of the blade moving in feet per second? $\boxed{\text{C}}$

50. The earth is moving around the sun in a nearly circular orbit of radius 93,000,000 miles, making one revolution approximately every 365 days. What is the linear speed of the earth in its orbit, in miles per hour? $\boxed{\text{C}}$

51. If the wheels of a car have 14-inch radius, how many revolutions per second will each wheel make when the car is traveling 45 miles per hour? $\boxed{\text{C}}$

SECTION **30**
Trigonometric Functions of Angles

A. DEFINITIONS

As stated in the introduction to this chapter, at the heart of trigonometry are six special functions called *trigonometric functions*. In the next chapter these will be considered as functions of real numbers—that is, the domains will be sets of real numbers. Here we consider them as functions of angles. The connection will be explained in the next chapter.

The **trigonometric functions** of an angle θ are the **sine** (sin θ), **cosine** (cos θ), **tangent** (tan θ), **cotangent** (cot θ), **secant** (sec θ), and **cosecant** (csc θ), defined as follows.

Definition   Given an angle θ, place it in standard position and then choose any point $P(x, y)$, other than the origin, on the terminal side of the angle (Figure 30.1). Let r denote the distance between the origin and $P(x, y)$. Then the trigonometric functions of θ are given by the following equations.

$$\sin \theta = y/r \qquad \csc \theta = r/y$$

$$\cos \theta = x/r \qquad \sec \theta = r/x \qquad (30.1)$$

$$\tan \theta = y/x \qquad \cot \theta = x/y$$

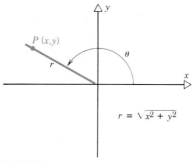

FIGURE 30.1

Since these six functions are to be functions of θ, we must verify that they depend only on θ and are independent of the particular choice made for $P(x, y)$ on the terminal side of θ. To this end, let $P'(x', y')$ be another point on the terminal side of θ. Figure 30.2 shows an example with an angle in the second quadrant. Let $OP = r$ and $OP' = r'$. Triangles OAP and $OA'P'$ are similar, so

$$\frac{AP}{OP} = \frac{A'P'}{OP'} \text{ and } \frac{y}{r} = \frac{y'}{r'}.$$

Therefore, the value obtained for $\sin \theta$ in (30.1) will be the same, whether it is computed with $P(x, y)$ or $P'(x', y')$.

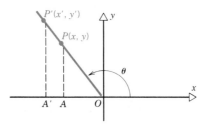

FIGURE 30.2

Similar arguments apply for other functions and for angles in other quadrants. The only difference in other quadrants is that if $y < 0$, for example, then the length of the corresponding side of the triangle will be $|y|$ rather than y. However, $r > 0$ and $r' > 0$ in every case, and $y > 0$ iff $y' > 0$. Thus y/r and

y'/r' are equal not only in absolute value but also in sign. For the functions that involve x a similar argument applies since $x > 0$ iff $x' > 0$.

Example 30.1 Assume that the terminal side of an angle θ in standard position passes through the point $(-3, 4)$. Determine the six trigonometric functions of the angle.

Solution See Figure 30.3. In this case $r = \sqrt{(-3)^2 + 4^2} = 5$.

$$\sin \theta = 4/5 \qquad\qquad \csc \theta = 5/4$$

$$\cos \theta = (-3)/5 = -3/5 \qquad \sec \theta = 5/(-3) = -5/3$$

$$\tan \theta = 4/(-3) = -4/3 \qquad \cot \theta = (-3)/4 = -3/4 \qquad \blacksquare$$

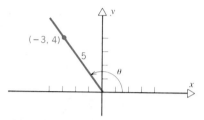

FIGURE 30.3

Angles will generally be denoted by Greek letters. A list of these appears inside the back cover. When no confusion seems likely, we often use the same symbol to denote both an angle and its measure.

B. REMARKS AND MORE EXAMPLES

Since each angle measure determines a unique angle in standard position, the symbol θ in the definitions can be replaced by any angle measure.

Example 30.2 Find the six trigonometric functions of $\pi/2$.

Solution The terminal side of an angle of measure $\pi/2$ in standard position lies along the positive y-axis. We can choose any point along this terminal side and then use (30.1). If we choose the point $(0, 1)$, we have $x = 0$, $y = 1$, and $r = 1$.

$$\sin \frac{\pi}{2} = \frac{1}{1} = 1 \qquad\qquad \csc \frac{\pi}{2} = \frac{1}{1} = 1$$

$$\cos \frac{\pi}{2} = \frac{0}{1} = 0 \qquad\qquad \sec \frac{\pi}{2} = \frac{1}{0} \quad \text{undefined}$$

$$\tan \frac{\pi}{2} = \frac{1}{0} \quad \text{undefined} \qquad \cot \frac{\pi}{2} = \frac{0}{1} = 0 \qquad \blacksquare$$

The cases $\tan \pi/2$ and $\sec \pi/2$ illustrate that a trigonometric function is undefined for any angle that yields a zero denominator. If the terminal side of an angle is along the x-axis, then $y = 0$ and the cosecant and cotangent of the angle are undefined.

The following reciprocal relationships follow directly from the definitions.

For example, $\csc \theta = r/y = 1/(y/r) = 1/\sin \theta$. Each equation is true for every angle θ for which both sides are defined.

$$\csc \theta = \frac{1}{\sin \theta} \qquad \sec \theta = \frac{1}{\cos \theta} \qquad \cot \theta = \frac{1}{\tan \theta} \qquad\qquad (30.2)$$

The sign of a function of an angle is determined by the quadrant of the angle. For any point in the first quadrant, $x > 0$, $y > 0$, and $r > 0$, and thus all six functions of a first quadrant angle are positive. To determine the signs of functions of angles in the other quadrants, we simply observe that $x > 0$ in quadrants I and IV, $x < 0$ in quadrants II and III, $y > 0$ in quadrants I and II, $y < 0$ in quadrants III and IV, and $r > 0$ in every quadrant. The conclusions about sines, cosines, and tangents are shown in Figure 30.4. Since a nonzero number and its reciprocal have the same sign, the conclusions for the other follow from (30.2).

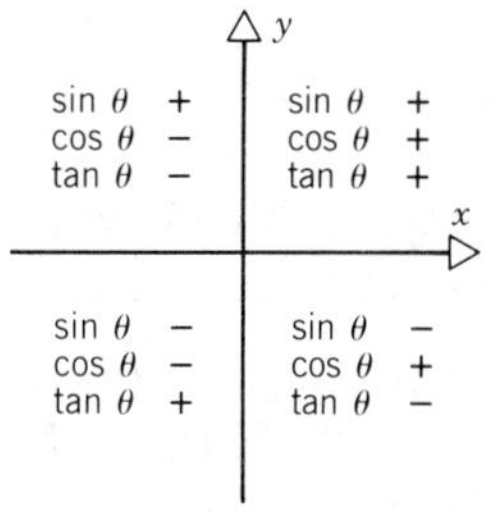

FIGURE 30.4

C. FUNCTIONS FROM RIGHT TRIANGLES

Many applications of trigonometry depend on relationships between the angles and sides of a right triangle. These relationships are simply the trigonometric functions of the acute angles (angles less than 90°) of the triangle. For this reason it is useful to express the functions directly in terms of the sides of the triangle. This is done as follows.

Let ABC be a right triangle with right angle at C, as shown in Figure 30.5. In considering the angle α, the two legs of the triangle are called the side **opposite** α and the side **adjacent** to α, as shown in the figure. Now place a coordinate system with its origin at A and its x-axis along the side AC (Figure 30.6). Then

$$x = \text{length of the side adjacent to } \alpha$$

$$y = \text{length of the side opposite } \alpha$$

$$r = \text{length of the hypotenuse.}$$

In this way we see that the six trigonometric functions of α can be expressed as follows, where, for simplicity, "length of the side adjacent to α" has been shortened to "adjacent," and similarly for the other sides.

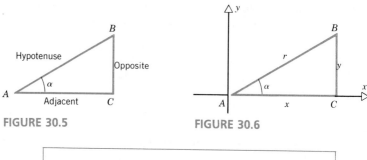

FIGURE 30.5   FIGURE 30.6

$$\sin \alpha = \frac{\text{opposite}}{\text{hypotenuse}} \qquad \csc \alpha = \frac{\text{hypotenuse}}{\text{opposite}}$$

$$\cos \alpha = \frac{\text{adjacent}}{\text{hypotenuse}} \qquad \sec \alpha = \frac{\text{hypotenuse}}{\text{adjacent}} \qquad (30.3)$$

$$\tan \alpha = \frac{\text{opposite}}{\text{adjacent}} \qquad \cot \alpha = \frac{\text{adjacent}}{\text{opposite}}$$

D. MORE EXAMPLES

The next three examples show how to compute the functions of 45°, 30°, and 60° using only elementary geometry.

Example 30.3 To compute the functions of 45° (or, in radians, $\pi/4$) we consider an isosceles right triangle as shown in Figure 30.7. Each of the equal legs has length 1, so the Pythagorean Theorem implies that the hypotenuse has length $\sqrt{2}$. Applying (30.3) to either 45° angle, we get the following results.

$$\sin 45° = \frac{1}{\sqrt{2}} = \frac{\sqrt{2}}{2} \qquad \csc 45° = \frac{\sqrt{2}}{1} = \sqrt{2}$$

$$\cos 45° = \frac{1}{\sqrt{2}} = \frac{\sqrt{2}}{2} \qquad \sec 45° = \frac{\sqrt{2}}{1} = \sqrt{2}$$

$$\tan 45° = \frac{1}{1} = 1 \qquad \cot 45° = \frac{1}{1} = 1 \qquad ▪$$

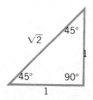

FIGURE 30.7

We can compute the trigonometric functions of 30° and 60° (or, in radians, $\pi/6$ and $\pi/3$) by considering an equilateral triangle, two units on each side (Figure 30.8). Each angle of the triangle is 60°. Therefore, if we draw an altitude of the triangle we divide the triangle into two right triangles whose acute angles are 30° and 60°. The hypotenuse of the right triangle ABD has length 2, and the leg AD has length 1. Thus the Pythagorean Theorem implies that the leg

BD has length $\sqrt{2^2 - 1^2}$, which is $\sqrt{3}$. The essential information from Figure 30.8 is shown in Figure 30.9. This figure should be memorized, because all of the functions of both 30° and 60° can be computed from it.

FIGURE 30.8   FIGURE 30.9

Example 30.4 Figure 30.9 and the relationships in (30.3) yield the following results.

$$\sin 30° = 1/2 \qquad\qquad \csc 30° = 2/1 = 2$$
$$\cos 30° = \sqrt{3}/2 \qquad\qquad \sec 30° = 2/\sqrt{3} = 2\sqrt{3}/3$$
$$\tan 30° = 1/\sqrt{3} = \sqrt{3}/3 \qquad \cot 30° = \sqrt{3}/1 = \sqrt{3}$$

Example 30.5 Figure 30.9 shows that the functions of 60° are as follows.

$$\sin 60° = \sqrt{3}/2 \qquad\qquad \csc 60° = 2/\sqrt{3} = 2\sqrt{3}/3$$
$$\cos 60° = 1/2 \qquad\qquad \sec 60° = 2/1 = 2$$
$$\tan 60° = \sqrt{3}/1 = \sqrt{3} \qquad \cot 60° = 1\sqrt{3} = \sqrt{3}/3$$

Because $30° = \pi/6$, $45° = \pi/4$, and $60° = \pi/3$, Examples 30.3, 30.4, and 30.5 also give the functions of any angle of radian measure $\pi/6$, $\pi/4$, or $\pi/3$.

Example 30.6 (a) $\sin \pi/6 = \sin 30° = \tfrac{1}{2}$

(b) $\tan \pi/4 = \tan 45° = 1$

(c) $\csc \pi/3 = \csc 60° = 2\sqrt{3}/3$

Example 30.7 If θ is an acute angle such that $\sin \theta = \tfrac{1}{5}$, what are the other trigonometric functions of θ?

Solution If θ is thought of as an angle of a right triangle, then the ratio *opposite/hypotenuse* must be $\sin \theta$, which is $\tfrac{1}{5}$. Figure 30.10 shows such a triangle. In this case the side opposite θ has been given length 1, and thus the length of the hypotenuse must be 5. Any other choices could be made for the lengths of these two sides *provided* the ratio is $\tfrac{1}{5}$.

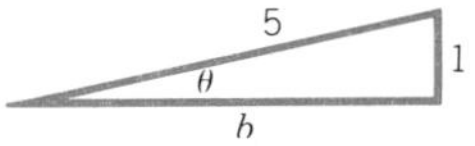

$$b^2 + 1^2 = 5^2$$
$$b = \sqrt{24} = 2\sqrt{6}$$

FIGURE 30.10

The calculation in Figure 30.10 shows that the length of the side adjacent to θ is $2\sqrt{6}$. Thus the functions of θ are as follows.

$$\sin \theta = 1/5 \qquad\qquad \csc \theta = 5/1 = 5$$
$$\cos \theta = 2\sqrt{6}/5 \qquad\qquad \sec \theta = 5/2\sqrt{6}$$
$$\tan \theta = 1/2\sqrt{6} \qquad\qquad\qquad = (5/2\sqrt{6})(\sqrt{6}/\sqrt{6}) = 5\sqrt{6}/12$$
$$= (1/2\sqrt{6})(\sqrt{6}/\sqrt{6}) = \sqrt{6}/12 \quad \cot \theta = 2\sqrt{6}/1 = 2\sqrt{6}$$

E. COFUNCTIONS

It is not accidental that the names of the six trigonometric functions are related in pairs: sine and *co*sine, secant and *co*secant, tangent and *co*tangent. The two functions in each pair are called **cofunctions,** and they satisfy the following relationships. (Recall from geometry that acute angles are **complementary** if the sum of their measures is 90°. In the following statement and its proof we continue our practice of using the same symbol to denote an angle and its measure.)

> Cofunctions of complementary angles are equal. That is, if α and β are acute angles such that $\alpha + \beta = 90°$, then
> $$\sin \alpha = \cos \beta, \quad \sec \alpha = \csc \beta, \quad \text{and} \quad \tan \alpha = \cot \beta.$$
(30.4)

Proof From plane geometry, the sum of the interior angles of a triangle is 180° (or π radians). Thus the sum of the two acute angles of a right triangle is 90°. It follows that α and β are related as in Figure 30.11. Therefore, the relationships from (30.3) can be applied as follows.

$$\sin \alpha = \frac{\text{opposite } \alpha}{\text{hypotenuse}} = \frac{\text{adjacent to } \beta}{\text{hypotenuse}} = \cos \beta$$

$$\sec \alpha = \frac{\text{hypotenuse}}{\text{adjacent to } \alpha} = \frac{\text{hypotenuse}}{\text{opposite } \beta} = \csc \beta$$

$$\tan \alpha = \frac{\text{opposite } \alpha}{\text{adjacent to } \alpha} = \frac{\text{adjacent to } \beta}{\text{opposite } \beta} = \cot \beta \qquad \square$$

FIGURE 30.11

Example 30.8 It can be shown that, to four decimal places, sin 25° = 0.4226, cos 25° = 0.9063, and tan 25° = 0.4663. Determine sin 65°, cos 65°, and cot 65°.

Solution Since 25° + 65° = 90°, the angles 25° and 65° are complementary. Thus we have the following answers.

$$\sin 65° = \cos 25° = 0.9063$$

$$\cos 65° = \sin 25° = 0.4226$$

$$\cot 65° = \tan 25° = 0.4663 \qquad \blacksquare$$

Example 30.9 An angle of 45° is its own complement, because 45° + 45° = 90°. Thus (30.4) implies that sin 45° = cos 45°, sec 45° = csc 45°, and tan 45° = cot 45°. These equations are also a consequence of Example 30.3. $\qquad \blacksquare$

In the next chapter we extend (30.4) and show that cofunctions of two angles are equal whenever the sum of the two angles is 90°, even if one of the angles is negative. For example, tan 150° = cot(−60°) because 150° + (−60°) = 90°.

At this point you should make sure that you know the definitions of the trigonometric functions from (30.1), and the relationships for the functions for acute angles from (30.3). You should also be able to determine quickly from Figures 30.7 and 30.9 the functions of 45°, 30°, and 60°.

EXERCISES FOR SECTION 30

In each of Exercises 1–6, a point is given on the terminal side of an angle in standard position. Find the six trigonometric functions of the angle.

1. (1, 2)   **2.** (5, 12)

3. (4, 3)   **4.** (− 12, 5)

5. (− 2, − 3)   **6.** (15, − 8)

7. Find all θ such that sec θ is undefined and $0 \le \theta < 2\pi$.

8. Find all θ such that tan θ is undefined and $0° \le \theta < 360°$.

9. Find all θ such that cot θ is undefined and $0 \le \theta < 2\pi$.

10. What is the quadrant of θ if sin θ < 0 and tan θ > 0?

11. What is the quadrant of θ if cos θ > 0 and csc θ < 0?

12. What is the quadrant of θ if cot θ < 0 and sec θ < 0?

13. Find the sine and cosecant of 135°. [You can work with the point (− 1, 1). Why?]

14. Find the cosine and secant of 225°. [You can work with the point (− 1, − 1). Why?]

15. Find the tangent and cotangent of 315°. [You can work with the point (1, − 1). Why?]

Give the exact value in each case, as in Examples 30.3–30.5.

16. cos π/4   **17.** sin π/3

18. tan π/6   **19.** csc π/6

20. cot π/4   **21.** sec π/3

22. tan π/3   **23.** sec π/6

24. sin π/4

In Exercises 25–27, find cos α and tan α.

25.   **26.**   **27.**

 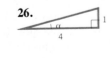 

In Exercises 28–30, find csc β and cot β.

28.   **29.**   **30.**

 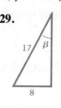 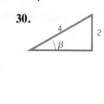

Ⓒ

31.   **32.**   **33.**

 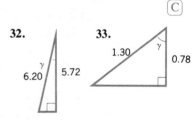

34. Find sin β and cos β if β is an acute angle such that cot β = 0.5.

35. Find sec γ and csc γ if γ is an acute angle such that $\tan \gamma = \dfrac{4}{3}$.

36. Find tan α and cot α if α is an acute angle such that $\csc \alpha = \dfrac{3}{2}$.

Determine the functional values in Exercises 37–45 by using the following functional values and the fact that cofunctions of complementary angles are equal.

sin 20° = 0.3420	sin 70° = 0.9397
tan 20° = 0.3640	tan 70° = 2.7475
sec 30° = 1.1547	sec 60° = 2.0000
cos 30° = 0.8660	cos 60° = 0.5000
cot 40° = 1.1918	cot 50° = 0.8391
csc 40° = 1.5557	csc 50° = 1.3054

37. csc 60°   **38.** sin 30°

39. cot 20°   **40.** tan 40°

41. cot 70°   **42.** csc 30°

43. cos 70°   **44.** sec 50°

45. tan 50°

46. If the terminal side of θ lies in the first quadrant, then csc θ > 0. In which other quadrant(s) is csc θ > 0?

47. Same as Exercise 46 with sec θ in place of csc θ.

48. Same as Exercise 46 with cot θ in place of csc θ.

49. If the terminal side of a first quadrant angle θ lies along the line $y = 2x$, what are cos θ and cot θ?

50. If the terminal side of a second quadrant angle θ lies along the line $y = -3x$, what are sin θ and sec θ?

51. If the terminal side of a third quadrant angle θ lies along the line $y = 2x$, what are csc θ and tan θ?

SECTION **31**
Calculators. Tables. Applications

A. CALCULATORS

To apply trigonometry we need a convenient source for the trigonometric functions of angles given in degrees or radians. In a few cases we can compute these functions directly—this was done for 30°, 45°, and 60° in Section 30. For most angles, however, we rely on either calculators or tables. We first consider calculators.

To find values of trigonometric functions with a scientific calculator, first place the calculator in the appropriate mode, either degrees, radians, or grads. Then, to find sin θ, for example, simply enter θ and press [sin]. The answer should be accurate to within the limits allowed by the display. The same idea applies to cos θ and tan θ, for which we use [cos] and [tan].

To find cotangents, secants, and cosecants we use [tan], [cos], and [sin], together with the reciprocal relationships cot θ = 1/tan θ, sec θ = 1/cos θ, and csc θ = 1/sin θ. The following example gives illustrations. Each computation has been taken from a calculator with a ten-digit display. Because of rounding, the answers may differ slightly from those found using other calculators or tables.

Example 31.1 (a) sin 251° = −0.945518576. The calculator must be in the degree mode.

(b) cos −92° = −0.034899497. Make sure you enter −92, not just 92, before pressing [cos].

(c) csc 68° = $\dfrac{1}{\sin 68°}$ = 1.078534743. With the calculator in the degree mode, enter 68, press [sin], and then press [1/x].

(d) sec 3 = $\dfrac{1}{\cos 3}$ = −1.010108666. Notice that this involves 3 radians, not 3°. Use the radian mode, enter 3, press [cos], and then press [1/x].

(e) tan π/2 is undefined. Some calculators will display ERROR in this case. However, the calculator must use a rational approximation for the irrational number π, so it may give tan θ for θ close to, but unequal to, π/2. One calculator gives −4,878,048,780. An answer like this should be a signal to think about the problem without using a calculator.

(f) cot 5π/3 = $\dfrac{1}{\tan 5\pi/3}$ = −0.577350270. Use the calculator to compute

$5\pi/3$ (approximately), press $\boxed{\text{tan}}$, and then press $\boxed{1/x}$. The exact value of cot $5\pi/3$ is $-\sqrt{3}/3$, or approximately -0.577350269. ▪

B. TABLES

From Table 31.1 we can read all of the functions of the angles at 10° intervals from 10° through 80°. Each entry has been rounded to the four figures shown. Since cofunctions of complementary angles are equal (Section 30E), it suffices to give the functions of 10°, 20°, 30°, and 40°; once we know these we automatically know the functions of their complements, 80°, 70°, 60°, and 50°, respectively. The table is arranged to take advantage of this. To find a function of either 10°, 20°, 30°, or 40°, read the angle from the column on the *left* and read the name of the function from the *top*. To find a function of either 80°, 70°, 60°, or 50°, read the angle from the column on the *right* and the name of the function from the *bottom*.

Example 31.2 To find cos 20° from Table 31.1, look in the row with 20° at the left and the column with cos θ at the top. The result is cos 20° = 0.9397. ▪

TABLE 31.1

θ	sin θ	cos θ	tan θ	cot θ	sec θ	csc θ	
10°	0.1736	0.9848	0.1763	5.671	1.015	5.759	80°
20°	0.3420	0.9397	0.3640	2.747	1.064	2.924	70°
30°	0.5000	0.8660	0.5774	1.732	1.155	2.000	60°
40°	0.6428	0.7660	0.8391	1.192	1.305	1.556	50°
	cos θ	sin θ	cot θ	tan θ	csc θ	sec θ	θ

Example 31.3 To find sec 60° from Table 31.1, look in the row with 60° at the right and the column with sec θ at the bottom. The result is sec 60° = 2.000. Notice that this entry is the same as csc 30°: again, cofunctions (such as secant and cosecant) of complementary angles (such as 60° and 30°) are equal. ▪

Table 31.1 is a condensed version of Table V in Appendix D, which gives the trigonometric functions of angles at 0.1° intervals from 0° through 90°. The following examples will serve as a check on how to read Table V. The idea is the same as that used for Table 31.1.

Example 31.4 Each illustration is from Table V in Appendix D.

(a) sin 37° = 0.6018

(b) cos 47° = 0.6820

(c) tan 2° = 0.0349

(d) cot 65.1° = 0.4642

(e) sec 45.6° = 1.4293

(f) ′csc 21° 24′ = csc 21.4° = 2.7407 ▪

Table VI in Appendix D gives functions of angles expressed in radians. It will be used in the next chapter.

C. REFERENCE ANGLES

This subsection shows how to express the functions of any angle in terms of the functions of an acute angle. This idea is essential if functions are to be found for nonacute angles using a table of functions of acute angles. It is also important to understand the idea even if you find functional values using a calculator. This subsection will use the facts from Section 30B about the signs of the functions for angles from different quadrants.

If θ is any angle, then the **reference angle** θ_r of θ is the smallest positive angle formed by the x-axis and the terminal side of θ when θ is in standard position, provided the terminal side is not along a coordinate axis.

Example 31.5 Figure 31.1 shows the following illustrations.

(a) If $\theta = 135°$, then $\theta_r = 45°$.

(b) If $\theta = -5\pi/3$, then $\theta_r = 2\pi - 5\pi/3 = \pi/3$.

(c) If $\theta = 250°$, then $\theta_r = 250° - 180° = 70°$.

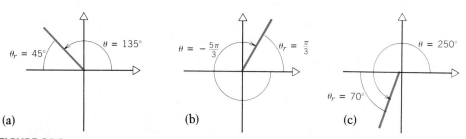

(a) (b) (c)

FIGURE 31.1

If the terminal side of θ coincides with either the positive or negative x-axis, then $\theta_r = 0°$ (or 0 radians). If the terminal side of θ coincides with either the positive or negative y-axis, then $\theta_r = 90°$ (or $\pi/2$ radians). If θ is any angle, then $0° \le \theta_r \le 90°$ (or $0 \le \theta_r \le \pi/2$).

Here is the essential fact about trigonometric functions and reference angles.

> Let θ be an angle with reference angle θ_r. Then
> $$\sin \theta = \sin \theta_r \quad \text{if} \quad \sin \theta > 0$$
> and
> $$\sin \theta = -\sin \theta_r \quad \text{if} \quad \sin \theta < 0.$$
> The statement is also true if sine is replaced by any other trigonometric function.

(31.1)

Proof Figure 31.2 shows an example with θ in the third quadrant. Using the point $P(x, y)$, we have $\sin \theta = y/r$. And from the right triangle OAP, we have $\sin \theta_r = |y|/r$. Therefore, since $y < 0$,

$$\sin \theta = \frac{y}{r} = -\frac{|y|}{r} = -\sin \theta_r,$$

as claimed in (31.1). A similar argument applies for other quadrants and other functions. □

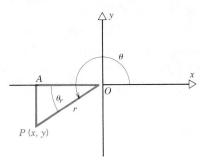

FIGURE 31.2

Example 31.6 Find each functional value.

    (a) tan 135°

    (b) sin 250°

    (c) sec $-\pi/3$

Solution (a) Figure 31.1 shows that the terminal side of 135° is in the second quadrant, so tan 135° < 0. The reference angle is 45°. From Example 30.3 we know that tan 45° = 1. Thus

$$\tan 135° = -\tan 45° = -1.$$

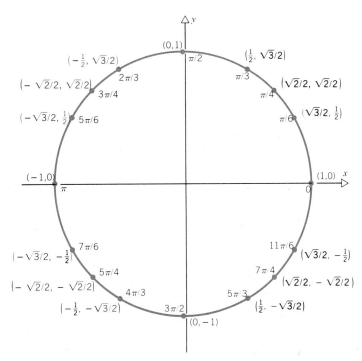

FIGURE 31.3

(b) Figure 31.1 shows that the terminal side of 250° is in the third quadrant, so sin 250° < 0. The reference angle is 70°. From Table 31.1 we see that sin 70° = 0.9397. Thus

$$\sin 250° = -\sin 70° = -0.9397.$$

(c) Since $-\pi/3$ is a fourth quadrant angle, sec $-\pi/3 > 0$. (Remember that sec θ has the same sign as cos θ.) The reference angle is $\pi/3$. From Example 30.5, sec $\pi/3$ = sec 60° = 2. Thus

$$\sec -\pi/3 = \sec \pi/3 = 2. \qquad \blacksquare$$

Remark Figure 31.3 shows the coordinates of all the points on a unit circle that correspond to angles that have reference angles of either 0, $\pi/6$, $\pi/4$, $\pi/3$, or $\pi/2$. If you can reproduce any part of the figure quickly, then you can easily find the functions of the corresponding angles.

D. APPLICATIONS

To **solve** a triangle means to find the lengths of its sides (relative to some unit of length) and the measures of its angles (in either degrees or radians). We now look at how to use the trigonometric functions to help solve right triangles. (Oblique triangles will be considered in Chapter IX.) We use the notational conventions shown in Figure 31.4. The angle at C is a right angle. The interior angles at the vertices A, B, and C will be denoted α, β, and γ, respectively. The lengths of the sides opposite A, B, and C will be denoted a, b, and c, respectively.

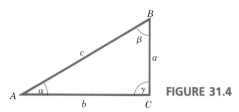

FIGURE 31.4

For a right triangle as in Figure 31.4:

* Given either one of α or β, the other can be determined from $\alpha + \beta = 90°$.
* Given any two of a, b, and c, the third can be determined from $a^2 + b^2 = c^2$.
* Given any two sides, any angle can be determined with an appropriate trigonometric function.
* Given any side and either acute angle, any other side can be determined with an appropriate trigonometric function.

The next two examples illustrate the third and fourth of the preceding statements.

Example 31.7 Find α if $b = 2$ and $c = 5$ (Figure 31.4).

Solution Pick a function that involves α, b, and c. There are two from which to choose:

either $\cos \alpha = b/c$ or $\sec \alpha = c/b$. From $\cos \alpha = b/c$, we have

$$\cos \alpha = \frac{2}{5} = 0.4.$$

That is, α is an acute angle whose cosine is 0.4.

Finding α with a scientific calculator. With the calculator in the degree mode, enter 0.4 and then press $\boxed{\cos^{-1}}$ (or press $\boxed{\text{inv}}$ and then $\boxed{\cos}$, where $\boxed{\text{inv}}$ is the inverse function key). The answer is $\alpha = 66.42°$. (More will be said about inverse trigonometric functions in Chapter VIII.)

Finding α with a table. Search through the cosine columns of Table V until you find 0.4. To the nearest 0.1°, $\alpha = 66.4°$.

We would arrive at the same answer from $\sec \alpha = \frac{5}{2} = 2.5$. However, we use the sine, cosine, and tangent unless there is a good reason to do otherwise. This is partly because these are the trigonometric functions on a scientific calculator. ∎

Example 31.8 Find c if $a = 10$ and $\beta = 30°$ (Figure 31.4).

Solution We need a function that involves c, a, and β. We can use either $\cos \beta = a/c$ or $\sec \beta = c/a$. Since we are to determine c in this case, we use $\sec \beta = c/a$, from which $c = a \sec \beta$. [The choice $\cos \beta = a/c$ gives $c = a/\cos \beta$, which offers no problem with a calculator. However, without a calculator multiplication (as in $c = a \sec \beta$) is easier than division (as in $c = a/\cos \beta$).] From Example 30.4, $\sec 30° = 2\sqrt{3}/3$. Therefore,

$$c = a \sec \beta = 10 \cdot 2\sqrt{3}/3 = 20\sqrt{3}/3. \quad ∎$$

Example 31.9 The hypotenuse of a right triangle has length 5 centimeters and one of the acute angles is 36°. Solve the triangle.

Solution We can take $c = 5$ and $\alpha = 36°$ in Figure 31.4. Then $\sin \alpha = a/c$. Using either a calculator or Table V, we find

$$a = c \sin \alpha = 5 \sin 36° = 5(0.5878) = 2.94 \text{ cm}.$$

We could now get b with the Pythagorean Theorem. Or we can use $\cos \alpha = b/c$:

$$b = c \cos \alpha = 5 \cos 36° = 5(0.8090) = 4.05 \text{ cm}.$$

The other acute angle is the complement of α:

$$\beta = 90° - \alpha = 90° - 36° = 54°. \quad ∎$$

An angle formed with the horizontal by a line of sight *above* the horizontal is called an **angle of elevation** (Figure 31.5). An angle formed with the horizontal by a line of sight *below* the horizontal is called an **angle of depression** (Figure 31.6). Positive angles are used in both cases—the words *elevation* and *depression* indicate the direction.

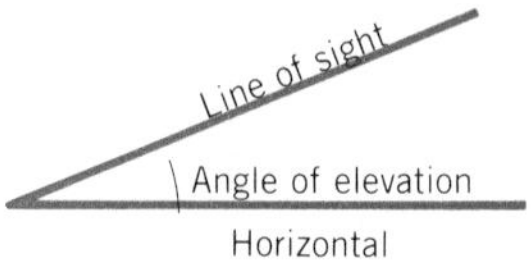

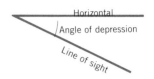

FIGURE 31.5

FIGURE 31.6

Example 31.10 The angle of elevation to the top of the Washington Monument from a point on the same level as the base and 1000 feet away is 29°. How tall is the monument?

Solution Figure 31.7 shows the essentials. From tan 29° = $h/1000$ and a calculator or Table V we have

$$h = 1000 \tan 29° = 1000(0.5543) \approx 554 \text{ ft.}$$

(The actual height is close to 555 ft.)

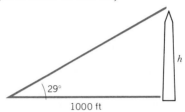

FIGURE 31.7

EXERCISES FOR SECTION 31

Some of these exercises specify the use of a table, some specify the use of a calculator, and others leave a choice. In any event, in this section and those that follow you are encouraged to use a calculator whenever nontrivial computation is required. This applies especially to the applied problems, which are usually not marked with a calculator symbol.

Find each functional value by using Table V.

1. tan 17°

2. cos 31°

3. sin 44°

4. sin 81.3°

5. csc 49.9°

6. cot 63.7°

7. sec 46.1°

8. tan 7.6°

9. sec 87.6°

10. cos 2° 12′

11. cot 72° 48′

12. csc 12° 36′

Find the reference angle θ, for each angle θ.

13. θ = −200°

14. θ = −π/4

15. θ = 95°

16. θ = 17π/6

17. θ = 610°

18. θ = −5π/4

Find each functional value using a reference angle. Show each step.

19. csc −200°

20. cos −π/4

21. sin 95°

22. tan 17π/6

23. cot 610°

24. sec −5π/4

25. cos 192°

26. csc −306°

27. cot 509°

Find each functional value using a calculator. [C]

28. tan 203°

29. cos −51°

30. sin 182°

31. sin 7

32. cot 12π/5

33. tan −95°

34. sec −π/8

35. sin 168°

36. csc −3π

37. cot −160.25°

38. csc −307.03°

39. cot 3415.82°

There is a unique acute angle α, β, or θ satisfying each equation. Find it, to the nearest 0.1°, using Table V. (See the solution of Example 31.7.)

40. csc α = 3.072

41. sin α = 0.8746

42. cos α = 0.9962

43. cot β = 0.1890

44. tan β = 0.9523

45. tan β = 3.867

46. sin θ = 0.3795

47. sec θ = 1.081

48. csc θ = 1.395

There is a unique acute angle α, β, or θ satisfying

each equation. Find it to the nearest 0.01° using a calculator. (See the solution of Example 31.7. To find θ given csc θ, first compute sin θ = 1/csc θ. A similar remark applies to sec θ and cot θ.) Ⓒ

49. $\sin \theta = 0.3777$

50. $\tan \alpha = 2.515$

51. $\cos \beta = 0.0642$

52. $\cos \beta = 0.9761$

53. $\sin \theta = 0.7098$

54. $\tan \alpha = 1.821$

55. $\cot \alpha = 28.49$

56. $\sec \beta = 1.019$

57. $\csc \theta = 2.753$

The notation in Exercises 58–69 is the same as that in Figure 31.4. In 58–63 give each answer to the nearest 0.1°.

58. Find α if a = 2 and b = 6.

59. Find β if b = 5 and c = 7.

60. Find α if a = 14 and b = 7.

61. Find β if b = 1.2 and c = 4.2.

62. Find α if $a = \frac{1}{2}$ and $b = \frac{1}{3}$.

63. Find β if a = 3 and c = 15.

64. Find a if c = 20 and α = 55°.

65. Find b if a = 100 and α = 70°.

66. Find c if a = 40 and β = 20°.

67. Find c if b = 0.1 and β = 5°.

68. Find a if $c = \frac{2}{3}$ and β = 45°.

69. Find b if a = π and α = 55°.

In Exercises 70–72, solve each right triangle from what is given. Determine each angle to the nearest 0.1°.

70. One leg has length 10 inches and the opposite acute angle is 3°.

71. One leg has length 2 meters and the hypotenuse has length 5 meters.

72. The hypotenuse has length 5 centimeters and one of the acute angles is 30°.

73. The Eiffel Tower is approximately 985 feet tall, measured from the top of the base. What is the angle of elevation to the top of the tower from a point on the same level as the top of the base and 300 feet away?

74. The angle of depression from the top of Trajan's Column to a point on the same level as the base and 150 feet away from the base is approximately 40°. How tall is the column? (Trajan's Column was erected in Rome in the second century A.D. to celebrate Emperor Trajan's victories in the Dacian Wars.)

75. Each outside edge of the Pentagon in Arlington, Virginia, is 921 feet long. How far is it from the center of the building to each of the five vertices?

76. Figure 31.8 shows two points on one bank of a river and one point on the opposite bank. The angle at B is a right angle. One other angle and one length have been measured and the results are shown in the figure. How wide is the river?

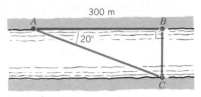

FIGURE 31.8

77. The Great Pyramid of Khufu, near Cairo, was built with a square base 230 meters on each side. The faces made an angle of 51.8° with the horizontal. How tall was the pyramid? (See Figure 31.9.)

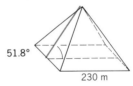

FIGURE 31.9

78. An astronaut is 100 miles above the earth's surface. What is the angle between the vertical and the line of sight to the horizon? (In Figure 31.10, the angle in question is θ. Use 3960 miles for the radius of the earth. The figure is not to scale.)

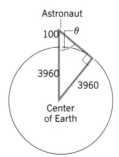

FIGURE 31.10

79. A basketball player's eyes are 6 feet above the floor and 15 feet from the point that is 6 feet above the floor and directly beneath the front of the rim of the goal. What is the angle of elevation from the player's eyes to the front of the rim, which is 10 feet above the floor?

80. The tallest known tree is a California redwood

having a height of approximately 366 feet. How far from the base of the tree is a point on the same level as the base if the angle of elevation to the top of the tree is 36°?

81. A vertical pole casts a 20-meter shadow on level ground when the angle of elevation to the sun is 29°. How tall is the pole?

82. It has often been claimed that a rectangle of length L and width W will have the most pleasing proportions if $L = (1 + \sqrt{5})W/2$. (Compare Exercise 56 of Section 9.) Determine the acute angles of a right triangle formed by a diagonal and two adjacent sides of such a rectangle.

83. A 12-foot ladder makes an angle of 22° 30′ with a vertical wall. How far is the bottom of the ladder from the wall?

84. The Gateway Arch in St. Louis is 630 feet high and has a span at the base that is also 630 feet (Exercise 36 of Section 28). What is the angle of elevation from an end of the base to the top?

85. In Figure 31.11, the circle has radius r and OC

is perpendicular to AB. Prove that $\overline{CO} = r \cos \dfrac{\alpha}{2}$ and $\overline{CB} = r \sin \dfrac{\alpha}{2}$.

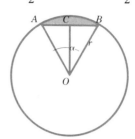

FIGURE 31.11

86. The shaded area in Figure 31.11 is a segment of the circle. Prove that its area is $r^2 \left(\dfrac{\alpha}{2} - \sin \dfrac{\alpha}{2} \cos \dfrac{\alpha}{2} \right)$ (Exercise 48 of Section 29 and Exercise 85 above will help.)

87. Prove that the perimeter of an n-sided regular polygon inscribed in a circle of radius r is $2nr \sin \pi/n$. (Exercise 85 will help.)

REVIEW EXERCISES FOR CHAPTER VI

1. Convert 5° 42′ to degrees, without using minutes and seconds in the answer.

2. Convert 21.82° to degrees, minutes, and seconds.

Convert from radians to degrees.

3. $2\pi/5$ **4.** $-8\pi/3$

Convert from degrees to radians.

5. $-200°$ **6.** $18°$

7. A 75° angle subtends a 10 centimeter arc when the vertex of the angle is at the center of a circle. What is the circle's radius?

8. An angle subtends an arc of length 3 inches when its vertex is at the center of a circle whose circumference is 2 feet. Find the measure of the angle in degrees.

9. When an angle α is placed in standard position its terminal side contains the point $(-2, 4)$. Find $\sin \alpha$, $\tan \alpha$, and $\sec \alpha$.

10. When an angle β is placed in standard position its terminal side contains the point $(1, -5)$. Find $\cos \beta$, $\cot \beta$, and $\csc \beta$.

Give the exact value in each case.

11. $\tan \pi/3$ **12.** $\cos \pi/4$

13. $\sin \pi/6$ **14.** $\sec 225°$

15. $\csc 330°$ **16.** $\cot(-120°)$

17. $\sin(-5\pi/6)$ **18.** $\tan(-4\pi/3)$

19. $\cos(5\pi/4)$ **20.** $\cot 270°$

21. $\sec 180°$ **22.** $\tan(-\pi/2)$

23. Find $\cos \beta$, $\tan \beta$, and $\csc \beta$.

24. Find $\sin \alpha$, $\cot \alpha$, and $\sec \alpha$.

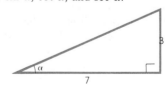

25. Find sin γ if γ is an acute angle such that tan γ = 0.8.

26. Find cot θ if θ is an acute angle such that cos θ = $\frac{5}{13}$.

27. If $0 \le \alpha < \pi/2$ and sin α = t, what is cos ($\pi/2$ − α)?

28. If $0 \le \beta < 90°$ and cot(90° − β) = u, what is tan β?

29. If θ is a second quadrant angle (in standard position) with reference angle $\theta_r = 20°$, what is θ?

30. What is the reference angle for θ = $10\pi/3$?

Find each functional value. Do not use a calculator.

31. tan(−51°)   **32.** cos 230°

33. sec 519°   **34.** csc(−212°)

Find each functional value using a calculator. Ⓒ

35. sin 22°   **36.** sec 85°

37. cot(−140°)   **38.** sin(−800°)

39. Use sin 15° = 0.2588 and cos 15° = 0.9659 to determine sin 75°, cos 165°, sin(−105°), and cos(−555°).

40. Compute the perimeter of a regular pentagon inscribed in a circle of radius 10 centimeters.

41. The interior of the Pantheon in Rome consists of a dome in the shape of a hemisphere resting on a drum in the shape of a right circular cylinder. Both the dome and the drum have a diameter of 143 feet, and the distance from the top of the dome to the floor is also 143 feet.

What is the angle of elevation to the top of the dome from a point on the outside edge of the floor?

42. The angle of depression from the observation platform on the Statue of Liberty to the water's edge at the tip of Manhattan Island is approximately 1.61°. If the platform is 260 feet above the surface of New York Harbor, what is the distance in miles from the statue to the tip of the island?

43. A football is on the 15 yard line (which is 25 yards from the goal posts) and centered between the sidelines of a football field. The goal posts are 23 feet 4 inches apart. What angle do the goal posts subtend at the football? (The angle of interest is α in the figure.)

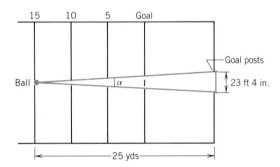

44. The angle of elevation to the top of a tower from a point 150 meters from the base and on the same level as the base is 35°. How tall is the tower?

CHAPTER VII

TRIGONOMETRIC FUNCTIONS

In the preceding chapter the trigonometric functions were considered as functions of *angles*. In this chapter we shift our viewpoint and consider them as functions of *real numbers*. This new way of viewing the functions leads to many important applications that do not involve angles. The goal is to study the general properties of these functions so that you will be able to handle the functions wherever they may arise.

SECTION **32**
Definitions. Basic Identities

A. DEFINITIONS AND EXAMPLES

Recall that the unit circle is the graph of $x^2 + y^2 = 1$, which is the circle with center at the origin and radius 1. To consider the trigonometric functions as functions of real numbers, we first associate a point $P(t)$ on the unit circle with each real number t, as follows.

If $t \geq 0$, then $P(t)$ is the point obtained when we start at $(1, 0)$ and move around the unit circle t units of arc length in the *counterclockwise* direction [Figure 32.1*a*].

If $t < 0$, then $P(t)$ is the point obtained when we start at $(1, 0)$ and move around the unit circle $|t|$ units of arc length in the *clockwise* direction [Figure 32.1*b*].

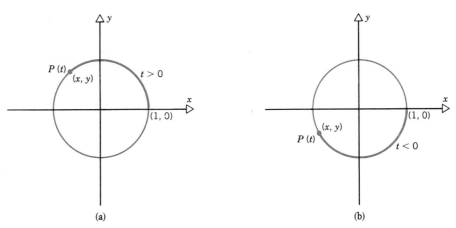

(a)   (b)

FIGURE 32.1

Definition For each real number t, let $P(t)$ denote the point on the unit circle as defined above. The **trigonometric functions** of t are defined in terms of the coordinates (x, y) of $P(t)$, as follows.

$$\sin t = y \qquad \csc t = 1/y$$
$$\cos t = x \qquad \sec t = 1/x \qquad (32.1)$$
$$\tan t = y/x \qquad \cot t = x/y$$

The trigonometric functions of t are also called **circular functions** of t.

Example 32.1   Figure 32.2 shows a real number t such that $P(t) = (-0.6, 0.8)$. Here are the corresponding trigonometric functions.

$$\sin t = 0.8 \qquad\qquad \csc t = 1/0.8 = 1.25$$

$$\cos t = -0.6 \qquad\qquad \sec t = 1/(-0.6) = -1.\overline{6}$$

$$\tan t = 0.8/(-0.6) = -1.\overline{3} \qquad \cot t = (-0.6)/0.8 = -0.75 \qquad \blacksquare$$

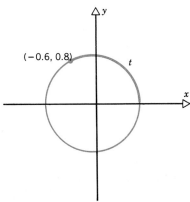

FIGURE 32.2

　　The connection between the definitions in (32.1) and the definitions of the trigonometric functions in Section 30 can be seen as follows. First, given a real number t, draw a ray from the origin through the point $P(t)$ (Figure 32.3). This produces an angle in standard position, and, because the unit circle has radius 1, the radian measure of this angle is t (Section 29B). With t thought of as the radian measure of the angle, Section 30 would yield $\sin t = y$ [take $r = 1$ in (30.1)]. This is precisely what we find in (32.1). The definitions in (32.1) are consistent with those in Section 30 for the other functions as well as for the sine function. Thus we can draw the following conclusion.

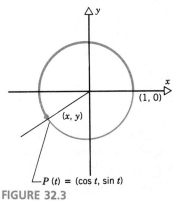

FIGURE 32.3

If t is a real number, then $\sin t$ as defined in (32.1) is equal to $\sin t$ as defined in (30.1), where, in the latter case, t is taken to denote t radians. The statement is also true if sine is replaced by any other trigonometric function.　　　　　　　　　　　　　　　　　　　　　　　　(32.2)

Because of the consistency guaranteed by Statement (32.2), it may seem an unnecessary distraction for us to have made the two different definitions of trigonometric functions given by (30.1) and (32.1). However, the definition given by (30.1) is well-justified by the great importance of trigonometric functions of angles. And the definition given by (32.1) is essential because t sometimes measures quantities that have no direct connection with angles—in some physical applications, for example, trigonometric functions of t are used with t representing time. You may rest assured that it is important to understand thoroughly both approaches—t representing the measure of an angle and t representing a real number.

Even though the definitions in (32.1) are independent of angles, Statement (32.2) allows us to draw on what we know about functions of angles whenever it can be useful. We do that in the next example.

Example 32.2 In Example 30.3 we computed all of the functions of 45°. Therefore, since $45° = \pi/4$ radians, we also know the functions of $\pi/4$ in the sense of (32.1). For example,

$$\sin \pi/4 = \sqrt{2}/2 \quad \text{and} \quad \cot \pi/4 = 1. \qquad ■$$

In the same way, Examples 30.4 and 30.5 provide the functions of $\pi/6$ and $\pi/3$. *It will be assumed that you remember or can compute all of the functions of $\pi/4$, $\pi/6$ and $\pi/3$.* You should also be able to use reference angles whenever necessary.

Example 32.3 Since $P(0) = (1, 0)$, we have the following functions of 0.

$$\sin 0 = 0 \qquad \qquad \csc 0 = 1/0 \text{ is undefined}$$

$$\cos 0 = 1 \qquad \qquad \sec 0 = 1/1 = 1$$

$$\tan 0 = 0/1 = 0 \qquad \cot 0 = 1/0 \text{ is undefined} \qquad ■$$

Example 32.4 For which values of t is $\csc t$ undefined?

Solution From $\csc t = 1/y$ we see that $\csc t$ is undefined for all t such that $y = 0$. This means all t such that $P(t)$ has coordinates $(1,0)$ or $(-1,0)$. This includes $t = 0$ and $t = \pi$, and all values differing from one of these by an integral multiple of 2π. Thus, $\csc t$ is undefined for

$$\{t: t = n\pi \text{ for } n \text{ an integer}\}. \qquad ■$$

By Statement (32.2) we can compute trigonometric functions of a real number t with a calculator or a table provided we regard t as representing radians. In particular, to compute a trigonometric function of a real number with a calculator, be sure to place the calculator in the *radian* mode.

B. BASIC IDENTITIES

Recall that an equation with *one* variable is an *identity* if every real number in the domain of the equation is a solution of the equation (Section 7A). An equation with *two* variables is an *identity* if every pair of numbers for which both sides of the equation are defined is a solution of the equation. For example, $x^2 - y^2 = (x + y)(x - y)$ is an identity, but $2x - y = x + y$ is not. (Identities with more than two variables are defined similarly, but they will not be needed.)

A **trigonometric identity** is an identity that involves trigonometric functions of one or more variables. The equations in (32.3)–(32.8), which follow, are all identities; each equation is true for all values of t for which both sides are defined. The identities in (32.3) were stated for functions of angles in (30.2). Section 34 will introduce identities with two variables.

Reciprocal identities

$$\csc t = \frac{1}{\sin t} \qquad \sec t = \frac{1}{\cos t} \qquad \cot t = \frac{1}{\tan t} \qquad (32.3)$$

Proof These follow directly from the definitions of the functions given in (32.1):

$$\csc t = 1/y = 1/(\sin t),$$

$$\sec t = 1/x = 1/(\cos t),$$

$$\cot t = x/y = 1/(y/x) = 1/(\tan t). \qquad \square$$

Quotient identities

$$\tan t = \frac{\sin t}{\cos t} \qquad \cot t = \frac{\cos t}{\sin t} \qquad (32.4)$$

Proof From the definitions of the functions,

$$\tan t = y/x = \sin t/\cos t$$

and

$$\cot t = x/y = \cos t/\sin t. \qquad \square$$

Even-odd identities

$$\sin(-t) = -\sin t \qquad \csc(-t) = -\csc t$$

$$\cos(-t) = \cos t \qquad \sec(-t) = \sec t \qquad (32.5)$$

$$\tan(-t) = -\tan t \qquad \cot(-t) = -\cot t$$

Proof Figure 32.4 will help in comparing the functions of t and $-t$ when $P(t)$ is in the second quadrant. We see that $P(t)$ and $P(-t)$ have equal x-coordinates, and the y-coordinates are equal in absolute value but opposite in sign. The same relationships will hold between the x- and y-coordinates of $P(t)$ and $P(-t)$ for any value of t. That is, for any value of t,

$$\text{if } P(t) = (x, y), \text{ then } P(-t) = (x, -y).$$

Therefore,

$$\sin(-t) = -y = -\sin t,$$

$$\cos(-t) = x = \cos t, \text{ and}$$

$$\tan(-t) = -y/x = -(y/x) = -\tan t.$$

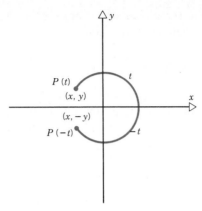

FIGURE 32.4

The other three identities in (32.5) now follow by using (32.3). For example,

$$\csc(-t) = \frac{1}{\sin(-t)} = \frac{1}{-\sin t} = -\csc t. \qquad \square$$

In the terminology of Section 20, $\sin t$, $\csc t$, $\tan t$, and $\cot t$ are *odd* functions of t, and $\cos t$ and $\sec t$ are *even* functions of t.

In the next identities the notation $\sin^2 t$ means $(\sin t)^2$, $\cos^2 t$ means $(\cos t)^2$, and so on. We do not write $\sin t^2$, because it might be taken to mean either $(\sin t)^2$ or $\sin(t^2)$.

Pythagorean identities

$$\sin^2 t + \cos^2 t = 1 \qquad (32.6)$$

$$\tan^2 t + 1 = \sec^2 t \qquad (32.7)$$

$$1 + \cot^2 t = \csc^2 t \qquad (32.8)$$

Proof In the definitions of the trigonometric functions of a real number t, in (32.1), x and y denote the coordinates of a point on the unit circle, so that $x^2 + y^2 = 1$. Since $\sin t = y$ and $\cos t = x$, this implies that $\cos^2 t + \sin^2 t = 1$, which yields Identity (32.6).

To prove (32.7), divide both sides of (32.6) by $\cos^2 t$ and then use (32.3) and (32.4).

$$\frac{\sin^2 t}{\cos^2 t} + \frac{\cos^2 t}{\cos^2 t} = \frac{1}{\cos^2 t}$$

$$\left(\frac{\sin t}{\cos t}\right)^2 + 1 = \left(\frac{1}{\cos t}\right)^2$$

$$\tan^2 t + 1 = \sec^2 t$$

Division by $\sin^2 t$ rather than $\cos^2 t$ will produce Identity (32.8). $\qquad \square$

Identities (32.6)–(32.8) are called *Pythagorean identities* because they are consequences of $x^2 + y^2 = 1$, which is a consequence of the Pythagorean theorem.

The identities in (32.3)–(32.8) should all be memorized. For (32.5), it suffices to remember what happens with the sine, cosine, and tangent; their reciprocals behave in the same way.

C. EXAMPLES

With the identities in (32.3)–(32.8) we can express all of the trigonometric functions in terms of any one of the functions, except for possible ambiguities in sign. The next example illustrates this for the sine function. The other cases are left as exercises.

Example 32.5 Express each of the other trigonometric functions of t in terms of $\sin t$.

Solution (a) From (32.6), $\cos^2 t = 1 - \sin^2 t$. Therefore,

$$\cos t = \pm\sqrt{1 - \sin^2 t}.$$

The choice of sign depends on the quadrant. If $P(t)$ is in Quadrant I or IV, then $\cos t > 0$ so we choose the positive sign. If $P(t)$ is in Quadrant II or III, then $\cos t < 0$ so we choose the negative sign.

(b) From (32.4), $\tan t = \dfrac{\sin t}{\cos t}$. Therefore, using part (a),

$$\tan t = \frac{\sin t}{\pm\sqrt{1 - \sin^2 t}}.$$

If $P(t)$ is in Quadrant I or IV, we choose the positive sign in the denominator. If $P(t)$ is in Quadrant II or III, we choose the negative sign. (Exercise 39. Remember to take the sign of $\sin t$ into account as well as the sign of $\tan t$.)

(c) From (32.3), $\csc t = \dfrac{1}{\sin t}$.

(d) From (32.3), $\sec t = \dfrac{1}{\cos t}$. Therefore, by part (a),

$$\sec t = \frac{1}{\pm\sqrt{1 - \sin^2 t}}.$$

(e) From (32.3), $\cot t = \dfrac{1}{\tan t}$. Therefore, by part (b),

$$\cot t = \frac{\pm\sqrt{1 - \sin^2 t}}{\sin t}.$$

To prove that an equation with one variable is *not* an identity, it suffices to find one value of the variable (in the domain of the equation) for which the equation is not true. Any such value of t provides a **counterexample** to the claim that the equation is an identity.

Example 32.6 Prove that $\sin 2t = 2 \sin t$ is not an identity.

Solution If $t = \pi/2$, then $\sin 2t = 2 \sin t$ becomes $\sin \pi = 2 \sin \pi/2$, or $0 = 2$, which is false. Therefore $\sin 2t = 2 \sin t$ is not an identity. [When we consider the solution of trigonometric equations in the next chapter, we shall see that, in fact, $\sin 2t \neq 2 \sin t$ for all t in the interval $[0, 2\pi)$ except $t = 0$ and $t = \pi$.]

EXERCISES FOR SECTION 32

1. What is $P(\pi)$?

2. What is $P(3\pi/2)$?

3. What is $P(-7\pi/2)$?

Find the six trigonometric functions of t *if* P(t) *has the given coordinates.*

4. $(5/13, -12/13)$

5. $(-0.8, -0.6)$

6. $(-1/3, 2\sqrt{2}/3)$

7. Find the six trigonometric functions of $-\pi/6$.

8. Find the six trigonometric functions of $3\pi/4$.

9. Find the six trigonometric functions of $4\pi/3$.

Find each functional value with the help of Table VI and, if necessary, Equations (32.5).

10. $\tan 0.49$ **11.** $\cos 1.14$

12. $\sin 0.36$ **13.** $\sin -1.48$

14. $\tan -0.1$ **15.** $\cos -1.50$

Find each functional value using a calculator. ⓒ

16. $\cos 9$ **17.** $\sin -1.234$

18. $\tan 8.72$ **19.** $\cot -0.05$

20. $\csc 2$ **21.** $\sec -58$

22. $\sec \pi/8$ **23.** $\cot -29$

24. $\csc \pi/12$

25. If $P(t)$ (on the unit circle) is in Quadrant IV and its x-coordinate is 0.4, what is its y-coordinate?

26. If $P(t)$ (on the unit circle) is in Quadrant III and its x- and y-coordinates are equal, what are they?

27. If $P(t)$ (on the unit circle) is in Quadrant II and the absolute value of its x-coordinate is twice its y-coordinate, what are the coordinates?

28. Determine the sign of each trigonometric function of t if $\pi < t < 3\pi/2$.

29. Determine the sign of each trigonometric function of t if $3\pi/2 < t < 2\pi$.

30. In which quadrant is $P(t)$ if $\sin t < 0$ and $\tan t > 0$?

31. In which quadrant is $P(t)$ if $\cos t > 0$ and $\tan t < 0$?

32. Find the six trigonometric functions of $7\pi/2$.

33. Find the six trigonometric functions of -13π.

34. Find the six trigonometric functions of $-11\pi/2$.

35. (a) Explain why $\sec t$ is undefined iff $\cos t = 0$.

    (b) There are infinitely many real numbers t for which $\cos t = 0$. What are they?

36. (a) Explain why $\cot t$ is undefined iff $\tan t = 0$.

    (b) There are infinitely many real numbers t for which $\tan t = 0$. What are they?

37. Find all t such that $0 \leq t < 2\pi$ and $\tan t = \cot t$.

38. Find all t such that $0 \leq t < 2\pi$ and $\sin t = \csc t$.

39. In (b) of the solution of Example 32.5 it is claimed that if $P(t)$ is in Quadrant II or III then the negative sign must be chosen in the expression there for $\tan t$. Explain why.

40. Express each of the other trigonometric functions of t in terms of $\cos t$.

41. Express each of the other trigonometric functions of t in terms of $\tan t$.

42. Express each of the other trigonometric functions of t in terms of $\sec t$.

43. Prove that $\cos^2 t - \sin^2 t = 1$ is not an identity.

44. Prove that $\sin t + \sec t = \tan t$ is not an identity.

45. Prove that $\tan 2t = 2 \tan t$ is not an identity.

46. If $\sin t = a$ and $\cos u = b$, what is $\sin(-t) - \cos(-u)$?

47. If $\cos t = c$ and $\cot(-u) = d$, what is $\cos(-t) - \tan u$?

48. Assume $f(u) = \cos u$ and $g(u) = 3u$.

    (a) What is $g(f(\pi))$?

    (b) What is $f(g(\pi))$?

SECTION **33**
Proving Identities

A. INTRODUCTION AND EXAMPLES

The basic identities in Section 32 can be used to prove a great many other identities, as the examples and exercises in this section will show. The primary reason for proving these other identities is to increase your familiarity with the basic identities and how to manipulate them. This will be useful for a number of reasons, including applications to calculus.

One way to prove that an equation is an identity is to begin with the expression on one side of the equation and transform it one step at a time until arriving at the expression on the other side of the equation. At each step we use either an algebraic property or a known trigonometric identity. If we show that $A = B$, $B = C$, and $C = D$ are all identities, for example, then we will know that $A = D$ is an identity. The important point is to ensure that each equation in the sequence (such as $B = C$) is itself an identity.

Example 33.1 Prove that $\sin t = \dfrac{\tan t}{\sec t}$ is an identity.

Solution This example illustrates two principles. First, if one side is simpler than the other, then try to reduce the more complicated side to the simpler side. Second, to simplify a trigonometric expression, if nothing else suggests itself then express each part in terms of sines and cosines and try to use algebra to simplify the result.

$$\frac{\tan t}{\sec t} = \frac{\dfrac{\sin t}{\cos t}}{\dfrac{1}{\cos t}} \qquad \text{Quotient and reciprocal identities.}$$

$$= \frac{\sin t}{\cos t} \cdot \frac{\cos t}{1} \qquad \text{Algebra.}$$

$$= \sin t \qquad \text{Algebra.}$$

Example 33.2 Prove the identity

$$\frac{1}{\sec u - \tan u} - \frac{1}{\sec u + \tan u} = 2 \tan u.$$

Solution

$$\frac{1}{\sec u - \tan u} - \frac{1}{\sec u + \tan u}$$

$$= \frac{1}{\sec u - \tan u} \cdot \frac{\sec u + \tan u}{\sec u + \tan u} \qquad \text{Get common denominators on the left.}$$

$$- \frac{1}{\sec u + \tan u} \cdot \frac{\sec u - \tan u}{\sec u - \tan u}$$

$$= \frac{(\sec u + \tan u) - (\sec u - \tan u)}{(\sec u - \tan u)(\sec u + \tan u)} \qquad \text{Subtract the fractions on the left.}$$

$$= \frac{2 \tan u}{\sec^2 u - \tan^2 u} \qquad \text{Algebra.}$$

$$= \frac{2 \tan u}{1} \qquad \text{Pythagorean identity.}$$

$$= 2 \tan u$$

This example shows the importance of being proficient with algebra; in this case being able to subtract fractions with different denominators. It also shows the need to learn the fundamental identities thoroughly in all of their different forms. For example, we used (32.7) in the alternate form $\sec^2 u - \tan^2 u = 1$. ■

Example 33.3 Prove the identity $\dfrac{\sin^3 t + \csc^3 t}{\sin t + \csc t} = \csc^2 t - \cos^2 t$.

Solution This example shows again the importance of knowing algebra well; in this case, the factorization formula $u^3 + v^3 = (u + v)(u^2 - uv + v^2)$ [Equation (5.4)].

$$\frac{\sin^3 t + \csc^3 t}{\sin t + \csc t}$$

$$= \frac{(\sin t + \csc t)(\sin^2 t - \sin t \csc t + \csc^2 t)}{\sin t + \csc t} \qquad \text{Factor.}$$

$$= \sin^2 t - \sin t \csc t + \csc^2 t \qquad \text{Algebra.}$$

$$= \sin^2 t - 1 + \csc^2 t \qquad \text{Reciprocal identity.}$$

$$= \csc^2 t - \cos^2 t \qquad \text{Pythagorean identity.} \quad ■$$

Example 33.4 Prove the identity $\dfrac{\cos \theta}{1 - \sin \theta} = \dfrac{1 + \sin \theta}{\cos \theta}$.

Solution This example requires a good idea. We multiply the expression on the left by 1 in the form $(1 + \sin \theta)/(1 + \sin \theta)$. This will introduce the factor $1 + \sin \theta$ into the numerator on the left (which we need, since it is also on the right). The method used here is similar to rationalizing a denominator.

$$\frac{\cos \theta}{1 - \sin \theta} = \frac{\cos \theta}{1 - \sin \theta} \cdot \frac{1 + \sin \theta}{1 + \sin \theta}$$

$$= \frac{\cos \theta (1 + \sin \theta)}{1 - \sin^2 \theta} \qquad \text{Algebra.}$$

$$= \frac{\cos \theta (1 + \sin \theta)}{\cos^2 \theta} \qquad \text{Pythagorean identity.}$$

$$= \frac{1 + \sin \theta}{\cos \theta} \qquad \text{Algebra.} \quad ■$$

In the next example we use the fact that if $A = C$ and $B = C$ are both identities, then $A = B$ is also an identity. We transform the left side of the original equation to a certain form, and then transform the right side of the original equation to that same form. In doing this it is important to work with only one side at a time.

Example 33.5 Prove the identity

$$\sec \alpha \csc \alpha + \cot \alpha = \tan \alpha + 2 \cos \alpha \csc \alpha.$$

Solution First express the left side in terms of $\sin \alpha$ and $\cos \alpha$ and simplify.

$$\sec \alpha \csc \alpha + \cot \alpha = \frac{1}{\cos \alpha} \cdot \frac{1}{\sin \alpha} + \frac{\cos \alpha}{\sin \alpha}$$

$$= \frac{1}{\cos \alpha} \cdot \frac{1}{\sin \alpha} + \frac{\cos \alpha}{\sin \alpha} \cdot \frac{\cos \alpha}{\cos \alpha}$$

$$= \frac{1 + \cos^2 \alpha}{\sin \alpha \cos \alpha}$$

Now express the right side in terms of $\sin \alpha$ and $\cos \alpha$ and simplify.

$$\tan \alpha + 2 \cos \alpha \csc \alpha = \frac{\sin \alpha}{\cos \alpha} + 2 \cos \alpha \cdot \frac{1}{\sin \alpha}$$

$$= \frac{\sin \alpha}{\cos \alpha} \cdot \frac{\sin \alpha}{\sin \alpha} + 2 \frac{\cos \alpha}{\sin \alpha} \cdot \frac{\cos \alpha}{\cos \alpha}$$

$$= \frac{\sin^2 \alpha + 2 \cos^2 \alpha}{\sin \alpha \cos \alpha}$$

$$= \frac{\sin^2 \alpha + \cos^2 \alpha + \cos^2 \alpha}{\sin \alpha \cos \alpha}$$

$$= \frac{1 + \cos^2 \alpha}{\sin \alpha \cos \alpha}$$

Both sides of the original equation have been transformed to the same form, so the original equation is an identity. ◾

B. SUMMARY AND REMARK

Here is a summary of the ideas that were brought out by Examples 33.1–33.5.

* Know the fundamental identities well in all their different forms. (For example, be able to recognize that $1 - \sin^2 t$ is $\cos^2 t$, as well as that $\sin^2 t + \cos^2 t$ is 1.)
* Make sure you are proficient with algebra. If necessary, review how to work with fractions and the formulas for factoring.
* If one side of an identity is simpler than the other side, try to reduce the more complicated side to the simpler side.
* If nothing else suggests itself, express one side (or, if necessary, both sides) in terms of sines and cosines and use algebra to simplify the result.

- If you cannot transform one side to the other side, try to transform both sides (separately) to some common form.
- Be alert to the possibility of multiplying an expression by 1 in a helpful form.

To see why it is important to work with only one side at a time when proving that an equation is an identity, consider the equation

$$\cos \theta = \sqrt{1 - \sin^2 \theta}. \tag{33.1}$$

This is *not* an identity, since the right side is never negative but the left side is negative for all θ in Quadrants II and III. However, if we square both sides of (33.1), we obtain

$$\cos^2 \theta = 1 - \sin^2 \theta, \tag{33.2}$$

which is an identity. The point to notice is that the step that took us from (33.1) to (33.2) is not reversible: (33.1) implies (33.2), but (33.2) does not imply (33.1). If we transform a given equation to an identity, then we can conclude that the given equation is an identity only if we know the steps we used are all reversible.

EXERCISES FOR SECTION 33

Prove that each equation is an identity.

1. $\cos t \tan t = \sin t$

2. $\sin t \cot t \sec t = 1$

3. $\dfrac{\sec t}{\csc t} = \tan t$

4. $\sin^2 \theta \,(1 + \cot^2 \theta) = 1$

5. $\tan^2 \theta = \dfrac{1}{\cos^2 \theta} - 1$

6. $(1 - \sin^2 \theta)(\sec^2 \theta) = 1$

7. $\dfrac{1}{\sec u + 1} - \dfrac{1}{\sec u - 1} = -2 \cot^2 u$

8. $\sec \alpha - \cos \alpha = \sin \alpha \tan \alpha$

9. $\dfrac{1}{1 + \cos u} + \dfrac{1}{1 - \cos u} = 2 \csc^2 u$

10. $\cot \alpha + \tan \alpha = \sec \alpha \csc \alpha$

11. $\dfrac{1 + \cos u}{\sin u} + \dfrac{\sin u}{1 + \cos u} = 2 \csc u$

12. $\dfrac{1 + \tan^2 \beta}{\tan^2 \beta} = \csc^2 \beta$

13. $\dfrac{1 - \sin^2 \beta}{\sin^2 \beta} = \cot^2 \beta$

14. $\dfrac{1 - \sin^3 \gamma}{1 - \sin \gamma} = 1 + \sin \gamma + \sin^2 \gamma$

15. $\sin \alpha - \csc \alpha = -\cos \alpha \cot \alpha$

16. $\dfrac{\tan^3 \gamma - \cot^3 \gamma}{\tan \gamma - \cot \gamma} = \sec^2 \gamma + \cot^2 \gamma$

17. $\dfrac{\sec^2 \gamma - 1}{\sec^2 \gamma} = \sin^2 \gamma$

18. $\dfrac{\sin^4 \gamma - \cos^4 \gamma}{\sin \gamma - \cos \gamma} = \sin \gamma + \cos \gamma$

19. $\dfrac{(1 + \cos A)(1 - \cos A)}{\sin A} = \sin A$

20. $\dfrac{\csc^2 A - \sin^2 A}{\csc A + \sin A} = \cos A \cot A$

21. $\dfrac{\sec^2 A - \tan^2 A}{\sin A} = \csc A$

22. $\sin(-x) \sec(-x) = -\tan x$

23. $\dfrac{\tan(-x)}{\sin x} = -\sec(-x)$

24. $\cot(-x) \sin(-x) = \cos(-x)$

25. $\dfrac{1 - \cos \alpha}{\sin \alpha} = \dfrac{\sin \alpha}{1 + \cos \alpha}$

26. $\dfrac{\sec \beta - 1}{\tan \beta} = \dfrac{\tan \beta}{\sec \beta + 1}$

27. $\csc \gamma - \cot \gamma = \dfrac{1}{\csc \gamma + \cot \gamma}$

28. $\dfrac{\tan t}{\tan t + \sec t} = \dfrac{1}{1 + \csc t}$

29. $\dfrac{\sec u}{\sec^2 u + \sec^3 u} = \dfrac{\cos u \cot u}{\cot u + \csc u}$

30. $\dfrac{\tan v \sec v - 1}{\tan v \sec v + \sec^2 v} = \dfrac{1 - \cot v \cos v}{1 + \csc v}$

31. $(1 - \cos t)^2 = 2 - 2 \cos t - \sin^2 t$

32. $(\sec t + \cos t)(\sec t - \cos t) =$
$(\sec^4 t - 1) \cos^2 t$

33. $\sin^2 t(\sin t + \sec t)^2 = \sin^4 t + 2 \sin^3 t \sec t + \tan^2 t$

34. $\dfrac{\cos(-\alpha)}{1 + \tan \alpha} - \dfrac{\sin(-\alpha)}{1 - \tan \alpha} = \dfrac{\sec \alpha}{1 - \tan^2 \alpha}$

35. $\dfrac{1}{\sin(-\beta)} + \dfrac{1}{\csc(-\beta)} = -\sin \beta - \csc \beta$

36. $\cos(-\gamma) = \dfrac{2}{1 + \cos(-\gamma)} -$
$\dfrac{1 - \cos \gamma + \sin^2 \gamma}{1 + \cos \gamma}$

37. $\dfrac{\sec A}{\cot^2 A} = \dfrac{1 - \cos^2 A}{\cos^3 A}$

38. $\dfrac{\sec B}{\sec B + \csc B} = \dfrac{1}{1 + \cot B}$

39. $\csc^4 C - \cot^4 C = \csc^2 C + \cot^2 C$

40. $\sin^3 \theta + \cos^3 \theta =$
$(\sin \theta + \cos \theta)(1 - \sin \theta \cos \theta)$

41. $\dfrac{1}{\tan \alpha + \cot \alpha} = \sin \alpha \cos \alpha$

42. $\dfrac{1 + \sin t + \cos t}{1 + \sin t - \cos t} = \dfrac{1 + \cos t}{\sin t}$

43. $\ln|\sin \theta| - \ln|\cos \theta| = \ln|\tan \theta|$

44. $\log|\csc t| = -\log|\sin t|$

45. $2 \log|\sin t| + 2 \log|\cos t| = \log \left(\dfrac{\sin t}{\sec t} \right)^2$

46. $10^{\sin^2 t} \cdot 10^{\cos^2 t} = (10^{\tan t})^{\cot t}$

47. $10^{-\log \sin^2 t} = \csc^2 t$

48. $[\ln e^{(\sec t - \tan t)}][\ln e^{(\sec t + \tan t)}] = 1$

The equations in Exercises 49–54 are not *identities. However, each equation is true for all values of the variable in precisely two quadrants. Decide which two in each case.*

49. $\sqrt{1 - \sin^2 \alpha} = \cos \alpha$

50. $\sqrt{1 - \cos^2 \beta} = -\sin \beta$

51. $\sqrt{\sec^2 \gamma - 1} = \tan \gamma$

52. $\tan t = |\tan(-t)|$

53. $\sin u = -|\sin u|$

54. $\cot v = \dfrac{\cos v}{\sqrt{1 - \cos^2 v}}$

SECTION **34**
Addition and Subtraction Formulas

A. FORMULA FOR $\cos(u - v)$

In this section we derive formulas (identities) for trigonometric functions of sums or differences of numbers in terms of functions of the individual numbers. For example, (34.1) expresses $\cos(u - v)$ in terms of functions of u and v. Such formulas are called **addition formulas** (for functions of $u + v$) or **subtraction formulas** (for functions of $u - v$). The proof of the first formula requires the most work. After it, the others will come easily.

$$\cos(u - v) = \cos u \cos v + \sin u \sin v \tag{34.1}$$

Proof We first restrict our attention to the case $0 \le v \le u < 2\pi$. If we measure v units counterclockwise from $(1, 0)$ along the unit circle, we reach the point with coordinates (cos v, sin v), labeled A in Figure 34.1a. If we measure u units, we reach B(cos u, sin u), also shown in Figure 34.1a. If we measure $u - v$ units, we reach D(cos($u - v$), sin($u - v$)) Figure 34.1b). With C denoting (1, 0) (Figure 34.1b) we see that the arc lengths $\overset{\frown}{AB}$ and $\overset{\frown}{CD}$ are equal: $\overset{\frown}{AB} = u - v = \overset{\frown}{CD}$. It follows that the chord lengths $\overline{AB}$ and $\overline{CD}$ must also be equal, which implies $\overline{AB}^2 = \overline{CD}^2$. To prove (34.1) we use the distance formula to compute $\overline{AB}^2$ and $\overline{CD}^2$, set $\overline{AB}^2 = \overline{CD}^2$, and simplify.

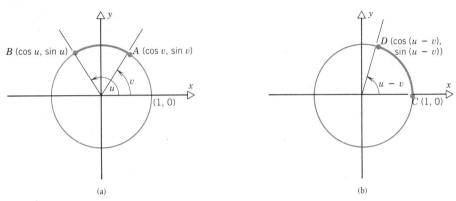

FIGURE 34.1

$$\overline{AB}^2 = (\cos u - \cos v)^2 + (\sin u - \sin v)^2$$

$$= \cos^2 u - 2 \cos u \cos v + \cos^2 v + \sin^2 u - 2 \sin u \sin v + \sin^2 v$$

$$= (\sin^2 u + \cos^2 u) + (\sin^2 v + \cos^2 v) - 2 \cos u \cos v - 2 \sin u \sin v$$

$$= 2 - 2 \cos u \cos v - 2 \sin u \sin v$$

$$\overline{CD}^2 = [\cos(u - v) - 1]^2 + [\sin(u - v) - 0]^2$$

$$= \cos^2(u - v) - 2 \cos(u - v) + 1 + \sin^2(u - v)$$

$$= [\cos^2(u - v) + \sin^2(u - v)] + 1 - 2 \cos(u - v)$$

$$= 2 - 2 \cos(u - v)$$

Since $\overline{AB}^2 = \overline{CD}^2$, we have

$$2 - 2 \cos u \cos v - 2 \sin u \sin v = 2 - 2 \cos(u - v)$$

$$\cos(u - v) = \cos u \cos v + \sin u \sin v,$$

which completes the proof for $0 \le v \le u < 2\pi$.

Because $\sin(t + 2\pi) = \sin t$ and $\cos(t + 2\pi) = \cos t$ for all t, a proof for $0 \le v < 2\pi$ and $0 \le u < 2\pi$ applies in all cases. Thus it only remains to consider $0 \le u < v < 2\pi$. However, this case follows from the case $0 \le v \le u < 2\pi$, already proved, since $\cos(u - v) = \cos[-(v - u)] = \cos(v - u)$ for all u and v. □

As with other trigonometric identities, (34.1) applies whether u and v represent real numbers, radian measures of angles, or degree measures of angles.

Example 34.1 Compute an exact value for cos 15°.

Solution   Use Identity (34.1) with $u = 45°$ and $v = 30°$. The result is

$$\cos 15° = \cos(45° - 30°)$$

$$= \cos 45° \cos 30° + \sin 45° \sin 30°$$

$$= \frac{\sqrt{2}}{2} \cdot \frac{\sqrt{3}}{2} + \frac{\sqrt{2}}{2} \cdot \frac{1}{2}$$

$$= \frac{1}{4}(\sqrt{6} + \sqrt{2}).$$

You can check this answer (approximately) by finding both $\cos 15°$ and $\frac{1}{4}(\sqrt{6} + \sqrt{2})$ with a calculator or tables. ◼

B. OTHER FORMULAS

The other formulas in this section are all consequences of (34.1) and identities proved earlier.

$$\boxed{\cos(u + v) = \cos u \cos v - \sin u \sin v} \qquad (34.2)$$

Proof   Use $-v$ in place of v in (34.1), and also $\cos(-v) = \cos v$ and $\sin(-v) = -\sin v$.

$$\cos(u + v) = \cos[u - (-v)]$$

$$= \cos u \cos (-v) + \sin u \sin(-v)$$

$$= \cos u \cos v + \sin u (-\sin v)$$

$$= \cos u \cos v - \sin u \sin v \qquad ◼$$

The next two identities should be compared with (30.4), "cofunctions of complementary angles are equal." Because $(\pi/2 - u) + u = \pi/2$, these next two identities are equivalent to (30.4) if $0 < u < \pi/2$. However, in contrast to (30.4), there is no restriction on u here.

$$\boxed{\cos \left(\frac{\pi}{2} - u \right) = \sin u} \qquad (34.3)$$

$$\boxed{\sin \left(\frac{\pi}{2} - u \right) = \cos u} \qquad (34.4)$$

Proof   For (34.3), apply (34.1) with $\pi/2$ in place of u, and u in place of v.

$$\cos \left(\frac{\pi}{2} - v \right) = \cos \frac{\pi}{2} \cos v + \sin \frac{\pi}{2} \sin v$$

$$= 0 \cdot \cos v + 1 \cdot \sin v$$

$$= \sin v$$

For (34.4), we can write

$$\cos u = \cos\left[\frac{\pi}{2} - \left(\frac{\pi}{2} - u\right)\right] = \sin\left(\frac{\pi}{2} - u\right),$$

where the second equality is (34.3) with u replaced by $\frac{\pi}{2} - u$. □

$$\boxed{\sin(u - v) = \sin u \cos v - \cos u \sin v} \qquad (34.5)$$

Proof

$$\sin(u - v) = \cos\left[\frac{\pi}{2} - (u - v)\right] \qquad \text{By (34.3)}.$$

$$= \cos\left[\left(\frac{\pi}{2} - u\right) + v\right]$$

$$= \cos\left(\frac{\pi}{2} - u\right)\cos v$$

$$\quad - \sin\left(\frac{\pi}{2} - u\right)\sin v \qquad \text{By (34.2)}.$$

$$= \sin u \cos v - \cos u \sin v \qquad \text{By (34.3) and (34.4)}. \quad □$$

The next identity follows from (34.5) in the same way that (34.2) follows from (34.1).

$$\boxed{\sin(u + v) = \sin u \cos v + \cos u \sin v} \qquad (34.6)$$

$$\boxed{\tan(u - v) = \frac{\tan u - \tan v}{1 + \tan u \tan v}} \qquad (34.7)$$

Proof

$$\tan(u - v) = \frac{\sin(u - v)}{\cos(u - v)} \qquad \text{Quotient identity.}$$

$$= \frac{\sin u \cos v - \cos u \sin v}{\cos u \cos v + \sin u \sin v} \qquad \text{By (34.5) and (34.1).}$$

$$= \frac{\dfrac{\sin u \cos v}{\cos u \cos v} - \dfrac{\cos u \sin v}{\cos u \cos v}}{\dfrac{\cos u \cos v}{\cos u \cos v} + \dfrac{\sin u \sin v}{\cos u \cos v}} \qquad \begin{array}{l}\text{Divide both} \\ \text{numerator and} \\ \text{denominator by} \\ \cos u \cos v.\end{array}$$

$$= \frac{\tan u - \tan v}{1 + \tan u \tan v} \qquad □$$

Exercise 51 asks you to prove the next identity with the idea in the proof of (34.2).

$$\tan(u + v) = \frac{\tan u + \tan v}{1 - \tan u \tan v} \qquad (34.8)$$

C. MORE EXAMPLES

Example 34.2 Express $\sin 4A \cos A - \cos 4A \sin A$ in terms of a single function.

Solution The given expression has the form of the right side of (34.5), with $u = 4A$ and $v = A$. Therefore,

$$\sin 4A \cos A - \cos 4A \sin A = \sin (4A - A) = \sin 3A. \qquad \blacksquare$$

Example 34.3 Assume that $\cos u = \frac{4}{5}$ and $\sin v = -\frac{5}{13}$, with u in Quadrant I and v in Quadrant III. Find $\cos(u + v)$.

Solution By (34.2),

$$\cos(u + v) = \cos u \cos v - \sin u \sin v.$$

From what is given we can construct the two parts of Figure 34.2. Thus we see that $\cos v = -\frac{12}{13}$ and $\sin u = \frac{3}{5}$. Therefore,

$$\cos(u + v) = (\tfrac{4}{5})(-\tfrac{12}{13}) - (\tfrac{3}{5})(-\tfrac{5}{13})$$

$$= -\tfrac{33}{65}. \qquad \blacksquare$$

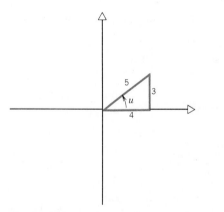
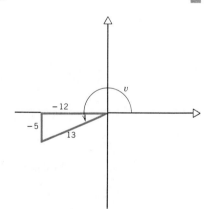

FIGURE 34.2

D. APPLICATION: INCLINATION AND PERPENDICULAR LINES

The following statement was made without proof in Section 13.

Two lines with nonzero slopes m_1 and m_2 are perpendicular iff $m_1 = -1/m_2$. $\qquad (34.9)$

We can prove (34.9) using identities in this section, but first we need some important facts about slope. The **angle of inclination** of a nonhorizontal line L is the smallest counterclockwise angle α from the positive x-axis to L; the angle of inclination of a horizontal line is defined to be zero. Thus $0 \le \alpha < \pi$. Figure 34.3 shows two examples.

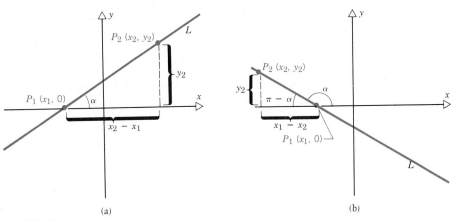

(a)   (b)

FIGURE 34.3

> If m is the slope of a nonvertical line with angle of inclination α, then $m = \tan \alpha$.

(34.10)

Proof of (34.10) If L is horizontal, then $m = 0$ and $\tan \alpha = 0$, so $m = \tan \alpha$. Figure 34.3a shows a case with $0 < \alpha < \pi/2$. With P_1 and P_2 as in the figure, the definition of slope (Section 13B) yields

$$m = \frac{y_2 - y_1}{x_2 - x_1} = \frac{y_2}{x_2 - x_1}.$$

But the fraction on the right equals $\tan \alpha$, so $m = \tan \alpha$.

Figure 34.3b shows a case with $\pi/2 < \alpha < \pi$. In this case $y_2 > 0$ and $x_1 - x_2 > 0$, so

$$\tan(\pi - \alpha) = \frac{y_2}{x_1 - x_2}.$$

But $\tan \alpha = -\tan(\pi - \alpha)$ [from (34.7) with $u = \pi$ and $v = \alpha$]. Thus

$$m = \frac{y_2 - y_1}{x_2 - x_1} = -\frac{y_2}{x_1 - x_2} = -\tan(\pi - \alpha) = \tan \alpha. \qquad \square$$

Proof of (34.9) Assume that L_1 and L_2 are perpendicular with $\alpha_1 < \alpha_2$, as shown in Figure 34.4. Then $\alpha_2 = \alpha_1 + \pi/2$. Therefore, by the identity in Exercise 30,

$$m_2 = \tan \alpha_2 = \tan\left(\alpha_1 + \frac{\pi}{2}\right) = -\cot \alpha_1 = \frac{-1}{m_1}.$$

To prove the converse, assume $m_2 = -1/m_1$. Then $m_1 \ne m_2$ so $\alpha_1 \ne \alpha_2$. Suppose $\alpha_2 > \alpha_1$. Then $0 < \alpha_1 < \pi/2$ and $\pi/2 < \alpha_2 < \pi$, because $\tan \alpha_1$ and

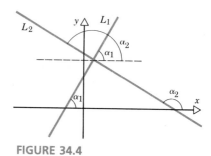

FIGURE 34.4

$\tan \alpha_2$ have opposite signs and both α_1 and α_2 are positive but less than π. Since $m_2 = -1/m_1$, we have

$$\tan \alpha_2 = -\frac{1}{\tan \alpha_1} = -\cot \alpha_1.$$

But $\tan(\alpha_1 + \pi/2) = -\cot \alpha_1$ (Exercise 30), so $\tan \alpha_2 = \tan(\alpha_1 + \pi/2)$. Because $\pi/2 < \alpha_2 < \pi$ and $\pi/2 < \alpha_1 + \pi/2 < \pi$, it follows that $\alpha_2 = \alpha_1 + \pi/2$, so L_1 and L_2 are perpendicular. (At the last step we have used the fact that the tangent function is one-to-one when its domain is restricted to the interval $(\pi/2, \pi)$, a fact that will become clearer in Section 39.) ☐

EXERCISES FOR SECTION 34

In Exercises 1–9 find the exact value. Use functions of $\pi/6$, $\pi/4$, and $\pi/3$.

1. $\tan 75°$

2. $\sin 15°$

3. $\cos 75°$

4. $\cos 7\pi/12$

5. $\tan \pi/12$

6. $\sin 7\pi/12$

7. $\sin 5\pi/12$

8. $\cos \pi/12$

9. $\tan 7\pi/12$

In Exercises 10–16, write each expression in terms of a single function.

10. $\cos t \cos 2t - \sin t \sin 2t$

11. $\dfrac{\tan 4t + \tan t}{1 - \tan 4t \tan t}$

12. $\sin at \cos bt - \cos at \sin bt$

13. $\dfrac{\tan(A/2) - \tan(A/3)}{1 + \tan(A/2) \tan(A/3)}$

14. $\sin(A - B) \cos B + \cos(A - B) \sin B$

15. $\cos B \cos 5B + \sin B \sin 5B$

16. $\sin(\theta - \varphi) \cos(2\varphi - \theta) + \cos(\theta - \varphi) \sin(2\varphi - \theta)$

17. Find $\tan v$ if $\tan u = \frac{2}{3}$ and $\tan(u + v) = \frac{5}{4}$.

18. Find $\tan \beta$ if $\tan \alpha = \frac{1}{5}$ and $\alpha - \beta = \pi/4$.

19. Assume that $\sin u = \frac{1}{3}$ and $\cos v = -\frac{3}{5}$, with u in Quadrant I and v in Quadrant II. Find $\sin(u - v)$.

20. Assume that $\sin u = -\frac{12}{13}$ and $\sin v = \frac{3}{5}$, with u in Quadrant IV and v in Quadrant I. Find $\cos(u - v)$.

21. Assume that $\cos u = \frac{5}{13}$ and $\sin v = \frac{1}{4}$, with u in Quadrant IV and v in Quadrant II. Find $\tan(u + v)$.

Prove each of the following identities. [See the proof of (34.3) for an example like 22–36.]

22. $\cos(u + \pi) = -\cos u$

23. $\sin(u - \pi) = -\sin u$

24. $\cos\left(u + \dfrac{\pi}{2}\right) = -\sin u$

25. $\cos\left(x - \dfrac{\pi}{2}\right) = \sin x$

26. $\sin\left(x + \dfrac{\pi}{2}\right) = \cos x$

27. $\sin\left(x - \dfrac{\pi}{2}\right) = -\cos x$

28. $\tan(\pi - x) = -\tan x$

29. $\tan(x - \pi) = \tan x$

30. $\tan\left(x + \dfrac{\pi}{2}\right) = -\cot x$

31. $\sec\left(\dfrac{\pi}{2} - u\right) = \csc u$

32. $\csc\left(\dfrac{\pi}{2} - u\right) = \sec u$

33. $\cot\left(\dfrac{\pi}{2} - u\right) = \tan u$

34. $\sin\left(\dfrac{3\pi}{2} + x\right) = -\cos x$

35. $\tan\left(x + \dfrac{\pi}{4}\right) = \dfrac{1 + \tan x}{1 - \tan x}$

36. $\tan\left(x - \dfrac{\pi}{4}\right) = \dfrac{\tan x - 1}{\tan x + 1}$

37. $\sin(u + v) + \sin(u - v) = 2 \sin u \cos v$

38. $\cos(u - v) - \cos(u + v) = 2 \sin u \sin v$

39. $(\cos \alpha \cos \beta - \sin \alpha \sin \beta)^2 +$ $(\sin \alpha \cos \beta + \cos \alpha \sin \beta)^2 = 1$

40. $\sin \gamma + \cos \gamma = \sqrt{2} \cos\left(\gamma - \dfrac{\pi}{4}\right)$

41. $\sin\left(\theta + \dfrac{\pi}{3}\right) - \cos\left(\theta + \dfrac{\pi}{6}\right) = \sin \theta$

42. $\sin\left(\theta + \dfrac{\pi}{6}\right) + \cos\left(\theta + \dfrac{\pi}{3}\right) = \cos \theta$

43. $\sec(u - v) = \dfrac{\sec u \sec v}{1 + \tan u \tan v}$

44. $\dfrac{\sin(\alpha + \beta)}{\sin(\alpha - \beta)} = \dfrac{\tan \alpha + \tan \beta}{\tan \alpha - \tan \beta}$

45. $\dfrac{\cos(A - B)}{\cos(A + B)} = \dfrac{1 + \tan A \cdot \tan B}{1 - \tan A \cdot \tan B}$

46. $\tan u + \tan v = \dfrac{\sin(u + v)}{\cos u \cos v}$

47. $\sqrt{2} \cos\left(x - \dfrac{\pi}{4}\right) = \cos x + \sin x$

48. $\sqrt{2} \cos\left(x + \dfrac{\pi}{4}\right) = \cos x - \sin x$

49. Derive the formula
$$\cot(u + v) = \dfrac{\cot u \cot v - 1}{\cot u + \cot v}.$$

50. Express $\sin(u + v + w)$ in terms of functions of u, v, and w. [Suggestion: Begin by expressing $\sin((u + v) + w)$ in terms of functions of $u + v$ and w.]

51. Prove Identity (34.8) by using $-v$ in place of v in Identity (34.7).

52. Prove that if α, β, and γ denote the interior angles of a triangle then $\sin \alpha = \sin \beta \cos \gamma + \cos \beta \sin \gamma$.

53. Prove that $\sec(u + v) = \sec u + \sec v$ is not an identity.

54. Prove that $\csc(u - v) = \csc u - \csc v$ is not an identity.

55. Prove that for all real numbers x and h, with $h \neq 0$,
$$\dfrac{\sin(x + h) - \sin x}{h} =$$
$$\sin x \left(\dfrac{\cos h - 1}{h}\right) + \cos x \left(\dfrac{\sin h}{h}\right).$$

56. Prove that for all real numbers x and h, with $h \neq 0$,
$$\dfrac{\cos(x + h) - \cos x}{h} =$$
$$\cos x \left(\dfrac{\cos h - 1}{h}\right) - \sin x \left(\dfrac{\sin h}{h}\right).$$

57. Find the tangent of the angle of inclination of the line whose equation is $2x - 3y + 5 = 0$.

58. What is the slope of a line whose angle of inclination is $2\pi/3$?

59. Assume that nonvertical lines L_1 and L_2 have angles of inclination α_1 and α_2, respectively, with $\alpha_1 < \alpha_2$ (Figure 34.4). Define the angle θ between L_1 and L_2 by $\theta = \alpha_2 - \alpha_1$. Prove that if L_1 and L_2 are not perpendicular, then
$$\tan \theta = \dfrac{m_2 - m_1}{1 + m_1 m_2},$$
where m_1 and m_2 denote the slopes of L_1 and L_2, respectively.

60. Prove: If the vertex of an angle θ ($0 \leq \theta \leq \pi$) is placed at the center of a circle of unit radius, and d denotes the length of the subtended chord (Figure 34.5), then
$$d = \sqrt{2 - 2 \cos \theta}.$$
[Suggestion: If θ is in standard position, the endpoints of the chord will be at $(1, 0)$ and $(\cos \theta, \sin \theta)$.]

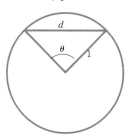

FIGURE 34.5

SECTION **35**
Double- and Half-Angle Formulas

A. DOUBLE-ANGLE FORMULAS

A **double-angle formula** is an identity that expresses a trigonometric function of twice a variable in terms of functions of the variable. Our first double-angle formula expresses $\sin 2u$ in terms of $\sin u$ and $\cos u$.

$$\sin 2u = 2 \sin u \cos u \tag{35.1}$$

Proof Apply Identity (34.6) with $v = u$:

$$\sin 2u = \sin(u + u) = \sin u \cos u + \cos u \sin u = 2 \sin u \cos u. \qquad \square$$

Example 35.1 Find $\sin 2\alpha$ if $\sin \alpha = \frac{4}{5}$ and $\pi/2 < \alpha < \pi$.

Solution Figure 35.1 shows the angle α such that $\sin \alpha = \frac{4}{5}$ and $\pi/2 < \alpha < \pi$. Identity (35.1) yields

$$\sin 2\alpha = 2 \sin \alpha \cos \alpha = 2(\tfrac{4}{5})(-\tfrac{3}{5}) = -\tfrac{24}{25}. \qquad \blacksquare$$

$$\cos 2u = \cos^2 u - \sin^2 u \tag{35.2}$$
$$\cos 2u = 1 - 2 \sin^2 u \tag{35.3}$$
$$\cos 2u = 2 \cos^2 u - 1 \tag{35.4}$$

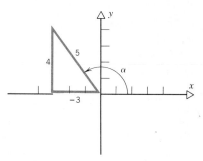

FIGURE 35.1

Proof For (35.2), apply Identity (34.2) with $v = u$:

$$\cos 2u = \cos(u + u) = \cos u \cos u - \sin u \sin u = \cos^2 u - \sin^2 u.$$

For (35.3), replace $\cos^2 u$ by $1 - \sin^2 u$ in (35.2). For (35.4), replace $\sin^2 u$ by $1 - \cos^2 u$ in (35.2). $\qquad \square$

$$\tan 2u = \frac{2 \tan u}{1 - \tan^2 u} \tag{35.5}$$

Proof Apply Identity (34.8) with $v = u$:

$$\tan 2u = \tan(u + u) = \frac{\tan u + \tan u}{1 - \tan u \tan u} = \frac{2 \tan u}{1 - \tan^2 u}. \qquad \square$$

B. IDENTITIES

To prove an identity that has trigonometric functions with different input elements (such as u and $2u$), it is generally best to begin by expressing both sides in terms of functions with the same input, if possible.

Example 35.2 Prove the identity $\dfrac{\sin 2\theta}{1 + \cos 2\theta} = \tan \theta.$

Solution

$$\frac{\sin 2\theta}{1 + \cos 2\theta} = \frac{2 \sin \theta \cos \theta}{1 + (2 \cos^2 \theta - 1)} \qquad \text{By (35.1) and (35.4).}$$

$$= \frac{2 \sin \theta \cos \theta}{2 \cos^2 \theta}$$

$$= \frac{\sin \theta}{\cos \theta}$$

$$= \tan \theta \qquad \blacksquare$$

In the preceding case it sufficed to replace $\sin 2\theta$ and $\cos 2\theta$ using double-angle formulas, and then simplify. To decide which of (35.2)–(35.4) to use for $\cos 2\theta$, think ahead to see which will do the most good; you may need more than one try.

C. HALF-ANGLE FORMULAS

A **half-angle formula** is an identity that expresses a trigonometric function of half of a variable in terms of trigonometric functions of the variable.

$$\sin \frac{u}{2} = \pm \sqrt{\frac{1 - \cos u}{2}} \qquad (35.6)$$

The sign on the right in (35.6) must be chosen to match the sign of $\sin \frac{1}{2}u$: If $\sin \frac{1}{2}u < 0$, choose the negative sign. Otherwise, choose the positive sign.

Proof First, for convenience, we rewrite the double-angle formula (35.3) with v in place of u.

$$\cos 2v = 1 - 2 \sin^2 v$$

Solve this equation for $\sin v$.

$$2 \sin^2 v = 1 - \cos 2v$$

$$\sin^2 v = \frac{1 - \cos 2v}{2}$$

$$\sin v = \pm \sqrt{\frac{1 - \cos 2v}{2}}$$

Now let $2v = u$, so that $v = u/2$. Then the last equation becomes (35.6). $\square$

Example 35.3 Compute the exact value of sin 22.5°.

Solution Use (35.6) with $u = 45°$. Since 22.5° is in Quadrant I, sin 22.5° > 0. Thus we choose the positive sign.

$$\sin 22.5° = \sqrt{\frac{1 - \cos 45°}{2}}$$

$$= \sqrt{\frac{1 - \dfrac{\sqrt{2}}{2}}{2}}$$

$$= \frac{\sqrt{2 - \sqrt{2}}}{2}$$

To compare, a calculator gives sin 22.5° = 0.382683432 and $\sqrt{2 - \sqrt{2}}/2$ = 0.382683433.

$$\boxed{\cos \frac{u}{2} = \pm\sqrt{\frac{1 + \cos u}{2}}} \qquad (35.7)$$

Proof Rewrite (35.4) with v in place of u.

$$\cos 2v = 2 \cos^2 v - 1$$

Solve for cos v.

$$2 \cos^2 v = 1 + \cos 2v$$

$$\cos^2 v = \frac{1 + \cos 2v}{2}$$

$$\cos v = \pm\sqrt{\frac{1 + \cos 2v}{2}}$$

Now let $2v = u$, so that $v = u/2$. Then the last equation becomes (35.7). $\square$

Example 35.4 Compute cos 195°.

Solution Use (35.7) with $u = 390°$. Since 195° is in Quadrant III, cos 195° < 0. Thus we choose the negative sign.

$$\cos 195° = -\sqrt{\frac{1 + \cos 390°}{2}}$$

$$= -\sqrt{\frac{1 + \cos 30°}{2}}$$

$$= -\sqrt{\frac{1 + \dfrac{\sqrt{3}}{2}}{2}}$$

$$= -\frac{\sqrt{2 + \sqrt{3}}}{2} \qquad \blacksquare$$

In the first of the following identities for $\tan \frac{1}{2}u$, the sign must be chosen to match that of $\tan \frac{1}{2}u$. No choice is required for (35.9) and (35.10).

$$\tan \frac{u}{2} = \pm\sqrt{\frac{1 - \cos u}{1 + \cos u}} \qquad (35.8)$$

$$\tan \frac{u}{2} = \frac{\sin u}{1 + \cos u} \qquad (35.9)$$

$$\tan \frac{u}{2} = \frac{1 - \cos u}{\sin u} \qquad (35.10)$$

Proof For (35.8), use $\tan \frac{1}{2}u = \dfrac{\sin \frac{1}{2}u}{\cos \frac{1}{2}u}$ and Identities (35.6) and (35.7).

$$\tan \tfrac{1}{2}u = \frac{\sin \frac{1}{2}u}{\cos \frac{1}{2}u}$$

$$= \frac{\pm\sqrt{\dfrac{1 - \cos u}{2}}}{\pm\sqrt{\dfrac{1 + \cos u}{2}}}$$

$$= \pm\sqrt{\frac{1 - \cos u}{1 + \cos u}}$$

To prove (35.9), use $\theta = u/2$ in Example 35.2. The proof of (35.10) is left as an exercise. $\qquad \square$

Example 35.5 Find $\tan \frac{1}{2}t$ if $\cos t = -\frac{4}{5}$ and $\pi < t < 3\pi/2$.

Solution From $\pi < t < 3\pi/2$ we have $\pi/2 < t/2 < 3\pi/4$. Therefore, $\frac{1}{2}t$ is in Quadrant II and $\tan \frac{1}{2}t < 0$. Thus (35.8) gives

$$\tan \frac{t}{2} = -\sqrt{\frac{1 - \cos t}{1 + \cos t}}$$

$$= -\sqrt{\frac{1 - (-\frac{4}{5})}{1 + (-\frac{4}{5})}}$$

$$= -\sqrt{\frac{\frac{9}{5}}{\frac{1}{5}}}$$

$$= -3.$$

For an alternative solution, use either (35.9) or (35.10), and $\sin t = -\frac{3}{5}$ (which follows since $\cos t = -\frac{4}{5}$ and t is in Quadrant III). ∎

EXERCISES FOR SECTION 35

Use half-angle formulas to compute the exact value of each of the following.

1. $\sin \pi/8$ **2.** $\cos \pi/8$

3. $\tan \pi/8$ **4.** $\cos 165°$

5. $\tan 165°$ **6.** $\sin 195°$

7. $\tan 13\pi/12$ **8.** $\sin 105°$

9. $\cos 75°$

Find sin 2u, cos 2u, and tan 2u under the given conditions.

10. $\sin u = \frac{3}{5}$ and $0 < u < \pi/2$

11. $\tan u = -\frac{1}{3}$ and $\pi/2 < u < \pi$

12. $\cos u = \frac{3}{4}$ and $-\pi/2 < u < 0$

Find sin ½θ, cos ½θ, and tan ½θ under the given conditions.

13. $\cos \theta = -\frac{3}{5}$ and $90° < \theta < 180°$

14. $\cot \theta = \frac{4}{3}$ and $540° < \theta < 630°$

15. $\csc \theta = -\frac{3}{2}$ and $270° < \theta < 360°$

Prove each of the following identities.

16. $2 \csc 2t = \sec t \csc t$

17. $\cos^4 \alpha - \sin^4 \alpha = \cos 2\alpha$

18. $(\sin \theta + \cos \theta)^2 = 1 + \sin 2\theta$

19. $\cos 2\beta = \dfrac{1 - \tan^2 \beta}{1 + \tan^2 \beta}$

20. $\sin 2\alpha = \dfrac{2 \tan \alpha}{1 + \tan^2 \alpha}$

21. $\sec 2\gamma = \dfrac{\sec^2 \gamma}{2 - \sec^2 \gamma}$

22. $\cot 2\theta = \dfrac{\cot^2 \theta - 1}{2 \cot \theta}$

23. $\sin 10t = 2 \sin 5t \cos 5t$

24. $\tan 3A = \dfrac{3 \tan A - \tan^3 A}{1 - 3 \tan^2 A}$

25. $\cos^2 2B - \sin^2 2B = \cos 4B$

26. $\tan 2C = \dfrac{1 - \cos 4C}{\sin 4C}$

27. $\dfrac{2 \tan 3v}{1 - \tan^2 3v} = \tan 6v$

28. $\dfrac{1 + \cos 2u}{\cos u - \sin 2u} = \dfrac{2 \cos u}{1 - 2 \sin u}$

29. $\tan \dfrac{u}{2} = \dfrac{1}{\csc u + \cot u}$

30. $\tan 2u = \dfrac{2 \sin u}{2 \cos u - \sec u}$

31. Write $\sin 3u$ as a function of $\sin u$. (Begin with $3u = 2u + u$.)

32. Write $\cos 3u$ as a function of $\cos u$. (Begin with $3u = 2u + u$.)

33. Write $\cos 4u$ as a function of $\cos u$. [Begin with $4u = 2(2u)$.]

34. Compute $\cos 2t$ if $\sin t = -0.1145$. Ⓒ

35. Compute $\tan 2t$ if $\tan t = 3.1039$. Ⓒ

36. Compute $\sin 2t$ if $\sin t = 0.1010$ and $\pi/2 < t < \pi$. Ⓒ

37. Compute $\tan \frac{1}{2}\alpha$ if $\sin \alpha = -0.7654$ and $7\pi/2 < \alpha < 4\pi$. Ⓒ

38. Compute $\sin \frac{1}{2}\beta$ if $\sin \beta = -0.7654$ and $\pi < \beta < 3\pi/2$. Ⓒ

39. Compute $\cos \frac{1}{2}\gamma$ if $\sec \gamma = 4.5017$ and $2\pi < \gamma < 5\pi/2$. Ⓒ

40. Compute $\sin 165°$ exactly using (a) $165° = 330°/2$ and a half-angle formula, and (b) $165° = 120° + 45°$ and an addition formula. Use a calculator or tables to compare the answers.

41. Prove Identity (35.10) by first using double-angle formulas to prove the identity

$$\frac{1 - \cos 2\theta}{\sin 2\theta} = \tan \theta.$$

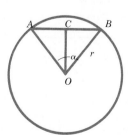

FIGURE 35.2

42. In Figure 35.2, the circle has radius r and OC is perpendicular to AB.

(a) Prove that the area of triangle OAB is $(r^2/2) \sin \alpha$. (Suggestion: See Exercise 85 of Section 31.)

(b) Prove that the area of an n-sided regular polygon inscribed in a circle of radius r is $(nr^2/2) \sin(2\pi/n)$.

SECTION **36**
Product, Sum, and Difference Formulas

A. PRODUCT FORMULAS

Each identity in this subsection expresses a product of two trigonometric functions as a sum or difference of trigonometric functions. Each identity in the next subsection expresses a sum or a difference as a product.

Product formulas

$$\sin u \cos v = \tfrac{1}{2}[\sin(u + v) + \sin(u - v)] \qquad (36.1)$$

$$\cos u \sin v = \tfrac{1}{2}[\sin(u + v) - \sin(u - v)] \qquad (36.2)$$

$$\cos u \cos v = \tfrac{1}{2}[\cos(u + v) + \cos(u - v)] \qquad (36.3)$$

$$\sin u \sin v = \tfrac{1}{2}[\cos(u - v) - \cos(u + v)] \qquad (36.4)$$

Only (36.1) will be proved here. The proofs of (36.2)–(36.4) are similar and are left as exercises.

Proof of (36.1) We begin by recalling (34.6) and (34.5).

$$\sin(u + v) = \sin u \cos v + \cos u \sin v$$

$$\sin(u - v) = \sin u \cos v - \cos u \sin v$$

If we add these two equations, we get

$$\sin(u + v) + \sin(u - v) = 2 \sin u \cos v.$$

Division of both sides by 2 yields (36.1). $\qquad \square$

Example 36.1 Express $\sin 4t \sin 3t$ as a sum or difference.

Solution Use (36.4) with $u = 4t$ and $v = 3t$.

$$\sin 4t \sin 3t = \tfrac{1}{2}[\cos(4t - 3t) - \cos(4t + 3t)]$$

$$= \tfrac{1}{2}(\cos t - \cos 7t)$$

B. SUM AND DIFFERENCE FORMULAS

Sum and difference formulas

$$\sin u + \sin v = 2 \sin \left(\frac{u+v}{2}\right) \cos \left(\frac{u-v}{2}\right) \qquad (36.5)$$

$$\sin u - \sin v = 2 \cos \left(\frac{u+v}{2}\right) \sin \left(\frac{u-v}{2}\right) \qquad (36.6)$$

$$\cos u + \cos v = 2 \cos \left(\frac{u+v}{2}\right) \cos \left(\frac{u-v}{2}\right) \qquad (36.7)$$

$$\cos u - \cos v = -2 \sin \left(\frac{u+v}{2}\right) \sin \left(\frac{u-v}{2}\right) \qquad (36.8)$$

Proof of (36.5)   For convenience, we first rewrite (36.1) with x in place of u and y in place of v:

$$\sin x \cos y = \tfrac{1}{2}[\sin(x+y) + \sin(x-y)]. \qquad (36.9)$$

Now let $u = x + y$ and $v = x - y$. Then

$$\tfrac{1}{2}(u+v) = \tfrac{1}{2}[(x+y) + (x-y)] = x$$

and

$$\tfrac{1}{2}(u-v) = \tfrac{1}{2}[(x+y) - (x-y)] = y.$$

Thus (36.9) becomes

$$\sin \tfrac{1}{2}(u+v) \cos \tfrac{1}{2}(u-v) = \tfrac{1}{2}[\sin u + \sin v]$$

from which

$$\sin u + \sin v = 2 \sin \tfrac{1}{2}(u+v) \cos \tfrac{1}{2}(u-v). \qquad \square$$

The proofs of (36.6)–(36.8) are left as exercises.

Example 36.2   Prove the identity $\dfrac{\sin 5\theta - \sin \theta}{\cos 5\theta + \cos \theta} = \tan 2\theta$.

Solution

$$\frac{\sin 5\theta - \sin \theta}{\cos 5\theta + \cos \theta} = \frac{2 \cos \left(\dfrac{5\theta + \theta}{2}\right) \sin \left(\dfrac{5\theta - \theta}{2}\right)}{2 \cos \left(\dfrac{5\theta + \theta}{2}\right) \cos \left(\dfrac{5\theta - \theta}{2}\right)}$$

$$= \frac{2 \cos 3\theta \sin 2\theta}{2 \cos 3\theta \cos 2\theta}$$

$$= \tan 2\theta$$

C. A SPECIAL SUM FORMULA

Identities such as the following are used in the study of mechanical vibrations, electrical circuits, and sound waves. Examples are given in the exercises for Sections 37 and 40.

> Assume that m, n, and b denote real numbers wtih $m \neq 0$ or $n \neq 0$. Then, for every real number t,
>
> $$m \cos bt + n \sin bt = a \cos(bt - c), \qquad (36.10)$$
>
> where $a = \sqrt{m^2 + n^2}$, $\cos c = m/a$, and $\sin c = n/a$.

Proof By the subtraction formula for cosine,

$$\cos(bt - c) = \cos bt \cos c + \sin bt \sin c. \qquad (36.11)$$

Therefore, Equation (36.10) will be an identity iff

$$m \cos bt + n \sin bt = a \cos bt \cos c + a \sin bt \sin c \qquad (36.12)$$

for every real number t. With $t = 0$, (36.12) implies

$$m = a \cos c \quad \text{and} \quad \cos c = m/a.$$

With $t = \pi/2b$, (36.12) implies

$$n = a \sin c \quad \text{and} \quad \sin c = n/a.$$

Finally,

$$m^2 + n^2 = (a \cos c)^2 + (a \sin c)^2$$
$$= a^2(\cos^2 c + \sin^2 c)$$
$$= a^2.$$

For convenience, we choose $a = \sqrt{m^2 + n^2}$ rather than $a = -\sqrt{m^2 + n^2}$. Then there will always be a unique value of c such that $\cos c = m/a$, $\sin c = n/a$, and $-\pi < c \leq \pi$. $\quad\square$

Example 36.3 Write $\cos 2t - \sqrt{3} \sin 2t$ in the form $a \cos(bt - c)$.

Solution Use Identity (36.10) with $m = 1$, $n = -\sqrt{3}$, and $b = 2$. Then $a = \sqrt{1^2 + (-\sqrt{3})^2} = 2$. Also, $\cos c = \frac{1}{2}$ and $\sin c = -\sqrt{3}/2$, so $c = -\pi/3$ (Figure 36.1). Thus

$$\cos 2t - \sqrt{3} \sin 2t = 2 \cos\left(2t + \frac{\pi}{3}\right).$$

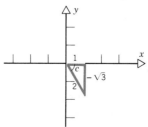

FIGURE 36.1

This can be checked by expanding the right side using the cosine addition formula. ∎

EXERCISES FOR SECTION 36

Express each of the following products as a sum or difference.

1. $\cos 4r \cos r$

2. $\sin 3s \cos 2s$

3. $\sin 5t \sin 6t$

4. $\cos \pi x \sin \pi y$

5. $\sin mx \sin nx$

6. $\sin 3\alpha \cos 3\beta$

Express each of the following sums or differences as a product.

7. $\cos 3\alpha + \cos 4\alpha$

8. $\sin \pi x + \sin \pi y$

9. $\cos 3\theta + \cos \theta$

10. $\sin 5x - \sin 2x$

11. $\cos(x + h) - \cos x$

12. $\sin(x + h) - \sin x$

Evaluate each expression without using a calculator or table.

13. $\cos 75° + \cos 15°$

14. $\sin 165° - \sin 105°$

15. $\sin 195° + \sin 105°$

16. $\sin \dfrac{\pi}{12} \cos \dfrac{5\pi}{12}$

17. $\cos \dfrac{7\pi}{12} \cos \dfrac{11\pi}{12}$

18. $\sin \dfrac{\pi}{12} \sin \dfrac{7\pi}{12}$

Prove each of the following identities.

19. $\dfrac{\sin 3u - \sin u}{\cos 3u - \cos u} = -\cot 2u$

20. $\dfrac{\cos 7v - \cos v}{\cos 7v + \cos v} = -\tan 4v \tan 3v$

21. $\dfrac{\sin w + \sin 3w}{\sin w - \sin 3w} = -\tan 2w \cot w$

22. $\sin(x + y) \sin(x - y) = \sin^2 x - \sin^2 y$

23. $\cos(x + y) \cos(x - y) = \cos^2 x - \sin^2 y$

24. $\sin(x + y) \sin(x - y) = \cos^2 y - \cos^2 x$

25. $\sin u + \sin 2u + \sin 3u = (1 + 2 \cos u) \sin 2u$

26. $\cos u + \cos 2u + \cos 3u = (1 + 2 \cos u) \cos 2u$

27. $4 \sin u \sin 2u \sin 3u = \sin 2u + \sin 4u - \sin 6u$

28. $\dfrac{\sin u + \sin v}{\cos u + \cos v} = \tan \dfrac{1}{2} (u + v)$

29. $\dfrac{\sin u + \sin v}{\cos u - \cos v} = -\cot \dfrac{1}{2} (u - v)$

30. $\dfrac{\cos u - \cos v}{\sin u - \sin v} = -\tan \dfrac{1}{2} (u + v)$

31. Prove Identity (36.2).

32. Prove Identity (36.3).

33. Prove Identity (36.4).

34. Prove Identity (36.6).

35. Prove Identity (36.7).

36. Prove Identity (36.8).

Express each sum in the form $a \cos (bt - c)$.

37. $\sqrt{3} \cos t + \sin t$

38. $2 \cos t - 2 \sin t$

39. $\cos t + \sqrt{3} \sin t$

40. $4 \sin 5t - 3 \cos 5t$

41. $5 \sin 2t + 12 \cos 2t$

42. $\sqrt{2} \sin 3t - \sqrt{2} \cos 3t$

43. Write $2 \cos \dfrac{m\pi x}{d} \cos \dfrac{n\pi x}{d}$ as a sum of two cosine functions. [This problem arises in the theory of Fourier series, which were used by the French mathematician Joseph Fourier (1768–1830) in the study of heat conduction.]

44. Show that $m \cos bt + n \sin bt$ can be expressed in the form $a \sin(bt - c)$. Determine the conditions that a and c must satisfy in terms of m and n.

45. Prove that

$$2 \sin \frac{x}{2} (\sin x + \sin 2x + \sin 3x) =$$

$$\cos \frac{x}{2} - \cos \frac{7x}{2}.$$

Find a similar identity involving

$$2 \sin \frac{x}{2} (\sin x + \sin 2x + \sin 3x + \sin 4x).$$

Do you see a pattern?

SECTION **37**
Trigonometric Equations

A. INTRODUCTION

A **trigonometric equation** is an equation involving one or more trigonometric functions. In particular, each trigonometric identity is a trigonometric equation. If a trigonometric equation is not an identity, then it is a *conditional* trigonometric equation. We now consider how to solve such conditional equations.

Example 37.1 Find all solutions of $\cos t = -1$.

Solution Let $P(t)$ denote the point on the unit circle corresponding to t, as in Section 32. Then $\cos t = -1$ iff the x-coordinate of $P(t)$ is -1, that is, iff $P(t) = (-1, 0)$. It follows that the only solution of $\cos t = -1$ in the interval $[0, 2\pi)$ (that is, for $0 \le t < 2\pi$) is $t = \pi$. However, any value of t that differs from π by an integral multiple of 2π is also a solution, because $P(\pi + k2\pi) = P(\pi)$ for every integer k. Therefore, the set of all solutions is

$$\{\pi + 2k\pi: \quad k \text{ is an integer}\}. \qquad \blacksquare$$

Solutions can be checked by substitution. This step is usually omitted in this book for the sake of brevity.

We shall often need to identify t when we are given $\sin t$, $\cos t$, or $\tan t$ and $t = 0$, $\pi/6$, $\pi/4$, $\pi/3$, $\pi/2$, π, or $3\pi/2$. Thus you may want to review those special cases before doing the exercises.

B. MORE EXAMPLES

If the variable in a trigonometric equation is t, then the equation may contain only trigonometric functions of t (as in Example 37.1), or it may contain trigonometric functions of multiples of t such as $2t$ or $3t$ or $\frac{1}{2}t$. Each equation in this subsection will contain only trigonometric functions of the equation's variable (such as t). For equations of this type it suffices to find all solutions in the interval $[0, 2\pi)$; the other solutions are the numbers obtained from these by adding integral multiples of 2π. Subsection C will treat equations with trigonometric functions of $2t$, $3t$, and so on. Subsection D will treat equations with trigonometric functions of $\frac{1}{2}t$, $\frac{1}{3}t$, and so on.

Example 37.2 Find all solutions of $2 \sin t - 1 = 0$.

Solution First, $2 \sin t - 1 = 0$ iff $\sin t = \frac{1}{2}$. This equation has solutions in each quadrant where $\sin t > 0$. In Quadrant I we have the solution $t = \pi/6$, since $\sin \pi/6 = \frac{1}{2}$. In Quadrant II we have the solution $t = 5\pi/6$, since that is the number whose reference angle is $\pi/6$.

The only two solutions in $[0, 2\pi)$ are $\pi/6$ and $5\pi/6$. The set of all solutions is

$$\left\{\frac{\pi}{6} + 2k\pi: k \text{ is an integer}\right\} \cup \left\{\frac{5\pi}{6} + 2k\pi: k \text{ is an integer}\right\}. \qquad \blacksquare$$

If an equation can be arranged so that one side is 0 and the other side is a product, then we can solve the equation by treating each factor separately, just as with algebraic equations.

Example 37.3 Find the solutions of $\sin x \tan x + \sin x = 0$ in the interval $[0, 2\pi)$.

Solution

$$\sin x \tan x + \sin x = 0$$

$$\sin x(\tan x + 1) = 0$$

$$\sin x = 0 \quad \text{or} \quad \tan x + 1 = 0$$

The solutions of $\sin x = 0$ in the interval $[0, 2\pi)$ are 0 and π. The solutions of $\tan x + 1 = 0$ in the interval $[0, 2\pi)$ are $3\pi/4$ and $7\pi/4$. (For $\tan x = -1$, x must be in Quadrant II or Quadrant IV and the reference angle must be $\pi/4$.) Therefore, the solutions of the original equation in $[0, 2\pi)$ are 0, $3\pi/4$, π, and $7\pi/4$.

If both sides of the equation in Example 37.3 had been divided by $\sin x$ after factoring, then we would have lost the solutions 0 and π; that is, we would have found the solutions of only $\tan x + 1 = 0$. See Example 7.8 and the remarks that accompany it concerning the effect of multiplying or dividing both sides of an equation by an expression that contains a variable. $\qquad \blacksquare$

Example 37.4 Find all solutions of $\sin \theta(\csc \theta - 1) = 0$.

Solution If $\sin \theta = 0$, then $\csc \theta$ is undefined. Therefore, the only solutions are those of $\csc \theta - 1 = 0$. The set of all solutions is

$$\left\{\frac{\pi}{2} + 2k\pi: k \text{ is an integer}\right\}.$$

The solutions of $\sin \theta = 0$ are *extraneous solutions* of the original equation. This example shows why solutions should be checked. $\qquad \blacksquare$

If factoring will not reduce the problem of solving an equation to that of solving simpler equations with only one function each, then we try to use identities to eliminate one or more functions, as in the next example.

Example 37.5 Find the solutions of $2 \sec \theta - \tan^2 \theta = 1$ in the interval $[0, 2\pi)$.

Solution Use the identity $\tan^2 \theta + 1 = \sec^2 \theta$ to express $\tan^2 \theta$ in terms of $\sec \theta$.

$$2 \sec \theta - (\sec^2 \theta - 1) = 1$$

$$2 \sec \theta - \sec^2 \theta = 0$$

$$\sec \theta(2 - \sec \theta) = 0$$

$$\sec \theta = 0 \quad \text{or} \quad 2 - \sec \theta = 0$$

$$\sec \theta = 0 \quad \text{or} \quad \sec \theta = 2$$

The equation $\sec \theta = 0$ has no solution. The solutions of $\sec \theta = 2$ in the interval $[0, 2\pi)$ are $\pi/3$ and $5\pi/3$. Therefore, the solutions of the original equation in $[0, 2\pi)$ are $\pi/3$ and $5\pi/3$. ∎

Example 37.6 Find the solutions of $\sin^2 \alpha - 3 \sin \alpha - 3 = 0$ in the interval $[0, 2\pi)$.

Solution It is not obvious how to factor the left side, so we apply the quadratic formula to find $\sin \alpha$.

$$\sin \alpha = \frac{-(-3) \pm \sqrt{(-3)^2 - 4(1)(-3)}}{2(1)}$$

$$= \frac{3 \pm \sqrt{21}}{2}$$

$$\approx 3.791, \ -0.7913$$

Since $\sin \alpha \leq 1$ for all α, the first value yields no solution. The second value yields solutions in Quadrants III and IV, since that is where $\sin \alpha < 0$. To find the reference angle α_r for these solutions, we use either a calculator or Table VI to find the acute angle α_r such that $\sin \alpha_r = 0.7913$. To the nearest 0.01 radian, $\alpha_r = 0.91$. This gives the solutions $\alpha \approx \pi + 0.91 \approx 4.05$ and $\alpha \approx 2\pi - 0.91 \approx 5.37$.

Remark I. To find α_r with a scientific calculator, enter 0.7913 and then press $\boxed{\sin^{-1}}$ (or press $\boxed{\text{inv}}$ and then $\boxed{\sin}$, where $\boxed{\text{inv}}$ is the inverse function key). If the answer is to be a real number, or in radians, then the calculator must be in the radian mode.

Remark II. If the solution is to be in degrees, then to find α_r, use Table V rather than Table VI (or, with a calculator, the degree mode rather than the radian mode). For $\sin \alpha_r = 0.7913$, either Table V or a calculator will give $\alpha_r \approx 52.3°$. ∎

C. EQUATIONS WITH FUNCTIONS OF *nt*

If an equation contains only functions of nt (n an integer), then we can get all solutions by first finding all solutions for nt and then dividing each of those by n.

Example 37.7 Find all solutions of $\cot 3t + \sqrt{3} = 0$.

Solution First find all solutions for $3t$ in the interval $[0, 2\pi)$.

$$\cot 3t = -\sqrt{3}$$

$$3t = 5\pi/6, \ 11\pi/6$$

The set of all solutions for $3t$ is

$$\left\{ \frac{5\pi}{6} + 2k\pi: \quad k \text{ is an integer} \right\} \cup \left\{ \frac{11\pi}{6} + 2k\pi: \quad k \text{ is an integer} \right\}.$$

Therefore, since

$$\frac{1}{3}\left(\frac{5\pi}{6} + 2k\pi \right) = \frac{5\pi + 12k\pi}{18} \quad \text{and} \quad \frac{1}{3}\left(\frac{11\pi}{6} + 2k\pi \right) = \frac{11\pi + 12k\pi}{18},$$

the set of all solutions is

$$\left\{\frac{5\pi + 12k\pi}{18}: \quad k \text{ is an integer}\right\} \cup \left\{\frac{11\pi + 12k\pi}{18}: \quad k \text{ is an integer}\right\}.$$

Notice that to find all solutions (for t) in the interval $[0, 2\pi)$, we can first find all solutions for $3t$ in the interval $[0, 6\pi)$ and then divide each of those by 3. The result is $5\pi/18$, $11\pi/18$, $17\pi/18$, $23\pi/18$, $29\pi/18$, and $35\pi/18$. These can also be obtained by using $k = 0$, $k = 1$, and $k = 2$ in both parts of the expression giving all solutions. ■

The next example illustrates a way to handle some of the equations that contain both functions of nt ($n \neq 1$) and functions of t. (For variety, x is used in place of t).

Example 37.8 Find all solutions of $\sin 2x = \sin x$ in the interval $[0, 2\pi)$.

Solution Use $\sin 2x = 2 \sin x \cos x$.

$$2 \sin x \cos x = \sin x$$

$$2 \sin x \cos x - \sin x = 0$$

$$\sin x(2 \cos x - 1) = 0$$

$$\sin x = 0 \quad \text{or} \quad \cos x = \tfrac{1}{2}$$

In the interval $[0, 2\pi)$ the solutions of $\sin x = 0$ are 0 and π, and the solutions of $\cos x = \tfrac{1}{2}$ are $\pi/3$ and $5\pi/3$. Therefore, the required solution set is $\{0, \pi/3, \pi, 5\pi/3\}$. ■

It is not always necessary to transform an equation so that functions of only a single variable are involved. To solve $\sin 3t \cos t = 0$, for example, we can solve $\sin 3t = 0$ and $\cos t = 0$, and then form the union of the two solution sets.

D. EQUATIONS WITH FUNCTIONS OF t/n

If an equation contains only functions of t/n (n is a positive integer), then we can get all solutions by first finding all solutions for t/n and then multiplying each of those by n.

Example 37.9 Find all solutions of $\tan \dfrac{t}{10} = \sqrt{3}$.

Solution The solutions for $t/10$ in the interval $[0, 2\pi)$ are $\pi/3$ and $4\pi/3$. Thus the set of all solutions for $t/10$ is

$$\left\{\frac{\pi}{3} + 2k\pi: \quad k \text{ is an integer}\right\} \cup \left\{\frac{4\pi}{3} + 2k\pi: \quad k \text{ is an integer}\right\}.$$

Since

$$10\left(\frac{\pi}{3} + 2k\pi\right) = \frac{10\pi + 60k\pi}{3} \quad \text{and} \quad 10\left(\frac{4\pi}{3} + 2k\pi\right) = \frac{40\pi + 60k\pi}{3},$$

the set of all solutions for t is

$$\left\{ \frac{10\pi + 60k\pi}{3} : \quad k \text{ is an integer} \right\} \cup \left\{ \frac{40\pi + 60k\pi}{3} : \quad k \text{ is an integer} \right\}.$$

Notice that there are no solutions (for t) in the interval $[0, 2\pi)$. ■

EXERCISES FOR SECTION 37

Find all solutions in the interval $[0, 2\pi)$.

1. $\sin t + 1 = 0$

2. $\cot t = 0$

3. $\tan t - 1 = 0$

4. $\csc t + 2 = 0$

5. $2 \cos t + 1 = 0$

6. $\sec t + 1 = 0$

7. $\sec^2 x - 1 = 0$

8. $\tan^2 x - 3 = 0$

9. $\cos^2 x - 4 = 0$

10. $\sin y + \cos y = 0$

11. $\csc^2 y - \csc y - 2 = 0$

12. $\sec(-y) + \sec y = 4$

13. $\sec^2 \theta - 16 = 0$

14. $\tan^2 \theta - 4 \tan \theta + 3 = 0$

15. $2 \csc^2 \theta + \csc \theta - 3 = 0$

16. $\sin u = \cos^2 u$

17. $2 \cos^2 u + 2 \sin u - 1 = 0$

18. $5 \cos u \sin u = \sin u$

19. $2 \sin 2t - 1 = 0$

20. $\cos 3u = 0$

21. $\tan 4v = 1$

22. $\sqrt{3} - 2 \cos \dfrac{\alpha}{2} = 0$

23. $\sqrt{3} \tan \dfrac{\beta}{2} + 1 = 0$

24. $2 \sin \dfrac{\gamma}{3} - 1 = 0$

25. $\sin 2\alpha = \cos \alpha$

26. $\cos 2\beta = \cos \beta$

27. $\cos 2\gamma = \sin \gamma$

28. $\sin^2 \alpha + \sin \alpha = 0$

29. $2 \sec^2 \alpha - 1 = 0$

30. $\tan \alpha \cos \alpha = \tan \alpha$

31. $\tan \beta = 3 \cot \beta$

32. $\csc \beta \cos \beta + \sqrt{2} \cos \beta = 0$

33. $3 \sec 3\beta = 1$

34. $\cot^2 t \cos t - 3 \cos t = 0$

35. $\sin^2 t = \cos^2 t$

36. $\csc^2 t + 2 \csc t + 1 = 0$

37. $2 \cos 4\theta + 1 = 0$

38. $\tan 3\theta = 1$

39. $\csc 4\theta + 2 = 0$

Find all solutions.

40. $3 \sin \dfrac{t}{3} = 0$

41. $2 \cos \dfrac{t}{5} = 1$

42. $2 \sin^2 \dfrac{t}{6} = 1$

43. $\sin(x - \pi) = \cos x$

44. $\tan(x + \pi) + \tan x + 2 = 0$

45. $\sin^3 x + \sin(-x) = 0$

46. $\tan t(1 + \sin t) = 0$

47. $\sin^2 t + \cos^2 t = 0$

48. $\cos t(2 \sec t - 1) = 0$

49. $\tan 3\gamma \left(2 - \csc \dfrac{\gamma}{3} \right) = 0$

50. $\cos \dfrac{\alpha}{3} \cot \dfrac{\alpha}{3} = \cot \dfrac{\alpha}{3}$

51. $2 \sin 4\beta = \sin 4\beta \sec \dfrac{\beta}{4}$

52. $\sin 3t \cos t + \cos 3t \sin t = 1$

53. $\cos 4u \cos u + \sin 4u \sin u = 0$

54. $\dfrac{\tan 3v + \tan v}{1 - \tan 3v \tan v} + \sqrt{3} = 0$

55. $\dfrac{2 \sin x}{\sin \left(x + \dfrac{\pi}{2}\right)} = 1$

56. $2 \cos x \sin \left(x + \dfrac{\pi}{2}\right) - 1 = 0$

57. $4 - 6 \cos 3 \left(x - \dfrac{\pi}{4}\right) = 1$

58. $\sin \theta + \cos \theta = 1$

59. $\sec \alpha = 1 + \tan \alpha$

60. $\sin \beta - \cos \beta = \sqrt{2}$

*Exercises 61–63 concern **Snell's law** (the law of refraction), which asserts that if light passes from one homogeneous medium to another, then the path followed will be made up of two straight line segments, meeting at the boundary between the media in such a way that*

$$\frac{\sin \alpha_1}{\sin \alpha_2} = \frac{c_1}{c_2}, \qquad (37.1)$$

where α_1 is the angle of incidence, α_2 *is the* angle of refraction, c_1 *is the velocity of light in the first medium, and c_2 is the velocity of light in the second medium (Figure 37.1). The ratio $\mu = \sin \alpha_1 / \sin \alpha_2$ is called the* index of refraction *of the second medium relative to the first. The index of refraction for water relative to air is approximately 1.3, and for diamond relative to air is approximately 2.4. [Snell's law explains why a straight object partially submerged in water appears to be bent at the surface. The law was shown by Pierre Fermat (1601–1669) to result from the assumption that the path followed by the light will make the total time required as small as possible.]*

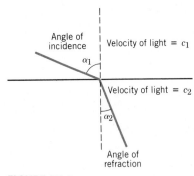

FIGURE 37.1

61. Assume that a beam of light passes from air into

water at an angle of incidence of 30°. Compute α_2, the angle of refraction.

62. Assume that a beam of light passes from water into air at an angle of incidence of 45°. Compute α_2, the angle of refraction.

63. The same as Exercise 61 with water replaced by diamond.

Exercises 64–66 concern oscillatory motion. If an object is attached to a vertical spring, then the spring and the weight of the object combine to determine an equilibrium postion from which the object will not move without being perturbed. We indicate the position of the object by its distance "d" below (positive) or above (negative) its equilibrium position (Figure 37.2). If the object is perturbed (vertically), then the

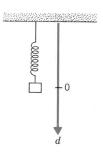

FIGURE 37.2

resulting oscillatory motion of the object is described by an equation of the form

$$d = m \cos bt + n \sin bt, \qquad (37.2)$$

where b is determined by the spring and the weight, and m and n are determined by the position and velocity of the object at the time $t = 0$. [In using (37.2) we are ignoring all forces other than those due to gravity and the spring.] Consider a case where

$$d = \frac{1}{4} \cos 8t + \frac{1}{2} \sin 8t. \qquad (37.3)$$

64. Find the smallest positive value of t for which $d = 0$ in Equation (37.3). What is the physical significance of this value of t?

65. Use Identity (36.10) to write Equation (37.3) in the form $d = a \cos(bt - c)$.

66. (a) Use the answer to Exercise 65 to determine the maximum displacement of the object above the equilibrium position. (Remember that $|\cos x| \le 1$ for all x.)

  (b) Find the smallest positive value of t for which the object reaches its maximum displacement above equilibrium.

REVIEW EXERCISES FOR CHAPTER VII

Find the six trigonometric functions of t *if* P(t) *has the given coordinates.*

1. $\left(-\dfrac{12}{13}, -\dfrac{5}{13}\right)$

2. $\left(-\dfrac{1}{4}, \dfrac{\sqrt{15}}{4}\right)$

3. Find the six trigonometric functions of $2\pi/3$.

4. Find the six trigonometric functions of $-3\pi/4$.

5. Find the six trigonometric functions of $5\pi/6$.

6. Find the six trigonometric functions of $-5\pi/2$.

Find each functional value
(a) with the help of Table VI, and
(b) using a calculator.

7. cos 1.31

8. tan(−0.6)

9. sin 2.81

10. cot 1.06

11. csc(−0.85)

12. sec 4

13. Determine the sign of each trigonometric function of t if $-3\pi/2 < t < -\pi$.

14. In which quadrant is $P(t)$ if $\tan t < 0$ and $\csc t < 0$?

15. Find the six trigonometric functions of 5π.

16. Find all real numbers t such that $-2\pi < t \le 2\pi$ and $\tan t$ is undefined.

17. Express $\tan t$ and $\csc t$ in terms of $\cos t$ if $\pi/2 < t < \pi$.

18. Express $\cot t$ and $\sec t$ in terms of $\sin t$ if $\pi < t < 3\pi/2$.

19. Find the domain of f if $f(t) = \tan\left(2t - \dfrac{\pi}{2}\right)$.

20. Find the angle of inclination of the line whose equation is $4x + y - 1 = 0$.

Prove that each equation in 21–37 is an identity.

21. $\dfrac{\sec t}{\sin t} - \dfrac{\sin t}{\cos t} = \cot t$

22. $1 - \tan^3 u = (1 - \tan u)(\sec^2 u + \tan u)$

23. $\sec^2 \alpha - \csc^2 \alpha = \tan^2 \alpha - \cot^2 \alpha$

24. $(\sec \beta - \tan \beta)^2 = \dfrac{1 - \sin \beta}{1 + \sin \beta}$

25. $\dfrac{\cot \gamma - 1}{\cot \gamma + 1} = \dfrac{\cos 2\gamma}{1 + \sin 2\gamma}$

26. $\tan \dfrac{\theta}{2} = \csc \theta - \cot \theta$

27. $\dfrac{1 - \cos 2t}{1 + \cos 2t} = \tan^2 t$

28. $\cos A - \sin \dfrac{A}{2} = \left(1 - 2\sin \dfrac{A}{2}\right)\left(1 + \sin \dfrac{A}{2}\right)$

29. $\sin^3 t = \dfrac{1}{2}\sin t\,(1 - \cos 2t)$

30. $\sec \dfrac{t}{2} + \csc \dfrac{t}{2} = \dfrac{2\left(\sin \dfrac{t}{2} + \cos \dfrac{t}{2}\right)}{\sin t}$

31. $\sec(\pi + \theta) = -\sec \theta$

32. $\tan(\pi - \theta) = -\tan \theta$

33. $\dfrac{\cos u + \cos v}{\cos u - \cos v} =$
$-\cot \dfrac{1}{2}(u + v)\cot \dfrac{1}{2}(u - v)$

34. $\dfrac{\sin u + \sin v}{\sin u - \sin v} = \dfrac{\tan \dfrac{1}{2}(u + v)}{\tan \dfrac{1}{2}(u - v)}$

35. $\cos u + \tan u = \sin u(\cot u + \sec u)$.

36. $\sin 4t = 4(\sin t \cos^3 t - \sin^3 t \cos t)$.
[Begin with $4t = 2(2t)$.]

37. $\dfrac{\tan \alpha}{1 - \cot \alpha} + \dfrac{\cot \alpha}{1 - \tan \alpha} =$
$\tan \alpha + \cot \alpha + 1$.

38. Prove that if $f(x) = \tan x$, then
$\dfrac{f(x + h) - f(x)}{h} =$

$\dfrac{\tan h}{h} \cdot \dfrac{1}{1 - \tan x \tan h} \cdot \sec^2 x$

for all real numbers x and h for which both sides are defined.

39. Find the exact value of cos 15° with a half-angle formula and then with a subtraction formula. Use a calculator or table to compare decimal approximations for the answers.

40. Find the exact value of tan 105° with a half-angle formula and then with an addition formula. Verify that the answers are equal.

41. Find $\sin(u - v)$ if $\sin u = -3/5$ with u in Quadrant IV and tan $v = 12/5$ with v in Quadrant III.

42. Find $\cos(u + v)$ if $\sin u = 4/5$ and $\cos v = -12/13$, with u in Quadrant I and v in Quadrant II.

43. Find $\sin(t/2)$ if $\cos t = -1/3$ and $\pi/2 < t < \pi$.

44. If $\cos t = -a$ and $\tan(-t) = b$, what is $\tan t - \sec(-t)$?

Express each sum in the form $a \cos(bt - c)$.

45. $2 \cos 3t - \sin 3t$

46. $\sqrt{3} \cos 5t + \sin 5t$

Find all solutions in the interval $[0, 2\pi)$.

47. $2 \cos^2 t - 3 \sin t - 3 = 0$

48. $\sin u + \sin(u/2) = 0$

49. $\tan 3\alpha = 1$

50. $2 \sin 3\alpha = 1$

51. $\sec(\beta/3) + 1 = 0$

52. $\cos(\beta/2) + \sqrt{3}/2 = 0$

Find all solutions.

53. $\sin^2 t - \sin t - 2 = 0$

54. $\cos^2 3u - \sin^2 3u = 0$

55. $\csc 2t = -1$

56. $\cot (t/5) = \sqrt{3}$

57. The angle of elevation to the top of a tree from a point 200 feet from the base and on the same level as the base is 24°. How tall is the tree? Ⓒ

58. Find the acute angles of a right triangle if the lengths of the two legs are 10 centimeters and 15 centimeters, respectively. Ⓒ

CHAPTER VIII

GRAPHS AND INVERSES OF TRIGONOMETRIC FUNCTIONS

By the *graph* of the sine function we mean the graph of the equation $y = \sin x$, where x (as well as y) denotes a real number. The graphs of the other trigonometric functions are defined similarly. In addition to graphs of equations like $y = \sin x$, this chapter also analyzes graphs of equations like $y = d + a \sin(bx - c)$, which arise naturally in a number of applications.

In the last section of this chapter we study inverse trigonometric functions. These functions could have been presented earlier, but they are more easily understood with the aid of graphs.

It is suggested that you quickly review the general principles about graphs from Section 20 A–C.

SECTION **38**
Graphs Involving the Sine and Cosine

A. PERIODIC FUNCTIONS

Loosely, behavior is periodic if it repeats itself at regular intervals. This notion applies to functions as follows. Suppose that f is a function and that p is a positive real number. We say that f is **periodic** with **period** p if

$$f(x + p) = f(x) \tag{38.1}$$

whenever x and $x + p$ are in the domain of f. If a nonconstant function is periodic, then the smallest positive number p satisfying (38.1) is called the **fundamental period** of the function.

Example 38.1 Figure 38.1 shows the graph of a function f that is periodic with fundamental period 2. For example, $f(1) = 1$ and $f(3) = 1$, so that $f(1 + 2) = f(1)$. More generally, $f(x + 2) = f(x)$ for every x.

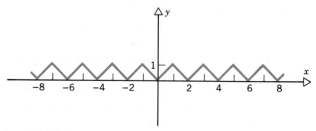

FIGURE 38.1

Notice that *any* given integer is a period for f. For example, 4 is a period because $f(x + 4) = f((x + 2) + 2) = f(x + 2) = f(x)$ for every x. However, the *fundamental period* is 2. ■

The graph in Figure 38.1 is typical of the graphs of periodic functions in that each such graph consists of a repeating pattern. This applies, in particular, to the graphs of the trigonometric functions: we shall show now that the trigonometric functions are periodic, and then we shall consider their graphs. The graphs will show that the periods given by the following identities are actually the fundamental periods—that is, π for the tangent and cotangent, and 2π for the other functions.

$$
\begin{array}{ll}
\sin(t + 2\pi) = \sin t & \csc(t + 2\pi) = \csc t \\
\cos(t + 2\pi) = \cos t & \sec(t + 2\pi) = \sec t \\
\tan(t + \pi) = \tan t & \cot(t + \pi) = \cot t
\end{array} \tag{38.2}
$$

Proof The circumference of the unit circle is 2π, so $P(t + 2\pi) = P(t)$ for every t. Since each function is defined in terms of the coordinates of $P(t)$, it follows that each function is periodic with period 2π. This justifies the equations involving sine, cosine, secant, and cosecant in (38.2).

For any t, $P(t)$ and $P(t + \pi)$ are located symmetrically with respect to the origin (Figure 38.2). Therefore, if $P(t)$ has coordinates (x, y), then $P(t + \pi)$ has coordinates $(-x, -y)$. Thus

$$\tan(t + \pi) = (-y)/(-x) = y/x = \tan t$$

and

$$\cot(t + \pi) = 1/\tan(t + \pi) = 1/\tan t = \cot t. \qquad \square$$

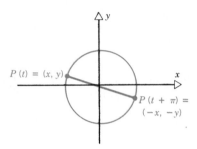

FIGURE 38.2

B. THE GRAPH OF $y = \sin x$

If f is any function whose domain and range are sets of real numbers, then the graph of f is the graph of the equation $y = f(x)$. In particular, the graph of the sine function is the graph of $y = \sin x$.

From (38.2) the sine function is periodic with period 2π. Therefore, it suffices to determine the graph for x in the interval $0 \le x < 2\pi$. We call this a **cycle** of the graph, because 2π is the fundamental period. The graph will repeat itself for x outside of the interval $0 \le x < 2\pi$. Table 38.1 shows some coordinate pairs for points on the graph. More points can be determined with either a calculator or Table VI. As such points are plotted, they will slowly reveal the graph, which is shown in Figure 38.3 for $-2\pi \le x < 4\pi$ (three fundamental periods).

TABLE 38.1

x	0	$\dfrac{\pi}{4}$	$\dfrac{\pi}{2}$	$\dfrac{3\pi}{4}$	π	$\dfrac{5\pi}{4}$	$\dfrac{3\pi}{2}$	$\dfrac{7\pi}{4}$
$y = \sin x$	0	0.707	1	0.707	0	-0.707	-1	-0.707

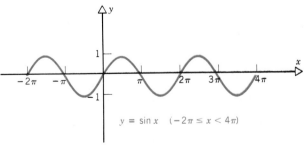

$$y = \sin x \quad (-2\pi \le x < 4\pi)$$

FIGURE 38.3

A graph with the general shape of the graph of $y = \sin x$ is said to be **sinusoidal.** In addition to being sinusoidal, the graph of $y = \sin x$ has the following important features.

- The fundamental period is 2π.
- The x-intercepts occur at the integral multiples of π: 0, $\pm\pi$, $\pm 2\pi$, ...
- The maximum y-value, which is 1, occurs at $x = \pi/2$ and at the other x-values different from $\pi/2$ by integral multiples of 2π, that is, at $\pi/2 + 2n\pi$ ($n = 0$, ± 1, ± 2, . . .).
- The minimum y-value, which is -1, occurs at $x = 3\pi/2$ and at the other x-values different from $3\pi/2$ by integral multiples of 2π, that is, at $3\pi/2 + 2n\pi$ ($n = 0$, ± 1, ± 2, . . .).
- The graph is symmetric with respect to the origin, because the sine function is odd, that is, $\sin(-x) = -\sin x$ [by (32.5)].

C. THE GRAPH OF $y = \cos x$

Figure 38.4 shows the graph of $y = \cos x$ for $-2\pi \leq x < 4\pi$. Table 38.2 shows some of the coordinate pairs; a calculator or Table VI will give others. If you compare Figure 38.4 with Figure 38.3, you will see that the graph of $y = \cos x$ can be obtained by translating the graph of $y = \sin x$ to the left by $\pi/2$ units. This is also a consequence of Statement (20.2) and the identity $\cos x = \sin(x + \pi/2)$ (Exercise 26 of Section 34). The graph of $y = \cos x$ is symmetric with respect to the y-axis, because the cosine function is even, that is, $\cos(-x) = \cos x$ [by (32.5)].

TABLE 38.2

x	0	$\dfrac{\pi}{4}$	$\dfrac{\pi}{2}$	$\dfrac{3\pi}{4}$	π	$\dfrac{5\pi}{4}$	$\dfrac{3\pi}{2}$	$\dfrac{7\pi}{4}$
$y = \cos x$	1	0.707	0	-0.707	-1	-0.707	0	0.707

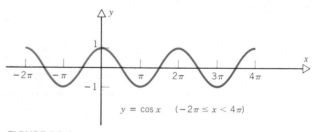

$y = \cos x \quad (-2\pi \leq x < 4\pi)$

FIGURE 38.4

D. THE GRAPHS OF $y = a \sin x$ and $y = a \cos x$

Example 38.2 Sketch the graph of $y = 3 \sin x$ for $0 \leq x < 4\pi$.

Solution Table 38.3 shows coordinate pairs for both $y = \sin x$ and $y = 3 \sin x$. Figure 38.5 shows the graphs. Each point on the graph of $y = 3 \sin x$ is obtained by simply multiplying the y-coordinate of a corresponding point on the graph of $y = \sin x$ by 3. For instance, instead of $(\pi/2, 1)$ (on the graph of $y = \sin x$), we get $(\pi/2, 3)$ (on the graph of $y = 3 \sin x$). The x-intercepts for $y = 3 \sin x$ are the same as those for $y = \sin x$. The graph is sinusoidal—it is just the graph of $y = \sin x$ stretched vertically by a factor of 3. ■

TABLE 38.3

x	0	$\dfrac{\pi}{4}$	$\dfrac{\pi}{2}$	$\dfrac{3\pi}{4}$	π	$\dfrac{5\pi}{4}$	$\dfrac{3\pi}{2}$	$\dfrac{7\pi}{4}$
$y = \sin x$	0	0.707	1	0.707	0	-0.707	-1	-0.707
$y = 3 \sin x$	0	2.121	3	2.121	0	-2.121	-3	-2.121

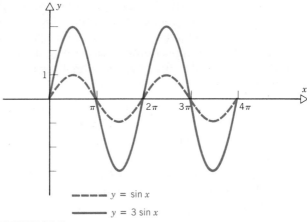

FIGURE 38.5

To sketch a sinusoidal graph, determine the x-intercepts and the maximum and minimum points and then rely on Figure 38.3 to give the general shape. Do not plot a large number of points. A similar remark will apply to graphs involving other functions.

Example 38.3   Sketch the graph of $y = -2 \cos x$ for $0 \le x < 4\pi$.

Solution   Multiply each y-coordinate in Table 38.2 by -2. The graph is shown in Figure 38.6. We can think of this graph as the result of modifying the graph of $y =$

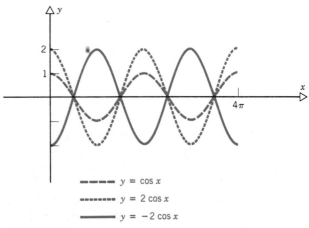

FIGURE 38.6

cos x in two steps: First, stretch $y = \cos x$ vertically by a factor of 2 to get the graph of $y = 2 \cos x$. Second, reflect the graph of $y = 2 \cos x$ through the x-axis; that is, replace each point (x, y) by the point $(x, -y)$. To illustrate, Figure 38.6 shows the graphs of $y = \cos x$ and $y = 2 \cos x$ along with the graph of $y = -2 \cos x$. ∎

If we replace the coefficient 3 in Example 38.2 by $\frac{1}{2}$, or any other positive number less than 1, then the effect will be to compress the graph in the vertical direction rather than to stretch it. The maximum y-value for the graph of $y = \frac{1}{2} \sin x$ will be $y = \frac{1}{2}$. The number $|a|$ is the **amplitude** of $y = a \sin x$ or $y = a \cos x$; it is the maximum distance of the graph from the x-axis. The amplitude of $y = -2 \cos x$ is 2. More generally, the **amplitude** of a sinusoidal graph is one-half of the absolute value of the difference of the maximum and minimum y-values of the graph. (This definition applies to sinusoidal graphs that have been translated upward or downward from the x-axis, as well as to cases like $y = a \sin x$ and $y = a \cos x$. Section 40 gives examples.) Our observations lead to the following principles.

- If $a > 0$ then the graph of $y = a \sin x$ has the same general shape as the graph of $y = \sin x$, except if $a > 1$ then the graph is stretched vertically, and if $a < 1$ then the graph is compressed vertically. A similar statement applies to $y = a \cos x$.

- If $a < 0$, then the graph of $y = a \sin x$ is obtained from the graph of $y = |a| \sin x$ by reflection through the x-axis. A similar statement applies to $y = a \cos x$.

E. THE GRAPHS OF $y = \sin bx$ and $y = \cos bx$

Example 38.4 Sketch the graph of $y = \sin 2x$ for $0 \le x < 2\pi$.

Solution Figure 38.7 shows the graph. The fundamental period of $y = \sin x$ is 2π. Therefore, as we have seen, the graph of $y = \sin x$ goes through one complete cycle as x increases from 0 to 2π. In the same way, the graph of $y = \sin 2x$ goes through one complete cycle as $2x$ increases from 0 to 2π, which means as x increases from 0 to π.

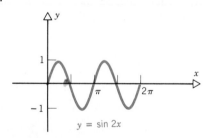

$y = \sin 2x$

FIGURE 38.7

The graph of $y = \sin x$ has x-intercepts $0, \pm\pi, \pm2\pi, \ldots$, so the graph of $y = \sin 2x$ has x-intercepts $0, \pm\pi/2, \pm\pi, \ldots$. A maximum of $y = \sin x$ occurs at $x = \pi/2$; the corresponding maximum of $y = \sin 2x$ occurs at $x = \pi/4$ since $\sin 2(\pi/4) = \sin \pi/2 = 1$.

Briefly, the graph of $y = \sin 2x$ is obtained from the graph of $y = \sin x$ by compressing in the horizontal direction by a factor of 2. ∎

Example 38.4 is a special case of $y = \sin bx$ with $b > 0$. The graph of $y = \sin bx$ goes through one complete cycle as x varies from 0 to $2\pi/b$, because $0 \leq bx < 2\pi$ iff $0 \leq x < 2\pi/b$. That is:

> The fundamental period of $y = \sin bx$ $(b > 0)$ is $2\pi/b$.

Example 38.5 Sketch the graph of $y = \sin \dfrac{\pi}{5} x$ for $-p \leq x < p$, where p is the fundamental period.

Solution By the remark above, the fundamental period is $2\pi/(\pi/5) = 10$. Figure 38.8 shows the graph. ▪

Examples 38.4 and 38.5 should convince you that if $b > 1$, then the graph of $y = \sin bx$ results from the graph of $y = \sin x$ by horizontal compression. And if $0 < b < 1$, then the graph of $y = \sin bx$ results from the graph of $y = \sin x$ by horizontal stretching. Similar statements are true for $y = \cos bx$.

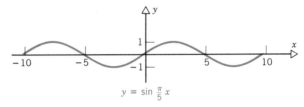

$y = \sin \frac{\pi}{5} x$

FIGURE 38.8

Now consider $y = \sin(-bx)$ with $b > 0$. By (32.5), $\sin(-bx) = -\sin bx$. Thus the graph of $y = \sin(-bx)$ results from the graph of $y = \sin bx$ by reflection through the x-axis.

On the other hand, (32.5) implies that $\cos(-bx) = \cos bx$. Thus the graph of $y = \cos(-bx)$ is the same as the graph of $y = \cos bx$.

F. THE GRAPHS OF $y = a \sin bx$ AND $y = a \cos bx$

To graph an equation of the form $y = a \sin bx$, we combine the methods already considered. Since $a \sin(-bx) = -a \sin bx$, we can always write the equation so that $b > 0$. [For example, $2 \sin(-3x) = -2 \sin 3x$.]

To graph $y = a \sin bx$ $(b > 0)$

- The graph is sinusoidal. Sketch one cycle and then continue as far as needed by using periodicity.
- The fundamental period is $2\pi/b$. One cycle begins at $x = 0$ and ends at $x = 2\pi/b$*.
- The x-intercepts are at $x = 0$ and $x = 2\pi/b$, and also at $x = \pi/b$, one-half of the way from 0 to $2\pi/b$.

*Technically, the cycle begins at $x = 0$ and goes up to but does not include $x = 2\pi/b$.

- If $a > 0$, a maximum y-value occurs for x one-fourth of the way from 0 to $2\pi/b$. The maximum y-value is the amplitude, a, if $a > 0$.
- If $a > 0$, a minimum y-value occurs for x three-fourths of the way from 0 to $2\pi/b$. The minimum y-value is $-a$ if $a > 0$.
- If $a < 0$, sketch the graph of $y = |a|\sin bx$ and then reflect it through the x-axis.

Example 38.6 Sketch the graph of $y = \frac{3}{2}\sin(-x/2)$ for $0 \le x < p$, where p is the fundamental period.

Solution First, $\frac{3}{2}\sin(-x/2) = -\frac{3}{2}\sin(x/2)$. Therefore, we graph $y = \frac{3}{2}\sin(x/2)$ and then reflect through the x-axis.

The period is $2\pi/(1/2) = 4\pi$. The amplitude is $\frac{3}{2}$. There is a maximum [for $y = \frac{3}{2}\sin(x/2)$] at $x = \pi$, and a minimum at $x = 3\pi$. Figure 38.9 shows the results. ∎

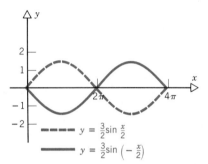

FIGURE 38.9

To graph an equation of the form $y = a\cos bx$, follow the ideas given for obtaining the graph of $y = a\sin bx$ from $y = \sin x$; proceed from $y = \cos x$ to $y = \cos bx$ to $y = a\cos bx$.

Because the cosine function is an even function, the graph of $y = a\cos(-bx)$ is the same as the graph of $y = a\cos bx$. Thus we can always write the equation so that $b > 0$.

$$\text{To graph } y = a\cos bx \ (b > 0)$$

- The graph is sinusoidal. Sketch one cycle and then continue as far as needed by using periodicity.
- The fundamental period is $2\pi/b$.
- If $a > 0$, the graph has a maximum y-value at $x = 0$ and another at $x = 2\pi/b$. A minimum y-value occurs at $x = \pi/b$, one-half of the way from $x = 0$ to $x = 2\pi/b$. The maximum y-value is a and the minimum y-value is $-a$.
- There are x-intercepts one-fourth and three-fourths of the way from $x = 0$ to $x = 2\pi/b$.
- If $a < 0$, sketch the graph of $y = |a|\cos bx$ and then reflect it through the x-axis.

EXERCISES FOR SECTION 38

Each figure in Exercises 1–3 shows the graph of a function for $0 \leq x < 2$. Assume that the function is periodic with fundamental period 2 and then draw the graph for $-2 \leq x < 6$.

1.

2.

3.

4. Suppose that f is periodic with fundamental period 3 and that $f(x) = 4 - x$ for $0 \leq x < 3$.
(a) What is $f(5)$?
(b) What is $f(10)$?
(c) What is $f(-6)$?
[Suggestion: $f(10) = f(7) = f(4) = f(1)$.]

5. Suppose that f is periodic with fundamental period 5 and that $f(x) = x^2$ for $-2 \leq x < 3$.
(a) What is $f(6)$?
(b) What is $f(13)$?
(c) What is $f(-3)$?

6. Suppose that f is periodic with fundamental period 10 and that $f(x) = x + 2$ for $10 \leq x < 20$.
(a) What is $f(5)$?
(b) What is $f(21)$?
(c) What is $f(-37)$?

7. Find all x such that $-8\pi \leq x < 8\pi$ and $\sin x = 0$.

8. Find all x such that $-4\pi \leq x < 4\pi$ and $\sin x = -1$.

9. Find all x such that $-4\pi \leq x < 4\pi$ and $\sin x = 1$.

10. Find all x such that $-8\pi \leq x < 8\pi$ and $\cos x = 1$.

11. Find all x such that $-4\pi \leq x < 4\pi$ and $\cos x = 0$.

12. Find all x such that $-4\pi \leq x < 4\pi$ and $\cos x = -1$.

Sketch the graph of each equation for $0 \leq x < 4\pi$.

13. $y = 2 \sin x$
14. $y = 1.5 \cos x$
15. $y = 0.5 \sin x$
16. $y = -3 \cos x$
17. $y = -5 \sin x$
18. $y = -4 \cos x$

Sketch the graph of each equation for $0 \leq x < p$, where p is the fundamental period.

19. $y = \sin(x/3)$
20. $y = \sin \pi x$
21. $y = \sin 4x$
22. $y = \cos 2\pi x$
23. $y = \cos 4x$
24. $y = \cos(x/2)$

25. Determine $a > 0$ and $b > 0$ so that $y = a \cos bx$ has fundamental period $\pi/2$ and amplitude 6.

26. Determine $a > 0$ and $b > 0$ so that $y = a \cos bx$ has fundamental period 10π and amplitude 0.1.

27. Determine $a > 0$ and $b > 0$ so that $y = a \cos bx$ has fundamental period 6 and amplitude 5.

28. Determine $a > 0$ and $b < 0$ so that $y = a \sin bx$ has fundamental period 5π and amplitude 0.4.

29. Determine $a > 0$ and $b > 0$ so that $y = a \sin bx$ has fundamental period 4 and amplitude 10.

30. Determine $a > 0$ and $b > 0$ so that $y = a \sin bx$ has fundamental period $\pi/6$ and amplitude $\sqrt{2}$.

Sketch the graph of each equation for $0 \leq x < p$, where p is the fundamental period.

31. $y = 3 \sin\left(-\dfrac{3}{2}x\right)$

32. $y = -2 \sin\dfrac{1}{4}x$

33. $y = -2.5 \sin\left(-\dfrac{3\pi}{2}x\right)$

34. $y = -\cos 3x$

35. $y = 5 \cos \pi x$

36. $y = -1.5 \cos(-2x)$

Prove each of the following identities using the ideas in the proof of (38.2). Then give another proof using an addition formula (from Section 34).

37. (a) $\sin(t + \pi) = -\sin t$

    (b) $\csc(t + \pi) = -\csc t$

38. (a) $\cos(t + \pi) = -\cos t$

    (b) $\sec(t + \pi) = -\sec t$

39. Suppose that $y = f(x)$ is periodic with fundamental period p, and that b is a positive real number. What is the fundamental period of $y = f(bx)$?

When a tuning fork vibrates to produce a pure tone, the motion of each of its tines about its equilibrium position varies sinusoidally with respect to time. Such motion is called **simple harmonic motion.** *The* **frequency,** *denoted by* f, *is the number of cycles per second. The unit of measure for frequency is the* **hertz,** *abbreviated Hz. Thus 1 cycle/sec = 1 Hz. Exercises 40–42 involve simple harmonic motion de-scribed by an equation*

$$x = a \sin bt \qquad (38.3)$$

where t *denotes time in seconds and* x *denotes distance from the equilibrium position measured in centimeters.*

40. (a) Explain why the period is the reciprocal of the frequency of $a \sin bt$. (Assume $b > 0$).

    (b) Use part (a) to explain why $f = b/2\pi$.

    (c) What value of b in Equation (38.3) corresponds to a frequency of 440 Hz (concert A)?

41. Sketch the graph describing the simple harmonic motion given by (38.3) with amplitude 0.02 cm and frequency 200 Hz. [By part (b) of Exercise 40, $f = b/2\pi$.]

42. Sketch the graph describing the simple harmonic motion produced by a synthesizer with amplitude 0.05 cm and period 0.004 sec/cycle. [Use (38.3).] What is the frequency?

SECTION **39**

Graphs Involving the Tangent, Cotangent, Secant, and Cosecant

A. GRAPHS INVOLVING THE TANGENT

Figure 39.1 shows the graph of $y = \tan x$ for $-3\pi/2 < x < 3\pi/2$. Table 39.1 shows some coordinate pairs; a calculator or Table VI will give others. The graph

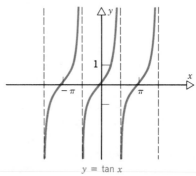

FIGURE 39.1

reveals that the fundamental period of the tangent function is π, as claimed in Section 38.

TABLE 39.1

x	0	0.2	0.4	0.6	0.8	1.0	1.2	1.4
$y = \tan x$	0	0.203	0.423	0.684	1.03	1.56	2.57	5.80

Because $\tan(-x) = -\tan x$ for all x, the tangent function is odd and the graph is symmetric with respect to the origin. The x-intercepts are at the integral multiples of π. The tangent function is undefined at $\pi/2$ and at other values that differ from $\pi/2$ by integral multiples of π, that is, at $\pm\pi/2, \pm3\pi/2, \ldots$. (These correspond to the points on the unit circle where the first coordinate is 0.) The values where $\tan x$ is undefined also yield vertical asymptotes for the graph. For example, in the notation of Section 22,

$$\sin x \to 1 \quad \text{as} \quad x \to \frac{\pi^-}{2}$$

and

$$\cos x \to 0^+ \quad \text{as} \quad x \to \frac{\pi^-}{2},$$

so that

$$\tan x = \frac{\sin x}{\cos x} \to \infty \quad \text{as} \quad x \to \frac{\pi^-}{2}.$$

The asymptotes are shown as dashed lines.

The graph of $y = a \tan bx$, with $b > 0$, goes through one complete cycle as x varies from $-\pi/(2b)$ to $\pi/(2b)$, because $-\pi/2 < bx < \pi/2$ iff $-\pi/(2b) < x < \pi/(2b)$. That is:

The fundamental period of

$$y = \tan bx \ (b > 0) \text{ is } \pi/b.$$

To graph $y = a \tan bx \ (b > 0)$

- Sketch one cycle and then continue as far as needed by using periodicity.
- The fundamental period is π/b.
- There are vertical asymptotes at $x = \pm\pi/(2b)$.
- There is an x-intercept at $x = 0$ (midway between the nearest vertical asymptotes).
- If $a > 0$, the function is increasing for all x between adjacent asymptotes.
- If $a < 0$, sketch the graph of $y = |a| \tan bx$ and then reflect it through the x-axis.

To graph $y = a \tan bx \ (b < 0)$, we can use $a \tan(-bx) = -a \tan bx$ (since the tangent function is odd). This reduces the problem to the type treated above.

Example 39.1 Sketch the graph of $y = \tan \frac{1}{2}x$ for $-p/2 < x < p/2$, where p is the fundamental period.

Solution The fundamental period is $\pi/\frac{1}{2} = 2\pi$. There are vertical asymptotes at $x = \pm\pi$, and there is an x-intercept at $x = 0$. Figure 39.2 shows the graph. ■

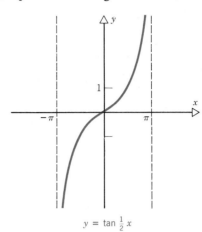

$y = \tan \frac{1}{2} x$

FIGURE 39.2

B. GRAPHS INVOLVING THE COTANGENT

Figure 39.3 shows the graph of $y = \cot x$ for $-\pi < x < \pi$. The general characteristics of the graph can be obtained using the graph of $y = \tan x$ and the identity $\cot x = 1/(\tan x)$. The fundamental period for $y = \cot x$ is π.

The graph of $y = \cot x$ has vertical asymptotes where $\tan x = 0$. And $\cot x = 0$ where $\tan x$ is undefined. Finally, $\tan x > 0$ iff $\cot x > 0$. Exercise 4 asks you to rewrite the six remarks preceding Example 39.1 so that they apply to $y = a \cot bx$ rather than $y = a \tan bx$.

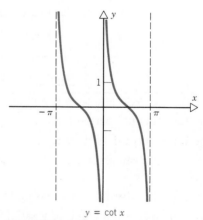

$y = \cot x$

FIGURE 39.3

Example 39.2 Sketch the graph of $y = 4 \cot \pi x$ for $0 < x < 2p$, where p is the fundamental period.

Solution The fundamental period is $\pi/\pi = 1$. We first sketch the cycle where $0 < x < 1$. There are vertical asymptotes at $x = 0$ and $x = 1$. There is an x-intercept at $x = \frac{1}{2}$, midway between the nearest vertical asymptotes. The factor 4 causes vertical stretching.

For $1 < x < 2$, we merely repeat the graph from $0 < x < 1$. Figure 39.4 shows the result. ◾

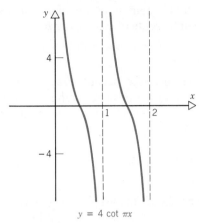

$y = 4 \cot \pi x$

FIGURE 39.4

C. GRAPHS INVOLVING THE SECANT

Figure 39.5 shows the graph of $y = \sec x$ for $-3\pi/2 < x < 5\pi/2$. The general characteristics of the graph can be obtained using the graph of $y = \cos x$ and the identity $\sec x = 1/(\cos x)$. To help make the connection clear, Figure 39.5

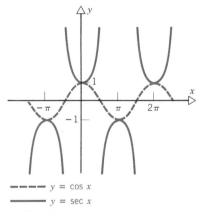

‑ ‑ ‑ ‑ $y = \cos x$

——— $y = \sec x$

FIGURE 39.5

also shows the graph of $y = \cos x$. The fundamental period of $y = \sec x$ is 2π. The graph of $y = \sec x$ has vertical asymptotes where $\cos x = 0$. Also, $\sec x > 0$ where $\cos x > 0$, and $\sec x < 0$ where $\cos x < 0$. The points where $\cos x = 1$ are minimum points for branches of $y = \sec x$; the points where $\cos x = -1$ are maximum points for branches of $y = \sec x$. The secant function is even, so the graph of $y = \sec x$ is symmetric with respect to the y-axis.

$$\text{To graph } y = a \sec bx \ (b > 0)$$

- Sketch the graph of $y = a \cos bx$ using the method in Section 38.
- Sketch the graph of $y = a \sec bx$ using the relationships suggested in Figure 39.5.

- In particular, minimum points for branches of $y = a \sec bx$ occur at maximum points of $y = a \cos bx$, and maximum points for branches of $y = a \sec bx$ occur at minimum points of $y = a \cos bx$. Asymptotes for $y = a \sec bx$ occur at x-intercepts of $y = a \cos bx$.
- If $a < 0$, sketch the graph of $y = |a| \sec bx$ and then reflect it through the x-axis.

Because the secant function is an even function, the graph of $y = a \sec(-bx)$ is the same as the graph of $y = a \sec bx$. For example, the graph of $y = 4 \sec(-3x)$ is the same as the graph of $y = 4 \sec 3x$, which is the next example.

Example 39.3 Sketch the graph of $y = 4 \sec 3x$ for $-p < x < p$, where p is the fundamental period.

Solution We first sketch the graph of $y = 4 \cos 3x$. The period is $2\pi/3$ and the amplitude is 4. Figure 39.6 shows the graph of $y = 4 \cos 3x$ as a dashed line. From that, we can sketch the graph of $y = 4 \sec 3x$. ■

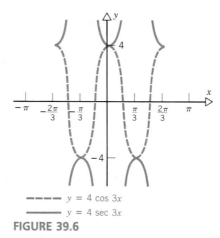

$$----\ y = 4 \cos 3x$$
$$———\ y = 4 \sec 3x$$

FIGURE 39.6

D. GRAPHS INVOLVING THE COSECANT

Figure 39.7 shows the graph of $y = \csc x$ for $-2\pi < x < 2\pi$. The graph of $y = a \csc bx$ is related to the graph of $y = a \sin bx$ in the same way the graph of $y = a \sec bx$ is related to the graph of $y = a \cos bx$.

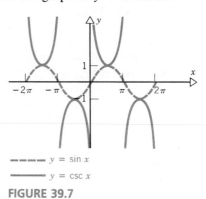

$$----\ y = \sin x$$
$$———\ y = \csc x$$

FIGURE 39.7

Example 39.4 Sketch the graph of $y = -\csc\dfrac{\pi}{3}x$ for $-p < x < p$, where p is the fundamental period.

Solution We first draw the graph of $y = -\sin(\pi/3)x$. The period is $2\pi/(\pi/3) = 6$. Figure 39.8 shows this graph as a dashed line. The graph of $y = -\csc(\pi/3)x$ has been sketched from that. ∎

Because the cosecant is an odd function, the graph of $y = a\csc(-bx)$ is the same as the graph of $y = -a\csc bx$. Thus, for example, the graph of $y = \csc[-(\pi/3)x]$ is the same as the graph of $y = -\csc(\pi/3)x$, which is Example 39.4.

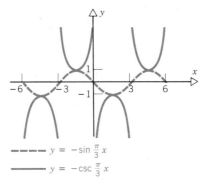

$$----\ \ y = -\sin\tfrac{\pi}{3}x$$

$$\underline{\hspace{2em}}\ \ y = -\csc\tfrac{\pi}{3}x$$

FIGURE 39.8

The graphs of the six trigonometric functions make it easy to see the facts about the domains and ranges that are summarized in Table 39.2. You should convince yourself that the table is correct.

TABLE 39.2

Function	Domain	Range
$\sin x$	All real numbers	$[-1, 1]$
$\cos x$	All real numbers	$[-1, 1]$
$\tan x$	All real numbers except $\left\{\pm\dfrac{\pi}{2},\ \pm\dfrac{3\pi}{2},\ \pm\dfrac{5\pi}{2},\ \ldots\right\}$	$(-\infty, \infty)$
$\cot x$	All real numbers except $\{0,\ \pm\pi,\ \pm2\pi,\ \pm3\pi,\ \ldots\}$	$(-\infty, \infty)$
$\sec x$	All real numbers except $\left\{\pm\dfrac{\pi}{2},\ \pm\dfrac{3\pi}{2},\ \pm\dfrac{5\pi}{2},\ \ldots\right\}$	$(-\infty, -1] \cup [1, \infty)$
$\csc x$	All real numbers except $\{0,\ \pm\pi,\ \pm2\pi,\ \pm3\pi,\ \ldots\}$	$(-\infty, -1] \cup [1, \infty)$

EXERCISES FOR SECTION 39

1. Make a table like Table 39.1 for $y = \cot x$, including all values of x in the set $\{\pm0.2, \pm0.4, \ldots, \pm1.4\}$, as well as $x = \pm\pi/2$. Then draw the graph of $y = \cot x$ for $-\pi/2 \le x \le \pi/2$.

2. Make a table like Table 39.1 for $y = \sec x$, including all values of x in the set $\{0, \pm0.2, \pm0.4, \ldots, \pm1.4\}$. Then draw the graph of $y = \sec x$ for $-\pi/2 < x < \pi/2$.

3. Make a table like Table 39.1 for $y = \csc x$, including all values of x in the set $\{\pm 0.2, \pm 0.4, \ldots, \pm 1.4\}$, as well as $x = \pm \pi/2$. Then draw the graph of $y = \csc x$ for $-\pi/2 \le x \le \pi/2$.

4. Rewrite the six remarks (indicated by •) preceding Example 39.1 so that they apply to $y = a \cot bx$ rather than $y = a \tan bx$.

5. Rewrite the four remarks (indicated by •) preceding Example 39.3 so that they apply to $y = a \csc bx$ rather than $y = a \sec bx$.

6. The function $\sec x$ is increasing for x in the interval $[0, \pi/2)$. For which other intervals such that $-3\pi/2 < x < 5\pi/2$ is $\sec x$ increasing?

7. Find all x such that $-2\pi < x < 2\pi$ and $\csc x = 1$.

8. Find all x such that $-2\pi < x < 2\pi$ and $\cot x$ is undefined.

9. Find all x such that $-2\pi < x < 2\pi$ and $\sec x = -1$.

Sketch the graph of each equation for $-\pi < x < 2\pi$.

10. $y = -\cot x$

11. $y = 2 \csc x$

12. $y = -2 \sec x$

13. $y = -0.5 \csc x$

14. $y = -2 \tan x$

15. $y = \dfrac{1}{4} \cot x$

Sketch the graph of each equation for $0 < x < p$, where p is the fundamental period.

16. $y = \sec \dfrac{\pi}{2} x$

17. $y = \cot 2x$

18. $y = \csc 0.5x$

19. $y = -0.25 \tan(-4x)$

20. $y = 3 \sec(-\pi x)$

21. $y = \tan(-0.25\pi x)$

22. Determine $b > 0$ so that $y = \tan bx$ has fundamental period $\pi/6$.

23. Determine $b > 0$ so that $y = \sec bx$ has fundamental period 10.

24. Determine $b > 0$ so that $y = \csc bx$ has fundamental period $\pi\sqrt{2}$.

25. If the domain of $f(x) = a \sec bx$ is the set of all real numbers except $\{\pm 2\pi, \pm 4\pi, \pm 6\pi, \ldots\}$, and the range is $(-\infty, -3] \cup [3, \infty)$, what are a and b?

26. If the graph of $y = a \tan bx$ has adjacent asymptotes $x = -\pi/8$ and $x = \pi/8$, and passes through the point $(\pi/16, 4)$, what are a and b?

27. A rotating light is on a ship at point S, two miles out from a straight shore line (Figure 39.9). The light is rotating at a rate of one revolution every six seconds. Let t denote time in seconds and assume that when $t = 0$ the light is shining at a point P such that SP is perpendicular to the shore. For $0 \le t < 1.5$, let x denote the distance in miles from P to the point T where the light shines on the shore. Find the function f such that $x = f(t)$ and then graph $x = f(t)$ for $0 \le t < 1.5$. (Suggestion: First express x as a function of θ and then express θ as a function of t.)

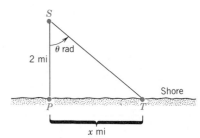

FIGURE 39.9

SECTION **40**

Translations

A. THE GRAPH OF $y = a \sin(bx - c)$

We know that changing b in $y = \sin bx$ changes the period. And changing a in $y = a \sin x$ changes the amplitude. We shall see now that the effect of changing c in $y = \sin(bx - c)$ is to translate (shift) the graph to the right or left, depending on the sign of c/b. By combining these results, we shall know that the graph of an equation of the form $y = a \sin(bx - c)$ differs from the graph of $y = \sin x$ only by a possible change in period or amplitude, or translation to the right or left.

The variable x is called the **argument** of $\sin x$. As the argument x increases from 0 to 2π, the graph of $y = \sin x$ goes through one cycle—the y-coordinate changes from 0 to 1 to 0 to -1 and back to 0 (Figure 40.1a). The **argument** of $y = \sin(bx - c)$ is $bx - c$, and the graph of $y = \sin(bx - c)$ will go through one cycle as $bx - c$ increases from 0 to 2π. By solving $bx - c = 0$ for x we shall know where a cycle begins, and by solving $bx - c = 2\pi$ for x we shall know where the cycle ends. If we use this with what we already know about graphs, we can reach the following conclusions.

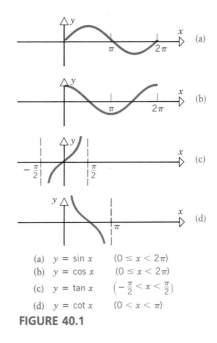

(a) $y = \sin x$ $(0 \le x < 2\pi)$
(b) $y = \cos x$ $(0 \le x < 2\pi)$
(c) $y = \tan x$ $\left(-\frac{\pi}{2} < x < \frac{\pi}{2}\right)$
(d) $y = \cot x$ $(0 < x < \pi)$

FIGURE 40.1

To graph $y = a \sin(bx - c)$ $(b > 0)$

- The graph is sinusoidal. Sketch one cycle and then continue as far as needed by using periodicity.
- Solve $bx - c = 0$ for x to determine where a cycle begins (in the sense of Figure 40.1a). Solve $bx - c = 2\pi$ for x to determine where the cycle ends.
- There are x-intercepts at the beginning, middle, and end of the cycle.
- If $a > 0$, a maximum occurs one-fourth of the way through the cycle. The maximum y-value is a if $a > 0$.
- If $a > 0$, a minimum occurs three-fourths of the way through the cycle. The minimum y-value is $-a$ if $a > 0$.
- If $a < 0$, sketch the graph of $y = |a| \sin(bx - c)$ and then reflect it through the x-axis.

Because $\sin(-bx - c) = \sin[-(bx + c)] = -\sin(bx + c)$, we can always arrange so that the coefficient of x is positive. Thus the preceding summary covers all cases of the form $y = a \sin(bx - c)$.

Example 40.1 Sketch one cycle of $y = \sin\left(x + \dfrac{\pi}{4}\right)$.

Solution If $x + \pi/4 = 0$, then $x = -\pi/4$, so a cycle begins at $x = -\pi/4$. If $x + \pi/4 = 2\pi$, then $x = 7\pi/4$, so the cycle ends at $x = 7\pi/4$. The amplitude is 1. Figure 40.2 shows the graph.

　　Notice that the middle of the cycle (where there is an x-intercept) is at the average of the x-coordinates of the endpoints of the cycle: $3\pi/4 = \frac{1}{2}[(-\pi/4) + (7\pi/4)]$ (Exercise 51 of Section 13). ■

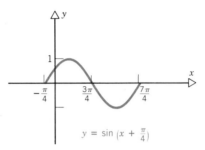

FIGURE 40.2

　　From Figure 40.2 we see that the graph of $y = \sin(x + \pi/4)$ is the graph of $y = \sin x$ translated $\pi/4$ units to the left. In general, the graph of $y = \sin(x - c)$ is the graph of $y = \sin x$ translated $|c|$ units to the right if $c > 0$, and $|c|$ units to the left if $c < 0$. The amount of the translation is called the **phase shift.** [You may want to verify that the remarks here are consistent with Statement (20.2). If $f(x) = \sin x$ then $f(x - c) = \sin(x - c)$.] The phase shift for $y = \sin(bx - c)$, with b perhaps different from 1, is discussed in Subsection D.

Example 40.2 Sketch two cycles of $y = 3 \sin(2x - \pi)$.

Solution If $2x - \pi = 0$, then $x = \pi/2$, so a cycle begins at $x = \pi/2$. If $2x - \pi = 2\pi$, then $x = 3\pi/2$, so the cycle ends at $3\pi/2$. The amplitude is 3. The graph in Figure 40.3 shows a second cycle from $3\pi/2$ to $5\pi/2$. ■

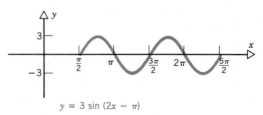

$y = 3 \sin (2x - \pi)$

FIGURE 40.3

B. OTHER GRAPHS WITH HORIZONTAL TRANSLATIONS

Graphs of equations of the form $y = a \cos(bx - c)$, $y = a \tan(bx - c)$, and $y = a \cot(bx - c)$ can be handled using the same principles as those for $y = a \sin(bx - c)$. We work from the cycles shown in Figure 40.1. For graphs involving the secant or cosecant we use what we know about graphs involving sine or cosine, together with the identities sec $x = 1/\cos x$ and csc $x = 1/\sin x$.

Example 40.3 Sketch one cycle of $y = -2 \cos \left(\dfrac{1}{2} x - \dfrac{\pi}{4} \right)$.

Solution First consider $y = 2 \cos(\frac{1}{2}x - \pi/4)$. If $\frac{1}{2}x - \pi/4 = 0$, then $x = \pi/2$, so a cycle begins at $x = \pi/2$. If $\frac{1}{2}x - \pi/4 = 2\pi$, then $x = 9\pi/2$, so the cycle ends at $x = 9\pi/2$. The amplitude is 2, so the cycle begins and ends with $y = 2$. The middle of the cycle is at $x = \frac{1}{2}(\pi/2 + 9\pi/2) = 5\pi/2$; there the graph has a minimum of $y = -2$. Figure 40.4 shows the graph of $y = 2 \cos(\frac{1}{2}x - \pi/4)$ as a dashed curve. By reflection through the x-axis we arrive at the graph of $y = -2 \cos(\frac{1}{2}x - \pi/4)$, which is shown as a solid curve. ∎

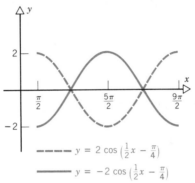

$$---- \quad y = 2 \cos \left(\tfrac{1}{2} x - \tfrac{\pi}{4} \right)$$

$$\text{———} \quad y = -2 \cos \left(\tfrac{1}{2} x - \tfrac{\pi}{4} \right)$$

FIGURE 40.4

Example 40.4 Sketch three cycles of $y = \cot \left(\dfrac{\pi}{3} x - \dfrac{\pi}{6} \right)$.

Solution Work from the model in Figure 40.1d, where the cycle occurs as the argument changes from 0 to π. If $(\pi/3)x - \pi/6 = 0$ then $x = \frac{1}{2}$, so for the present example a cycle starts at $x = \frac{1}{2}$. If $(\pi/3)x - \pi/6 = \pi$ then $x = \frac{7}{2}$, so the cycle ends at $x = \frac{7}{2}$.

To sketch the graph, we follow the pattern in Figure 40.1d and draw asymptotes at the beginning and end of the cycle, $x = \frac{1}{2}$ and $x = \frac{7}{2}$. An x-intercept occurs midway between the asymptotes, at $x = \frac{1}{2}(\frac{1}{2} + \frac{7}{2}) = 2$. See Figure 40.5.

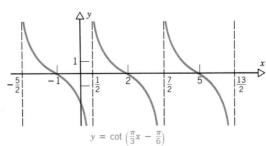

$$y = \cot \left(\tfrac{\pi}{3} x - \tfrac{\pi}{6} \right)$$

FIGURE 40.5

The instructions ask for three cycles. One has been sketched to the right of our first cycle and another is to the left. The distance between any pair of adjacent asymptotes is 3 units, the same as the distance between $x = \frac{1}{2}$ and $x = \frac{7}{2}$ (which is also the period). ∎

C. VERTICAL TRANSLATIONS

To get the graph of $y = 4 + \sin x$ from the graph of $y = \sin x$, we simply add 4 to each y-value; that is, we translate the graph upward by 4 units. Similarly, to get the graph of $y = -4 + \sin x$ we translate downward by 4 units. Examples involving other functions are similar.

Example 40.5 Sketch one cycle of $y = 3 + \sec\left(-x - \dfrac{\pi}{2}\right)$.

Solution We begin with $y = \cos(-x - \pi/2)$ and then use the fact that the secant is the reciprocal of the cosine. After that we use a vertical translation.

First, $\cos(-x - \pi/2) = \cos[-(x + \pi/2)] = \cos(x + \pi/2)$, because cosine is an even function. Thus we can work with $\cos(x + \pi/2)$. If $x + \pi/2 = 0$, then $x = -\pi/2$, so a cycle begins at $x = -\pi/2$. If $x + \pi/2 = 2\pi$, then $x = 3\pi/2$, so the cycle ends at $x = 3\pi/2$. The graph of $y = \cos(-x - \pi/2)$ is shown as a dashed curve in Figure 40.6.

To get the graph of $y = \sec(-x - \pi/2)$ we now use the ideas in Section 39C. The result is the second dashed curve in Figure 40.6. For the graph of $y = 3 + \sec(-x - \pi/2)$ we translate upward by 3 units. The result is the solid curve. ∎

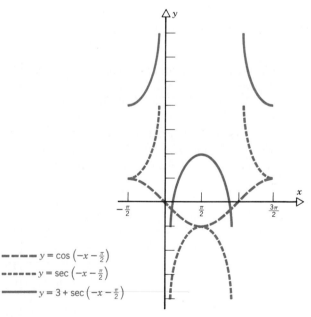

$$- - - - \; y = \cos\left(-x - \tfrac{\pi}{2}\right)$$
$$\cdots\cdots\; y = \sec\left(-x - \tfrac{\pi}{2}\right)$$
$$\text{———}\; y = 3 + \sec\left(-x - \tfrac{\pi}{2}\right)$$

FIGURE 40.6

D. GENERAL REMARKS

The key used throughout this section has been to find where one cycle of a graph begins and ends, and then follow an appropriate pattern from Figure 40.1. If we apply these ideas to a general equation such as $y = d + a \sin(bx - c)$, then we can determine the amplitude, period, and vertical and horizontal translations directly in terms of the constants a, b, c, and d. Some typical results will be stated here and the details will be left to the exercises.

Consider

$$y = d + a \sin(bx - c). \tag{40.1}$$

For convenience, we rewrite this as

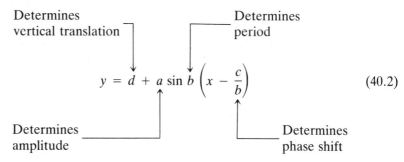

$$y = d + a \sin b\left(x - \frac{c}{b}\right) \tag{40.2}$$

Exercise 22 asks you to verify the statements concerning period and phase shift that follow.

- The amplitude is $|a|$.
- The period is $2\pi/|b|$.
- The phase shift is $|c/b|$ units to the right if $c/b > 0$, and $|c/b|$ units to the left if $c/b < 0$.
- The term d induces translation $|d|$ units upward if $d > 0$, and $|d|$ units downward if $d < 0$.

The results for the other functions are similar, except the period of $y = d + a \tan(bx - c)$ is $\pi/|b|$ rather than $2\pi/|b|$; the same remark applies with the cotangent in place of the tangent.

To sketch the graph of a specific equation the method in Subsections A–C is recommended, rather than reliance on the more general remarks in this subsection.

EXERCISES FOR SECTION 40

Sketch two cycles of the graph of each equation.

1. $y = \cos\left(x - \dfrac{\pi}{4}\right)$

2. $y = \sin\left(x + \dfrac{3}{2}\pi\right)$

3. $y = \tan\left(x + \dfrac{\pi}{3}\right)$

4. $y = \sin\left(2x - \dfrac{\pi}{2}\right)$

5. $y = \tan\left(\dfrac{1}{2}x - \dfrac{\pi}{4}\right)$

6. $y = \cos\left(\dfrac{\pi}{2}x + \dfrac{\pi}{4}\right)$

7. $y = \csc\left(\dfrac{\pi}{2}x + \pi\right)$

8. $y = \sec(2x - \pi)$

9. $y = \cot\left(\dfrac{1}{3}x - \dfrac{\pi}{3}\right)$

10. $y = 2\sin\left(-x + \dfrac{\pi}{4}\right)$

11. $y = -\tan\left(-x - \dfrac{\pi}{2}\right)$

12. $y = -3\cos(-2x + \pi)$

13. $y = 1 + \tan\dfrac{1}{3}x$

14. $y = 2 - \csc\left(\pi x + \dfrac{\pi}{2}\right)$

15. $y = -3 + \sec\left(-\pi x + \dfrac{\pi}{3}\right)$

The equations in Exercises 16–21 are identities. To illustrate this in each case, sketch the graphs of the functions represented by the two sides of the equation and observe that they coincide. Explain clearly how you obtain each graph.

16. $\cos(x - \pi/2) = \sin x$

17. $\cot(x + \pi/2) = -\tan x$

18. $\cos(x + \pi) = -\cos x$

19. $\sin(x - \pi) = -\sin x$

20. $\cos(x + \pi/2) = -\sin x$

21. $\sin(\pi/2 - x) = \cos x$

22. Use the method in Subsection A to determine where one cycle begins and ends for $y = d + a\sin(bx - c)$. Use your answer to verify that the period is $2\pi/|b|$, and the phase shift is $|c/b|$ units to the right if $c/b > 0$ and $|c/b|$ units to the left is $c/b < 0$.

23. Use the method of Subsection A to determine where one cycle begins and ends for $y = d + a\tan(bx - c)$. Use your answer to determine the period and phase shift of the graph.

24. Determine the period and phase shift for $y = d + a\csc(bx - c)$.

Without sketching the graph, use the information in Subsection D to determine the period and phase shift of the graph of each function. In 25–27, also determine the amplitude.

25. $y = -2\sin(4x - \pi)$

26. $y = 5 + 5\cos\left(\pi x + \dfrac{\pi}{8}\right)$

27. $y = \sqrt{2}\sin(0.5x - \pi)$

28. $y = 3 - \tan\pi\left(\dfrac{1}{2}x + \dfrac{1}{3}\right)$

29. $y = \tan 3(-x - \pi)$

30. $y = -0.4 - 2\tan 0.5(2x - 1)$

Sketch the graph of each equation in Exercises 31–36 by first applying Identity (36.10) to write the equation in the form y = a cos(bt − c).

31. $y = \cos t + \sqrt{3}\sin t$

32. $y = \sqrt{3}\cos t - \sin t$

33. $y = 5\cos t - 5\sin t$

34. $y = 12\cos 2t - 5\sin 2t$

35. $y = \sqrt{2}\cos 4t - \sqrt{2}\sin 4t$

36. $y = 3\cos 3t + 4\sin 3t$

Alternating current in an electrical circuit varies with time according to an equation of the form

$$I = a\sin(bt - c), \qquad (40.3)$$

*where I is measured in amperes (abbreviated amp), t is measured in seconds, and b > 0. The **frequency**, denoted by f, is the number of cycles per second. The unit for f is the **hertz** (abbreviated Hz), defined by 1 cycle/sec = 1 Hz. Exercises 37–39 relate to a circuit satisfying Equation (40.3).*

37. (a) Explain why the frequency is the reciprocal of the period of $a\sin(bt - c)$.

   (b) Use part (a) to explain why $f = b/2\pi$.

   (c) Assume $I = 10\sin(120\pi t - 30\pi)$. Determine the amplitude, period, frequency, and phase shift, and sketch the graph of I as a function of t.

38. Write an equation of the form $I = a\sin(bt - c)$ for a circuit in which the maximum value of I is 160 amp, the frequency is 60 Hz, and the phase shift (c/b) is zero. Sketch the graph of I as a function of t.

39. Write an equation of the form $I = a\sin(bt - c)$ for a circuit in which the maximum value of I is 20 amp, the frequency is 120 Hz, and the phase shift c/b is 1/6. Sketch the graph of I as a function t.

SECTION **41**
Inverse Trigonometric Functions

A. THE INVERSE SINE

Recall from Section 17 that a function has an inverse iff the function is one-to-one. It might appear, then, that trigonometric functions cannot have inverse functions, since they are not one-to-one. We met a similar problem with the function $f(x) = x^2$ (for all x), which also is not one-to-one. We overcame this problem for $f(x) = x^2$ in Example 17.6, by restricting the domain of the function to nonnegative values of x: the function $f(x) = x^2$ ($x \geq 0$) does have an inverse—the function $f^{-1}(x) = \sqrt{x}$ ($x \geq 0$). This section introduces the *inverse trigonometric functions* by using this same idea of restricting a domain.

Figure 41.1 shows the graph of $y = \sin x$ for $-\pi/2 \leq x \leq \pi/2$. Each horizontal line intersects the graph at most once, so the function represented by the graph has an inverse (Section 17).

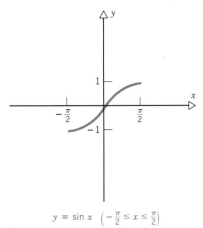

$$y = \sin x \quad \left(-\frac{\pi}{2} \leq x \leq \frac{\pi}{2}\right)$$

FIGURE 41.1

Definition   The inverse of the function $f(x) = \sin x$ ($-\pi/2 \leq x \leq \pi/2$) is called the **inverse sine function.** The inverse sine of x is denoted by $\sin^{-1} x$. Thus

$$y = \sin^{-1} x \quad \text{iff} \quad x = \sin y \text{ and } -\frac{\pi}{2} \leq y \leq \frac{\pi}{2}. \qquad (41.1)$$

Note well that $\sin^{-1} x$ *does not* mean $1/\sin x$. This use of -1 as a superscript is not the same as its use as an exponent. The notation **arcsin x** is also used to denote $\sin^{-1} x$. Both $\sin^{-1} x$ and arcsin x will be used in this book, since you may encounter either notation elsewhere.

Example 41.1   (a) $\sin^{-1} \sqrt{2}/2 = \pi/4$ because $\pi/4$ is the unique value of y such that $\sin y = \sqrt{2}/2$ and $-\pi/2 \le y \le \pi/2$.

(b) $\arcsin(-1) = -\pi/2$ because $-\pi/2$ is the unique value of y such that $\sin y = -1$ and $-\pi/2 \le y \le \pi/2$.

(c) $\sin^{-1} 0 = 0$ because $\sin 0 = 0$ and $-\pi/2 \le 0 \le \pi/2$.   ■

Example 41.2   Compute $\arcsin 0.5646$.

Solution with a calculator   To compute $\sin^{-1} x$ on a calculator with an (sin⁻¹) key, place the calculator in the radian mode, enter x, press (sin⁻¹), and read the answer in the display. On some calculators the correct sequence is to enter x, press (inv) (the inverse function key), press (sin), and then read the answer in the display. In this example the answer is $\arcsin 0.5646 = 0.60$ (rounded to the nearest 0.01).

Solution with a table   Use Table VI in reverse. That is, look in the sine column to find the input 0.5646, and then read the answer 0.60 in the outside column, rather than vice versa. The accuracy of any such answer is determined by the accuracy of the table, of course.   ■

In this book, unless explicitly stated to the contrary, $\sin^{-1} x$ will always represent a real number. Given this, we can think of $\sin^{-1} x$ as $\sin^{-1} x$ *radians*, but *not* as $\sin^{-1} x$ *degrees*. Thus $\sin^{-1} \frac{1}{2}$ equals $\pi/6$, not 30°. We can think of $\sin^{-1} x$ as "the angle between $-\pi/2$ and $\pi/2$ (inclusive) whose sine is x," provided we interpret *angle* to mean the angle's radian measure.

Recall that if f^{-1} is the inverse of a function f, then the domain of f^{-1} is the range of f, and the range of f^{-1} is the domain of f. Since the range of the sine function is the interval $[-1, 1]$, and the inverse sine is obtained by restricting the domain of the sine function to the interval $[-\pi/2, \pi/2]$, we have the following facts about the inverse sine function.

The domain of the inverse sine is $[-1, 1]$.

The range of the inverse sine is $[-\pi/2, \pi/2]$.

The graph of a function and its inverse are located symmetrically about the line $y = x$ (Section 17 again). Thus the graph of $y = \sin^{-1} x$ is obtained by reflecting the graph in Figure 41.1 through the line $y = x$. The result is shown in Figure 41.2.

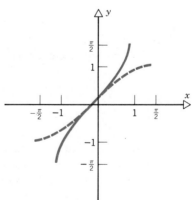

----- $y = \sin x \left(-\frac{\pi}{2} \le x \le \frac{\pi}{2}\right)$

——— $y = \sin^{-1} x \ (-1 \le x \le 1)$

FIGURE 41.2

Example 41.3 (a) $\sin(\sin^{-1} 0.5) = \sin \pi/6 = 0.5$

(b) $\arcsin[\sin(-\pi/3)] = \arcsin(-\sqrt{3}/2) = -\pi/3$

(c) $\sin^{-1}(\sin 4\pi) = \sin^{-1} 0 = 0$ ■

Parts (a) and (b) of Example 41.3 illustrate the following identities, which restate that the sine (with restricted domain) and the inverse sine are inverse functions.

$$\begin{array}{lll} \sin(\sin^{-1} x) = x & \text{for} & -1 \le x \le 1 \\ \sin^{-1}(\sin x) = x & \text{for} & -\pi/2 \le x \le \pi/2 \end{array} \qquad (41.2)$$

Part (c) of Example 41.3 illustrates the following important fact.

If x is not in the interval $[-\pi/2, \pi/2]$, then $\sin^{-1}(\sin x) \ne x$.

B. THE INVERSE COSINE

The dashed curve in Figure 41.3 shows the graph of $y = \cos x$ for $0 \le x \le \pi$. Each horizontal line intersects the graph at most once, so the function represented by the curve has an inverse. It is the inverse cosine function.

Definition The inverse of the function $f(x) = \cos x$ $(0 \le x \le \pi)$ is called the **inverse cosine function.** The inverse cosine of x is denoted by either **cos**$^{-1}$ x or **arccos** x.

$$y = \cos^{-1} x \quad \text{iff} \quad x = \cos y \text{ and } 0 \le y \le \pi. \qquad (41.3)$$

The domain of the inverse cosine is $[-1, 1]$.
The range of the inverse cosine is $[0, \pi]$.

Example 41.4 (a) $\cos^{-1} 0 = \pi/2$ because $\pi/2$ is the unique value of y such that $\cos y = 0$ and $0 \le y \le \pi$.

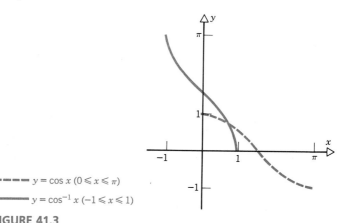

$----$ $y = \cos x$ $(0 \le x \le \pi)$

$\rule{1cm}{0.5pt}$ $y = \cos^{-1} x$ $(-1 \le x \le 1)$

FIGURE 41.3

(b) $\arccos(-\sqrt{3}/2) = 5\pi/6$ because $5\pi/6$ is the unique value of y such that $\cos y = -\sqrt{3}/2$ and $0 \le y \le \pi$.

(c) $\cos(\arccos 1) = \cos 0 = 1$

(d) $\cos^{-1}(\cos 3\pi) = \cos^{-1}(-1) = \pi$

Here are the cosine equivalents of the statements in (41.2).

$$\cos(\cos^{-1} x) = x \qquad \text{for} \qquad -1 \le x \le 1$$
$$\cos^{-1}(\cos x) = x \qquad \text{for} \qquad 0 \le x \le \pi$$

(41.4)

Part (d) of Example 41.4 illustrates the following statement.

If x is not in the interval $[0, \pi]$, then $\cos^{-1}(\cos x) \neq x$.

C. INVERSES OF THE OTHER FUNCTIONS

The inverses of the other trigonometric functions are defined in the same manner as the inverse sine and the inverse cosine. Table 41.1 gives the domains and ranges of all of the inverse trigonometric functions. Figure 41.4 shows the graph of the inverse tangent function. Exercises 70–72 ask you to sketch the graphs of the inverse cosecant, secant, and cotangent.

TABLE 41.1

Function	Domain	Range*
$y = \sin^{-1} x$	$[-1, 1]$	$[-\pi/2, \pi/2]$
$y = \cos^{-1} x$	$[-1, 1]$	$[0, \pi]$
$y = \tan^{-1} x$	All real numbers.	$(-\pi/2, \pi/2)$
$y = \cot^{-1} x$	All real numbers.	$(0, \pi)$
$y = \sec^{-1} x$	$(-\infty, -1] \cup [1, \infty)$	$[0, \pi/2) \cup (\pi/2, \pi]$
$y = \csc^{-1} x$	$(-\infty, -1] \cup [1, \infty)$	$[-\pi/2, 0) \cup (0, \pi/2]$

*Some books use other choices for the ranges of the inverse secant and inverse cosecant. There are advantages to the choices in Table 41.1, but in calculus there are advantages to the other choices.

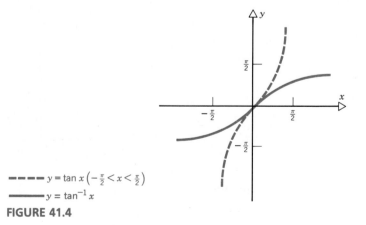

$- - -\;y = \tan x \left(-\frac{\pi}{2} < x < \frac{\pi}{2}\right)$

$\underline{\qquad}\,y = \tan^{-1} x$

FIGURE 41.4

D. MORE EXAMPLES INVOLVING INVERSE FUNCTIONS

Example 41.5 Find $\tan(\arccos x)$ if $0 < x < 1$.

Solution Let $\theta = \arccos x$. Then $\cos \theta = x$, and θ can be thought of as an acute angle because $0 < x < 1$. Figure 41.5 shows this angle with the hypotenuse chosen as

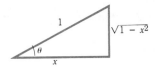

FIGURE 41.5

1 and the side adjacent to θ chosen as x, so that $\cos \theta = x$, as required. The length of the side opposite θ is $\sqrt{1 - x^2}$, by the Pythagorean Theorem. It follows that

$$\tan \theta = \sqrt{1 - x^2}/x,$$

so that

$$\tan(\arccos x) = \sqrt{1 - x^2}/x.$$

Example 41.6 Find $\cos[\sin^{-1}(-\frac{1}{4})]$.

Solution Let $\theta = \sin^{-1}(-\frac{1}{4})$. Then θ can be thought of as the unique angle such that $-\pi/2 \le \theta \le \pi/2$ and $\sin \theta = -\frac{1}{4}$. Figure 41.6 shows this angle, where the value $\sqrt{15}$ has been computed by the Pythagorean Theorem. From the figure, we see that $\cos \theta = \sqrt{15}/4$, so that $\cos[\sin^{-1}(-\frac{1}{4})] = \sqrt{15}/4$.

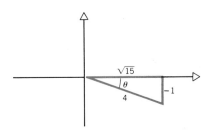

FIGURE 41.6

Example 41.7 Find $\sin[2 \sin^{-1}(-\frac{1}{4})]$.

Solution Let $\theta = \sin^{-1}(-\frac{1}{4})$. Then the functions of θ can be read from Figure 41.6. Using the identity $\sin 2\theta = 2 \sin \theta \cos \theta$, we have

$$\sin\left[2 \sin^{-1}\left(-\frac{1}{4}\right)\right] = \sin 2\theta$$

$$= 2 \sin \theta \cos \theta$$

$$= 2 \left(\frac{-1}{4}\right)\left(\frac{\sqrt{15}}{4}\right)$$

$$= \frac{-\sqrt{15}}{8}.$$

Example 41.8 Find $\tan[\arccos \frac{1}{3} + \arcsin(-\frac{1}{4})]$.

Solution Use Identity (34.8) with $u = \arccos \frac{1}{3}$ and $v = \arcsin(-\frac{1}{4})$:

$$\tan(u + v) = \frac{\tan u + \tan v}{1 - \tan u \tan v}.$$

From Figure 41.5 with $x = \frac{1}{3}$, we have

$$\tan u = \tan (\arccos \tfrac{1}{3}) = \frac{\sqrt{1 - (\frac{1}{3})^2}}{\frac{1}{3}} = \sqrt{8} = 2\sqrt{2}.$$

From Figure 41.6,

$$\tan v = \tan [\arcsin (-\tfrac{1}{4})] = -1/\sqrt{15} = -\sqrt{15}/15.$$

Thus

$$\tan(u + v) = \frac{2\sqrt{2} + (-\sqrt{15}/15)}{1 - (2\sqrt{2})(-\sqrt{15}/15)}$$

$$= \frac{30\sqrt{2} - \sqrt{15}}{15 + 2\sqrt{30}} \quad \text{(exact answer)}$$

$$\approx 1.485 \quad \text{(approximate answer).}$$

Ⓒ

The approximate answer can also be found using a calculator or table without using the formula for $\tan(u + v)$. Compute $\arccos(\frac{1}{3})$, then $\arcsin(-\frac{1}{4})$, then add, and then find the tangent. ■

EXERCISES FOR SECTION 41

Find each functional value without using a table or calculator.

1. $\sin^{-1} 1$
2. $\cos^{-1}(-1)$
3. $\tan^{-1} 0$
4. $\arctan(-1)$
5. $\arcsin(-\sqrt{3}/2)$
6. $\arccos(-1/2)$
7. $\arccos\sqrt{2}/2$
8. $\arctan \sqrt{3}$
9. $\arcsin 0.5$
10. $\sec^{-1}(-1)$
11. $\csc^{-1} 2$
12. $\cot^{-1}(-\sqrt{3}/3)$

Find each functional value with the help of Table VI.

13. $\cot^{-1} 0.5090$
14. $\sin^{-1}(-0.1987)$
15. $\cos^{-1} 0.7317$
16. $\arccos(-0.8253)$
17. $\text{arccot } 0.0208$
18. $\arctan(-2.912)$

Find each functional value with a calculator. Ⓒ

19. $\arcsin 0.7174$
20. $\arccos(-0.4447)$
21. $\arctan 0.8595$
22. $\tan^{-1}(-0.7602)$

23. $\sin^{-1} 0.0998$
24. $\sin^{-1}(-0.8866)$

Find each functional value without using a table or calculator.

25. $\cos(\cos^{-1} 0.3)$
26. $\sin[\sin^{-1}(-0.8)]$
27. $\tan(\tan^{-1} 5)$
28. $\arctan(\tan 0.6)$
29. $\arccos(\cos 0.75)$
30. $\arcsin(\sin 1.5)$
31. $\sin^{-1}(\sin 5\pi/6)$
32. $\tan^{-1}(\tan 2\pi/3)$
33. $\cos^{-1}[\cos(-\pi/6)]$
34. $\text{arccot}[\cot(-\pi/4)]$
35. $\text{arcsec}(\sec 3\pi)$
36. $\text{arccsc}(\csc 3\pi/2)$

37. $\cos^{-1}(\sin 2\pi/3)$

38. $\cos^{-1}[\sin(-\pi/4)]$

39. $\sin^{-1}[\cos(5\pi/6)]$

40. $\tan[\arccos(-2/3)]$

41. $\sin[\arctan(4/3)]$

42. $\cos[\arcsin(-3/5)]$

Write each of the following expressions without trigonometric or inverse trigonometric functions. Assum $0 < x < 1$.

43. $\sin(\cos^{-1} x)$

44. $\cos(\tan^{-1} x)$

45. $\tan(\sin^{-1} x)$

Find the exact value. (Do not use a table or calculator.)

46. $\cos(2 \arcsin 0.1)$

47. $\tan\left[2 \cos^{-1}\left(\dfrac{2}{3}\right)\right]$

48. $\sin(2 \tan^{-1} 0.2)$

49. $\tan\left[\dfrac{1}{2} \cos^{-1}\left(\dfrac{1}{4}\right)\right]$

50. $\sin\left[\dfrac{1}{2} \arcsin(-0.6)\right]$

51. $\cos\left[\dfrac{1}{2} \arccos(-0.1)\right]$

52. $\sin\left[\arccos\dfrac{3}{5} + \arctan\left(-\dfrac{3}{4}\right)\right]$

53. $\cos[\sin^{-1}(-0.8) - \sin^{-1}(0.6)]$

54. $\tan[\tan^{-1} 2.5 - \tan^{-1} 3]$

Compute with a calculator.   Ⓒ

55. $\cos(\arcsin 0.3415)$

56. $\tan(\arccos 0.0193)$

57. $\sin(\arccos 0.1916)$

58. $\tan(2 \sin^{-1} 0.3712)$

59. $\sin(2 \tan^{-1} 42.17)$

60. $\cos(2 \cos^{-1} 0.0901)$

61. $\sin(\arccos 0.52 + \arctan 0.25)$

62. $\cos(\tan^{-1} 9.071 - \tan^{-1} 3.125)$

63. $\tan(\sin^{-1} 0.7365 + \cos^{-1} 0.3248)$

64. Show that if $0 <, a < 1$, then $\cos^{-1} a = \sec^{-1}(1/a)$.

65. Show that if $a > 0$, then $\arctan a = \arcsin(a/\sqrt{a^2 + 1})$.

66. Show that if $0 < a < 1$, then $\sin^{-1} a = \cos^{-1} \sqrt{1 - a^2}$.

67. (a) For which values of x is $\tan(\arctan x) = x$?

    (b) For which values of x is $\arctan(\tan x) = x$?

68. (a) For which values of x is $\cot(\text{arccot } x) = x$?

    (b) For which values of x is $\text{arccot}(\cot x) = x$?

69. (a) For which values of x is $\sec(\text{arcsec } x) = x$?

    (b) For which values of x is $\text{arcsec}(\sec x) = x$?

70. Sketch the graphs of $y = \csc x$ ($-\pi/2 \le x \le \pi/2$, $x \ne 0$) and $y = \csc^{-1} x$ in the same figure.

71. Sketch the graphs of $y = \sec x$ ($0 \le x \le \pi$, $x \ne \pi/2$) and $y = \sec^{-1} x$ in the same figure.

72. Sketch the graphs of $y = \cot x$ ($0 < x < \pi$) and $y = \cot^{-1} x$ in the same figure.

73. Solve $y = d + a \sin(bx - c)$ for x, assuming $-\pi/2 \le bx - c \le \pi/2$. Why is the condition on $bx - c$ necessary?

74. Snell's law for light passing from one homogeneous medium to another states that

$$\frac{\sin \alpha_1}{\sin \alpha_2} = \mu, \qquad (41.5)$$

where α_1 is the angle of incidence, α_2 is the angle of refraction, and μ is the index of refraction of the second medium relative to the first. (See Exercises 61–63 in Section 37). Solve Equation (41.5) for α_2.

75. A painting a feet fall is hung with its bottom b feet above the level of a viewer's eye. The viewer is x feet from the wall. Let α denote the angle subtended by the painting at the viewer's eye (Figure 41.7).

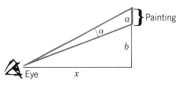

FIGURE 41.7

(a) Show that

$$\alpha = \arctan \frac{a + b}{x} - \arctan \frac{b}{x}.$$

(b) Find $\tan \alpha$ in terms of a, b, and x.

REVIEW EXERCISES FOR CHAPTER VIII

1. If f is periodic with fundamental period 4 and $f(x) = x^2 + 1$ for $0 \le x < 4$, what is $f(11)$?

2. If f is periodic with fundamental period 5 and $f(x) = |x - 1|$ for $0 \le x < 5$, what is $f(-9.5)$?

Sketch the graph of each equation for $0 \le$ x $<$ p, where p *is the fundamental period.*

3. $y = 3 \cos \pi x$

4. $y = -\tan x$

5. $y = -2 \sin (3x/2)$

6. $y = 2 \tan (x/5)$

7. $y = 0.5 \cot 2x$

8. $y = -\csc 2\pi x$

9. $y = \sec(x + \pi/4)$

10. $y = -\cos(x - \pi/4)$

11. $y = -1 + 2 \sin(\tfrac{1}{2}x - \pi/3)$

12. $y = 3 - 0.5 \cos(2x + \pi/4)$

Find each functional value without using a table or calculator.

13. $\arcsin(-\sqrt{3}/2)$

14. $\arctan\sqrt{3}/3$

15. $\arccos(-0.5)$

16. $\cos^{-1}[\sin(5\pi/6)]$

17. $\sin^{-1}[\sin(7\pi/2)]$

18. $\tan^{-1}(\cos \pi)$

Find the exact value. (Do not use a table or calculator.)

19. $\sin[2 \arccos(3/5)]$

20. $\cos 2[\arctan(-2)]$

Find each functional value (a) with the help of Table VI, and (b) with a calculator.

21. $\tan^{-1}(-0.8771)$

22. $\sin^{-1}(-0.8134)$

CHAPTER IX

APPLICATIONS OF TRIGONOMETRY

In Section 31 we used trigonometry to solve problems involving right triangles. The first two sections of this chapter show how to use trigonometry to handle oblique triangles—those without a right angle. The third section introduces vectors and their application to velocity and force problems. The last section, which is independent of the preceding sections of the chapter, discusses polar coordinates.

SECTION **42**
Law of Cosines

A. STATEMENT AND PROOF

Throughout this chapter, A, B, and C will denote the vertices of a triangle. The interior angles at A, B, and C will be denoted α, β, and γ, respectively. The lengths of the sides opposite A, B, and C will be denoted a, b, and c, respectively. Figure 42.1 gives an example. We shall frequently denote an angle and its measure with the same letter.

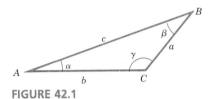

FIGURE 42.1

Very often we know some of the six quantities α, β, γ, a, b, and c, and then must determine one or more of the others. The Law of Cosines (which follows) and the Law of Sines (in Section 43) are two keys to the solution of such problems. For simplicity, we shorten the phrase "length of a side" to just "side" in the following statement.

Law of Cosines

$$a^2 = b^2 + c^2 - 2bc \cos \alpha$$

$$b^2 = a^2 + c^2 - 2ac \cos \beta$$

$$c^2 = a^2 + b^2 - 2ab \cos \gamma$$

In words, the square of any side of a triangle equals the sum of the squares of the other two sides minus twice the product of the other two sides and the cosine of the angle between them.

Proof We prove the law in the form of the first of the three equations. To this end, assume a given triangle ABC, and assume a coordinate system placed so that A is at the origin and the side AB is along the positive x-axis. Figure 42.2a shows a case where α is obtuse, and Figure 42.2b shows two typical cases where α is acute. The computations that follow provide a proof whether α is obtuse or acute.

If (x, y) denotes the coordinates of C, then

$$\cos \alpha = \frac{x}{b} \quad \text{and} \quad \sin \alpha = \frac{y}{b}$$

so

$$x = b \cos \alpha \quad \text{and} \quad y = b \sin \alpha.$$

Thus C has coordinates $(b \cos \alpha, b \sin \alpha)$, as shown in the figure. We can

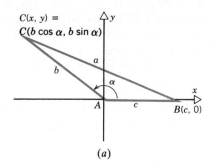

(a)

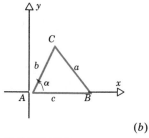

(b)

FIGURE 42.2

compute a^2 by the distance formula. It is the square of the distance between the points $B(c, 0)$ and $C(b \cos \alpha, b \sin \alpha)$.

$$a^2 = (b \cos \alpha - c)^2 + (b \sin \alpha - 0)^2$$

$$= b^2 \cos^2 \alpha - 2bc \cos \alpha + c^2 + b^2 \sin^2 \alpha$$

$$= b^2(\sin^2 \alpha + \cos^2 \alpha) + c^2 - 2bc \cos \alpha$$

$$= b^2 + c^2 - 2bc \cos \alpha$$

This proves the law in the first of the stated forms. The forms giving b^2 and c^2 are proved in the same way; only the letters are different. ☐

The Pythagorean Theorem is a special case of the Law of Cosines. For example, if there is a right angle at C, then $\cos \gamma = \cos 90° = 0$, and $c^2 = a^2 + b^2 - 2ab \cos \gamma$ becomes $c^2 = a^2 + b^2$.

B. APPLICATIONS

In Section 31D we considered how to solve right triangles—that is, how to find all of α, β, γ, a, b, and c given some subset that determines the triangle uniquely. With the Law of Cosines, and the fact that $\alpha + \beta + \gamma = 180°$, we can solve any triangle if we are given either all three sides or any two sides and the angle between them. (We could also handle the case of any two sides and an angle not between them, but that will be easier with the Law of Sines.) Examples 42.1 and 42.2 illustrate the main ideas. Remember that when an exercise asks you to solve a triangle it means to find all of the angles and all of the sides. For emphasis, here are two types of problems the Law of Cosines can help solve.

SAS Given two sides of a triangle and the angle included between them, solve the triangle.

SSS Given three sides of a triangle, solve the triangle.

The notation SAS for the first case stands for side-angle-side, representing the information given in that case (two sides and the angle between them). Similarly, SSS stands for side-side-side.

In most of the examples of this chapter the following conventions on accuracy will be used.

Lengths will be given to four significant figures.

Angles will be given to the nearest 0.01°.

If Table V is used for the exercises, give the answers for angles to the nearest 0.1°.

Example 42.1 Find a for the triangle in Figure 42.3, where, as indicated, $b = 2$ cm, $c = 5$ cm, and $\alpha = 120°$.

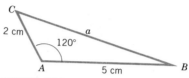

FIGURE 42.3

Solution

$$a^2 = b^2 + c^2 - 2bc \cos \alpha$$

$$= 2^2 + 5^2 - 2(2)(5) \cos 120°$$

$$= 4 + 25 - 20(-\tfrac{1}{2})$$

$$= 39$$

Therefore, $a = \sqrt{39} \approx 6.245$ cm.

Now that we know all three sides of this triangle, we could find β by the method of the next example, and then find γ by using $\alpha + \beta + \gamma = 180°$. ■

Example 42.2 Find γ for the triangle in Figure 42.4.

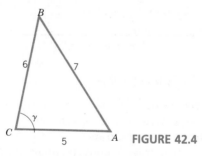

FIGURE 42.4

Solution Use the form of the Law of Cosines that involves γ.

$$c^2 = a^2 + b^2 - 2ab \cos \gamma$$

$$7^2 = 6^2 + 5^2 - 2(6)(5) \cos \gamma$$

$$60 \cos \gamma = 36 + 25 - 49$$

$$\cos \gamma = 0.2$$

From a calculator, $\gamma = 78.46°$.

Example 42.3 To determine the distance between points B and C on opposite shores of a lake, a surveyor measures the distances to B and C from a third point A and finds them to be 400 meters and 250 meters, respectively. The angle between AB and AC is 27°. Compute the distance between B and C (Figure 42.5).

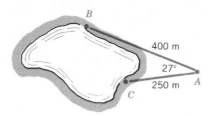

FIGURE 42.5

Solution By the Law of Cosines (with the notation of Figure 42.1),

$$a^2 = b^2 + c^2 - 2bc \cos \alpha$$

$$= 250^2 + 400^2 - 2(250)(400) \cos 27°$$

$$\approx 44{,}299$$

$$a \approx 210.5 \text{ m.}$$

Ⓒ ■

EXERCISES FOR SECTION 42

Although these exercises can be solved with a table of trigonometric functions and a table of square roots, you are encouraged to use a calculator as much as possible.

1. Find a if $b = 12$, $c = 7$, and $\alpha = 13°$.
2. Find c if $a = 9$, $b = 20$, and $\gamma = 110°$.
3. Find b if $a = 13$, $c = 3$, and $\beta = 2.5$ radians.
4. Find β if $a = 6$, $b = 7$, and $c = 8$.
5. Find a if $b = 20$, $c = 40$, and $\alpha = \pi/3$.
6. Find α if $a = 10$, $b = 20$, and $c = 15$.
7. Find c if $a = 25$, $b = 15$, and $\gamma = 5\pi/6$.
8. Find α if $a = 3$, $b = 9$, and $c = 11$.
9. Find b if $a = 14$, $c = 2$, and $\beta = 20.3°$.
10. Find γ if $a = 6$, $b = 6$, and $c = 10$.
11. Find β if $a = 10$, $b = 15$, and $c = 6$.
12. Find γ if $a = 2$, $b = 3$, and $c = 4$.

Solve the triangle determined by the given information.

13. $a = 7$, $b = 8$, $c = 9$
14. $a = 18$, $c = 15$, $\beta = 2\pi/3$

15. $a = 50$, $b = 20$, $\gamma = 1$ radian
16. $b = 10$, $c = 5$, $\alpha = 40°$
17. $a = 6$, $b = 10$, $c = 6$
18. $a = 11$, $b = 6$, $c = 7$
19. The sides of a triangular lot are 200 feet, 175 feet, and 140 feet. Find the angle between the shortest two sides.
20. Two of the adjacent sides of a parallelogram have lengths 3 centimeters and 5 centimeters, and the angle between them is 60°. Find the length of the *longest* diagonal of the parallelogram.
21. A baseball diamond is a square 90 feet on each side. The pitcher's mound is 60.5 feet away from home plate toward second base. Find the distance from the pitcher's mound to first base.
22. A hill makes a 15° angle with the horizontal. A wire extends from a 10-meter tall vertical pole to a point on the hill 20 meters downhill from the base of the pole. How long is the wire?
23. The three mutually tangent circles in Figure 42.6 have radii 3 centimeters, 4 centimeters, and 5 centimeters. Find the smallest of the three angles at A, B, and C.

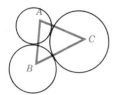

FIGURE 42.6

24. A diagonal of a rhombus has length 5 centimeters and each angle of the rhombus opposite that diagonal is three times as large as each of the other two angles of the rhombus. Find the length of each edge of the rhombus.

25. The Great Pyramid of Khufu was built with a square base 230 meters on each side. The distance from the top to each corner of the base was 219 meters. Use the Law of Cosines to compute the angle at the top of each face.

26. The opening of an ice-hockey goal is 6 feet across. The distance from the red spot at the center of a face-off area to the nearer front edge of a goal is 25.9 feet, and the distance from the red spot to the farther front edge is 30.1 feet. Compute the angle subtended at the red spot by the opening of the goal. (The angle is labeled α in Figure 42.7.)

27. A quadrilateral lot has three known sides and two known angles as shown in Figure 42.8. Compute the length of the fourth side and the measure of the other two angles. (Suggestion: Divide the figure into triangles.)

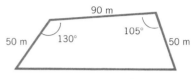

FIGURE 42.8

28. A central angle of 50° subtends a chord of length 3 centimeters on a circle. Use the Law of Cosines to compute the circle's radius.

29. A central angle of measure α ($0 < \alpha < \pi$) subtends a chord of length d on a circle of radius r. Use the Law of Cosines to prove that $d = r\sqrt{2(1 - \cos \alpha)}$.

30. Use Exercise 29 to prove that the perimeter of an n-sided regular polygon inscribed in a circle of radius r is $nr\sqrt{2[1 - \cos(2\pi/n)]}$. Use an identity to show that this agrees with the formula in Exercise 87 of Section 31.

31. Prove that $a^2 + b^2 + c^2 = 2(bc \cos \alpha + ac \cos \beta + ab \cos \gamma)$.

32. Prove that $\dfrac{\cos \alpha}{a} + \dfrac{\cos \beta}{b} + \dfrac{\cos \gamma}{c} = \dfrac{a^2 + b^2 + c^2}{2abc}$.

33. Solve $a^2 = b^2 + c^2 - 2bc \cos \alpha$ for b.

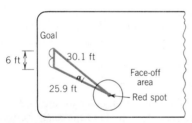

FIGURE 42.7

SECTION **43**

Law of Sines

A. STATEMENT AND PROOF

Law of Sines

$$\frac{\sin \alpha}{a} = \frac{\sin \beta}{b} = \frac{\sin \gamma}{c}$$

Proof The law actually consists of three equations,

$$\frac{\sin \alpha}{a} = \frac{\sin \beta}{b}, \quad \frac{\sin \beta}{b} = \frac{\sin \gamma}{c}, \quad \text{and} \quad \frac{\sin \alpha}{a} = \frac{\sin \gamma}{c}.$$

We now prove the first of these equations. The other two then follow simply by changing the letters appropriately.

Since $\alpha + \beta + \gamma = 180°$, not both α and β can be obtuse (greater than 90°). We assume $\alpha < 90°$ and then consider two cases: $\beta \leq 90°$ and $\beta > 90°$.

Figure 43.1a illustrates the case $\beta < 90°$. (The equations that follow are also true if $\beta = 90°$.) The altitude CD has length h.

From right triangle ADC, $\dfrac{h}{b} = \sin \alpha$ so $h = b \sin \alpha$.

From right triangle BDC, $\dfrac{h}{a} = \sin \beta$ so $h = a \sin \beta$.

Therefore,

$$b \sin \alpha = a \sin \beta \quad \text{so} \quad \frac{\sin \alpha}{a} = \frac{\sin \beta}{b}.$$

Figure 43.1b illustrates the case $\beta > 90°$.

From right triangle ADC, $h = b \sin \alpha$.

From right triangle BDC, $h = a \sin(180° - \beta)$.

However, $\sin(180° - \beta) = \sin \beta$ [Identity (34.5) with $u = 180°$ and $v = \beta$], so $h = a \sin \beta$. Therefore, as before,

$$\frac{\sin \alpha}{a} = \frac{\sin \beta}{b}. \qquad \Box$$

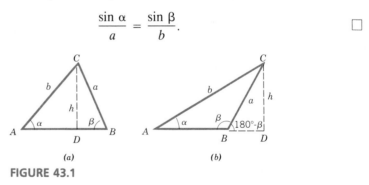

(a)   (b)

FIGURE 43.1

B. APPLICATIONS

To solve a triangle, we use trigonometric functions and one or more of these four basic results: the Law of Cosines, the Law of Sines, $\alpha + \beta + \gamma = 180°$, and (for a right triangle) the Pythagorean Theorem. Which results are used depends on what information is given. Rather than trying to memorize a method for each different situation, try to analyze the problems individually. Remember the four basic results listed above, and also remember that when you know all but one variable in an equation then you should be able to solve for that variable (at least for the types of equations that arise in solving triangles).

In Section 42 we used the Law of Cosines to treat the case SAS (where two sides and the included angle are given) and the case SSS (where all three sides are given). There is a unique solution in each case, provided, in the SSS case, that the sum of every pair of sides is greater than the third side. The Law of Sines is used in the following cases.

SAA Given two angles and a side of a triangle, solve the triangle. (There is a unique solution.)

SSA Given two sides of a triangle, and an angle opposite one of them, solve the triangle. (There may be no solution, one solution, or two solutions.)

When using the Law of Sines to compute a side of a triangle, it is convenient to invert each member of the equation and begin with the equivalent form

$$\frac{a}{\sin \alpha} = \frac{b}{\sin \beta} = \frac{c}{\sin \gamma}.$$

A calculator has been used throughout the following examples. ©

Example 43.1 Solve triangle ABC if $\alpha = 77°$, $\beta = 38°$, and $c = 15$ (Figure 43.2).

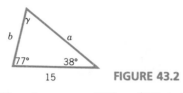

15   **FIGURE 43.2**

Solution From $\alpha + \beta + \gamma = 180°$ we have $\gamma = 180° - (77° + 38°) = 65°$. By the Law of Sines,

$$\frac{a}{\sin \alpha} = \frac{c}{\sin \gamma},$$

so

$$a = \frac{c \sin \alpha}{\sin \gamma} = \frac{15 \sin 77°}{\sin 65°} = \frac{15(0.9744)}{0.9063} \approx 16.13.$$

Also,

$$\frac{b}{\sin \beta} = \frac{c}{\sin \gamma},$$

so

$$b = \frac{c \sin \beta}{\sin \gamma} = \frac{15 \sin 38°}{\sin 65°} = \frac{15(0.6157)}{0.9063} \approx 10.19.$$ ∎

The case where we are given two sides of a triangle and an angle opposite one of them is called the **ambiguous case,** because there may be no solution, one solution, or two solutions. Figure 43.3 illustrates the possibilities for given α, a, and b.

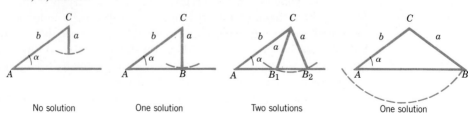

No solution   One solution   Two solutions   One solution

FIGURE 43.3

Example 43.2 Solve triangle ABC if $\alpha = 67°$, $a = 9$, and $b = 12$.

Solution From $\dfrac{\sin \alpha}{a} = \dfrac{\sin \beta}{b}$ we have

$$\sin \beta = \frac{b \sin \alpha}{a} = \frac{12 \sin 67°}{9} = \frac{12(0.9205)}{9} \approx 1.227.$$

Since $\sin \beta > 1$ is an impossibility, there is no solution for β. Thus there is no triangle with α, a, and b as given.

Example 43.3 Solve triangle ABC if $\gamma = 40°$, $a = 14$, and $c = 12$ (Figure 43.4).

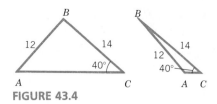

FIGURE 43.4

Solution From $\dfrac{\sin \alpha}{a} = \dfrac{\sin \gamma}{c}$ we have

$$\sin \alpha = \frac{a \sin \gamma}{c} = \frac{14 \sin 40°}{12} = \frac{14(0.6428)}{12} \approx 0.7499.$$

The acute angle satisfying $\sin \alpha = 0.7499$ is $48.58°$ (to the nearest $0.01°$). There is also an obtuse angle such that $\sin \alpha = 0.7499$, namely $180° - 48.58° = 131.42°$. Let

$$\alpha_1 = 48.58° \quad \text{and} \quad \alpha_2 = 131.42°.$$

For β, the choice α_1 gives

$$\beta_1 = 180° - (\alpha_1 + \gamma) = 180° - (48.58° + 40°) = 91.42°.$$

And the choice α_2 gives

$$\beta_2 = 180° - (\alpha_2 + \gamma) = 180° - (131.42° + 40°) = 8.58°.$$

There are two solutions in this case, one using α_1 and β_1, the other using α_2 and β_2. To complete each solution use the Law of Sines to determine the corresponding sides b_1 and b_2, respectively.

$$b_1 = \frac{c \sin \beta_1}{\sin \gamma} = \frac{12 \sin 91.42°}{\sin 40°} = \frac{12(0.9997)}{0.6428} \approx 18.66$$

$$b_2 = \frac{c \sin \beta_2}{\sin \gamma} = \frac{12 \sin 8.58°}{\sin 40°} = \frac{12(0.1492)}{0.6428} \approx 2.79$$

Thus the two solutions, each with $\gamma = 40°$, $a = 14$, and $c = 12$, are

$$\alpha_1 = 48.58°, \beta_1 = 91.42°, b = 18.66$$

and

$$\alpha_2 = 131.42°, \beta_2 = 8.58°, b_2 = 2.79.$$

Example 43.4 Solve triangle ABC if $\beta = 150°$, $b = 7.5$, and $c = 3.2$ (Figure 43.5).

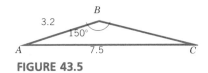

FIGURE 43.5

Solution From $\dfrac{\sin \beta}{b} = \dfrac{\sin \gamma}{c}$, we have

$$\sin \gamma = \frac{c \sin \beta}{b} = \frac{3.2 \sin 150°}{7.5} = \frac{3.2(0.5)}{7.5} \approx 0.2133.$$

Since $\beta = 150°$, we must have $\alpha + \gamma = 30°$, so, in particular, γ must be acute. The acute angle satisfying $\sin \gamma = 0.2133$ is $12.32°$ (the the nearest $0.01°$). This gives

$$\alpha = 180° - (\beta + \gamma) = 180° - (150° + 12.32°) = 17.68°.$$

To find a, we write

$$a = \frac{b \sin \alpha}{\sin \beta} = \frac{7.5 \sin 17.68°}{\sin 150°} = \frac{7.5(0.3037)}{0.5} \approx 4.56.$$

Thus the unique solution, with $\beta = 150°$, $b = 7.5$, and $c = 3.2$, is $\gamma = 12.32°$, $\alpha = 17.68°$, and $a = 4.56$. ∎

EXERCISES FOR SECTION 43

The use of a calculator to compute functions and perform arithmetic is encouraged on these exercises.

Solve the triangles determined by the given information. Give all solutions. If there is no solution, so state.

1. $\alpha = 20°$, $\gamma = 103°$, $a = 8$

2. $\alpha = 62°$, $\beta = 35°$, $b = 10$

3. $\beta = 114°$, $\gamma = 50°$, $c = 40$

4. $\alpha = 100°$, $a = 85$, $b = 60$

5. $\beta = 75°$, $b = 10$, $c = 14$

6. $\gamma = 115°$, $a = 8$, $c = 10$

7. $\beta = 40°$, $a = 25$, $b = 15$

8. $\gamma = 12°$, $b = 50$, $c = 16$

9. $\alpha = 25°$, $a = 5$, $c = 7$

10. $\gamma = 36°$, $a = 7$, $c = 5$

11. $\alpha = 30°$, $a = 4$, $b = 5$

12. $\beta = 110°$, $b = 25$, $c = 30$

13. $\alpha = 32.3°$, $a = 25.3$, $c = 36.2$

14. $\beta = 35.83°$, $a = 30.05$, $b = 17.52$

15. $\gamma = 27.17°$, $b = 17.14$, $c = 7.96$

The conditions given in each of Exercises 16–18 determine a unique triangle. Compute the area of the triangle in each case.

16. $\alpha = 40°$, $\beta = 60°$, $c = 10$

17. $a = 14$, $b = 40$, $c = 45$

18. $\alpha = 35°$, $b = 12$, $c = 15$

19. Points A and B are on one bank of a river and point C is on the opposite bank. Measurement reveals $\alpha = 70°$, $\beta = 62°$, and $c = 100$ meters (with our usual notation). Compute the distance between A and C.

20. Two observers five miles apart on level ground are directly beneath the path of a plane that is flying with constant altitude. What is the altitude of the plane, in feet, at an instant when the angles of elevation from the observers to the plane are $20°$ and $35°$, respectively?

21. From a point 50 feet down a hill from the base of a (vertical) tree, the angle of elevation to the top of the tree is 36°. The hill makes an angle of 15° with the horizontal. How tall is the tree?

22. To determine the distance between points C and D on opposite shores of a lake, a surveyor measures the distances and angles involving C, D and two other points A and B, obtaining the results shown in Figure 43.6. Compute the distance between C and D.

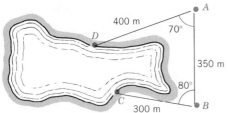

FIGURE 43.6

23. A vertical pole stands on a hill that makes an angle of 20° with the horizontal. When the angle of elevation to the sun is 45°, the pole casts a 20 meter shadow straight down the hill. How tall is the pole?

24. The angle of elevation to a plane from the 102nd floor observatory of the Empire State Building (1250 feet up) is 20°. At the same instant, the angle of elevation to the plane from the 86th floor observatory (1050 feet up) is 21.5°. What is the straight line distance from the 102nd floor to the plane?

Exercises 25–30 relate to Figure 43.7, which shows the relative locations of five points of interest in Washington, D.C. The line through the Lincoln Memorial, the Washington Monument, and the Capitol is the perpendicular bisector of the segment connecting the White House and the Jefferson Memorial, which lie on a north-south line. Except for Exercise 30, do each

exercise without using results from a previous exercise.

25. Find the distance between the White House and the Jefferson Memorial.

26. Find the distance between the White House and the Washington Monument.

27. Find the distance between the White House and the Capitol.

28. Find the distance between the Lincoln Memorial and the Capitol.

29. In the line from the White House through the Lincoln Memorial is extended 1.47 miles past the Memorial, we arrive at a point P that is just east of the Tomb of the Unknown Soldier. Find the distance from P to the Jefferson Memorial.

30. Robert F. Kennedy Stadium lies 1.98 miles due east of the Capitol. Find the distance between the White House and the Stadium. (Suggestion: Do Exercise 27 or Exercise 28 first.)

31. Prove that if A denotes the area of a triangle, then $A = \dfrac{1}{2} ab \sin \gamma$.

32. Prove that if A denotes the area of a triangle, then $A = \dfrac{b^2 \sin \alpha \sin \gamma}{\sin \beta}$. (Use Exercise 31.)

33. Prove that the area of a parallelogram equals the product of the lengths of any two adjacent sides and the sine of the angle between them. (Use Exercise 31.)

34. Prove the identity
$$\frac{\sin \alpha - \sin \beta}{\sin \alpha + \sin \beta} = \frac{a - b}{a + b}.$$

35. Prove the identity
$$\frac{2 \cos \dfrac{1}{2} (\alpha + \beta) \sin \dfrac{1}{2} (\alpha - \beta)}{2 \sin \dfrac{1}{2} (\alpha + \beta) \cos \dfrac{1}{2} (\alpha - \beta)} = \frac{a - b}{a + b}.$$

[Use Exercise 34 and Identities (36.5) and (36.6).]

36. Prove the **Law of Tangents:**
$$\frac{\tan \dfrac{1}{2} (\alpha - \beta)}{\tan \dfrac{1}{2} (\alpha + \beta)} = \frac{a - b}{a + b}.$$

(Use Exercise 35. The Law of Tangents has two other forms obtained by changing the variables in the obvious ways.)

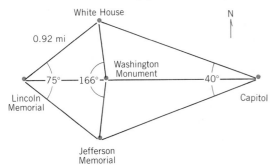

FIGURE 43.7

SECTION **44**
Vectors

A. INTRODUCTION

Some physical quantities, such as time, temperature, and length, can be measured with a real number (30 seconds of time, for example). Such quantities are called **scalar quantities.** Other physical quantities, such as velocity and force, require both a real number and a direction (wind velocity 30 miles per hour from the north, for example). These latter quantities are called **vector quantities.** In this section we study some of the elementary properties of vectors and look at several applications.

A directed line segment is called a **vector.** Thus a vector has both a **length** (or **magnitude**) and a **direction.** Figure 44.1 shows examples. The vector on the left in Figure 44.1 has **initial point** A and **terminal point** B. The notation $\overrightarrow{AB}$ will be used to denote such a vector. Vectors will also be denoted by boldface letters such as $\mathbf{u}$, $\mathbf{v}$, and so on. The lengths of vectors $\overrightarrow{AB}$ and $\mathbf{u}$ will be denoted by $|\overrightarrow{AB}|$ and $|\mathbf{u}|$, respectively. When working with vectors, we often refer to real numbers as **scalars.** Scalars will be denoted by italic letters such as a, b, and so on.

Vectors are defined to be **equal** if they have the same length and direction. For example, $\overrightarrow{AB} = \overrightarrow{EF}$ in Figure 44.1.

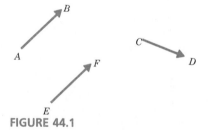

FIGURE 44.1

B. OPERATIONS INVOLVING VECTORS

To define the **vector sum** of vectors $\overrightarrow{AB}$ and $\overrightarrow{CD}$, place $\overrightarrow{CD}$ so that C coincides with B. (As a consequence of the definition of equality of vectors, this does not change $\overrightarrow{CD}$ provided its length and direction are not changed.) Then, by definition, $\overrightarrow{AB} + \overrightarrow{CD} = \overrightarrow{AD}$ (Figure 44.2).

A useful geometric interpretation of vector addition is shown in Figure 44.3. The sum $\mathbf{u} + \mathbf{v}$ is represented by a diagonal of a parallelogram formed with $\mathbf{u}$ and $\mathbf{v}$ as a pair of adjacent sides. Notice that the side opposite $\mathbf{u}$, given the same direction as $\mathbf{u}$, represents a vector equal to $\mathbf{u}$.

The vector with length zero is called the **zero vector** and will be denoted by $\mathbf{0}$. If $\mathbf{u}$ is any vector, then

$$\mathbf{u} + \mathbf{0} = \mathbf{0} + \mathbf{u} = \mathbf{u}. \tag{44.1}$$

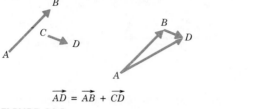

$$\overrightarrow{AD} = \overrightarrow{AB} + \overrightarrow{CD}$$

FIGURE 44.2

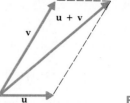

FIGURE 44.3

If **u** is a vector and c is a scalar, then the **product** of c and **u**, denoted c**u**, is the vector defined as follows.

The length of c**u** is $|c| \cdot |\mathbf{u}|$ (the absolute value of c times the length of **u**).

The direction of c**u** is the same as that of **u** if $c > 0$, and opposite to that of **u** if $c < 0$.

The negative of **u**, denoted $-\mathbf{u}$, is defined to be $(-1)\mathbf{u}$. Thus $-\mathbf{u}$ has the same length as **u** but the opposite direction. In particular,

$$\mathbf{u} + (-\mathbf{u}) = \mathbf{0} \quad \text{and} \quad (-\mathbf{u}) + \mathbf{u} = \mathbf{0}. \tag{44.2}$$

Notice that $\overrightarrow{AB} = -\overrightarrow{BA}$. Figure 44.4 shows several examples.

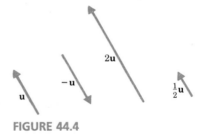

FIGURE 44.4

It can be proved that

$$\mathbf{u} + \mathbf{v} = \mathbf{v} + \mathbf{u} \tag{44.3}$$

for all vectors **u** and **v** (the *commutative property*) and

$$\mathbf{u} + (\mathbf{v} + \mathbf{w}) = (\mathbf{u} + \mathbf{v}) + \mathbf{w} \tag{44.4}$$

for all vectors **u**, **v**, and **w** (the *associative property*).

The difference $\mathbf{u} - \mathbf{v}$ of **u** and **v** (in that order) is defined to be $\mathbf{u} + (-\mathbf{v})$. By (44.4) and (44.2),

$$(\mathbf{u} - \mathbf{v}) + \mathbf{v} = [\mathbf{u} + (-\mathbf{v})] + \mathbf{v}$$
$$= \mathbf{u} + [(-\mathbf{v}) + \mathbf{v}]$$
$$= \mathbf{u} + \mathbf{0}$$
$$= \mathbf{u}.$$

The last equation provides a geometric interpretation of $\mathbf{u} - \mathbf{v}$: when $\mathbf{u} - \mathbf{v}$ is added to **v**, the result is **u**. See Figure 44.5.

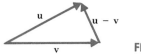

FIGURE 44.5

C. COMPONENTS

To work with vectors in a Cartesian plane we introduce the two **unit vectors i** and **j**. By definition, **i** and **j** are the vectors of length one unit pointing in the positive x and y directions, respectively (Figure 44.6).

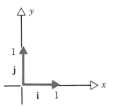

FIGURE 44.6

Consider a vector **u** in a Cartesian plane with its initial point at the origin (Figure 44.7). If its endpoint has coordinates (u_x, u_y), then

$$\mathbf{u} = u_x\mathbf{i} + u_y\mathbf{j}. \tag{44.5}$$

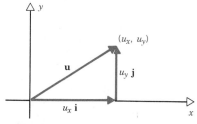

FIGURE 44.7

This vector will also be denoted by $\langle u_x, u_y \rangle$. The numbers u_x and u_y are called the **components** of **u**.

By the Pythagorean Theorem, the length of a vector $\langle u_x, u_y \rangle$ is given by

$$|\langle u_x, u_y \rangle| = \sqrt{u_x^2 + u_y^2}. \tag{44.6}$$

(Consider Figure 44.7.)

If the vector **u** is thought of as the terminal side of an angle θ in standard position (Figure 44.8), then

$$\frac{u_x}{|\mathbf{u}|} = \cos\theta \quad \text{and} \quad \frac{u_y}{|\mathbf{u}|} = \sin\theta.$$

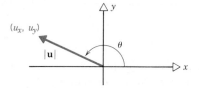

FIGURE 44.8

Thus $u_x = |\mathbf{u}| \cos \theta$ and $u_y = |\mathbf{u}| \sin \theta$, and

$$\mathbf{u} = (|\mathbf{u}| \cos \theta)\mathbf{i} + (|\mathbf{u}| \sin \theta)\mathbf{j}. \tag{44.7}$$

We call θ the **direction** of $\mathbf{u}$. The direction θ is uniquely determined by $\mathbf{u}$ if we make the restriction $0° \le \theta < 360°$.

Example 44.1   Write $\mathbf{u}$ in the form $u_x\mathbf{i} + u_y\mathbf{j}$ if $|\mathbf{u}| = 3$ and the direction of $\mathbf{u}$ is $120°$.

Solution   By Equation (44.7),

$$\mathbf{u} = (3 \cos 120°)\mathbf{i} + (3 \sin 120°)\mathbf{j} = -\frac{3}{2}\mathbf{i} + \frac{3\sqrt{3}}{2}\mathbf{j}. \qquad \blacksquare$$

Here is a summary, in terms of components, of some of the important vector properties.

$$\langle u_x, u_y \rangle + \langle v_x, v_y \rangle = \langle u_x + v_x, u_y + v_y \rangle$$

$$\langle u_x, u_y \rangle + \langle 0, 0 \rangle = \langle u_x, u_y \rangle$$

$$c\langle u_x, u_y \rangle = \langle cu_x, cu_y \rangle$$

$$-\langle u_x, u_y \rangle = \langle -u_x, -u_y \rangle$$

$$\langle u_x, u_y \rangle - \langle v_x, v_y \rangle = \langle u_x - v_x, u_y - v_y \rangle$$

$$|\langle u_x, u_y \rangle| = \sqrt{u_x^2 + u_y^2}$$

Example 44.2   Assume $\mathbf{u} = \langle 2, -1 \rangle$ and $\mathbf{v} = \langle 4, 3 \rangle$.

(a) $\mathbf{u} + \mathbf{v} = \langle 2, -1 \rangle + \langle 4, 3 \rangle = \langle 6, 2 \rangle$
(b) $\mathbf{u} - \mathbf{v} = \langle 2, -1 \rangle - \langle 4, 3 \rangle = \langle -2, -4 \rangle$
(c) $6\mathbf{v} = 6\langle 4, 3 \rangle = \langle 24, 18 \rangle$
(d) $-\mathbf{u} = -\langle 2, -1 \rangle = \langle -2, 1 \rangle$
(e) $|\mathbf{v}| = \sqrt{4^2 + 3^2} = 5$ $\blacksquare$

Vector properties are treated more fully in books on linear algebra and vector analysis. We move on now to two applications.

D. APPLICATION: VELOCITY

To describe the direction of vectors representing velocity, consider two perpendicular axes SN (representing south-north) and WE (representing west-east), as shown in Figure 44.9. Given a velocity vector, place its initial point at the intersection of these axes. If the vector is not directed along one of the axes, then it makes an acute angle θ with either the north or the south direction. Depending on the quadrant of the vector, the notation NθW, NθE, SθE, or SθW is used to denote the direction of the vector, as shown in Figure 44.9. See Figure 44.10 for two specific cases.

The next example illustrates the use of vectors to solve a velocity problem. The **speed** of an object is represented by the length of the corresponding velocity vector.

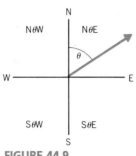

FIGURE 44.9

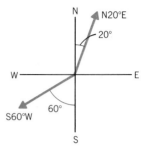

FIGURE 44.10

Example 44.3 A plane is headed N30°E at 150 mi/hr. The wind is blowing *from* N10°W at 20 mi/hr. Find the direction and speed of the plane's motion relative to the ground.

Solution In Figure 44.11, $\overrightarrow{AB}$ represents the plane's velocity ignoring the wind, and $\overrightarrow{BC}$ represents the velocity of the wind. The solution is the sum $\overrightarrow{AC}$ of $\overrightarrow{AB}$ and $\overrightarrow{BC}$. Using elementary geometry, we see that angle ABC is a 40° angle. (Lines SN and DB are parallel and the two 30° angles are opposite interior angles determined by AB.) Thus b, the length of $\overrightarrow{AC}$, can be computed by the Law of Cosines.

$$b^2 = 20^2 + 150^2 - 2(20)(150) \cos 40°$$

$$= 400 + 22500 - 6000(0.7660)$$

$$b = 135 \text{ mi/hr}$$

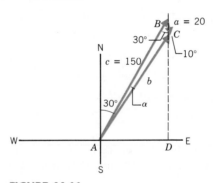

FIGURE 44.11

To determine the direction of $\overrightarrow{AC}$ we first compute α by the Law of Sines.

$$\frac{\sin \alpha}{a} = \frac{\sin 40°}{b}$$

$$\sin \alpha = \frac{20 \sin 40°}{135} = \frac{20(0.6428)}{135} = 0.0952$$

$$\alpha = 5.46°$$

Since $30° + 5.46° = 35.46°$, the motion of the plane relative to the ground is 135 mi/hr in the direction N35.46°E.

E. APPLICATION: FORCE

A force acting on an object has both a magnitude and a direction, and can, therefore, be represented by a vector. In solving force problems it is often convenient to work separately with components in appropriately chosen perpendicular directions. Thus we begin with an example showing how to resolve a force into such component forces. (We consider only forces in a plane. Similar ideas apply to forces acting in three dimensions.)

Example 44.4 A 50-kilogram crate is resting on an inclined plane with a 10° elevation. What force does the crate exert perpendicular to the plane? What force does the crate exert parallel to the plane?

Solution The force due to the weight is directed straight downward, as shown by **F** in Figure 44.12. The two component forces are $\mathbf{F_n}$ (for *normal,* or perpendicular, to the plane) and $\mathbf{F_p}$ (for *parallel* to the plane). The angle at A is given as 10°.

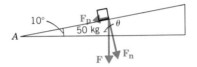

FIGURE 44.12

Therefore, θ is also a 10° angle, because its sides are perpendicular to those of the angle at A. It follows that

$$|\mathbf{F_n}| = |\mathbf{F}| \cos \theta = 50 \cos 10° = 50(0.9848) = 49.24$$

$$|\mathbf{F_p}| = |\mathbf{F}| \sin \theta = 50 \sin 10° = 50(0.1736) = 8.68.$$

That is, the crate exerts a force of 49.24 kg perpendicular to the plane and a force of 8.68 kg parallel to the plane. ▨

The vector sum of two or more forces acting on an object is called the **resultant** of the forces. The resultant force acting on the object has the same effect as the combination of the individual forces acting concurrently. The next example shows two ways to compute the resultant of two forces.

Example 44.5 Find the resultant **F** of the forces **P** and **Q** if **P** is a 10-lb force directed north and **Q** is a 30-lb force directed N70°E (Figure 44.13).

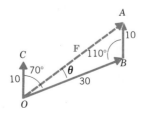

FIGURE 44.13

First Solution The magnitude of **F** is given by $\overline{OA}$. Since angle COB is $70°$, angle OBA is $110°$
(Law of (because OC and BA are parallel). Therefore, by the Law of Cosines,
Cosines and
Law of Sines)

$$|\mathbf{F}|^2 = |\mathbf{P}|^2 + |\mathbf{Q}|^2 - 2|\mathbf{P}| \cdot |\mathbf{Q}| \cos 110°$$

$$= 10^2 + 30^2 - 2(10)(30)(-0.3420)$$

$$= 1205.2$$

$$|\mathbf{F}| = 34.7.$$

By the Law of Sines,

$$\frac{\sin \theta}{10} = \frac{\sin 110°}{|\mathbf{F}|}$$

$$\sin \theta = \frac{10 \sin 110°}{34.7} = \frac{10(0.9397)}{34.7} = 0.2708$$

$$\theta = 15.7°.$$

Therefore, angle $COA = 70° - 15.7° = 54.3°$. Thus the resultant is a 34.7-lb
force directed N54.3°E.

Second The north component of **F** is the sum of the north components of **P** and **Q**.
Solution
(Components) $\mathbf{F_N}$ = north component of $\mathbf{F}$ = $10 + 30 \cos 70° = 10 + 30(0.3420) = 20.26$

The east component of **F** is the sum of the east components of **P** and **Q**.

$\mathbf{F_E}$ = east component of $\mathbf{F}$ = $0 + 30 \sin 70° = 30(0.9397) = 28.19$

Therefore (Figure 44.14),

$$|\mathbf{F}|^2 = |\mathbf{F_N}|^2 + |\mathbf{F_E}|^2 = 20.26^2 + 28.19^2 = 1205.1$$

$$|\mathbf{F}| = 34.7.$$

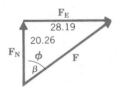

FIGURE 44.14

And the direction of **F** is NϕE, where

$$\phi = \arctan \frac{|\mathbf{F_E}|}{|\mathbf{F_N}|} = \arctan \frac{28.19}{20.26}$$

$$= \arctan 1.3914 = 54.3°.$$

EXERCISES FOR SECTION 44

*In each of Exercises 1–9, draw **u** and **v**, and find and*
simplify

1. u $= 3\mathbf{i} + \mathbf{j}$, **v** $= 3\mathbf{i} - 4\mathbf{j}$

2. u $= -2\mathbf{j}$, **v** $= 12\mathbf{i} + 5\mathbf{j}$

3. u $= 7\mathbf{i} + 10\mathbf{j}$, **v** $= \mathbf{i} - 6\mathbf{j}$

(a) u $+$ **v**   **(b)** $-3\mathbf{v}$ **4. u** $= \langle 2, -2 \rangle$, **v** $= \langle 0, 5 \rangle$

(c) v $- 2\mathbf{u}$ **(d)** $|\mathbf{v}|$.   **5. u** $= \langle 4, 4 \rangle$, **v** $= \langle -3, 10 \rangle$

6. $\mathbf{u} = \langle -6, 0 \rangle$, $\mathbf{v} = \langle 2, -3 \rangle$

7. $\mathbf{u} = 3(\mathbf{i} - \mathbf{j})$, $\mathbf{v} = -(4\mathbf{i} + 2\mathbf{j})$

8. $\mathbf{u} = -2(2\mathbf{i} - \mathbf{j})$, $\mathbf{v} = \mathbf{j} + (2\mathbf{i} - 2\mathbf{j})$

9. $\mathbf{u} = -\mathbf{i} + (\mathbf{i} - 2\mathbf{j})$, $\mathbf{v} = 4\mathbf{i} - (\mathbf{i} + 4\mathbf{j})$

*Use the given information to write **u** in the form $u_x\mathbf{i} + u_y\mathbf{j}$. As in (44.7), θ denotes the direction of **u**.*

10. $|\mathbf{u}| = 2$ and $\theta = 150°$

11. $|\mathbf{u}| = 0.5$ and $\theta = 300°$

12. $|\mathbf{u}| = 5$ and $\theta = 225°$

13. $u_x = 3$ and $\theta = 315°$

14. $u_y = -1$ and $\theta = 210°$

15. $u_x = -5$ and $\theta = 120°$

16. Use $c\langle u_x, u_y \rangle = \langle cu_x, cu_y \rangle$ and Equation (44.6) to verify $|c\mathbf{u}| = |c| \cdot |\mathbf{u}|$.

17. Determine u_x and u_y if **u** equals the vector with initial point at (a, b) and terminal point at (c, d).

18. The vertices of a parallelogram, in counterclockwise order starting at A, are A, B, C, and D. What is k if $(\overrightarrow{AB} + \overrightarrow{CB}) = k(\overrightarrow{AD} + \overrightarrow{CD})$? Justify your answer.

19. A plane is headed S45°W at 200 mi/hr. The wind is blowing due north at 15 mi/hr. Find the direction and speed of the plane's motion relative to the ground.

20. A plane is headed east at 175 mi/hr. Because of the wind, the plane's motion relative to the ground is 160 mi/hr N80°E. Find the speed of the wind and the direction from which it is blowing.

21. A pilot wants to fly N50°W at 200 mi/hr (relative to the ground). The wind is from due east at 10 mi/hr. What direction should she head and what should the speed be relative to the air?

22. A balloon is rising at a rate of 10 ft/sec and the wind is blowing (horizontally) at 20 ft/sec. What is the angle of inclination of the balloon's path?

23. A ship is headed due north at 30 km/hr. The ocean current is in the direction S80°W at 6 km/hr. Find the actual direction and speed of the ship's motion.

24. A one-quarter mile wide river flows from north to south with a current of 3 mi/hr. A man wants to row across the river on a straight east-west course in 20 minutes. In what direction and at what speed should he row?

25. A 3000-lb car is parked on an inclined plane with a 12° elevation. What force does the car exert perpendicular to the plane? What force does the car exert parallel to the plane?

26. A cable that makes a 15° angle with the horizontal is attached to the front of a car. The tension in the cable is 750 lb. (The tension is the force with which the cable is pulling on the car.) Determine the vertical and horizontal components of the tension.

27. A wind blowing from N22°W is exerting a 500-lb force on the sail of a boat. Determine the south-north and the west-east components of the force.

28. Find the resultant of the forces **P** and **Q** if **P** is a 15-kg force directed N50°W and **Q** is a 25-kg force directed S50°W.

29. A 75-kg weight is suspended as shown in Figure 44.15. What is the tension in the rope? [Suggestion: Think of three forces acting at point **P**, (1) a horizontal force due to the support, (2) a force downward due to the weight, and (3) the force (tension) exerted by the rope. The vector sum (and therefore the vertical and horizontal components of the vector sum) of the three forces is zero. Concentrate on the vertical components.]

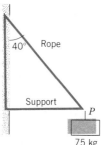

75 kg **FIGURE 44.15**

30. Forces $\mathbf{F_1}$ and $\mathbf{F_2}$ are acting together to exert a resultant 900-kg force in the direction S80°W. Force $\mathbf{F_1}$ is a 400-kg force in the direction N80°W. Determine the magnitude and direction of $\mathbf{F_2}$.

*The **inner product** (or **dot product**) of two vectors $\mathbf{u} = \langle u_x, u_y \rangle$ and $\mathbf{v} = \langle v_x, v_y \rangle$ is defined by*

$$\mathbf{u} \cdot \mathbf{v} = u_x v_x + u_y v_y. \qquad (44.8)$$

*Prove the properties in Exercises 31–34 for all vectors **u**, **v**, and **w** and every scalar c.*

31. $\mathbf{u} \cdot \mathbf{v} = \mathbf{v} \cdot \mathbf{u}$

32. $\mathbf{u} \cdot (\mathbf{v} + \mathbf{w}) = \mathbf{u} \cdot \mathbf{v} + \mathbf{u} \cdot \mathbf{w}$

33. $c(\mathbf{u} \cdot \mathbf{v}) = (c\mathbf{u}) \cdot \mathbf{v} = \mathbf{u} \cdot (c\mathbf{v})$

34. $\mathbf{u} \cdot \mathbf{u} = |\mathbf{u}|^2$

35. Use the Law of Cosines to prove that if θ is the angle between two nonzero vectors $\mathbf{u}$ and $\mathbf{v}$ (both with initial point at the origin), then

$$\cos \theta = \frac{\mathbf{u} \cdot \mathbf{v}}{|\mathbf{u}| \cdot |\mathbf{v}|}. \qquad (44.9)$$

[Suggestion: $\mathbf{u}$, $\mathbf{v}$, and $\mathbf{u} - \mathbf{v}$ represent the three sides of a triangle with $\mathbf{u} - \mathbf{v}$ opposite θ. Use (44.6).]

36. Use Exercise 35 to prove that two nonzero vectors are perpendicular iff their inner product is zero.

SECTION **45**
Polar Coordinates

A. INTRODUCTION

In this section we consider polar coordinate systems, which are more convenient for some purposes than the Cartesian coordinate systems we have used previously. Throughout the section all points and lines are assumed to lie in a given plane.

We begin with a point that we denote by O and call the **origin** (or **pole**). We choose a ray with endpoint O and call it the **polar axis.** (See Figure 45.1,

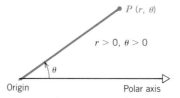

$P\ (r,\ \theta)$

$r > 0,\ \theta > 0$

θ

Origin　　　　Polar axis　　**FIGURE 45.1**

which follows the standard practice of directing the polar axis horizontally to the right.) We next choose a unit of length relative to which all distances will be measured. Finally, we say that an angle with its vertex at the origin is positive if it is generated by counterclockwise rotation from the polar axis and negative if it is generated by clockwise rotation.

Now consider an ordered pair $[r, \theta]$, where r is a positive real number and θ is the measure of an angle in either degrees or radians. From θ we obtain a unique ray with endpoint at O—namely, the terminal side of the angle with measure θ, vertex at O, and initial side along the polar axis. We associate with $[r, \theta]$ the point P that is r units from O measured along this ray. Figure 45.2 shows examples. We call $[r, \theta]$ **polar coordinates** of P. The notation $P[r, \theta]$ will denote the point with polar coordinates $[r, \theta]$. This square bracket notation will distinguish polar coordinates from Cartesian coordinates, which we shall continue to write with parentheses.

If $r < 0$, we associate with $[r, \theta]$ the point P that is $|r|$ units from O measured in the direction opposite to that determined by θ. Figure 45.3 shows the points with polar coordinates $[5, 135°]$ and $[-5, 135°]$ for comparison. For $r \neq 0$,

$$P[r, \theta] = P[-r, \theta + 180°] \qquad \text{(for } \theta \text{ in degrees)}$$

$$P[r, \theta] = P[-r, \theta + \pi] \qquad \text{(for } \theta \text{ in radians).}$$

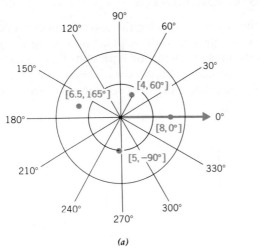
(a)

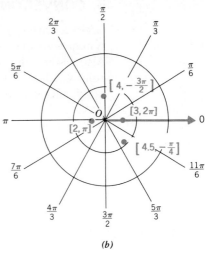
(b)

FIGURE 45.2

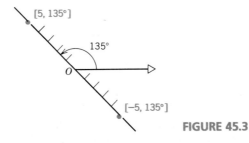

FIGURE 45.3

With each pair $[0, \theta]$ we associate the point O, the origin. Although each point in a Cartesian plane has a unique pair of Cartesian coordinates, relative to any polar coordinate system each point has many polar coordinates. Figure 45.4 shows three of the possibilities for $P(8, 90°)$. In general, if θ is in degrees, then

$$P[r, \theta] = P[r, \theta + k \cdot 360°] \qquad \text{for every integer } k,$$

$$- P[8, 90°] = P[8, 450°] = P[-8, 270°]$$

90°

O

FIGURE 45.4

and if θ is in radians, then

$$P[r, \theta] = P[r, \theta + 2k\pi] \qquad \text{for every integer } k.$$

With restrictions such as $r > 0$ and $0 \le \theta < 2\pi$, or $r > 0$ and $-\pi < \theta \le \pi$, each point other than the origin will have unique polar coordinates.

B. CONVERTING COORDINATES

Given a Cartesian coordinate system for the plane, consider a polar coordinate system with the same origin and with its polar axis along the positive x-axis. Also assume the same unit of length for both coordinate systems. Then each point will have both Cartesian and polar coordinates, and from Figure 45.5 we can deduce the following relationships between these coordinates.

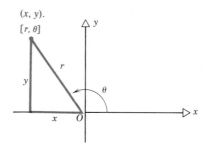

FIGURE 45.5

$$x = r \cos \theta \qquad\qquad y = r \sin \theta \qquad\qquad (45.1)$$

$$r^2 = x^2 + y^2 \qquad \tan \theta = \frac{y}{x} \quad (x \neq 0) \qquad\qquad (45.2)$$

With (45.1) we can compute the Cartesian coordinates if polar coordinates are known, and with (45.2) we can compute polar coordinates if the Cartesian coordinates are known. In the latter case the signs of x and y determine the quadrant of θ.

Example 45.1 Convert from polar to Cartesian coordinates.

(a) $[3, 150°]$ (b) $[-4.2, 2.7]$

Solution Use (45.1) in both cases. For (b) use a calculator (in the radian mode) or Table VI.

(a) $x = 3 \cos 150° = 3 \left(-\dfrac{\sqrt{3}}{2} \right) = \dfrac{-3\sqrt{3}}{2}$

$\quad\; y = 3 \sin 150° = 3 \left(\dfrac{1}{2} \right) = 1.5$

The exact answer is $(-3\sqrt{3}/2, 1.5)$. An approximate answer is $(-2.598, 1.5)$.

(b) $x = -4.2 \cos 2.7 = -4.2(-0.9041) = 3.797$

$\quad\; y = -4.2 \sin 2.7 = -4.2(0.4274) = -1.795$

The answer is $(3.797, -1.795)$. ∎

Example 45.2 Convert from Cartesian to polar coordinates, using degrees in part (a) and radians in part (b).

(a) $(3, 6)$ (b) $(-40, -30)$

Solution Use (45.2) in both cases.

(a) The point is in Quadrant I, so we can choose $r > 0$ and $0° < \theta < 90°$.

$$r = \sqrt{3^2 + 6^2} = \sqrt{45} = 3\sqrt{5}$$

$$\tan \theta = \frac{6}{3} = 2 \quad \text{and} \quad \theta = \arctan 2$$

An exact answer is $[3\sqrt{5}, \arctan 2]$. Using a calculator or Table V we get $3\sqrt{5} = 6.708$ and $\arctan 2 = 63.4°$. Thus an approximate answer is [6.708, 63.4°].

(b) The point is in Quadrant III, so we can choose $r > 0$ and $\pi < \theta < 3\pi/2$.

$$r = \sqrt{(-40)^2 + (-30)^2} = 50$$

$$\tan \theta = \frac{-30}{-40} = 0.75$$

The reference angle for θ is $\arctan 0.75$. Thus $\theta = \pi + \arctan 0.75$, and an exact answer is $[50, \pi + \arctan 0.75]$. From either a calculator or Table VI, $\pi + \arctan 0.75 \approx 3.14 + 0.64 = 3.78$, so an approximate answer is [50, 3.78].

Another form for the answer is $[-50, \arctan 0.75]$, or, approximately $[-50, 0.64]$. ◼

C. POLAR EQUATIONS AND THEIR GRAPHS

An equation involving one or both of the letters r and θ representing polar coordinates is called a **polar equation.** Given a plane with a polar coordinate system, the **graph** (or **polar graph**) of a polar equation is the set of all points whose coordinates satisfy the equation.

Example 45.3 The graph of any equation of the form $r = c$, where c is a positive constant, is a circle centered at the origin. Figure 45.6 shows the case $r = 6$. ◼

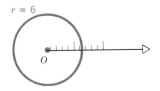

FIGURE 45.6

Example 45.4 The graph of any equation of the form $\theta = c$, where c is a constant angle measure in either degrees or radians, is a line through the origin. Figure 45.7 shows the case $\theta = \pi/4$. ◼

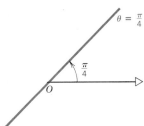

FIGURE 45.7

Example 45.5 Sketch the graph of the polar equation $r = 2(1 + \cos \theta)$.

Solution Table 45.1 shows some of the coordinate pairs. By plotting the corresponding points, along with others, we are led to the graph in Figure 45.8.

TABLE 45.1

θ	0	$\pi/4$	$\pi/2$	$3\pi/4$	π	$5\pi/4$	$3\pi/2$	$7\pi/4$
$r = 2(1 + \cos \theta)$	4	3.414	2	0.586	0	0.586	2	3.414

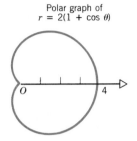

Polar graph of
$r = 2(1 + \cos \theta)$

FIGURE 45.8

It is instructive in this case to draw on what we know about the graphs of trigonometric equations in Cartesian planes. Figure 45.9 shows the graph of $r = 2(1 + \cos \theta)$ for $0 \le \theta < 2\pi$ with (r, θ) interpreted as Cartesian coordinates. We see that as θ increases from 0 to π, r decreases from 4 to 0. This same variation appears in Figure 45.8. And as θ increases from π to 2π, r increases from 0 to 4 on both graphs. The variation shown by the Cartesian graph does not reveal the exact shape of the polar graph, of course, but it does give a general idea of what to expect. *Remember, however, that unless specified otherwise a request to sketch the graph of an equation with* r *and* θ *will mean to sketch the polar graph, that is, the graph relative to a polar coordinate system.*

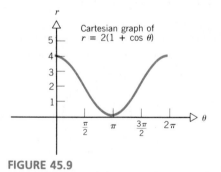

Cartesian graph of
$r = 2(1 + \cos \theta)$

FIGURE 45.9

The graph of any equation of the form $r = 2a(1 + \cos \theta)$ will have the same general shape as that in Figure 45.8. Such a graph is called a **cardioid.**

The equations in (45.1) and (45.2) can be used to convert any equation with x and y into an equation with r and θ, and, conversely, any equation with r and θ into an equation with x and y. Conversions of these kinds can sometimes make graphing easier. The next example gives an illustration.

Example 45.6 Find an equation with x and y that has the same graph as $r = 2a \cos \theta$. Sketch the graph.

Solution Since cos θ = x/r, the given equation is equivalent to

$$r = 2ax/r \quad \text{or} \quad r^2 = 2ax.$$

Using $r^2 = x^2 + y^2$ and completing the square, we can now write

$$x^2 + y^2 = 2ax$$

$$x^2 - 2ax + y^2 = 0$$

$$x^2 - 2ax + a^2 + y^2 = a^2$$

$$(x - a)^2 + y^2 = a^2.$$

We recognize this as an equation whose graph is a circle with radius |a| and center at the point with Cartesian coordinates (a, 0). (The polar coordinates of the center are also [a, 0].) The graph for a > 0 is shown in Figure 45.10.

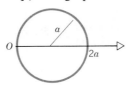

$r = 2a \cos θ \quad (a > 0)$ **FIGURE 45.10**

Exercise 58 asks you to show that the graph of $r = 2a \sin θ$ is a circle with radius |a| and center at the point with polar cordinates [a, π/2]. ■

Example 45.7 Sketch the graph of $r = a \sin 3θ \ (a > 0)$.

Solution Table 45.2 shows the variations in θ, 3θ, and r as θ increases from 0° to 180°. A calculator or Table V can be used to compute specific coordinate pairs. Figure

TABLE 45.2

θ	3θ	$r = a \sin 3θ$
0° to 30°	0° to 90°	0 to a
30° to 60°	90° to 180°	a to 0
60° to 90°	180° to 270°	0 to −a
90° to 120°	270° to 360°	−a to 0
120° to 150°	360° to 450°	0 to a
150° to 180°	450° to 540°	a to 0

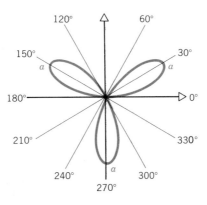

$r = a \sin 3θ \quad (a > 0)$

FIGURE 45.11

45.11 shows the graph. As θ increases from 180° to 360° the points obtained will repeat those determined by $0° \leq \theta < 180°$. ▪

The graph in Figure 45.11 is an example of a **rose curve.** Equations of either of the forms

$$r = a \sin n\theta, \; r = a \cos n\theta$$

produce such curves. If n is an odd integer, the curve will have n loops. If n is an even integer, the curve will have $2n$ loops.

EXERCISES FOR SECTION 45

For each exercise, plot the point with the given polar coordinates and give two other polar representations for the same point, one with $r < 0$ *and another with* $\theta < 0$.

1. [3, 100°]

2. [2.7, 205°]

3. [6.8, 10°]

4. [19.8, $\pi/5$]

5. [7, 11π/12]

6. [4, 9π/8]

Convert from polar to Cartesian coordinates.

7. [-3, 60°]

8. [5, 3π/4]

9. [10, $-2\pi/3$]

Convert from Cartesian to polar coordinates. Use radians as the angle measure in each case.

10. $(1, \sqrt{3})$

11. $(-5, 5)$

12. $(\sqrt{3}, -1)$

13. $(-6, -8)$

14. $(5, -12)$

15. $(-20, 15)$

Sketch the graph of each equation. In Exercises 34–39, θ must be in radians.

16. $\theta = 0$

17. $r = 5$

18. $\theta = 5\pi/6$

19. $r = -4.5$

20. $\theta = -100°$

21. $r^2 = 4$

22. $\theta^2 = \pi^2/16$

23. $r = 6 \sin \theta$

24. $r = 3 \cos \theta$

25. $2r = \cos \theta$

26. $r = 5(1 + \cos \theta)$

27. $r = 3(1 - \cos \theta)$

28. $r = 5(1 + \sin \theta)$

29. $r = 3(1 - \sin \theta)$

30. $r = 5 \sec \theta$

31. $r = 2 \cos \theta - 1$

32. $r = 1 + 2 \cos \theta$

33. $r = 3 + 2 \cos \theta$

34. $r = \theta \; (\theta > 0)$

35. $r = -2\theta \; (\theta > 0)$

36. $r\theta = 1 \; (\theta > 0)$

37. $r\theta = -2 \; (\theta > 0)$

38. $r = \ln \theta$

39 $\log r = \theta$ (Suggestion: First solve for r.)

40. $r = 3 \sin 2\theta$

41. $r = 2 \cos 3\theta$

42. $r = \cos 2\theta$

Find an equation with x and y that has the same graph as the given equation.

43. $r = 6 \sin \theta$

44. $r = 12 \cos \theta$

45. $r = 2$

46. $\theta = -\pi/6$

47. $r = \csc \theta$

48. $r = \sec \theta$

Find a polar equation that has the same graph as the given equation.

49. $x^2 + y^2 = a^2$

50. $y = x$

51. $x^2 + y^2 = ax$

52. $y = -3$

53. $x = 5$

54. $x^2 - y^2 = 25$

55. The graph of a polar equation is symmetric with respect to the line $\theta = 0$ if an equivalent equation is obtained when θ is replaced by $-\theta$.

(a) Find a corresponding statement relating to symmetry with respect to the origin.

(b) Find a corresponding statement relating to symmetry with respect to the line $\theta = \pi/2$.

56. Prove that the graph of $r = a \sin \theta + b \cos \theta$ is a circle with center at $(b/2, a/2)$. What is the radius? (Suggestion: Begin by multiplying both sides by r.)

57. Use the Law of Cosines to prove that if d is the

distance between $P_1[r_1, \theta_1]$ and $P_2[r_2, \theta_2]$, then
$$d^2 = r_1^2 + r_2^2 - 2r_1r_2\cos(\theta_2 - \theta_1).$$

58. By changing to x and y, show that the graph of $r = 2a\sin\theta$ is a circle with radius $|a|$ and center at the point with polar coordinates $[a, \pi/2]$.

59. (a) Prove that the polar equation of the circle with center at $P_1[r_1, \theta_1]$ and radius a is
$$r^2 + r_1^2 - 2rr_1\cos(\theta - \theta_1) = a^2.$$
$$\text{(45.3)}$$

(Suggestion: Refer to Figure 45.12 and use the Law of Cosines.)

(b) Verify that if $r_1 = a$ and $\theta_1 = 0$, then Equation 45.3 reduces to
$$r = 2a\cos\theta.$$

(c) Verify that if $r_1 = a$ and $\theta_1 = \pi/2$, then Equation 45.3 reduces to $r = 2a\sin\theta$.

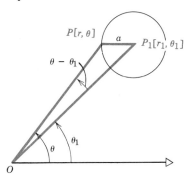

FIGURE 45.12

60. (a) Prove that if the perpendicular from O to a line L has length p and makes an angle of

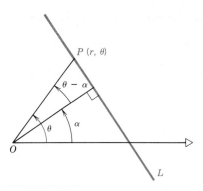

FIGURE 45.13

measure α with the polar axis, then an equation for L is
$$r = p\sec(\theta - \alpha). \quad \text{(45.4)}$$
(Suggestion: Use Figure 45.13.)

(b) Use Equation (45.4) to verify that every vertical line to the right of the origin is the graph of an equation of the form $r = p\sec\theta$. What about a vertical line to the left of the origin?

(c) Use Equation (45.4) to verify that every horizontal line above the origin is the graph of an equation of the form $r = p\csc\theta$. What about a horizontal line below the origin?

REVIEW EXERCISES FOR CHAPTER IX

The notation is that used throughout the chapter. Use a calculator as much as possible.

1. Find c if $a = 5$, $b = 14$, and $\gamma = 38°$.

2. Find b if $a = 8$, $c = 15$, and $\beta = 71°$.

3. Find c if $b = 20$, $\alpha = 68°$, and $\gamma = 42°$.

4. Find α if $b = 41$, $\beta = 105°$, and $a = 32$.

Solve the triangle determined by the given information.

5. $b = 6$, $c = 7$, $\alpha = 10°$

6. $a = 10$, $b = 9$, $\alpha = 80°$

7. $b = 30$, $c = 11$, $\beta = 1.83$

8. $a = 30$, $b = 25$, $\beta = 1.91$

In Exercises 9 and 10, compute and simplify

(a) $\mathbf{u} + \mathbf{v}$ (b) $2\mathbf{u} - 3\mathbf{v}$ (c) $|\mathbf{u}|$.

9. $\mathbf{u} = 4\mathbf{i} - \mathbf{j}$, $\mathbf{v} = 10\mathbf{i} + \mathbf{j}$

10. $\mathbf{u} = \langle -12, 5 \rangle$, $\mathbf{v} = \langle 8, -3 \rangle$

11. Write $\mathbf{u}$ in the form $u_x\mathbf{i} + u_y\mathbf{j}$ if $|\mathbf{u}| = 10$ and $\theta = 120°$, where θ is the direction of $\mathbf{u}$.

12. Write $\mathbf{u}$ in the form $u_x\mathbf{i} + u_y\mathbf{j}$ if $|\mathbf{u}| = 5$ and $\theta = 7\pi/6$, where θ is the direction of $\mathbf{u}$.

13. A plane is headed N20°E at 250 mi/hr. The wind is blowing due north at 10 mi/hr. Find the direction and speed of the plane's motion relative to the ground.

14. Find the resultant of the forces $\mathbf{P}$ and $\mathbf{Q}$ if $\mathbf{P}$ is a 20-kg force directed S10°W and **Q** is a 50-kg force directed S10°E.

Convert from polar to Cartesian coordinates.

15. $[7, 240°]$ **16.** $[15, 5\pi/6]$

Convert from Cartesian to polar coordinates.

17. $(-\sqrt{3}, -1)$ **18.** $(9, -12)$

19. Find an equation with x and y that has the same graph as $r = 5 \cos \theta$.

20. Find an equation with r and θ that has the same graph as $x^2 + y^2 = by$.

Sketch the graph of each equation.

21. $r = 3(1 + \cos \theta)$ **22.** $r^2 = 1$

23. $9\theta^2 = \pi^2$ **24.** $r = 2 + \cos \theta$

25. $r \sin 3\theta$ **26.** $r = 4 \cos \theta$

27. Prove that $\dfrac{a + b}{b} = \dfrac{\sin \alpha + \sin \beta}{\sin \beta}$.

28. (a) Solve $a^2 = b^2 + c^2 - 2bc \cos \alpha$ for b.

(b) In Section 43 we used the Law of Sines to solve triangles when we were given two sides and an angle opposite one of them (case SSA). How could we have done this case without the Law of Sines?

CHAPTER X

COMPLEX NUMBERS

Because $x^2 \geq 0$ for every real number x, the quadratic equation $x^2 = -1$ has no solution among the real numbers. The quadratic formula shows that, more generally, a quadratic equation has no solution among the real numbers if its discriminant is negative (Section 8E). The real numbers also fail to provide a solution for many polynomial equations of degree greater than two. Mathematicians have overcome this problem by extending the system of real numbers to a larger system—the system of *complex numbers*—in which every polynomial equation does have a solution. Complex numbers are indispensable in many parts of higher mathematics. They are also valuable in many applications, such as the study of alternating current electrical circuits.

Section 46 introduces the complex numbers and Section 47 shows how to compute powers and roots of complex numbers. Section 48 shows how the complex numbers help to complete the study of polynomial equations that was begun in Chapter IV.

SECTION **46**
Complex Numbers

A. INTRODUCTION

The system of complex numbers, to be introduced in this section, has the following fundamental properties:

The system contains the system of real numbers.

The system also contains other numbers, including a number i such that
$$i^2 = -1.$$

The system has operations $(+, -, \times, \div)$ that satisfy all of the properties listed in Section 1B. More precisely, the system satisfies the field axioms (see page 5) and thus all of the properties that can be proved from the field axioms.

If the system is to have the properties listed, then for each real number b the product bi must also be a complex number (because b and i are both complex numbers). For each real number a, then, the sum $a + bi$ must be a complex number (because a and bi are both complex numbers). It turns out that every complex number can be written in this form $a + bi$, where a and b are real numbers. In fact, this form $a + bi$ will be our starting point in studying the complex numbers.

B. DEFINITIONS AND OPERATIONS

Definition The set of **complex numbers** is the set of all expressions of the form $a + bi$, where a and b are real numbers and $i^2 = -1$.

Statement (46.1), which follows, summarizes the way in which complex numbers are combined to form a system with the properties listed in Subsection A. Equations (46.2)–(46.5), which give explicit formulas for the fundamental operations $(+, -, \times, \div)$ between complex numbers, are consequences of this statement.

The number i is combined with real numbers to form complex numbers in the same way that a variable x is combined with real numbers to form polynomials, except that i^2 can always be replaced by -1.	(46.1)

Example 46.1 (a) Each real number is a complex number because $a = a + 0i$ (just as $a = a + 0x$ for polynomials).

(b) i is a complex number because $i = 0 + 1i$ (just as $x = 0 + x$ for polynomials).

(c) $a - bi = a + (-b)i$.

(d) $0 = 0 + 0i$ and $1 = 1 + 0i$. ■

When we refer to the form $a + bi$ we always assume that a and b are real numbers. We call a the **real part** of $a + bi$ and bi the **imaginary part.** If $b \neq 0$, then $a + bi$ is said to be **imaginary.** (Thus a complex number is *imaginary* if it is *not real.*) Both $\sqrt{2} + 4i$ and $-2i$ are imaginary. If $b \neq 0$ and $a = 0$, then $a + bi$ is said to be a **pure imaginary number.** Thus $-2i$ is pure imaginary but $\sqrt{2} + 4i$ is not.

Figure 46.1 summarizes the relation between different sets of complex numbers; each set contains the sets that appear beneath it.

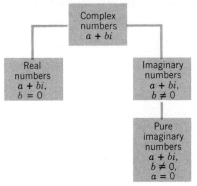

FIGURE 46.1

Two complex numbers are **equal** iff their real parts are equal and their imaginary parts are equal. Thus

$$a + bi = c + di$$

iff

$$a = c \text{ and } b = d.$$

Compare the following equations for addition and subtraction with the corresponding equations for polynomials: for example $(a + bx) + (c + dx) = (a + c) + (b + d)x$.

Addition

$$(a + bi) + (c + di) = (a + c) + (b + d)i \tag{46.2}$$

Subtraction

$$(a + bi) - (c + di) = (a - c) + (b - d)i \tag{46.3}$$

The next computation is a direct consequence of Statement (46.1) and will give the equation for multiplication:

$$(a + bi)(c + di) = a(c + di) + bi(c + di)$$
$$= ac + adi + bci + bdi^2$$
$$= (ac - bd) + (ad + bc)i.$$

Multiplication

$$(a + bi)(c + di) = (ac - bd) + (ad + bc)i \qquad (46.4)$$

Instead of using (46.4) when multiplying complex numbers, it is generally better to use the ideas in its derivation.

Example 46.2 Write each expression in the form $a + bi$ and simplify.

(a) $(2 + i) + (3 + 6i) = (2 + 3) + (1 + 6)i = 5 + 7i$
(b) $(2 + i) - (8 - 3i) = (2 - 8) + (1 + 3)i = -6 + 4i$
(c) $(3 + 2i)(1 + 4i) = (3 - 8) + (12 + 2)i = -5 + 14i$
(d) $(7 - i)(2 - 5i) = (14 - 5) + (-35 - 2)i = 9 - 37i$ ◼

The equations $i^2 = -1$ and $(-i)^2 = (-1)^2(i)^2 = -1$ show that both i and $-i$ are square roots of -1. We choose i as the principal square root. That is,

$$\boxed{\sqrt{-1} = i.}$$

More generally, the principal square root of any negative real number is defined as follows.

$$\boxed{\sqrt{-a} = \sqrt{a}\sqrt{-1} = \sqrt{a}\,i \text{ for each positive real number } a.}$$

Example 46.3 (a) $\sqrt{-9} = \sqrt{9}\sqrt{-1} = 3i$

(b) $(\sqrt{2} + \sqrt{-4}) - (1 + 5i) = (\sqrt{2} + \sqrt{4}\sqrt{-1}) - (1 + 5i) = (\sqrt{2} + 2i) - (1 + 5i) = (\sqrt{2} - 1) - 3i$ ◼

The law $\sqrt{ab} = \sqrt{a}\sqrt{b}$, which holds when both a and b are positive and when one is positive and the other is negative, does not hold when both a and b are negative. For example,

$$\sqrt{(-4)(-4)} = \sqrt{16} = 4 \text{ but } \sqrt{-4}\sqrt{-4} = 2i \cdot 2i = 4i^2 = -4,$$

so that

$$\sqrt{(-4)(-4)} \neq \sqrt{-4}\sqrt{-4}.$$

Thus in simplifying a product of imaginary numbers (such as $\sqrt{-4}\sqrt{-4}$), it is important to reduce each factor to the form $a + bi$ before multiplying.

The powers in the sequence i, i^2, i^3, . . . can be simplified easily by using

$$i^2 = -1,$$

$$i^3 = i^2 i = (-1)i = -i,$$

and

$$i^4 = i^3 i = (-i)i = -i^2 = -(-1) = 1.$$

Specifically, any power i^n can be reduced to 1, i, -1, or $-i$ by separating the highest multiple of 4 in the exponent, as in the following example.

Example 46.4 (a) $i^5 = i^4 i = 1i = i$

(b) $i^{20} = (i^4)^5 = 1^5 = 1$

(c) $i^{1935} = i^{4(483)+3} = (i^4)^{483} i^3 = 1(-i) = -i$

(d) $(2 + i)^3 = 2^3 + 3(2)^2(i) + 3(2)(i)^2 + i^3$ [by Equation (5.6)]

$$= 8 + 12i - 6 - i$$

$$= 2 + 11i$$

∎

To see how to divide complex numbers we need the following idea.

> The **conjugate** of a complex number $a + bi$ is $a - bi$.

Thus the conjugate of $2 + 3i$ is $2 - 3i$, the conjugate of $4 - i$ is $4 + i$, the conjugate of i is $-i$, and the conjugate of 6 is 6.

The product of a complex number and its conjugate is always a real number:

$$(c + di)(c - di) = c^2 - (di)^2$$

$$= c^2 - d^2 i^2$$

$$= c^2 - d^2(-1)$$

$$= c^2 + d^2.$$

If the real number $c^2 + d^2$ is nonzero, then we can rewrite the last result as

$$\frac{1}{c^2 + d^2} (c - di)(c + di) = 1.$$

Since $c^2 + d^2 \neq 0$ whenever $c + di \neq 0$ (that is, whenever $c \neq 0$ or $d \neq 0$), we see that every nonzero complex number has a multiplicative inverse (reciprocal):

$$(c + di)^{-1} = \frac{1}{c^2 + d^2} (c - di) \quad \text{if} \quad c + di \neq 0.$$

We can now define **quotients** of complex numbers as follows:

$$\frac{a + bi}{c + di} = (a + bi)(c + di)^{-1} \quad \text{if} \quad c + di \neq 0.$$

The following steps show how to determine the real and imaginary parts of such a quotient.

$$\frac{a + bi}{c + di} = \frac{a + bi}{c + di} \cdot \frac{c - di}{c - di}$$

$$= \frac{a(c - di) + bi(c - di)}{c^2 + d^2}$$

$$= \frac{ac - adi + bci - bdi^2}{c^2 + d^2}$$

$$= \frac{ac + bd}{c^2 + d^2} + \frac{bc - ad}{c^2 + d^2} i.$$

Division

$$\frac{a + bi}{c + di} = \frac{ac + bd}{c^2 + d^2} + \frac{bc - ad}{c^2 + d^2} i \qquad (46.5)$$

To simplify a fraction we use the idea behind Equation (46.5) rather than the equation itself; that is, if the denominator is $c + di$, multiply the fraction by $(c - di)/(c - di)$ and simplify. The idea is the same as that used to rationalize denominators in Section 3F.

Example 46.5 Write each number in the form $a + bi$ and simplify.

(a) $\dfrac{1}{1 + i} = \dfrac{1}{1 + i} \cdot \dfrac{1 - i}{1 - i} = \dfrac{1 - i}{1 + 1} = \dfrac{1}{2} - \dfrac{1}{2} i$

(b) $\dfrac{2}{i} = \dfrac{2}{i} \cdot \dfrac{-i}{-i} = \dfrac{-2i}{-i^2} = \dfrac{-2i}{1} = -2i$

(c) $\dfrac{2 + i}{3 - 2i} = \dfrac{2 + i}{3 - 2i} \cdot \dfrac{3 + 2i}{3 + 2i} = \dfrac{6 + 7i - 2}{9 + 4} = \dfrac{4 + 7i}{13} = \dfrac{4}{13} + \dfrac{7}{13} i$ ∎

C. SOLUTIONS OF QUADRATIC EQUATIONS

The quadratic formula (Section 8D) shows that if a quadratic equation has real coefficients and its discriminant is negative, then the solutions of the equation are complex conjugates. Here is an example.

Example 46.6 The discriminant of

$$x^2 + 2x + 5 = 0$$

is

$$2^2 - 4(1)(5) = 4 - 20 = -16 < 0.$$

The solutions of the equation are

$$x = \frac{-2 \pm \sqrt{-16}}{2} = \frac{-2 \pm \sqrt{16}\sqrt{-1}}{2} = \frac{-2 \pm 4i}{2} = -1 \pm 2i.$$ ∎

Table 46.1 gives a summary concerning the solutions of a quadratic equation $ax^2 + bx + c = 0$ with real coefficients a, b, and c. (Compare Table 8.1.)

TABLE 46.1

Discriminant	Character of the Solutions
$b^2 - 4ac < 0$	Two conjugate imaginary solutions
$b^2 - 4ac = 0$	One real solution
$b^2 - 4ac > 0$	Two unequal real solutions

D. COMPLEX PLANES

Real numbers are represented geometrically by using a real line (Section 1). To represent complex numbers geometrically we use a plane: Choose a Cartesian plane and assign the complex number $a + bi$ to the point with coordinates (a, b). This establishes a one-to-one correspondence between the set of complex numbers and the set of points in the plane. A plane together with such a correspondence is called a **complex plane.** Figure 46.2 shows examples of points with the corresponding complex numbers.

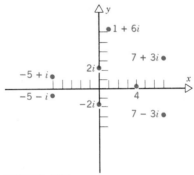

FIGURE 46.2

Notice that any complex number and its conjugate are located symmetrically with respect to the horizontal axis: compare the conjugate pairs $-5 \pm i$, $\pm 2i$, and $7 \pm 3i$ in Figure 46.2.

The usefulness of geometrical representations of complex numbers will be shown in the next section.

EXERCISES FOR SECTION 46

In each of Exercises 1–3 make three lists, one of the real numbers, one of the imaginary numbers, and one of the pure imaginary numbers. A number may be in more than one list.

1. $4 - i$, $\sqrt{3}$, πi, $2 + \sqrt{-4}$, 0

2. $5i$, 1, $\sqrt{-25}$, $5 + i$, $-\sqrt{2}$

3. $3 + \sqrt{-15}$, π, $\sqrt{2} + i$, -1, $4i$

Write each expression in the form a + bi and simplify.

4. $(9 + i) + (3 - 2i)$

5. $(7 - i) - (6 - i)$

6. $(14 + 5i) + (-8 + 2i)$

7. $5i - (4 - i)$ **8.** $\sqrt{-1} + \sqrt{25}$

9. $-2 + (7 - \sqrt{-2})$

10. $(1 + \sqrt{-16}) - (\pi + 4i)$

11. $10 - (1 - \sqrt{-9})$

12. $(\sqrt{2} + i) - (1 - 2\sqrt{-4})$

13. $i(2 + 3i)$ **14.** $2i(5 + i)$

15. $-3i(8 + i)$ **16.** $(2 + i)(1 - 2i)$

17. $(-4 + i)(2 + i)$ **18.** $(6 - i)(-1 - i)$

19. $(1 + i)(1 - i)$ **20.** $(5 - i)^2$

21. $(-1 - 3i)(1 - 3i)$ **22.** $(3 + i)^2$

23. $(-2 + i)(2 + i)$ **24.** $(-1 - 2i)^2$

25. i^{15} **26.** i^{18}

27. i^{32} **28.** $\dfrac{2}{2 + i}$

29. $\dfrac{1}{3 - i}$ **30.** $\dfrac{1}{1 + 2i}$

31. $\dfrac{i}{1 - i}$ **32.** $\dfrac{-2i}{1 + i}$

33. $\dfrac{-5i}{3 - 4i}$ **34.** $\dfrac{1 - i}{2 + i} + \dfrac{2i}{3 - i}$

35. $\dfrac{4}{i} - \dfrac{3 + i}{2 - i}$ **36.** $\dfrac{i}{(1 - i)^2} - \dfrac{1}{2i}$

37. $(1 + i)i^{-5}$ **38.** $(-i)^{-10}(2 - i)$

39. $(-i)^{-7}(-1 + i)$ **40.** $(1 - i)^3$

41. $(1 + 2i)^3$ **42.** $(3 - 2i)^3$

Solve for x.

43. $x^2 - x + 2 = 0$ **44.** $x^2 + 2x + 2 = 0$

45. $x^2 + x + 3 = 0$ **46.** $2x^2 + x = -1$

47. $3x^2 = 5x - 3$ **48.** $4x^2 + 1 = 0$

49. $2x^2 = -18$ **50.** $x^2 + 10 = 0$

51. $2x^2 - 3x = -2$

Solve for x *and* y, *assuming that both are real numbers.* (*Remember: For* a, b, c, *and* d *real,* a + bi = c + di *iff* a = c *and* b = d.)

52. $2x + i = 5 + yi$

53. $x + 4 + i = 2 - yi$

54. $(x - y) + (x + 2y)i = 1 + 10i$

55. Is i a solution of $x^4 + x^3 + x^2 + x + 1 = 0$? Justify your answer.

56. Is $-i$ a solution of $x^3 + x^2 + x + 1 = 0$? Justify your answer.

57. Is i a solution of $x^6 + x^4 + x^2 + 1 = 0$? Justify your answer.

58. Verify that both of the numbers $-\dfrac{1}{2} \pm \dfrac{1}{2}\sqrt{3}i$ are cube roots of 1, that is, that both are solutions of $x^3 = 1$.

59. Verify that both of the numbers $\dfrac{1}{2} \pm \dfrac{1}{2}\sqrt{3}i$ are cube roots of -1, that is, that both are solutions of $x^3 = -1$.

60. Verify that $-i, \dfrac{1}{2}\sqrt{3} + \dfrac{1}{2}i,$ and $-\dfrac{1}{2}\sqrt{3} + \dfrac{1}{2}i$ are cube roots of i, that is, that each number is a solution of $x^3 = i$.

61. (a) Show that if a and b are both real, then the sum of $a + bi$ and its conjugate is a real number.

     (b) Show that if a and b are both real and $b \neq 0$, then $a + bi$ minus its conjugate is a pure imaginary number.

62. If the sum of a complex number and its reciprocal is 1, what is the number?

63. If the sum of a complex number and its reciprocal is 0, what is the number?

Let $\bar{z}$ *denote the conjugate of a complex number* z. *That is, if* z = a + bi, *then* $\bar{z}$ = a − bi (*where* a *and* b *are real*). *Assume that* z = a + bi *and* w = c + di, *and then prove each statement in Exercises 64–72.*

64. $\bar{z} = z$ iff z is real (that is, iff $b = 0$)

65. $\bar{\bar{z}} = z$ [where $\bar{\bar{z}}$ means $\overline{(\bar{z})}$]

66. $\bar{z} + z = 0$ iff the real part of z is 0 (that is, iff $a = 0$)

67. $\overline{z + w} = \bar{z} + \bar{w}$

68. $\overline{z - w} = \bar{z} - \bar{w}$

69. $\overline{\left(\dfrac{z}{w}\right)} = \dfrac{\bar{z}}{\bar{w}}$

70. $\overline{zw} = \bar{z}\,\bar{w}$

71. $\overline{z^2} = (\bar{z})^2$

72. $\overline{z^n} = (\bar{z})^n$ (Use Exercise 71.)

In each of Exercises 73–75, draw a complex plane and label the points associated with the given complex numbers.

73. $3 + 2i, -2, 4i, -1 - 2i, -4 + i$

74. $1 - i, 6 + i, -2i, 5, -3 - i$

75. $-i, 2 - i, -4, -5 + i, 4 + 3i$

SECTION **47**

Trigonometric Form. De Moivre's Theorem

A. TRIGONOMETRIC FORM

In this section we develop a method to compute powers and roots of complex numbers. The method uses the *trigonometric* (or *polar*) *form* for the numbers, which is defined as follows.

Figure 47.1 shows the point in a complex plane corresponding to the complex number $a + bi$. Let r denote the distance between the point and the origin. Let

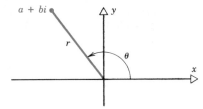

FIGURE 47.1

θ denote the angle from the positive x-axis to the ray from the origin through the point. Then

$$\frac{a}{r} = \cos \theta \quad \text{and} \quad \frac{b}{r} = \sin \theta,$$

so

$$a = r \cos \theta \quad \text{and} \quad b = r \sin \theta.$$

Therefore

$$a + bi = r(\cos \theta + i \sin \theta). \tag{47.1}$$

The expression $r(\cos \theta + i \sin \theta)$ is called the **trigonometric form** of $a + bi$. The nonnegative number r is called the **absolute value** (or **modulus**) of $a + bi$ and is denoted by $|a + bi|$. Thus

$$|a + bi| = \sqrt{a^2 + b^2}.$$

The angle θ is called the **argument** (or **amplitude**) of $a + bi$. The absolute value of $a + bi$ is uniquely determined by $a + bi$, but the argument is not: if θ is an argument of $a + bi$, then so is $\theta + 2n\pi$ for any integer n. We shall restrict θ so that $0 \leq \theta < 2\pi$; then each complex number will have a unique argument.

Example 47.1 Determine the absolute value, argument, and trigonometric form of each number.

(a) $3 + 3i$ (b) $-i$ (c) $-1 + \sqrt{3}i$

Solution (a) See Figure 47.2a.

$$|3 + 3i| = \sqrt{3^2 + 3^2} = \sqrt{18} = 3\sqrt{2}$$

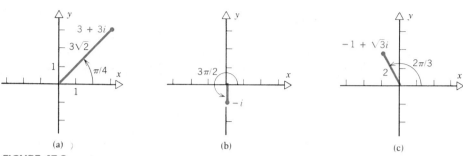

FIGURE 47.2

The argument is a first quadrant angle θ such that $\tan \theta = 3/3 = 1$. Thus $\theta = \pi/4$. The trigonometric form is

$$3\sqrt{2}\left(\cos \frac{\pi}{4} + i \sin \frac{\pi}{4}\right).$$

(b) See Figure 47.2b.

$$|-i| = \sqrt{0^2 + 1^2} = 1$$

The argument is $3\pi/2$. The trigonometric form is

$$\cos \frac{3\pi}{2} + i \sin \frac{3\pi}{2}.$$

(c) See Figure 47.2c.

$$|-1 + \sqrt{3}\,i| = \sqrt{(-1)^2 + \sqrt{3}^2} = \sqrt{4} = 2$$

The argument is a second quadrant angle θ such that $\tan \theta = \sqrt{3}/(-1) = -\sqrt{3}$. The reference angle is the acute angle θ_r such that $\tan \theta_r = \sqrt{3}$. Thus $\theta_r = \pi/3$ and $\theta = 2\pi/3$. The trigonometric form is

$$2\left(\cos \frac{2\pi}{3} + i \sin \frac{2\pi}{3}\right).$$

B. PRODUCTS AND QUOTIENTS

The trigonometric form is especially well-suited for computing products and quotients of complex numbers.

If $z = r(\cos \theta + i \sin \theta)$ and $w = s(\cos \phi + i \sin \phi)$, then

$$zw = rs[\cos(\theta + \phi) + i \sin(\theta + \phi)]. \qquad (47.2)$$

Thus the absolute value of a product is the product of the absolute values, and the argument of a product is the sum of the arguments.

Proof $zw = r(\cos \theta + i \sin \theta) \cdot s(\cos \phi + i \sin \phi)$

$= rs \cos \theta \cos \phi + rsi \cos \theta \sin \phi + rsi \sin \theta \cos \phi + rsi^2 \sin \theta \sin \phi$

$= rs[(\cos \theta \cos \phi - \sin \theta \sin \phi) + i(\sin \theta \cos \phi + \cos \theta \sin \phi)]$

$= rs[\cos(\theta + \phi) + i \sin(\theta + \phi)]$ by (34.2) and (34.6).   □

Example 47.2 Use Equation (47.2) to compute the product of $3 + 3i$ and $-i$.

Solution Using parts (a) and (b) of Example 47.1, we have

$$(3 + 3i)(-i) = 3\sqrt{2}\left(\cos\frac{\pi}{4} + i\sin\frac{\pi}{4}\right) \cdot \left(\cos\frac{3\pi}{2} + i\sin\frac{3\pi}{2}\right)$$

$$= 3\sqrt{2}\left[\cos\left(\frac{\pi}{4} + \frac{3\pi}{2}\right) + i\sin\left(\frac{\pi}{4} + \frac{3\pi}{2}\right)\right]$$

$$= 3\sqrt{2}\left(\cos\frac{7\pi}{4} + i\sin\frac{7\pi}{4}\right)$$

$$= 3\sqrt{2}\left(\frac{\sqrt{2}}{2} - i\frac{\sqrt{2}}{2}\right)$$

$$= 3 - 3i.$$

As a check,

$$(3 + 3i)(-i) = -3i - 3i^2 = -3i + 3 = 3 - 3i.$$

If $z = r(\cos\theta + i\sin\theta)$ and $w = s(\cos\phi + i\sin\phi) \neq 0$, then

$$\frac{z}{w} = \frac{r}{s}[\cos(\theta - \phi) + i\sin(\theta - \phi)]. \qquad (47.3)$$

The proof of (47.3) is left as an exercise.

Example 47.3 Express z/w in the form $a + bi$ if $z = 4(\cos 30° + i\sin 30°)$ and $w = 3(\cos 240° + i\sin 240°)$.

Solution

$$\frac{z}{w} = \frac{4}{3}[\cos(30° - 240°) + i\sin(30° - 240°)]$$

$$= \frac{4}{3}[\cos(-210°) + i\sin(-210°)]$$

$$= \frac{4}{3}\left(-\frac{\sqrt{3}}{2} + \frac{1}{2}i\right)$$

$$= -\frac{2\sqrt{3}}{3} + \frac{2}{3}i.$$

C. DE MOIVRE'S THEOREM

With $w = z$, Equation (47.2) yields

$$z^2 = r^2(\cos 2\theta + i\sin 2\theta). \qquad (47.4)$$

Now, using (47.4), and (47.2) with $w = z^2$, we have

$$z^3 = zz^2 = r(\cos\theta + i\sin\theta)r^2(\cos 2\theta + i\sin 2\theta)$$

$$= r^3(\cos 3\theta + i\sin 3\theta).$$

If you look at the pattern for z, z^2, and z^3, and the way in which they are derived, the following theorem should be evident. [A careful proof requires mathematical induction (Section 58).]

De Moivre's Theorem

If n is a positive integer and $z = r(\cos \theta + i \sin \theta)$, then

$$z^n = r^n(\cos n\theta + i \sin n\theta).$$

Example 47.4 Use De Moivre's Theorem to compute $(-1 + \sqrt{3}\, i)^5$.

Solution From Example 47.1c,

$$-1 + \sqrt{3}\, i = 2 \left(\cos \frac{2\pi}{3} + i \sin \frac{2\pi}{3} \right).$$

Therefore,

$$(-1 + \sqrt{3}\, i)^5 = 2^5 \left[\cos 5\left(\frac{2\pi}{3} \right) + i \sin 5\left(\frac{2\pi}{3} \right) \right]$$

$$= 32 \left(\cos \frac{10\pi}{3} + i \sin \frac{10\pi}{3} \right)$$

$$= 32 \left(-\frac{1}{2} - \frac{\sqrt{3}}{2} i \right)$$

$$= -16 - 16\sqrt{3}\, i.$$

A standard abbreviation for $\cos \theta + i \sin \theta$ is cis θ. For example, using this abbreviation De Moivre's Theorem becomes: If n is a positive integer and $z = r \cdot$ cis θ, then $z^n = r^n \cdot$ cis $n\theta$. Although the book does not use cis θ, it may be convenient to use when you do the exercises.

D. ROOTS OF COMPLEX NUMBERS

For each integer $n \geq 1$, there are exactly n complex numbers z such that $z^n = 1$. These numbers are called the complex **nth roots of unity.** They can be computed as follows.

For each integer $n \geq 1$, the n complex nth roots of unity are

$$\cos \frac{2k\pi}{n} + i \sin \frac{2k\pi}{n}, \quad k = 0, 1, \ldots, n - 1. \qquad (47.5)$$

Proof By De Moivre's Theorem, the absolute value of the nth power of each number in (47.5) is $1^n = 1$, and the argument is $n(2k\pi/n) = 2k\pi$, which is an argument of 1. Thus each of the numbers is an nth root of unity. The n numbers in (47.5) are distinct because the numbers $2k\pi/n$ are distinct and $0 \leq 2k\pi/n < 2\pi$ for $k = 0, 1, \ldots, n - 1$. There cannot be more than n complex nth roots of unity because, as will be seen in the next section, the equation $z^n = 1$ cannot have more than n solutions. $\square$

Example 47.5 Compute the complex 6th roots of unity and represent them geometrically.

Solution Compute $\cos \dfrac{2k\pi}{6} + i \sin \dfrac{2k\pi}{6}$ for $k = 0, 1, 2, 3, 4, 5$.

$$k = 0: \qquad \cos 0 + i \sin 0 = 1$$

$$k = 1: \qquad \cos \frac{\pi}{3} + i \sin \frac{\pi}{3} = \frac{1}{2} + \frac{\sqrt{3}}{2} i$$

$$k = 2: \qquad \cos \frac{2\pi}{3} + i \sin \frac{2\pi}{3} = -\frac{1}{2} + \frac{\sqrt{3}}{2} i$$

$$k = 3: \qquad \cos \pi + i \sin \pi = -1$$

$$k = 4: \qquad \cos \frac{4\pi}{3} + i \sin \frac{4\pi}{3} = -\frac{1}{2} - \frac{\sqrt{3}}{2} i$$

$$k = 5: \qquad \cos \frac{5\pi}{3} + i \sin \frac{5\pi}{3} = \frac{1}{2} - \frac{\sqrt{3}}{2} i$$

Figure 47.3 shows the geometrical representations.

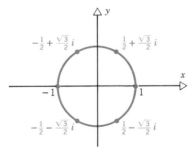

FIGURE 47.3

The next statement extends (47.5) to give a formula for the nth roots of any complex number. Exercise 31 asks for a proof. Geometrically, the n roots are represented by n equally spaced points on the circle with center at the origin and radius $r^{1/n}$, beginning at $r^{1/n}[\cos(\theta/n) + i \sin(\theta/n)]$.

For each integer $n > 1$, the n complex nth roots of

$$z = r(\cos \theta + i \sin \theta)$$

are

$$r^{1/n} \left(\cos \frac{\theta + 2k\pi}{n} + i \sin \frac{\theta + 2k\pi}{n} \right), \quad k = 0, 1, \ldots, n - 1, \quad (47.6)$$

where $r^{1/n}$ denotes the positive real nth root of r.

Example 47.6 Compute the complex 4th roots of $-1 + i$.

Solution In this case $r = \sqrt{2}$ and $\theta = 3\pi/4$. Thus we compute

$$\sqrt{2}^{1/4} \left(\cos \frac{\dfrac{3\pi}{4} + 2k\pi}{4} + i \sin \frac{\dfrac{3\pi}{4} + 2k\pi}{4} \right) \text{ for } k = 0, 1, 2, 3.$$

Notice that $\sqrt{2}^{1/4} = (2^{1/2})^{1/4} = 2^{1/8}$.

$$k = 0: \quad 2^{1/8} \left(\cos \frac{3\pi}{16} + i \sin \frac{3\pi}{16} \right)$$

$$k = 1: \quad 2^{1/8} \left(\cos \frac{11\pi}{16} + i \sin \frac{11\pi}{16} \right)$$

$$k = 2: \quad 2^{1/8} \left(\cos \frac{19\pi}{16} + i \sin \frac{19\pi}{16} \right)$$

$$k = 3: \quad 2^{1/8} \left(\cos \frac{27\pi}{16} + i \sin \frac{27\pi}{16} \right)$$

The four roots above are exact. For numerical approximations use a calculator or table. ◼

If θ is measured in degrees rather than radians, then replace $(\theta + 2k\pi)/n$ in (47.6) by $(\theta + k \cdot 360°)/n$. Similarly, in degrees, $2k\pi/n$ in (47.5) becomes $k \cdot 360°/n$.

EXERCISES FOR SECTION 47

Write each of the following complex numbers in the form a + bi.

1. The number with absolute value 2 and argument $\pi/6$.

2. The number with absolute value $\dfrac{1}{2}$ and argument $5\pi/3$.

3. The number with absolute value 5 and argument $7\pi/4$.

Determine the absolute value, argument, and trigonometric form of each number. For 13–15, use a calculator or Table VI.

4. $1 + i$

5. $\sqrt{3} - i$

6. -5

7. $-2i$

8. $2 - 2i$

9. $2\sqrt{3} + 2i$

10. $-1 + \sqrt{3}i$

11. $4i$

12. $-1 - i$

13. $2 - 3i$

14. $-5 + 2i$

15. $4 - 7i$

Use De Moivre's Theorem to write each of the following complex numbers in the form a + bi.

16. $(1 + i)^{10}$

17. $(\sqrt{3} + i)^5$

18. $(1 - i)^6$

19. $(-i)^{10}$

20. $\left(\dfrac{1}{2} - \dfrac{1}{2} i \right)^4$

21. $(-\sqrt{2} - \sqrt{2}i)^{12}$

22. Compute the complex cube roots of unity and represent them geometrically.

23. Compute the complex 8th roots of unity and represent them geometrically.

24. Draw a graph showing the complex 5th roots of unity.

25. Compute the complex 4th roots of i.

26. Compute the complex square roots of $-i$.

27. Compute the complex cube roots of -2.

28. Compute the complex cube roots of $1 + i$.

29. Compute the complex 4th roots of $\sqrt{3} - i$.

30. Prove that if $z = r(\cos \theta + i \sin \theta)$ then $z^{-1} = r^{-1}[\cos(-\theta) + i \sin(-\theta)]$.

31. Prove Statement (47.6).

32. Prove that if n is an integer, then
$$i^n + i^{-n} = 2 \cos n\pi/2.$$
Then compute $i^n + i^{-n}$ for $0 \le n \le 12$. (Suggestion: Use De Moivre's Theorem and appropriate trigonometric identities.)

33. Prove that if n is an integer and z is pure imaginary with $|z| = 1$, then $z^n - z^{-n}$ is either zero or pure imaginary.

Prove each statement in Exercises 34–39. (See the remarks preceding Exercise 64 in Section 46 for the meaning of $\bar{z}$.)

34. $|z| = |-z|$

35. $|z| = |\bar{z}|$

36. $|z| = 0$ iff $z = 0$

37. $|z| = \sqrt{z\bar{z}}$

38. $|zw| = |z| \cdot |w|$

39. $\left|\dfrac{z}{w}\right| = \dfrac{|z|}{|w|}$

SECTION **48**
Complex Zeros of Polynomials

A. COMPLEX POLYNOMIALS

Previously we have considered only **real polynomials,** that is, polynomials whose coefficients are real numbers. In this section we consider the more general class of **complex polynomials**—polynomials whose coefficients are complex numbers. (Every real polynomial is also complex, of course, since every real number is a complex number.)

All of the results about divisibility, factors, and zeros from Sections 18 and 19 hold for complex polynomials as well as for real polynomials. This subsection contains examples to illustrate the highlights, without proofs. For proofs it suffices to replace real numbers by complex numbers in the proofs in Sections 18 and 19.

Remember that to divide by $z - c$ using synthetic division the leading number in the first row must be c.

Example 48.1 Use synthetic division to determine the quotient and remainder when $z^4 - 3iz^2 + 6z + i$ is divided by $z + 2i$.

Solution

$$
\begin{array}{r|rrrrr}
-2i & 1 & 0 & -3i & 6 & i \\
 & & -2i & -4 & -6 + 8i & 16 \\
\hline
 & 1 & -2i & -4 - 3i & 8i & 16 + i
\end{array}
$$

The quotient is $z^3 - 2iz^2 + (-4 - 3i)z + 8i$. The remainder is $16 + i$. ∎

The Remainder Theorem states that if $f(z)$ is divided by $z - c$, then the remainder is $f(c)$.

Example 48.2 Use the Remainder Theorem and synthetic division to compute $f(1 + i)$ if $f(z) = iz^3 + z^2 - iz$.

Solution

$$1 + i \,\Big|\, \begin{array}{cccc} i & 1 & -i & 0 \\ & -1 + i & -1 + i & -1 - i \\ \hline i & i & -1 & -1 - i \end{array}$$

The remainder is $-1 - i$, so $f(1 + i) = -1 - i$.   ■

The Factor Theorem states that $z - c$ is a factor of $f(z)$ iff $f(c) = 0$.

Example 48.3 Use synthetic division and the Factor Theorem to show that $z - 2i$ is a factor of $f(z) = z^4 + 6z^2 + 8$.

Solution

$$2i \,\Big|\, \begin{array}{ccccc} 1 & 0 & 6 & 0 & 8 \\ & 2i & -4 & 4i & -8 \\ \hline 1 & 2i & 2 & 4i & 0 \end{array}$$

Thus $f(2i) = 0$, so $z - 2i$ is a factor of $f(z)$ by the Factor Theorem. The numbers in the third row of the synthetic division show that, in fact,

$$z^4 + 6z^2 + 8 = (z - 2i)(z^3 + 2iz^2 + 2z + 4i).$$ ■

If c is a zero of $f(z)$ and $(z - c)^m$ is the highest power of $z - c$ that divides $f(z)$, then c is said to be a zero of *multiplicity m*.

Example 48.4 Form a complex polynomial of lowest degree having 0, i, and $-2i$ as zeros of multiplicities 3, 2, and 1, respectively.

Solution The answer must have z^3, $(z - i)^2$, and $z + 2i$ as factors. Thus an acceptable answer is $z^3(z - i)^2(z + 2i)$. To expand this, write

$$z^3(z - i)^2(z + 2i) = z^3(z^2 - 2iz - 1)(z + 2i)$$
$$= z^3(z^3 + 3z - 2i)$$
$$= z^6 + 3z^4 - 2iz^3.$$ ■

B. THE FUNDAMENTAL THEOREM OF ALGEBRA

From Section 46 we know that every second-degree real polynomial has a zero among the complex numbers. The following important theorem extends that result from *real* polynomials of degree *two* to *complex* polynomials of *every* positive degree.

The Fundamental Theorem of Algebra

Every complex polynomial whose degree is at least one has a zero among the complex numbers.

Again, notice the implications of this theorem. To have zeros for all real polynomials we must go beyond the real numbers to the complex numbers. The Fundamental Theorem of Algebra asserts that to find zeros for complex polynomials there will be no need to go further: the complex numbers contain a zero not only for every real polynomial, but also for every complex polynomial.

Proofs of the Fundamental Theorem of Algebra lie outside the scope of this book. However, we can give a proof of a useful corollary based on the theorem. First, recall that in Section 19 we showed that each *real* polynomial of degree n has at most n real zeros. The argument used there can also be used to show that each *complex* polynomial of degree n has at most n *complex* zeros. If, as before, each zero of multiplicity m is counted m times, we can prove the following more precise result.

Corollary Each polynomial of degree $n \geq 1$ has exactly n complex zeros.

Proof Assume that

$$f(z) = a_n z^n + a_{n-1} z^{n-1} + \cdots + a_1 z + a_0 \text{ with } a_n \neq 0.$$

If $n = 1$, then $f(z) = a_1 z + a_0$. By inspection there is one zero, $-a_0/a_1$, as required.

Assume that $n > 1$. By the Fundamental Theorem of Algebra $f(z)$ has at least one zero. If c_1 is a zero, then $z - c_1$ is a factor of $f(z)$, and thus $f(z) = (z - c_1)f_1(z)$ for some complex polynomial $f_1(z)$ of degree $n - 1$. If $n = 2$, then $f_1(z)$ has degree one; therefore, as required, $f(z)$ has two zeros, c_1 together with the zero of $f_1(z)$. If $n > 2$, then $f_1(z)$ has degree greater than one. Thus $f_1(z)$ has a zero, say c_2, and $f(z) = (z - c_1)(z - c_2)f_2(z)$ for some polynomial $f_2(z)$ of degree $n - 2$. We can continue in this way until we arrive at

$$f(z) = a_n(z - c_1)(z - c_2) \cdots (z - c_n). \tag{48.1}$$

[The factor a_n appears because a_n is the coefficient of z^n in $f(z)$, and the coefficients of $f(z)$ on the two sides of the equation must be equal.] Equation (48.1) shows that $f(z)$ has at least the n complex numbers $c_1, c_2, \ldots, c_n$ as zeros. On the other hand, Equation (48.1) also shows that if $c \neq c_k$ so that $c - c_k \neq 0$ for $1 \leq k \leq n$, then $f(c) \neq 0$; thus $f(z)$ has no other zeros. $\square$

The following fact was used in Section 23E on partial fractions.

Corollary A complex polynomial in z is identically zero (that is, zero for every value of z) iff each of its coefficients is zero.

Proof It is obvious that if a polynomial $f(z)$ has only zero coefficients, then $f(c) = 0$ for every c. Suppose, on the other hand, that $f(z)$ is identically zero but has one or more nonzero coefficients. Then $f(z)$ has degree n for some $n \geq 1$, since a nonzero constant polynomial is not identically zero. But if $f(z)$ has degree n, then (by the preceding corollary) $f(z)$ has at most n zeros, which contradicts the fact that $f(z)$ is identically zero. This proves that if $f(z)$ is identically zero, then each of its coefficients must be zero. $\square$

In Section 5 we defined a polynomial to be *reducible* if it can be written as a product of two other polynomials that are both of positive degree; otherwise it is *irreducible*. A polynomial has been *factored completely* when it has been written as a product of irreducible factors. The reducibility or irreducibility of a polynomial depends on whether the factors are required to be real polynomials or whether they can be complex polynomials. By the Factor Theorem and the Fundamental Theorem of Algebra, a complex polynomial is irreducible relative to complex factors iff the polynomial is linear, that is, of degree one.

Example 48.5 (a) The polynomial $x^2 + 1$ is reducible with complex factors:
$$x^2 + 1 = (x + i)(x - i).$$
(b) The polynomial $x^2 + 1$ is irreducible if the factors are required to be real polynomials. ■

Example 48.6 Given that -3 is a zero of
$$f(z) = z^3 + 3z^2 + 4z + 12,$$
factor $f(z)$ completely using complex factors.

Solution To determine $q(z)$ such that $f(z) = (z + 3)q(z)$, use synthetic division.

$$
\begin{array}{r|rrrr}
-3 & 1 & 3 & 4 & 12 \\
   &   & -3 & 0 & -12 \\
\hline
   & 1 & 0 & 4 & 0
\end{array}
$$

Thus
$$f(z) = (z + 3)(z^2 + 4).$$
The zeros of $z^2 + 4$ are the solutions of $z^2 + 4 = 0$, which are $\pm 2i$. Thus
$$z^2 + 4 = (z - 2i)(z + 2i)$$
and
$$f(z) = (z + 3)(z - 2i)(z + 2i). \qquad ■$$

Example 48.7 A complex number z is a cube root of -1 iff $z^3 = -1$. One cube root of -1 is -1. Find the others.

Solution The answer will be the zeros of $g(z)$, where $z^3 + 1 = (z + 1)g(z)$. The computation

$$
\begin{array}{r|rrrr}
-1 & 1 & 0 & 0 & 1 \\
   &   & -1 & 1 & -1 \\
\hline
   & 1 & -1 & 1 & 0
\end{array}
$$

shows that $g(z) = z^2 - z + 1$. By the quadratic formula $g(z)$ has zeros
$$z = \frac{1 \pm \sqrt{1 - 4}}{2} = \frac{1 \pm \sqrt{3}i}{2}.$$
Thus the cube roots of -1 are -1 and $(1 \pm \sqrt{3}i)/2$. (You can check each one by cubing.) ■

C. COMPLEX ZEROS OF REAL POLYNOMIALS

The zeros of a real quadratic polynomial are either real, or imaginary conjugates (Table 46.1). Therefore, if $a + bi$ is a zero of such a polynomial, then $a - bi$ is also a zero. This is a special case of the following theorem.

Conjugate Zero Theorem

If $a + bi$ is a zero of a real polynomial $f(x)$, then its conjugate $a - bi$ is also a zero of $f(x)$.

To paraphrase the Conjugate Zero Theorem: *Imaginary zeros of real polynomials occur in conjugate pairs.* A proof of the Conjugate Zero Theorem is outlined in Exercise 30.

Example 48.8 Given that $1 + i$ is a zero of

$$f(x) = x^4 - 2x^3 + 5x^2 - 6x + 6,$$

factor $f(x)$ completely using complex factors.

Solution If $1 + i$ is a zero, then $1 - i$ must also be a zero, by the Conjugate Zero Theorem. Therefore,

$$[x - (1 + i)][x - (1 - i)]$$

must be a factor of $f(x)$. Since

$$[x - (1 + i)][x - (1 - i)] = [(x - 1) - i][(x - 1) + i] =$$
$$(x - 1)^2 + 1 = x^2 - 2x + 2,$$

there is a polynomial $g(x)$ such that $f(x) = (x^2 - 2x + 2)g(x)$. Division of $f(x)$ by $x^2 - 2x + 2$ yields $g(x) = x^2 + 3$. The zeros of $x^2 + 3$ are $\pm\sqrt{3}i$, so that $g(x) = (x - \sqrt{3}i)(x + \sqrt{3}i)$. Thus the complete factorization of $f(x)$ is

$$f(x) = (x - 1 - i)(x - 1 + i)(x - \sqrt{3}i)(x + \sqrt{3}i). \quad \blacksquare$$

Example 48.9 Form a *real* polynomial of lowest degree having $3i$ and -1 as zeros of multiplicities 1 and 2, respectively.

Solution The Conjugate Zero Theorem implies that if $3i$ is a zero then its conjugate $-3i$ must also be a zero. Thus the answer must have $x - 3i$, $x + 3i$, and $(x + 1)^2$ as factors. An acceptable polynomial, in factored form, is

$$(x - 3i)(x + 3i)(x + 1)^2.$$

To make it obvious that this is a real polynomial, we can write it as

$$(x^2 + 9)(x + 1)^2. \quad \blacksquare$$

We have seen that the only complex polynomials that are irreducible relative to complex factors are the linear polynomials. The next theorem gives the corresponding fact for real factors of real polynomials. The proof of this theorem will be omitted.

Irreducible Real Polynomials

A real polynomial is irreducible relative to real factors iff it is either linear or quadratic with a negative discriminant.

Example 48.10 Factor $f(x) = x^4 - 2x^3 + 5x^2 - 6x + 6$ completely using real factors.

Solution The complete factorization using complex factors was given in Example 48.8. The solution there also gives the complete factorization using real polynomials, namely,

$$f(x) = (x^2 - 2x + 2)g(x)$$
$$= (x^2 - 2x + 2)(x^2 + 3).$$

EXERCISES FOR SECTION 48

Use synthetic division to find the quotient and remainder when f(z) *is divided by* g(z).

1. $f(z) = z^3 - 2iz^2 + z + i$, $g(z) = z - i$

2. $f(z) = iz^4 + z^2 - 2iz$, $g(z) = z + i$

3. $f(z) = 2z^4 - z^3 + iz^2 - z + 1 + i$, $g(z) = z - 2i$

Use the Remainder Theorem and synthetic division to compute the indicated values $f(c)$.

4. $f(z) = (1 + i)z^4 - z^3 + z - 2i$; $f(2)$, $f(1 + i)$

5. $f(z) = 3z^3 - iz^2 + z - 1 - i$; $f(i)$, $f(1 + i)$

6. $f(z) = iz^3 - iz^2 - z + 3i$; $f(-i)$, $f(1 - i)$

Use synthetic division and the Factor Theorem to show that z − c *is a factor of* f(z). *Also determine* q(z) *such that* f(z) = (z − c)q(z).

7. $z - 2i$; $f(z) = z^4 + z^2 - 12$

8. $z - 3i$; $f(z) = z^3 - 3iz^2 + z - 3i$

9. $z + \sqrt{5}i$; $f(z) = z^4 + 4z^2 - 5$

Form a complex *polynomial of lowest degree having the specified zeros with the specified multiplicities.*

10. 0, $-i$, and $1 + i$ of multiplicities 2, 2, and 1, respectively.

11. 1, i, and $-2i$ of multiplicities 1, 1, and 2, respectively.

12. $1 - i$ and $2 + i$ of multiplicities 1 and 2, respectively.

Form a real *polynomial of lowest degree having the specified zeros with the specified multiplicities. Write the answer so that it is obvious that it has only real coefficients.*

13. $-2i$ and 2 of multiplicities 1 and 2, respectively.

14. $1 + i$ and 0 of multiplicities 1 and 3, respectively.

15. $1 - i$ and $3i$ of multiplicities 1 and 2, respectively.

For each polynomial in Exercises 16–27:
(a) *Determine the complex zeros, along with their multiplicities; one zero is given in each case.*
(b) *Factor the polynomial completely using complex factors.*
(c) *Factor the polynomial completely using real factors.*

16. $z^3 + 2z^2 + 5z + 10$; -2 is a zero.

17. $z^3 - 4z^2 - 4z - 5$; 5 is a zero.

18. $z^3 + z + 2$; -1 is a zero.

19. $z^4 + 5z^2 + 4$; i is a zero.

20. $z^4 + z^2 + 1$; $(-1 + \sqrt{3}i)/2$ is a zero.

21. $z^3 + z^2 - 4z + 6$; $1 + i$ is a zero.

22. $4x^3 - 7x - 3$; $\dfrac{3}{2}$ is a zero.

23. $2x^3 + x^2 + 6x + 3$; $-\dfrac{1}{2}$ is a zero.

24. $2 - 11x + 17x^2 - 6x^3$; 2 is a zero.

25. $2x^3 - x^2 + 8x - 4$; $\dfrac{1}{2}$ is a zero.

26. $-1 + 5x - x^2 + 5x^3$; $-i$ is a zero.

27. $x^4 - 5x^3 + 4x^2 - 20x$; $2i$ is a zero.

28. Prove that every real polynomial of odd degree has at least one real zero. (Use the Conjugate Zero Theorem.)

29. Give an example of a complex polynomial of odd degree that does not have a real zero. (Compare Exercise 28.)

30. Prove the Conjugate Zero Theorem. [Suggestion: Assume that z is a zero of

$$f(x) = a_n x^n + \cdots + a_1 x + a_0.$$

Then $f(z) = 0$. Use results from Exercises 64–72 of Section 46 to deduce that $f(\bar{z}) = 0$. For $n = 2$, the proof would look like this:

$$f(\bar{z}) = a_2\bar{z}^2 + a_1\bar{z} + a_0 = \overline{a_2}\overline{z^2} + \overline{a_1}\bar{z} + \overline{a_0}$$
$$= \overline{a_2 z^2 + a_1 z + a_0} = \overline{f(z)} = \bar{0} = 0.$$

Justify each step.]

REVIEW EXERCISES FOR CHAPTER X

Write each expression in the form a + bi *and simplify.*

1. $(5 + i) - 2(-1 + i)$

2. $(\sqrt{4} - \sqrt{-4}) - (\sqrt{9} + \sqrt{-9})$

3. $i(4 + i) - 4(-1 + i)$

4. $2i[(1 - 6i) - 2i(3 + i)]$

5. $(2 + i)(-3 + 2i)$

6. $(-1 + \sqrt{-9})(4 + 2i)$

7. $\dfrac{5i}{2 + i}$

8. $\dfrac{-3}{5 - i}$

9. $i^{10} - 3i^7 + i^3 + 5$

10. $i^5(2 - i)^3$

Write each complex number in the form a + bi.

11. The number with absolute value 3 and argument $2\pi/3$.

12. The number with absolute value 4.5 and argument $3\pi/4$.

Determine the absolute value, argument, and trigonometric form of each number.

13. $-5i$

14. $-5\sqrt{3} - 5i$

15. $-6 + 6i$

16. -24

Use De Moivre's Theorem to write each number in the form a + bi.

17. $(\sqrt{2} - \sqrt{2}i)^6$

18. $(1 + \sqrt{3}i)^{10}$

19. Compute the complex 12th roots of unity and represent them geometrically.

20. Compute the complex cube roots of i.

Solve for x.

21. $2x^2 - 2x + 1 = 0$ **22.** $x^2 + 3x = -3$

Form a complex *polynomial of lowest degree having the specified zeros with the specified multiplicities.*

23. 0, $2i$, and $-i$ of multiplicities 2, 1, and 1, respectively.

24. 3, $-i$, and i of multiplicities 1, 2, and 1, respectively.

Form a real *polynomial of lowest degree having the specified zeros with the specified multiplicities.*

25. $2 + 2i$ and 0 of multiplicities 1 and 2, respectively.

26. $-i$ and $3i$ of multiplicities 2 and 1, respectively.

For each polynomial in Exercises 27–28:
(a) *Determine the complex zeros, along with their multiplicities; one zero is given in each case.*
(b) *Factor the polynomial completely using complex factors.*
(c) *Factor the polynomial completely using real factors.*

27. $2x^3 - 3x^2 + 8x - 12$; $2i$ is a zero.

28. $4x^4 + 9x^2 + 2$; $- \sqrt{2}i$ is a zero.

CHAPTER XI

SYSTEMS OF EQUATIONS AND INEQUALITIES

The solution of many problems involves one equation with one variable. Other problems involve more than one variable; the solution of such problems generally requires that we write more than one equation or inequality involving those variables. This creates the need for a systematic analysis of *systems* (sets) of equations and inequalities. Such an analysis is the purpose of this chapter.

SECTION **49**
Systems of Equations with Two Variables

A. LINES AND SYSTEMS OF LINEAR EQUATIONS

A **solution** of a system of two linear equations such as

$$a_1 x + b_1 y = c_1$$
$$a_2 x + b_2 y = c_2$$

(49.1)

is an ordered pair of numbers (a, b) such that both equations are satisfied when $x = a$ and $y = b$.

Example 49.1 The pair $(2, -3)$ is a solution of the system

$$4x + y = 5$$
$$x + 2y = -4$$

because

$$4(2) + (-3) = 5$$

and

$$(2) + 2(-3) = -4.$$ ■

A system of two linear equations with two variables may have one solution, infinitely many solutions, or no solution. By thinking geometrically it will be easy for us to see why. First, recall that the graph of an equation in x and y is the set of all points whose coordinates satisfy the equation. Therefore, a system of two such equations will have a solution iff the graphs of the equations have at least one point in common; any point they have in common will provide a solution. For instance, the graphs for Example 49.1 are shown in Figure 49.1. The coordinates of the point of intersection, $(2, -3)$, give the solution of the system.

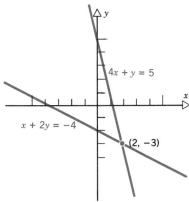

FIGURE 49.1

Since the graph of each linear equation in x and y is a straight line (Section 13), there are three possibilities for a system of two such equations:

I. The lines intersect in a single point. In this case there is exactly one solution.

II. The lines coincide. In this case there are infinitely many solutions.

III. The lines are distinct and parallel. In this case there is no solution.

If there is no solution the system is said to be **inconsistent** (possibility III). If there is at least one solution the system is said to be **consistent** (possibilities I and II). If there are infinitely many solutions the system is also said to be **dependent** (possibility II).

Example 49.2 The line in Figure 49.2 is the graph of both of the equations in the system

$$2x - 3y = -7$$
$$-4x + 6y = 14.$$

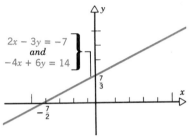

FIGURE 49.2

Therefore, the system has infinitely many solutions and is dependent. The coordinates of any point on the graph provide a solution. Some examples are $(-\frac{7}{2}, 0)$, $(0, \frac{7}{3})$, and $(1, 3)$. Notice that the second equation is the result of multiplying both sides of the first equation by -2. That is typical of a dependent system of two linear equations; either one will result from multiplying both sides of the other by an appropriate nonzero constant, followed, possibly, by a rearrangement of terms. ◼

Example 49.3 The lines in Figure 49.3 are the graphs of the equations in the system

$$x + 2y = 2$$
$$2x + 4y = -3.$$

The lines are parallel, because they both have slope $-\frac{1}{2}$. (Remember from

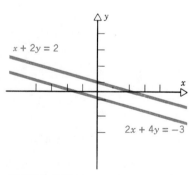

FIGURE 49.3

Section 13 that the slope is the coefficient of x after the equation has been solved for y.) The system has no solution; it is inconsistent. ◼

B. SOLUTION BY SUBSTITUTION

One way to solve a system like (49.1) is by **substitution**: Solve one of the equations for either variable, x or y. Substitute the answer in place of that variable in the other equation; this will give a linear equation with only one variable. Solve for that variable, and substitute the result in the answer from the first step to determine the value of the other variable.

Example 49.4 Solve the system

$$x - 2y = 6 \qquad \text{(A)}$$
$$2x + y = 7 \qquad \text{(B)}$$

by substitution.

Solution Solve Equation (A) for x:

$$x = 2y + 6. \qquad \text{(C)}$$

Use (C) to replace x in (B):

$$2(2y + 6) + y = 7.$$

Solve for y:

$$4y + 12 + y = 7$$
$$5y = -5$$
$$y = -1.$$

Now use $y = -1$ in (C) to get x:

$$x = 2(-1) + 6 = 4.$$

The system has one solution, $(4, -1)$.

Check Substitute $x = 4$ and $y = -1$ in both (A) and (B):

$$\text{(A)} \quad 4 - 2(-1) \overset{?}{=} 6 \qquad \text{(B)} \quad 2(4) + (-1) \overset{?}{=} 7$$
$$6 = 6 \qquad\qquad\qquad 7 = 7 \qquad ◼$$

Remark The only reason to check the solutions obtained from a linear system is to catch possible mistakes in computation; our methods will not lead to extraneous solutions. Hereafter, the step of checking a solution will not be shown.

For a dependent system the method of substitution will lead to the equation $0 = 0$ during the solution process, as in the next example.

Example 49.5 Solve the system

$$2x - 3y = -7 \qquad \text{(A)}$$
$$-4x + 6y = 14 \qquad \text{(B)}$$

by substitution. (This is the system in Example 49.2.)

Solution Solve Equation (A) for y:

$$3y = 2x + 7$$

$$y = \tfrac{1}{3}(2x + 7). \tag{C}$$

Use (C) to replace y in (B):

$$-4x + 6[\tfrac{1}{3}(2x + 7)] = 14$$

$$-4x + 4x + 14 = 14$$

$$0 = 0$$

The last equation is true for every value of x. To get a solution of the original system, assign x arbitrarily and determine the corresponding value of y from (C); any pair obtained in this way will be a solution. For example, if $x = 4$, then (C) gives

$$y = \tfrac{1}{3}(2 \cdot 4 + 7) = 5.$$

Thus $(4, 5)$ is one of the infinitely many solutions. In set-builder notation the set of all solutions is

$$\{(x, \tfrac{1}{3}(2x + 7)): \quad x \text{ is a real number}\}$$

For an inconsistent system the method of substitution will lead to a contradiction, as in the following example.

Example 49.6 Solve the system

$$x + 2y = 2 \tag{A}$$

$$2x + 4y = -3 \tag{B}$$

by substitution. (This is the system in Example 49.3)

Solution Solve Equation (A) for x:

$$x = -2y + 2. \tag{C}$$

Use (C) to replace x in (B).

$$2(-2y + 2) + 4y = -3$$

$$-4y + 4 + 4y = -3 \tag{D}$$

$$4 = -3.$$

Equation (D) is contradictory, so the given system has no solution.

C. INTERSECTIONS OF LINES AND CIRCLES

A **solution** of a system of equations (not necessarily linear) in x and y is defined as for linear systems: it is an ordered pair of numbers (a, b) such that all of the equations are satisfied when $x = a$ and $y = b$. A system of equations is said to be **nonlinear** if at least one of its equations is nonlinear. An attempt to find the points of intersection of a line and a circle (if there are any) will lead to such a system. [Recall that the equation of a circle can be written in either of the forms (12.4) or (12.5).]

Many two-variable systems composed of a linear equation and one other equation can be solved by substitution.

Example 49.7 Solve the system

$$x - 2y = 1 \tag{A}$$

$$x^2 + y^2 = 29 \tag{B}$$

by substitution.

Solution Solve the linear equation, (A), for x:

$$x = 2y + 1. \tag{C}$$

Substitute $2y + 1$ for x in (B):

$$(2y + 1)^2 + y^2 = 29.$$

Solve for y:

$$4y^2 + 4y + 1 + y^2 = 29$$

$$5y^2 + 4y - 28 = 0$$

$$(5y + 14)(y - 2) = 0$$

$$y = -\tfrac{14}{5} \quad \text{or} \quad y = 2.$$

Now use $y = -\tfrac{14}{5}$, and then $y = 2$ in (C) to determine the corresponding values of x:

If $y = -\tfrac{14}{5}$, then $x = 2\left(-\tfrac{14}{5}\right) + 1 = -\tfrac{23}{5}$.

If $y = 2$, then $x = 2(2) + 1 = 5$.

The solutions of the system are $\left(-\tfrac{23}{5}, -\tfrac{14}{5}\right)$ and $(5, 2)$. They can be checked by substitution in both of the original equations.

Figure 49.4 shows a geometrical interpretation of this example: the straight line is the graph of Equation (A) and the circle is the graph of Equation (B). The solutions of the system are the coordinates of the points where the line and the circle intersect. ■

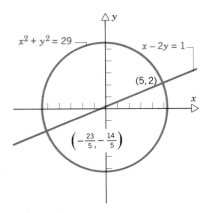

FIGURE 49.4

A line will intersect a circle in either two points, one point, or no point. It follows that a system composed of a linear equation and an equation of the form

$$x^2 + y^2 + ax + by + c = 0 \qquad (49.2)$$

will have either two solutions, one solution, or no solution. In a case with one solution, the line is tangent to the circle. A nonlinear system may have imaginary solutions (Chapter X) even if its coefficients are all real, but such imaginary solutions will not correspond to intersections of the corresponding graphs. Only real solutions are being considered in this section.

D. INTERSECTIONS OF CIRCLES

If two distinct circles intersect, then we can find the point or points of intersection by first eliminating the second-degree terms and then using substitution, as in the following example.

Example 49.8 Find all of the points of intersection of the circles represented by the following equations.

$$(x - 2)^2 + (y + 1)^2 = 4$$

$$x^2 + (y - 2)^2 = 9$$

Solution First, rewrite both equations in the form (49.2).

$$x^2 + y^2 - 4x + 2y + 1 = 0 \qquad (A)$$

$$x^2 + y^2 - 4y - 5 = 0 \qquad (B)$$

Now eliminate the second-degree terms by subtracting (B) from (A).

$$-4x + 6y + 6 = 0 \qquad (C)$$

Solve (C) for x and substitute in (B). [We could also substitute in (A). Or, we could solve (C) for y and substitute in either (A) or (B).]

$$x = \tfrac{3}{2}(y + 1) \qquad (D)$$

$$[\tfrac{3}{2}(y + 1)]^2 + y^2 - 4y - 5 = 0 \qquad (E)$$

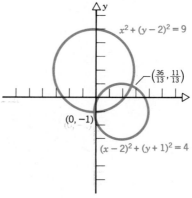

FIGURE 49.5

Simplify (E) and solve for y.

$$13y^2 + 2y - 11 = 0$$

$$(y + 1)(13y - 11) = 0$$

$$y = -1, \tfrac{11}{13}$$

If $y = -1$, then (D) yields $x = 0$. If $y = \tfrac{11}{13}$, then (D) yields $x = \tfrac{36}{13}$. Therefore, the points of intersection are $(0, -1)$ and $(\tfrac{36}{13}, \tfrac{11}{13})$. Figure 49.5 shows the graphs.

If Equation (E) had only one real solution, then the circles would intersect in a single point. If Equation (E) had no real solution, then the circles would not intersect. ◼

E. APPLICATIONS

Many applied problems can be solved using the ideas in this section. Here are three examples.

Example 49.9 We require 4 liters of mixture that is 15% alcohol. We can draw from two mixtures, one that is 12% alcohol and another that is 20% alcohol. How many liters of each should we use?

Solution From what is given we can write two equations, one involving the mixture and the other involving the alcohol. Let

$$x = \text{amount of 12\% mixture to be used}$$

$$y = \text{amount of 20\% mixture to be used.}$$

Then

$$x + \quad y = 4 \qquad \text{(mixture)}$$

$$0.12x + 0.20y = 0.15(4) \quad \text{(alcohol).}$$

The solution of this system is $x = 2.5$ and $y = 1.5$. Thus we should use 2.5 liters of 12% mixture and 1.5 liters of 20% mixture. ◼

Example 49.10 A plane flies 560 miles with the wind in 2 hours and 20 minutes. On its return trip, which is against the same wind but with the same constant air speed, the trip takes 2 hours and 40 minutes. Determine the air speed of the plane and the speed of the wind. (The air speed is the speed the plane would be traveling relative to the ground if there were no wind.)

Solution We use $D = RT$ (distance equals rate times time) and write two equations, one for the trip *with* the wind and the other for the trip *against* the wind. Let p denote the air speed of the plane and w the speed of the wind. Then the rate of the plane is $p + w$ when it flies with the wind, and $p - w$ when it flies against the wind. With the wind the time is $\tfrac{7}{3}$ hours; against the wind the time is $\tfrac{8}{3}$ hours. Therefore,

$$560 = (p + w)\tfrac{7}{3} \quad \text{(with the wind)}$$

$$560 = (p - w)\tfrac{8}{3} \quad \text{(against the wind).}$$

The solution of this system is $p = 225$ and $w = 15$. Thus the air speed is 225 mph and the wind speed is 15 mph. ◼

Example 49.11 The perimeter of a rectangle is 35 centimeters and the area is 49 square centimeters. What are the dimensions?

Solution Let x and y denote the dimensions. Then the area is xy and the perimeter is $2x + 2y$. This gives the system

$$xy = 49 \tag{A}$$

$$2x + 2y = 35. \tag{B}$$

This system can be solved by substitution. Solve (B) for y and substitute the result into (A). This leads to

$$x \left[\tfrac{35}{2} - x \right] = 49$$

$$2x^2 - 35x + 98 = 0$$

$$(2x - 7)(x - 14) = 0$$

$$x = 3.5, \ 14.$$

If $x = 3.5$, then either (A) or (B) yields $y = 14$. If $x = 14$, then either (A) or (B) yields $y = 3.5$. Either solution implies that the rectangle is 3.5 centimeters by 14 centimeters. ∎

EXERCISES FOR SECTION 49

Each system in Exercises 1–6 has exactly one solution. Find it by substitution. Also, draw the graphs of the two equations and use the point of intersection to check your answer.

1. $x - 3y = -1$
   $3x + 2y = 8$

2. $x + 2y = 3$
   $-2x - y = 6$

3. $x + 5y = 15$
   $4x - 2y = -6$

4. $3x - 5y = -1$
   $4x + 2y = 3$

5. $5x - 2y = -4$
   $3x - 2y = 0$

6. $2x - y = -10$
   $7x + 5y = -1$

Each system in Exercises 7–9 has infinitely many solutions. Find the solutions and express each answer in set-builder notation. (See the answer to Example 49.5.)

7. $3x - y = 1$
   $-6x + 2y = -2$

8. $4x - 2y = 4$
   $10x - 5y = 10$

9. $x + 2y + 3 = 0$
   $3x + 6y + 9 = 0$

Verify that each system in Exercises 10–12 is inconsistent.

10. $x - 3y = 2$
   $2x - 6y = 2$

11. $4x + 2y - 5 = 0$
   $6x + 3y + 4 = 0$

12. $x - 2y = 5$
   $-2x + 4y = -8$

In each of Exercises 13–21, find all of the points of intersection of the two lines whose equations are given.

13. $5x - 2y = 1$
   $-x + 3y = 5$

14. $2x - 4y + 1 = 0$
   $3x - 6y + 2 = 0$

15. $6x + 3y = 3$
   $8x + 4y = 4$

16. $10u - 5v + 5 = 0$
   $-4u + 2v - 2 = 0$

17. $4w - 7z = -1$
   $-2w + 5z = 1$

18. $-0.25p + 0.5q + 0.2 = 0$
   $p - 2q + 1 = 0$

19. $0.5x - 1.5y = 1.0$
   $-0.4x + 1.2y = -0.6$

20. $-\dfrac{1}{3}x + \dfrac{1}{2}y + \dfrac{1}{6} = 0$
   $\dfrac{1}{2}x - \dfrac{3}{4}y - \dfrac{1}{4} = 0$

21. $0.25x + 0.5y = 0$

$\dfrac{1}{6}x - \dfrac{1}{3}y = -4$

Each of Exercises 22–27 has an equation of a line and a circle. Find all of the points where the line intersects the circle. Also, draw the graphs to check your answer.

22. $x - 2y - 1 = 0$

$(x - 1)^2 + y^2 = 5$

23. $x + 4y = 17$

$x^2 + y^2 = 34$

24. $y - x = 6$

$x^2 + y^2 = 18$

25. $x + y = 3$

$x^2 + y^2 + 2y - 3 = 0$

26. $x = 2$

$x^2 + y^2 + 2x + 2y - 7 = 0$

27. $x - y - 1 = 0$

$x^2 + y^2 + 4x - 2y + 1 = 0$

In each of Exercises 28–30, find all of the points of intersection of the two circles whose equations are given.

28. $x^2 + y^2 + 2x + 4y - 4 = 0$

$x^2 + y^2 - 4x - 4y + 4 = 0$

29. $x^2 + y^2 - 4y = 0$

$x^2 + y^2 + 2y - 3 = 0$

30. $(x - 3)^2 + (y + 1)^2 = 5$

$(x + 1)^2 + (y - 1)^2 = 9$

31. Container *A* holds a solution that is 20% alcohol and container *B* holds a solution that is 60% alcohol. How much solution from each should we use to produce 2 liters of solution that is 50% alcohol?

32. Part of $8000 is invested in an account that pays 10% simple annual interest and the remainder is invested in an account that pays 8% simple annual interest. The total interest earned from the two accounts in one year is $669. How much is invested in each account?

33. Tickets to a theater cost $2.00 for children and $3.50 for adults. The receipts for one evening total $1075 from 335 total tickets sold. How many of the tickets were for children and how many were for adults?

34. A retiree wants to earn $5000 per year from investing $50,000. She can earn 7% annually from an insured savings account and 12% from slightly risky bonds. How much should she in-

vest in each so as to earn the $5000 but minimize the risk?

35. A plane flies 350 miles *against* the wind in 1 hour and 15 minutes. On its return trip, which is *with* the same wind, the trip takes 1 hour and 10 minutes. Determine the air speed of the plane and the speed of the wind.

36. A riverboat travels 6 miles downstream in 20 minutes. The return trip, upstream, takes 30 minutes. Determine the speed of the river current and the speed of the boat in still water.

37. When a walker and a runner start together on a quarter-mile track and go in the same direction, the runner passes the walker after three minutes. When they start together and go in opposite directions they meet after one minute. Determine the rate of each in miles per hour. Assume each has a constant rate.

38. Wilt Chamberlain once scored 100 points in a single National Basketball Association game, from a total of 64 field goals and free throws. How many field goals (worth two points each) and how many free throws (worth one point each) did he make? (Use linear equations, not trial and error).

39. The purity of gold is measured in *karats*. Pure (100%) gold is 24 karats; other degrees of purity are expressed as proportional parts of 24. Thus a 12-karat gold alloy is 50% gold and a 15-karat gold alloy is 62.5% gold. A jeweler wants to mix 14-karat gold with 24-karat gold to obtain 20 grams of 18-karat gold. How many grams each of 14- and 24-karat gold should the jeweler use?

40. Three times the tens digit of a two-digit integer is five more than the units digit. If the digits are reversed, the resulting integer exceeds by five the sum of the original integer added to twice the sum of the two digits. Find the original integer. (If *t* is the tens digit and *u* is the units digit, then the original integer is $10t + u$.)

41. If both the numerator and denominator of a fraction are increased by 1, the resulting fraction equals $\frac{2}{3}$. If both the numerator and denominator of the original fraction are decreased by 13, the resulting fraction equals $\frac{1}{2}$. Find the fraction.

42. In the 75 years from 1905 through 1979, American League teams won the World Series 13 years more than National League teams. How many years did each league have the winning team?

43. The sum of two real numbers is 4 and the difference of their squares is also 4. Find the numbers.

44. The perimeter of a rectangle is 17 inches and the area is 15 square inches. What are the dimensions?

45. The areas of two circles differ by 144π square centimeters and their circumferences differ by 12π centimeters. Find the radius of each circle.

46. The perimeter of a right triangle is 60 centimeters and the hypotenuse is 26 centimeters. How long are the legs? (Suggestion: $P = a + b + c$ and $a^2 + b^2 = c^2$, where a and b denote the lengths of the legs.)

47. If a plane's usual speed for the trip between two cities 420 miles apart is increased by 60 miles per hour, the trip will take 10 minutes less than usual. Determine the usual speed and flying time. [Suggestion: $420 = RT$ and $420 = (R + 60)(T - \frac{1}{6})$.]

48. Savings Accounts A and B each earn $100 simple interest annually. The rate for Account B is 2% less than the rate for Account A, and the principal for Account B is $250 more than the principal for Account A. Determine the principal and rate for each account. [Suggestion: If P and R denote the principal and rate for Account A, then $100 = PR$ and $100 = (P + 250)(R - 0.02)$.]

49. For which values of k does the system

$$x^2 + y^2 = 9$$
$$y = x + k$$

have two solutions? one solution? no solution? Interpret your answers geometrically.

50. For which values of k does the system

$$x^2 + (y - 3)^2 = 4$$
$$y = kx$$

have two solutions? one solution? no solution? Interpret your answers geometrically.

51. Two circles are to be placed inside a larger circle so that the centers are collinear and the three circles are tangent as shown in Figure 49.6. Show

that this can be done in such a way that the area of the larger circle is k times the sum of the areas of the smaller circles, provided $1 \le k \le 2$. (Allow 0 as a radius.) Determine a and b in terms of c and k. What does the figure look like in the case $k = 1$? $k = 2$?

Exercises 52–54 concern the system

$$a_1 x + b_1 y = c_1 \qquad (49.3)$$
$$a_2 x + b_2 y = c_2$$

where a_1, b_1, c_1, a_2, b_2, and c_2 are all assumed to be nonzero. To answer each exercise, argue geometrically after writing the equations in slope-intercept from [Equation (13.5)].

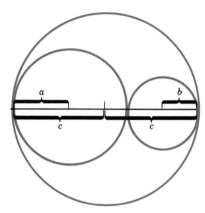

FIGURE 49.6

52. Explain why the system is inconsistent iff

$$\frac{a_1}{a_2} = \frac{b_1}{b_2} \neq \frac{c_1}{c_2}.$$

53. Explain why the system is dependent iff

$$\frac{a_1}{a_2} = \frac{b_1}{b_2} = \frac{c_1}{c_2}.$$

54. Explain why the system has exactly one solution iff

$$\frac{a_1}{a_2} \neq \frac{b_1}{b_2}.$$

SECTION **50**
Systems of Linear Equations

A. INTRODUCTION

The idea of a linear equation in two variables extends in a natural way to more than two variables. For example, a **linear equation** in the variables x, y, and z is an equation that can be written in the form

$$ax + by + cz = d$$

for some constant real numbers a, b, c, and d, with a, b, and c not all 0. In this section we consider systems of such linear equations.

A **solution** of a system such as

$$a_1x + b_1y + c_1z = d_1$$

$$a_2x + b_2y + c_2z = d_2 \qquad (50.1)$$

$$a_3x + b_3y + c_3z = d_3$$

is an ordered triple (a, b, c) of real numbers that satisfies each of the equations in the system. If necessary, we can be more explicit and write a solution in the form $x = a$, $y = b$, and $z = c$. The definition of a *solution* of a system with more variables or different variables, or with nonlinear equations, is defined in the same way.

Example 50.1 The triple $(4, 0, -3)$ is a solution of the system

$$x + 3y - z = 7$$

$$2x - y + 3z = -1$$

$$-x + 5y - 2z = 2.$$

If we substitute $x = 4$, $y = 0$, and $z = -3$ in the second equation, for instance, the result is

$$2(4) - (0) + 3(-3) \overset{?}{=} -1$$

$$-1 = -1.$$

You can check the first and third equations in the same way. ▪

We concentrate primarily on three linear equations with three variables, but the method we use will carry over to systems with more linear equations and variables. As with systems of two linear equations with two variables, any system of linear equations will have either one solution, infinitely many solutions, or no solution. If a system has at least one solution, then the system is said to be **consistent**. If there is no solution, then the system is said to be **inconsistent**.

For linear systems with three variables, the solutions have a geometrical interpretation similar to that in the case of two variables. It can be shown that relative to a three-dimensional coordinate system, each linear equation in x, y, and z is represented by a plane. A solution for a system corresponds to a point of intersection of the planes that represent the system. If there is a unique point of intersection, then the system has a unique solution. If the planes intersect in a line or a plane, then there are infinitely many solutions. If the planes have no point in common, then there is no solution.

B. SOLUTION BY ELIMINATION

One system of equations is **equivalent** to another if the two systems have the same solutions. In the method of **elimination** we solve a system by replacing it by a succession of equivalent systems until we arrive at a system for which the solutions are obvious. This method depends on the fact that we arrive at an equivalent system if we perform any combination of the following operations on a given system:

I. Interchange the equations.

II. Multiply an equation by a nonzero constant.

III. Add a constant multiple of one of the equations to the other equation.

In II, to multiply an equation by a constant means to multiply both sides of the equation by the constant. Operations I–III can be used to try to arrive at equations with only one variable each; that is, equations from which all but one variable has been *eliminated*. Before we look at an example, let us see why operations I–III produce equivalent systems. For simplicity, we consider two equations with two variables.

It is obvious that operation I will not change the solutions of a system. Operation II will not change the solutions because if $k \neq 0$, then

$$ax + by = c \quad \text{iff} \quad k(ax + by) = kc.$$

Finally, operation III will not change the solutions because if

$$a_1x + b_1y = c_1 \quad \text{and} \quad a_2x + b_2y = c_2 \tag{50.2}$$

and k is a real number, then

$$a_1x + b_1y = c_1 \quad \text{and} \quad k(a_1x + b_1y) + a_2x + b_2y = kc_1 + c_2; \tag{50.3}$$

and, conversely, if (50.3) is true, then (50.2) is true [add $-k$ times the first equation in (50.3) to the second equation in (50.3) to produce the second equation in (50.2)].

Example 50.2 Solve the system

$$4x + y = 5$$
$$x + 2y = -4$$

by elimination.

Solution First we replace the system by an equivalent system in which the coefficient of the first variable in the first equation is 1. This is not an essential step, but it fits a systematic pattern that will arise later when we consider systems with more equations and more variables. Interchange the two equations.

$$x + 2y = -4$$

$$4x + y = 5$$

Add -4 times the first equation to the second equation. This will eliminate x from the second equation.

$$x + 2y = -4$$

$$-7y = 21$$

Multiply the second equation by $-\frac{1}{7}$, so that the coefficient on y will be 1.

$$x + 2y = -4$$

$$y = -3$$

Add -2 times the second equation to the first equation. This will eliminate y from the first equation.

$$x \quad = 2$$

$$y = -3$$

Thus the solution is $(2, -3)$. ◼

The method of elimination will uncover dependent and inconsistent systems in the same way as the method of substitution (Section 49B): a dependent system will lead to the equation $0 = 0$, and an inconsistent system will lead to a contradiction.

C. GAUSSIAN ELIMINATION

We have solved linear systems by two methods, substitution and elimination. The method we use in this section is a combination of the two; we begin with elimination and then complete the process by substitution. This method is called **Gaussian elimination** [after Karl Friedrich Gauss (1777–1855)]. Operations I–III (Subsection B) are used for all but the last two steps in the following outline; the last two steps use substitution.

Outline of Gaussian Elimination

*The goal of the step on the left is to produce the form on the right, where each * denotes an unspecified constant.*

Obtain an equivalent system with 1 as the coefficient of x in the first equation. This may require no change or it may require the use of I and II.

$$x + {}^*y + {}^*z = {}^*$$
$${}^*x + {}^*y + {}^*z = {}^*$$
$${}^*x + {}^*y + {}^*z = {}^*$$

Eliminate x from the second and third equations by adding appropriate multiples of the first equation to the second and third equations.

$$x + {}^*y + {}^*z = {}^*$$
$${}^*y + {}^*z = {}^*$$
$${}^*y + {}^*z = {}^*$$

Obtain an equivalent system with 1 as the coefficient of y in the second equation, without changing the first equation.

$$\begin{aligned} x + {}^*y + {}^*z &= {}^* \\ y + {}^*z &= {}^* \\ {}^*y + {}^*z &= {}^* \end{aligned}$$

Eliminate y from the third equation by adding an appropriate multiple of the second equation to the third equation.

$$\begin{aligned} x + {}^*y + {}^*z &= {}^* \\ y + {}^*z &= {}^* \\ {}^*z &= {}^* \end{aligned}$$

Obtain an equivalent system with 1 as the coefficient of z in the third equation, without changing the first two equations. The value of z will now be obvious.

$$\begin{aligned} x + {}^*y + {}^*z &= {}^* \\ y + {}^*z &= {}^* \\ z &= {}^* \end{aligned}$$

Substitute the value of z into the second equation and solve for y.

$$\begin{aligned} x + {}^*y + {}^*z &= {}^* \\ y \qquad &= {}^* \\ z &= {}^* \end{aligned}$$

Substitute the values of y and z into the first equation and solve for x.

$$\begin{aligned} x \qquad &= {}^* \\ y \qquad &= {}^* \\ z &= {}^* \end{aligned}$$

Gaussian elimination can also be used for more than three variables and equations. For four linear equations in $w, x, y,$ and z, for example, work for the form

$$\begin{aligned} w + {}^*x + {}^*y + {}^*z &= {}^* \quad \text{and then} \\ x + {}^*y + {}^*z &= {}^* \\ y + {}^*z &= {}^* \\ z &= {}^* \end{aligned} \qquad \begin{aligned} w \qquad &= {}^* \\ x \qquad &= {}^* \\ y \qquad &= {}^* \\ z &= {}^*. \end{aligned}$$

The steps listed should be thought of as an outline that can be varied whenever it is convenient or necessary. For example, if a system is inconsistent, a contradiction will arise somewhere along the way, and then we simply stop. Or, for a system with infinitely many solutions the equation $0 = 0$ will arise somewhere along the way; Example 50.5 will show how to handle that.

Example 50.3 Solve the system

$$\begin{aligned} 3x - y - 2z &= 8 \\ x + 2y - z &= -1. \\ -2x - 5y + z &= -1. \end{aligned}$$

Solution We follow the plan outlined above, with a variation where noted. Interchange the first two equations.

$$\begin{aligned} x + 2y - z &= -1 \\ 3x - y - 2z &= 8 \\ -2x - 5y + z &= -1 \end{aligned}$$

Add -3 times the first equation to the second equation.

Add 2 times the first equation to the third equation.

$$x + 2y - z = -1$$
$$- 7y + z = 11$$
$$- y - z = -3$$

We could now multiply the second equation by $-\frac{1}{7}$ so that the coefficient on y would be 1. But that would introduce fractions, which can be avoided in this problem as follows. (Fractions cannot always be avoided, however.) Multiply the third equation by -1 to make 1 the coefficient of y; then interchange the second and third equations.

$$x + 2y - z = -1$$
$$y + z = 3$$
$$-7y + z = 11$$

Add 7 times the second equation to the third equation.

$$x + 2y - z = -1$$
$$y + z = 3$$
$$8z = 32$$

Multiply the third equation by $\frac{1}{8}$.

$$x + 2y - z = -1$$
$$y + z = 3$$
$$z = 4$$

Thus $z = 4$. With $z = 4$ the second equation becomes $y + 4 = 3$, so that $y = -1$. Finally, with $y = -1$ and $z = 4$ the first equation becomes $x + 2(-1) - 4 = -1$, so that $x = 5$. Therefore, the solution of the system is $(5, -1, 4)$. ◼

Example 50.4 Solve the system

$$x - 2y + 4z = 9$$
$$2x + y + 3z = 3$$
$$-x - 4y + 2z = 4.$$

Solution Again we follow the plan preceding Example 50.3. The first step there requires no change in this problem.

Add -2 times the first equation to the second equation.

Add the first equation to the third equation.

$$x - 2y + 4z = 9$$
$$5y - 5z = -15$$
$$-6y + 6z = 13$$

Multiply the second equation by $\frac{1}{5}$.

$$x - 2y + 4z = 9$$
$$y - z = -3$$
$$-6y + 6z = 13$$

Add 6 times the second equation to the third equation.

$$x - 2y + 4z = 9$$
$$y - z = -3$$
$$0 = -5$$

The equation $0 = -5$ is a contradiction, so the system is inconsistent. ▨

Example 50.5 Solve the system

$$x - y + 3z = 0$$
$$2x - y + 4z = 1$$
$$5x - 3y + 11z = 2.$$

Solution Add -2 times the first equation to the second equation. Add -5 times the first equation to the third equation.

$$x - y + 3z = 0$$
$$y - 2z = 1$$
$$2y - 4z = 2$$

Add -2 times the second equation to the third equation.

$$x - y + 3z = 0$$
$$y - 2z = 1$$
$$0 = 0$$

The last set of equations shows that if t is any real number then we get a solution with

$$z = t$$
$$y = 2t + 1 \quad \text{(from the second equation)}$$
$$x = y - 3z$$
$$= (2t + 1) - 3t$$
$$= -t + 1 \quad \text{(from the first equation).}$$

For example, $t = 0$ gives the solution $(1, 1, 0)$. And $t = -2$ gives the solution $(3, -3, -2)$. In set-builder notation, the set of all solutions is

$$\{(-t + 1, 2t + 1, t): \quad t \text{ is a real number}\}. \quad ▨$$

D. APPLICATIONS

Example 50.6 A farmer needs 1000 pounds of fertilizer that is 50% nitrogen, 15% phosphorous, and 35% potassium. He can draw from mixtures A, B, and C, whose compositions are shown in Table 50.1. How many pounds of each mixture should he use?

TABLE 50.1

	Nitrogen	Phosphorus	Potassium
A	40%	20%	40%
B	60%		40%
C	50%	20%	30%

Solution Let x, y, and z denote the required number of pounds of mixtures A, B, and C, respectively. We can write one equation for each nutrient. For example, the total amount of nitrogen, which is 50% of 1000 pounds, or 500 pounds, must equal the sum of the amounts from mixture A (40% of x, or $0.4x$), mixture B (60% of y, or $0.6y$), and mixture C (50% of z, or $0.5z$). The equations that arise in this way are

$$0.4x + 0.6y + 0.5z = 500 \quad \text{nitrogen}$$

$$0.2x \qquad\quad + 0.2z = 150 \quad \text{phosphorus}$$

$$0.4x + 0.4y + 0.3z = 350 \quad \text{potassium.}$$

The solution of this system is $x = 250$, $y = 250$, and $z = 500$, so the farmer should use 250 pounds each of mixtures A and B, and 500 pounds of mixture C. ∎

If two points are in a Cartesian plane but are not on the same vertical line, they determine a unique *linear* function—the function whose graph passes through the two points (Section 15). In the same way, if three noncollinear points are in a Cartesian plane but no two are on the same vertical line, they determine a unique *quadratic* function—the function whose graph passes through the three points. The following example shows how to find the quadratic function from the coordinates of the three points.

Example 50.7 There is a unique quadratic function

$$f(x) = ax^2 + bx + c$$

such that $f(-1) = 0$, $f(1) = 2$, and $f(2) = 9$. What is it? [Geometrically, the problem asks for the equation of the parabola through $(-1, 0)$, $(1, 2)$, and $(2, 9)$. See Figure 50.1.]

Solution If $f(-1) = 0$ then

$$0 = a(-1)^2 + b(-1) + c \text{ so that } a - b + c = 0.$$

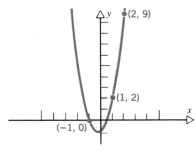

FIGURE 50.1

Similarly, $f(1) = 2$ implies $a + b + c = 2$, and $f(2) = 9$ implies $4a + 2b + c = 9$. Therefore, a, b, and c must satisfy each of the following equations.

$$a - b + c = 0$$

$$a + b + c = 2$$

$$4a + 2b + c = 9$$

The solution of this system is $a = 2$, $b = 1$, and $c = -1$, so the function is

$$f(x) = 2x^2 + x - 1.$$ ◾

Example 50.7 is related to the following general fact: If $b_1, b_2, \ldots, b_{n+1}$ are any real numbers and $a_1, a_2, \ldots, a_{n+1}$ are any distinct real numbers, then there is a unique polynomial function of degree not exceeding n such that $f(a_i) = b_i$ for $1 \le i \le n + 1$. Exercise 31 gives an illustration with $n = 3$.

EXERCISES FOR SECTION 50

Solve each system by Gaussian elimination. If there are infinitely many solutions, give the answer in set-builder notation. If a system is inconsistent, so state.

1. $\begin{aligned} x + 2y + z &= 1 \\ y + 3z &= -7 \\ 4x + 8y + 3z &= 6 \end{aligned}$

2. $\begin{aligned} x + 5y - 2z &= -5 \\ -3x + 2y - 4z &= 2 \\ y + z &= 4 \end{aligned}$

3. $\begin{aligned} x + 2y - 2z &= -2 \\ x - 4y + z &= 4 \\ 2x + 4y - z &= 5 \end{aligned}$

4. $\begin{aligned} x - 3y + 2z &= 1 \\ 2x - 5y + 6z &= 3 \\ 4x - 11y + 10z &= 5 \end{aligned}$

5. $\begin{aligned} 2u + 7v + 4w &= 2 \\ u + 4v + 3w &= 1 \\ u + 5v + 5w &= 1 \end{aligned}$

6. $\begin{aligned} 2r + + 6t &= -7 \\ s + 2t &= 2 \\ r + s + 5t &= 1 \end{aligned}$

7. $\begin{aligned} 3p + q + 14r &= 10 \\ q - 4r &= -12 \\ p + q + 2r &= 1 \end{aligned}$

8. $\begin{aligned} -x + 3y - 3z &= 2 \\ y - z &= 1 \\ 2x - y + 2z &= 7 \end{aligned}$

9. $\begin{aligned} 5x + 2y - z &= 0 \\ y + 2z &= -10 \\ x - y - 3z &= 14 \end{aligned}$

10. $\begin{aligned} 2a + 3b - c &= 0 \\ a + 2b + c - 7 &= 0 \\ 5a + 8b - c - 7 &= 0 \end{aligned}$

11. $\begin{aligned} 2a + + 36c &= 7 \\ 4a + 6b + 3c &= -1 \\ a + 2b - 5c &= 2 \end{aligned}$

12. $\begin{aligned} p + 9q + 3r - 5 &= 0 \\ p + + 6r - 6 &= 0 \\ 3p + 9q + 15r - 17 &= 0 \end{aligned}$

13. $\begin{aligned} x + 4y - z &= 2 \\ -2x + 3y + z &= 0 \\ 4x + 5y - 3z &= 4 \end{aligned}$

14. $\begin{aligned} x + y - 2z - 2 &= 0 \\ 4x + 3y - 4z - 7 &= 0 \\ 6x - y + 4z - 2 &= 0 \end{aligned}$

15. $\begin{aligned} -x + 4y + 2z &= -1 \\ 3x + 2y - 2z &= 2 \\ 9x - 8y - 10z &= 6 \end{aligned}$

16. $\begin{aligned} 9x - y + 5z &= 2 \\ 4x + 2y + 3z &= 2 \\ -2x + 3y + z &= 1 \end{aligned}$

17. $\begin{aligned} 2x + 2y - 2z &= 3 \\ -3x + y + 2z &= 1 \\ 7x + 3y - 6z &= 5 \end{aligned}$

18. $\begin{aligned} 2x - y + 2z &= 3 \\ -3x + 2y - z &= -3 \\ 4x + 2y - z &= -1 \end{aligned}$

19. $\begin{aligned} x + y + 2z - w &= 1 \\ y - z + w &= 8 \\ 2x + + 2w &= 10 \\ x - y + + w &= 1 \end{aligned}$

20. $\begin{aligned} w - 2x + 3y &= 4 \\ w - 3x + 4y - z &= 0 \\ 4x - 3y &= 2 \\ 5w - 4x + 11y - 2z &= 16 \end{aligned}$

21.
$$y - 2z + w = 3$$
$$x - y - 3z - w = 1$$
$$2x + 4y + 2w = 5$$
$$4x + 3y - 8z + w = 9$$

22. Suppose the farmer in Example 50.6 needs 1000 pounds of fertilizer that is 53% nitrogen, 10% phosphorous, and 37% potassium. How many pounds of each mixture (A, B, and C) should he use?

23. Suppose the farmer in Example 50.6 has the same requirements as in the example, except that he must draw from mixtures A and B (as in the example) and a mixture D that is 40% nitrogen, 30% phosphorous, and 30% potassium. How many pounds of each mixture (A, B, and D) should he use?

24. A total of $1000 is divided into three accounts, which pay 6%, 7%, and 9% simple annual interest, respectively. The total interest earned in 1 year is $75. The sum of the amounts in the 6% and 9% accounts is $100 more than the amount in the 7% account. How much is invested in each account?

25. In the 50 years from 1926 through 1975, National League teams won the World Series one more time than the Yankees (an American League team), and the Yankees won eight more times than the other American League teams combined. Determine how many times the Series was won by National League teams, by the Yankees, and by American League teams other than the Yankees. (It is irrelevant to the solution of this problem, but during 1926–1975 the Yankees lost the series seven times.)

26. Of the 50 states in the U.S., the number entering the Union from 1800 through 1850 was one less than the number entering before 1800, one more

than the number entering from 1851 through 1900, and three times the number entering after 1900. How many entered in each of the four periods: before 1800, 1800–1850, 1851–1900, and after 1900?

27. The normal annual precipitation amounts in Cleveland, Phoenix, and Mobile satisfy these relationships: the amount in Cleveland is five times the amount in Phoenix, the amount in Mobile is just three inches less than ten times the amount in Phoenix, and the total amount in the three cities is 109 inches. Find the amount in each city.

28. There is a unique quadratic function f such that $f(-1) = 0$, $f(1) = 1$, and $f(4) = 0$. What is it?

29. Find the equation of the parabola through the points $(-2, 6)$, $(1, 3)$, and $(2, 10)$.

30. Find the equation of the parabola that passes through the points $(0, 4)$ and $\left(1, \dfrac{3}{2}\right)$ and is symmetric with respect to the y-axis.

31. There is a unique polynomial function f of degree three such that $f(-1) = -2$, $f(1) = 2$, $f(2) = 4$, and $f(3) = 14$. What is it?

32. Prove that there is no quadratic function f such that $f(0) = -3$, $f(2) = 1$, and $f(3) = 3$. (Suggestion: The method of Example 50.7 will lead to $a = 0$.)

33. A small rocket is to be fired upward near the earth's surface with initial velocity v_0 feet per second and initial height s_0 feet, in such a way that its maximum altitude will be 225 feet and it will reach the ground 4.5 seconds after it is fired. Determine what v_0 and s_0 should be. (See Example 16.5. Use $g = 32$.)

SECTION **51**

Linear Systems and Row-Equivalent Matrices

A. AUGMENTED MATRICES

The essential information from the system

$$a_1 x + b_1 y + c_1 z = d_1$$

$$a_2 x + b_2 y + c_2 z = d_2$$

$$a_3 x + b_3 y + c_3 z = d_3$$

is contained in the rectangular array

$$\text{Row } 1 \rightarrow \begin{bmatrix} a_1 & b_1 & c_1 & d_1 \\ a_2 & b_2 & c_2 & d_2 \\ a_3 & b_3 & c_3 & d_3 \end{bmatrix}, \tag{51.1}$$

$$\underset{\text{Column 4}}{\uparrow}$$

which is called the **augmented matrix** of the system. In general, the term *matrix* is used to describe any rectangular array of numbers. Any matrix consists of (horizontal) **rows** and (vertical) **columns,** numbered respectively from top to bottom and from left to right as suggested in (51.1).

Example 51.1 The augmented matrix of the system

$$2x + y - 8z = 7$$
$$y - 2z = 2$$
$$x - 2y + 6z = 1$$

is

$$\begin{bmatrix} 2 & 1 & -8 & 7 \\ 0 & 1 & -2 & 2 \\ 1 & -2 & 6 & 1 \end{bmatrix}.$$

The 0 in the second row and first column is the coefficient of the missing x in the second equation. ∎

Augmented matrices for linear systems with more variables or different variables are defined in a similar way. Just remember that before writing the augmented matrix the variables must be in the same order in all of the equations. Also, the constant terms are to be on the right of the equality signs.

B. LINEAR SYSTEMS AND ROW-EQUIVALENCE

Rather than writing out all of the equations at each step in solving a system, we can simply write the corresponding augmented matrix. Each of the three types of operations used to solve a system by elimination produces a corresponding operation on the augmented matrix. If we interchange the first and third equations, for example, then we correspondingly interchange the first and third rows of the augmented matrix. The operations on matrices that can arise in this way are called **elementary row operations.** They are of three types (compare I, II, and III in Section 50B).

I. Interchange two rows.

II. Multiply a row by a nonzero constant.

III. Add a constant multiple of one row to another row.

To solve systems in this section we use I–III on the augmented matrix until the solution is obvious. This amounts to modifying Gaussian elimination so that substitution is not used. If possible, transform the augmented matrix as follows (for the case of three equations with three variables; other cases are similar).

$$\begin{bmatrix} * & * & * & * \\ * & * & * & * \\ * & * & * & * \end{bmatrix} \rightarrow \begin{bmatrix} 1 & * & * & * \\ * & * & * & * \\ * & * & * & * \end{bmatrix} \rightarrow \begin{bmatrix} 1 & * & * & * \\ 0 & * & * & * \\ 0 & * & * & * \end{bmatrix} \rightarrow \begin{bmatrix} 1 & * & * & * \\ 0 & 1 & * & * \\ 0 & * & * & * \end{bmatrix}$$

$$\rightarrow \begin{bmatrix} 1 & 0 & * & * \\ 0 & 1 & * & * \\ 0 & 0 & * & * \end{bmatrix} \rightarrow \begin{bmatrix} 1 & 0 & * & * \\ 0 & 1 & * & * \\ 0 & 0 & 1 & * \end{bmatrix} \rightarrow \begin{bmatrix} 1 & 0 & 0 & * \\ 0 & 1 & 0 & * \\ 0 & 0 & 1 & * \end{bmatrix}$$

In some cases the final form will necessarily be different, as Examples 51.3 and 51.4 will show. To keep track of the operations we use abbreviations like these:

(Row 1) $\leftrightarrow$ (Row 2) will mean to interchange Rows 1 and 2.

5(Row 1) will mean to multiply each element of Row 1 by 5.

(Row 3) $-$ 6(Row 1) will mean to add -6 times Row 1 to Row 3.

Example 51.2 Use augmented matrices to solve the system in Example 51.1.

Solution
$$\begin{bmatrix} 2 & 1 & -8 & 7 \\ 0 & 1 & -2 & 2 \\ 1 & -2 & 6 & 1 \end{bmatrix} \xrightarrow{\text{(Row 1)} \leftrightarrow \text{(Row 3)}} \begin{bmatrix} 1 & -2 & 6 & 1 \\ 0 & 1 & -2 & 2 \\ 2 & 1 & -8 & 7 \end{bmatrix}$$

$$\xrightarrow{\text{(Row 3)} - 2\text{(Row 1)}} \begin{bmatrix} 1 & -2 & 6 & 1 \\ 0 & 1 & -2 & 2 \\ 0 & 5 & -20 & 5 \end{bmatrix}$$

$$\xrightarrow[\text{(Row 3)} - 5\text{(Row 2)}]{\text{(Row 1)} + 2\text{(Row 2)}} \begin{bmatrix} 1 & 0 & 2 & 5 \\ 0 & 1 & -2 & 2 \\ 0 & 0 & -10 & -5 \end{bmatrix}$$

$$\xrightarrow{-\dfrac{1}{10}\text{(Row 3)}} \begin{bmatrix} 1 & 0 & 2 & 5 \\ 0 & 1 & -2 & 2 \\ 0 & 0 & 1 & \frac{1}{2} \end{bmatrix}$$

$$\xrightarrow[\text{(Row 2)} + 2\text{(Row 3)}]{\text{(Row 1)} - 2\text{(Row 3)}} \begin{bmatrix} 1 & 0 & 0 & 4 \\ 0 & 1 & 0 & 3 \\ 0 & 0 & 1 & \frac{1}{2} \end{bmatrix}$$

The last matrix is the augmented matrix of

$$\begin{aligned} x \qquad\quad &= 4 \\ y \quad &= 3 \\ z &= \tfrac{1}{2}. \end{aligned}$$

Thus the solution is $(4, 3, \tfrac{1}{2})$.

Example 51.3 Use augmented matrices to solve the system

$$x - 3y - 5z = 2$$
$$2x - 5y + z = 3$$
$$x - 2y + 6z = 2.$$

Solution
$$\begin{bmatrix} 1 & -3 & -5 & 2 \\ 2 & -5 & 1 & 3 \\ 1 & -2 & 6 & 2 \end{bmatrix} \xrightarrow[\text{(Row 3) - (Row 1)}]{\text{(Row 2) - 2(Row 1)}} \begin{bmatrix} 1 & -3 & -5 & 2 \\ 0 & 1 & 11 & -1 \\ 0 & 1 & 11 & 0 \end{bmatrix}$$

$$\xrightarrow[\text{(Row 3) - (Row 2)}]{\text{(Row 1) + 3(Row 2)}} \begin{bmatrix} 1 & 0 & 28 & -1 \\ 0 & 1 & 11 & -1 \\ 0 & 0 & 0 & 1 \end{bmatrix}$$

The third row of the last matrix corresponds to the equation $0x + 0y + 0z = 1$, which implies $0 = 1$, an impossibility. Therefore, the system is inconsistent. ▨

Example 51.4 Use augmented matrices to solve the system

$$x - 2y + z = 4$$
$$2x - 4y + 3z = 14$$
$$-x + 2y = 2.$$

Solution
$$\begin{bmatrix} 1 & -2 & 1 & 4 \\ 2 & -4 & 3 & 14 \\ -1 & 2 & 0 & 2 \end{bmatrix} \xrightarrow[\text{(Row 3) + (Row 1)}]{\text{(Row 2) - 2(Row 1)}} \begin{bmatrix} 1 & -2 & 1 & 4 \\ 0 & 0 & 1 & 6 \\ 0 & 0 & 1 & 6 \end{bmatrix}$$

No row operation will transform the last matrix to the form

$$\begin{bmatrix} 1 & * & * & * \\ 0 & 1 & * & * \\ 0 & * & * & * \end{bmatrix}.$$

But we can get the form

$$\begin{bmatrix} 1 & * & * & * \\ 0 & 0 & 1 & * \\ 0 & 0 & 0 & 0 \end{bmatrix}.$$

$$\begin{bmatrix} 1 & -2 & 1 & 4 \\ 0 & 0 & 1 & 6 \\ 0 & 0 & 1 & 6 \end{bmatrix} \xrightarrow[\text{(Row 3) - (Row 2)}]{\text{(Row 1) - (Row 2)}} \begin{bmatrix} 1 & -2 & 0 & -2 \\ 0 & 0 & 1 & 6 \\ 0 & 0 & 0 & 0 \end{bmatrix}$$

The last matrix translates to

$$x - 2y = -2$$
$$z = 6.$$

The first equation will give a value of x for each value of y. The second equation dictates $z = 6$. Therefore, the complete solution is

$$\{(2t - 2, t, 6): t \text{ is a real number}\}. ▨$$

The ideas in the preceding examples can be summarized as follows. If one matrix can be obtained from another by a finite sequence of elementary row operations, then the two matrices are said to be **row equivalent.** It follows that two systems of linear equations are equivalent iff their augmented matrices are row equivalent. Therefore, we can solve a system by transforming its augmented matrix by a sequence of elementary row operations until the solutions become obvious. An appropriate form is given in the following statement, which is presented here without proof.

Each matrix is row equivalent to a unique matrix in **reduced echelon form,** which is, by definition, a matrix such that

1. the first nonzero entry (the **leading entry**) of each row is 1;
2. the other entries in any column containing such a leading entry are 0;
3. the leading entry in each row is to the right of the leading entry in each preceding row; and
4. rows containing only 0's are below rows with nonzero entries.

Example 51.5 (a) For the reasons indicated, these matrices are not in reduced echelon form.

$$\begin{bmatrix} 1 & 0 & 4 & 3 \\ 0 & 2 & -4 & 0 \\ 0 & 0 & 0 & 0 \end{bmatrix} \quad \begin{bmatrix} 1 & 0 & 5 \\ 1 & 0 & 10 \end{bmatrix} \quad \begin{bmatrix} 1 & 0 & 0 & 8 \\ 0 & 0 & 1 & 4 \\ 0 & 1 & 0 & -7 \end{bmatrix} \quad \begin{bmatrix} 1 & 0 & 0 & 0 & -9 \\ 0 & 0 & 0 & 0 & 0 \\ 0 & 0 & 1 & 0 & 6 \\ 0 & 0 & 0 & 0 & 0 \end{bmatrix}$$

Violates 1 Violates 2 Violates 3 Violates 4

(b) Each of these matrices is in reduced echelon form and is equivalent to the corresponding matrix in part (a).

$$\begin{bmatrix} 1 & 0 & 4 & 3 \\ 0 & 1 & -2 & 0 \\ 0 & 0 & 0 & 0 \end{bmatrix} \quad \begin{bmatrix} 1 & 0 & 5 \\ 0 & 0 & 1 \end{bmatrix} \quad \begin{bmatrix} 1 & 0 & 0 & 8 \\ 0 & 1 & 0 & -7 \\ 0 & 0 & 1 & 4 \end{bmatrix} \quad \begin{bmatrix} 1 & 0 & 0 & 0 & -9 \\ 0 & 0 & 1 & 0 & 6 \\ 0 & 0 & 0 & 0 & 0 \\ 0 & 0 & 0 & 0 & 0 \end{bmatrix}$$

C. MORE EXAMPLES

Linear systems with two variables or with more than three variables can also be solved by the method of this section. The method applies as well when the number of equations is different from the number of variables, as illustrated by the following examples.

Example 51.6 Solve

$$2x \qquad + 5z = 6$$
$$3y \qquad = -1$$
$$4x - 3y + \quad z = 4$$
$$2x + 3y + \quad z = 1.$$

Solution The augmented matrix

$$\begin{bmatrix} 2 & 0 & 5 & 6 \\ 0 & 3 & 0 & -1 \\ 4 & -3 & 1 & 4 \\ 2 & 3 & 1 & 1 \end{bmatrix}$$

is row equivalent to the reduced echelon matrix

$$\begin{bmatrix} 1 & 0 & 0 & \frac{1}{2} \\ 0 & 1 & 0 & -\frac{1}{3} \\ 0 & 0 & 1 & 1 \\ 0 & 0 & 0 & 0 \end{bmatrix}$$

Therefore, the solution is $(\frac{1}{2}, -\frac{1}{3}, 1)$.

Example 51.7 Solve

$$
\begin{aligned}
w + 2x + 7y \qquad\quad - 1 &= 0 \\
x + 2y \qquad\quad - 1 &= 0 \\
-w + \quad x - \quad y + 5z - 6 &= 0.
\end{aligned}
$$

Solution The augmented matrix

$$\begin{bmatrix} 1 & 2 & 7 & 0 & 1 \\ 0 & 1 & 2 & 0 & 1 \\ -1 & 1 & -1 & 5 & 6 \end{bmatrix}$$

is row equivalent to the reduced echelon matrix

$$\begin{bmatrix} 1 & 0 & 3 & 0 & -1 \\ 0 & 1 & 2 & 0 & 1 \\ 0 & 0 & 0 & 1 & \frac{4}{5} \end{bmatrix}$$

The last row gives $z = \frac{4}{5}$. No leading coefficient occurs in the third column, so y can be assigned arbitrarily. If $y = t$, then the first and second rows give

$$w + 3t = -1 \quad \text{and} \quad x + 2t = 1.$$

Therefore,

$$w = -3t - 1 \quad \text{and} \quad x = -2t + 1,$$

so that

$$w = -3t - 1, \quad x = -2t + 1, \quad y = t, \quad \text{and } z = \frac{4}{5}$$

is a solution for each real number t. Or, the set of all solutions is

$$\{(-3t - 1, -2t + 1, t, \tfrac{4}{5}): t \text{ is a real number}\}.$$

The following statements can be proved about a system of m linear equations with n unknown variables. If $m > n$, then the system may have one solution, infinitely many solutions, or no solution, depending on the system. If $m < n$ then the system may have infinitely many solutions or no solution, depending on the system; there will never be just one solution in this case.

EXERCISES FOR SECTION 51

Each matrix in Exercises 1–9 is the reduced echelon matrix for a system with variables w, x, y and z (in that order). Write the solution or solutions, or indicate that the system is inconsistent.

1. $\begin{bmatrix} 1 & 0 & 0 & 0 & -5 \\ 0 & 1 & 0 & 0 & 0 \\ 0 & 0 & 1 & 0 & 6 \\ 0 & 0 & 0 & 0 & 1 \end{bmatrix}$

2. $\begin{bmatrix} 1 & 0 & 0 & 0 & 0 \\ 0 & 1 & 0 & 0 & 8 \\ 0 & 0 & 1 & 0 & -3 \\ 0 & 0 & 0 & 1 & 12 \end{bmatrix}$

3. $\begin{bmatrix} 1 & 0 & 0 & 0 & 4 \\ 0 & 1 & 0 & 3 & -1 \\ 0 & 0 & 1 & 2 & 7 \\ 0 & 0 & 0 & 0 & 0 \end{bmatrix}$

4. $\begin{bmatrix} 1 & 0 & 0 & 0 & -1 \\ 0 & 1 & 8 & 0 & -2 \\ 0 & 0 & 0 & 1 & 23 \\ 0 & 0 & 0 & 0 & 0 \end{bmatrix}$

5. $\begin{bmatrix} 1 & 0 & 0 & 0 & 13 \\ 0 & 1 & 0 & 0 & 4 \\ 0 & 0 & 0 & 0 & 1 \\ 0 & 0 & 0 & 0 & 0 \end{bmatrix}$

6. $\begin{bmatrix} 1 & 0 & 0 & 0 & 8 \\ 0 & 1 & 0 & 0 & -9 \\ 0 & 0 & 1 & 0 & 7 \\ 0 & 0 & 0 & 1 & 23 \end{bmatrix}$

7. $\begin{bmatrix} 1 & 0 & 0 & 0 & -3 \\ 0 & 1 & 0 & 0 & 8 \\ 0 & 0 & 1 & 0 & 6 \\ 0 & 0 & 0 & 1 & 6 \end{bmatrix}$

8. $\begin{bmatrix} 1 & 4 & 0 & 0 & 7 \\ 0 & 0 & 1 & 0 & 4 \\ 0 & 0 & 0 & 1 & -3 \\ 0 & 0 & 0 & 0 & 0 \end{bmatrix}$

9. $\begin{bmatrix} 1 & 2 & 0 & 0 & 11 \\ 0 & 0 & 1 & 0 & -2 \\ 0 & 0 & 0 & 0 & 1 \\ 0 & 0 & 0 & 0 & 0 \end{bmatrix}$

In Exercises 10–30 solve each system by reducing the augmented matrix to reduced echelon form.

10.
$$x - 3y = -6$$
$$3x + 2y = 26$$

11.
$$2x + y = -2$$
$$3x + 2y = 0$$

12.
$$x + 5y = -2$$
$$4x + 2y = -1$$

13.
$$x - 7y + 2z = -6$$
$$5x - 5y + 3z = 2$$
$$-2x + y - z = -2$$

14.
$$x + y + z = 3$$
$$x - y - z = -1$$
$$2x + 3y = 4$$

15.
$$x + 5y + 10z = 2$$
$$-3x + 15z = 4$$
$$-x + 10y + 35z = 8$$

16.
$$-a - 2b = 3$$
$$2a + 5b - c = -4$$
$$-3a - 5b - c = 12$$

17.
$$p - 4q + r = 0$$
$$3p - 11q + 4r = 1$$
$$2p - 7q + 3r = -2$$

18.
$$-t + 4u - v = 2$$
$$3t - u + 2v = 1$$
$$-4t + u - v = -1$$

19.
$$2x + y - z - 2 = 0$$
$$-5x + y + 3z = 0$$
$$3x + 5y - z - 8 = 0$$

20.
$$3x - 3y + z - 6 = 0$$
$$2x + 2z - 5 = 0$$
$$-4x + 6y + 7 = 0$$

21.
$$5x - 2y + z - 2 = 0$$
$$2x + y - z - 2 = 0$$
$$3x + 6y - 5z + 6 = 0$$

22.
$$w - 2x + 3y + 7z = 8$$
$$x + z = -3$$
$$w + y + 3z = 2$$
$$w + x + y + 4z = -1$$

23.
$$w + 2x - y = -6$$
$$w + 3x + 3z = -1$$
$$-2x + y = 2$$
$$2w - 2y + 5z = -7$$

24.
$$w + 2y = 1$$
$$x + z = -2$$
$$w + 2x + 2y + 2z = 0$$
$$-w + x - 2y = 0$$

25.
$$x + 5y = 8$$
$$2x + 4y = 7$$
$$-x + 3y = 4$$

26.
$$2x - y = 5$$
$$x + 2y = 4$$
$$-x + 8y = 3$$

27.
$$x - y = 4$$
$$-2x + 2y = -8$$
$$5x - 10y = 20$$

28.
$$p - q + 2r - 4 = 0$$
$$-2p + 2q - 4r + 7 = 0$$

29.
$$t - v = -3$$
$$u + 2v = 1$$
$$w = 2$$
$$2t + u = -5$$

30.
$$2a + b + 2c = 0$$
$$b + 2c = 1$$
$$2a + b + 2c - 3d = -1$$
$$2a + 2b + 4c = 1$$

Solve each exercise by writing a system of equations and then solving the system with augmented matrices.

31. The perimeter of a triangle is 45 centimeters. The length of the longest side is 3 centimeters less than the sum of the lengths of the other two sides. The longest side is 3 times as long as the shortest side. Find the length of each side.

32. The sum of three numbers is 85. The sum of 10% of the first, 20% of the second, and 25% of the third is 17. The sum of 25% of the first, 40% of the second, and 90% of the third is 51. Find the three numbers.

33. The sum of three real numbers is 3, the largest of the three is 5 times the smallest of the three, and the remaining number is 3 times the smallest of the three. Find the numbers.

34. The three largest lakes in the world, by area, are the Caspian Sea, Lake Superior, and Lake Victoria. The area of the Caspian Sea is 16 thousand square miles more than 4 times the area of Lake Superior and 9 thousand square miles more than 5 times the area of Lake Victoria. The area of Lake Superior is 5 thousand square miles more than the area of Lake Victoria. Find the area of each lake. (All areas are rounded to the nearest thousand square miles. The Caspian Sea is a saltwater lake; Superior and Victoria are both freshwater lakes.)

35. In the 50 years from 1926 through 1975, the number of seven-game World Series was four more than twice the number of four-game Series, the number of seven-game Series was six more than twice the number of six-game Series, and the number of Series that lasted either four or five games was two-thirds the number that lasted either six or seven games. Find the number of Series of each length. (A World Series lasts either four, five, six, or seven games.)

36. The population of the U.S. in 1950 was 1 million less than twice the population in 1900. The population east of the Mississippi River increased by 45 million from 1900 to 1950, and the population west of the river increased by 30 million from 1900 to 1950. In 1950 the population east of the river was 2 million less than twice the population west of the river. Determine the population both east and west of the river in 1900, and both east and west of the river in 1950. Then determine the percentage increases both east and west of the river between 1900 and 1950. (All populations are rounded to the nearest million.)

37. There is a unique polynomial function f of degree three such that $f(-2) = 10$, $f(-1) = 4$, $f(1) = 4$, and $f(2) = -2$. What is it?

38. There is exactly one circle that passes through the three points $(1, 3)$ $(3, -1)$, and $(4, 2)$. Find its equation. [Compare Example 50.7. Begin with Equation (12.5), $x^2 + y^2 + ax + by + c = 0$, and determine a, b, and c.]

39. Prove that no circle passes through the three points $(-3, 2)$, $(-2, 1)$, and $(1, -2)$. (The method suggested in Exercise 38 will lead to an inconsistent system.)

SECTION **52**
Systems of Inequalities

A. GRAPHS OF INEQUALITIES

The next section will show that applications can involve systems of inequalities as well as systems of equations. Before looking at such applications, however, we need more facts about inequalities that contain variables.

A **solution** of an inequality in variables x and y is an ordered pair (a, b) of numbers that satisfies the inequality; that is, such that a true statement results if x is replaced by a and y is replaced by b.

Example 52.1 Consider the inequality $2x - 3y < 6$.

(a) The pair $(0, 0)$ is a solution because $2(0) - 3(0) < 6$.

(b) The pair $(2, -1)$ is not a solution because $2(2) - 3(-1) \not< 6$. ▪

    The **graph** of an inequality in x and y is the set of all points in the Cartesian plane whose coordinates are solutions of the inequality. To graph an inequality we make use of the equation that results when the inequality sign ($<$, $>$, $\leq$, or $\geq$) is replaced by an equality sign; we call this equation the **associated equation** of the inequality. Thus the associated equation of $2x - 3y < 6$ is $2x - 3y = 6$. An inequality is said to be **linear** if its associated equation is linear; otherwise the inequality is **nonlinear.**

    The first step in graphing an inequality is to graph its associated equation. If the sign in the inequality is either $<$ or $>$, then the points on the graph of the associated equation *will not* belong to the graph of the inequality; we indicate this by graphing the associated equation as a *dashed* line or curve (as in Figure 52.1). If the sign in the inequality is either $\leq$ or $\geq$, then the points on the graph of the associated equation *will* belong to the graph of the inequality; in this case we graph the associated equation as a *solid* line or curve (as in Figure 52.2).

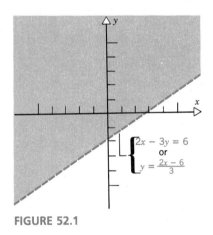

FIGURE 52.1

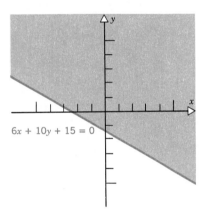

FIGURE 52.2

Example 52.2 Draw the graph of $2x - 3y < 6$.

Solution The graph of the associated equation is the dashed line in Figure 52.1. To graph the inequality we first solve it for y:

$$-3y < 6 - 2x$$

$$y > (6 - 2x)/-3$$

$$y > (2x - 6)/3.$$

For each x, a point with coordinates (x, y) will be

on the dashed line if $y = (2x - 6)/3$

and

above the dashed line if $y > (2x - 6)/3$.

Therefore, the graph of the inequality is the shaded region consisting of all of the points above the dashed line. ▪

Any line in a plane divides the points of the plane into three sets: the points on one side of the line, the points on the other side of the line, and the points on the dividing line itself. Each of the first two sets is called an **open half-plane.** The graph of any linear inequality with $<$ or $>$ (like Figure 52.1) will be an open half-plane. An open half-plane together with the points on the dividing line is called a **closed half-plane.** The graph of any linear inequality with $\leq$ and $\geq$ (like Example 52.3, which follows) will be a closed half-plane. The graph of any linear inequality can be drawn quickly with the following two steps.

Step I Draw the graph of the associated equation, making the appropriate choice of a dashed line or a solid line.

Step II Choose any point (a, b) not on the graph of the associated equation. If (a, b) *satisfies* the inequality, then the graph is the half-plane *containing* (a, b) (open or closed, as appropriate). If (a, b) *does not satisfy* the inequality, then the graph is the half-plane *not containing* (a, b) (again, open or closed, as appropriate).

For Step I, recall that the quickest way to graph most linear equations is to use the intercepts, but the important thing is to use points that are reasonably far apart. For Step II, the easiest point to check will be $(0, 0)$; if $(0, 0)$ satisfies the associated equation, then we must use some other point to help decide which half-plane to use.

Example 52.3 Draw the graph of $6x + 10y + 15 \geq 0$.

Solution The y-intercept of the associated equation is $(0, -1.5)$. The x-intercept is $(-2.5, 0)$. Draw the solid line through these points (Figure 52.2). The pair $(0, 0)$ satisfies the inequality, so the graph is the closed half-plane containing the origin, as shown. ◾

The two-step method given for graphing linear inequalities can be modified to handle many nonlinear inequalities, especially when the graph of the associated equation divides the other points of the plane into two easily identifiable parts. This includes most of the cases where the associated equation can be written as either $y = f(x)$ or $x = g(y)$ for a function f or g. Here is an example.

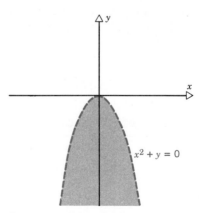

$x^2 + y = 0$

FIGURE 52.3

Example 52.4 Draw the graph of $x^2 + y < 0$.

Solution This can be rewritten as $y < -x^2$. The graph of the associated equation is the dashed parabola in Figure 52.3, which divides the plane into two parts. It is clear that the points for which $y < -x^2$ are those below the parabola; thus they make up the graph of the inequality. We could also determine which of the two parts of the plane to use by testing a pair such as (1, 1); since this pair does not satisfy $y < -x^2$, the graph is the part of the plane not containing (1, 1). ■

B. SYSTEMS OF INEQUALITIES

A **solution** of a system of inequalities in x and y is an ordered pair of numbers that satisfies every inequality in the system. The set of all solutions of such a system is called the **solution set** of the system.

In many cases this solution set can be described most easily with a graph. An element is in the **intersection** of a collection of sets iff it is in every set of the collection (Appendix B). It follows that the graph of the solution set of a system of inequalities is the intersection of the graphs of all of the inequalities in the system.

Example 52.5 Graph the solution set of the system

$$x^2 + y^2 < 21$$

$$4x - y^2 \geq 0.$$

Solution The graph of $x^2 + y^2 = 21$ is the dashed circle in Figure 52.4. The graph of $x^2 + y^2 < 21$ is the set of all points inside that circle.

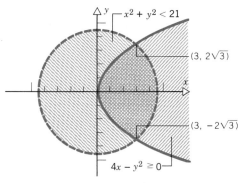

FIGURE 52.4

The graph of $4x = y^2$ is the parabola in Figure 52.4. The parabola divides the other points of the plane into two parts; the graph of $4x - y^2 \geq 0$ includes the part to the right of the parabola, as can be seen by verifying that the pair (1, 0) satisfies the inequality.

The graph of the solution set of the system is the intersection of the graphs of the two inequalities.

The coordinates of the points where the circle and the parabola intersect can be found by the methods of Section 49. Addition of the associated equations gives

$$x^2 + 4x = 21$$

$$x^2 + 4x - 21 = 0$$

$$(x + 7)(x - 3) = 0$$

$$x = -7 \quad \text{or} \quad x = 3.$$

If $x = -7$, the $4x - y^2 = 0$ implies $y^2 = -28$, which has no real solution. If $x = 3$, then $4x - y^2 = 0$ implies $y^2 = 12$, or $y = \pm2\sqrt{3}$. Thus the points of intersection are $(3, 2\sqrt{3})$ and $(3, -2\sqrt{3})$, as shown in the figure. ■

Example 52.6 Graph the solution set of the system

$$x \geq 0, \quad y \geq 0, \quad x + 2y \leq 4.$$

Solution The graph of $x \geq 0$ is the closed half-plane of all points on or to the right of the y-axis. The graph of $y \geq 0$ is the closed half-plane of all points on or above the x-axis. The graph of $x + 2y \leq 4$ is the closed half-plane of all points on or below the line $x + 2y = 4$. The intersection of these graphs is the triangular region in Figure 52.5. ■

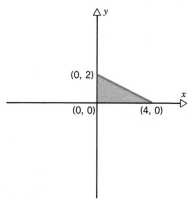

FIGURE 52.5

EXERCISES FOR SECTION 52

Draw the graph of each inequality.

1. $5x + 2y \geq 10$ **2.** $3x - 4y > 12$

3. $x - 2y \leq 4$ **4.** $-x + 2y < 6$

5. $2x + 2y \leq 5$ **6.** $4x + 3y > 8$

7. $x^2 + y^2 < 4$ **8.** $y - 2x^2 \geq 0$

9. $y + x^2 < 1$ **10.** $y \geq |x|$

11. $y < -|x|$ **12.** $x \geq |y|$

13. $y - x^2 \geq 0$
$\quad y < 2$

14. $x^2 + y^2 < 9$
$\quad x \geq 0$

15. $x^2 + y^2 \leq 25$
$\quad y - x \geq 0$

16. $5x - 6y > -3$
$\quad y > 0$
$\quad x \leq 0$

17. $-x + y \geq -2$
$\quad x > 0$
$\quad y < 0$

18. $x + y < 3$
$\quad y > 1$
$\quad x \geq 0$

19. $x^2 + y^2 > 2$
$\quad -3 \leq x \leq 3$
$\quad -3 \leq y \leq 3$

20. $y < -x^2 + 2$
$\quad y > x^2 - 2$

Graph the solution set of each system. Show the coordinates of the points where the graphs of the associated equations intersect.

21. $y - x^2 \geq 0$
$\quad x - y^2 \geq 0$

22. $y + 2x \geq 4$
$\quad x \geq 0$
$\quad y \geq 0$

23. $y - x^2 > 0$
$\quad\ x - y \geq -2$

24. $x^2 + y^2 > 4$
$\quad\ x^2 + y^2 < 9$
$\quad\ y \geq 0$

25. $y - x < 2$
$\quad\ y + x < 2$
$\quad\ y - x > -2$
$\quad\ y + x > -2$

26. $2y + 3x \geq 6$
$\quad\ 4y + x \geq 4$
$\quad\ 4y + 3x \leq 12$

27. $2x + 5y \geq 10$
$\quad\ 6x + 2y \geq 12$
$\quad\ x \geq 0$
$\quad\ y \geq 0$

28. $3x - y \geq 0$
$\quad\ x + 3y \geq 0$
$\quad\ x + y < 2$

29. $y < |x|$
$\quad\ y > |x| - 2$
$\quad\ |x| \leq 1$

30. $|y - x| \leq 1$
$\quad\ xy \leq 0$

In each of Exercises 31–36, write a system of in-equalities having the given set as its solution set. (*The points on the* x- *and* y-*axes are not in any quad-rant.*)

31. The points in the first quadrant.

32. The points in the second quadrant.

33. The points in the fourth quadrant.

34. The points in the third quadrant that are more than 10 units from the origin.

35. The points above the x-axis that are less than 5 units from the origin.

36. The points that are closer to the x-axis than to the y-axis.

SECTION **53**
Linear Programming

A. LINEAR PROGRAMMING THEOREM

Many applied problems involve maximizing or minimizing the output of a function subject to certain constraints on the input of the function. We now consider several examples of such problems that can be solved by a method known as *linear programming*. For motivation, before reading Example 53.1 you may want to glance at the problems in Examples 53.2 and 53.3; Example 53.1 is less interesting than the later examples, but it is free of the distractions of "word" problems.

Example 53.1 Consider the function of x and y defined by $f(x, y) = 3x + 5y + 2$ for each pair (x, y) in the solution set of Example 52.6 (Figure 52.5). For instance,

$$f(2.5, 0) = 3(2.5) + 5(0) + 2 = 9.5$$

and

$$f(2, 1) = 3(2) + 5(1) + 2 = 13.$$

Problem For which pair or pairs (x, y), if any, does $f(x, y)$ attain a maximum value? What about a minimum value?

Solution The triangular region in Figure 52.5 contains an infinite number of points, so we certainly cannot answer the question by checking every possible pair (x, y). The theorem following this example will tell us that, in fact, we need to check only the vertices of the triangular region; the function f will attain both its maximum and minimum values when it is evaluated at appropriate vertices. In the present case, the vertices are at $(0, 0)$, $(0, 2)$, and $(4, 0)$, and

$$f(0, 0) = 3(0) + 5(0) + 2 = 2$$
$$f(0, 2) = 3(0) + 5(2) + 2 = 12$$
$$f(4, 0) = 3(4) + 5(0) + 2 = 14.$$

Thus f attains its maximum value, which is 14, at $(4, 0)$. And f attains its minimum value, which is 2, at $(0, 0)$. ∎

Before we look at the theorem used in Example 53.1 we need several preliminary remarks. A **linear function** of x and y is a function of the form $f(x, y) = ax + by + c$, where a, b, and c are real numbers. Each linear programming problem that we consider will involve such a function and a corresponding **feasible set,** which will be the solution set of a finite system of linear inequalities in x and y. Geometrically, each feasible set will be the intersection of a finite number of closed half-planes; these half-planes are bounded by lines (the graphs of the associated equations), and the points where these lines intersect are called the **vertices** of the feasible set. Figures 52.5 and 53.1 show feasible sets with three vertices each; Figure 53.2 shows a feasible set with four vertices.

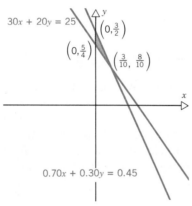

FIGURE 53.1

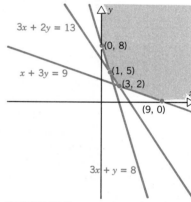

FIGURE 53.2

Linear Programming Theorem

Suppose that f is a linear function of x and y and that S is the feasible set of f. If f attains a maximum (minimum) value for (x, y) in S, then that maximum (minimum) value is attained at a vertex of S.

Subsection C will indicate why the theorem is true.

B. APPLICATIONS

Example 53.2 A hospital dietician must plan a day's menu using some combination of roasted chicken and canned tuna as a primary source of protein. Each person must receive at least 25 units of protein from the combination, and the cost must not exceed $0.45 per person. Subject to those restrictions the number of calories is to be made as small as possible. Use the data in Table 53.1 to help decide the amount of chicken and tuna that the dietician should choose.

TABLE 53.1

	Chicken (per unit)	Tuna (per unit)
Protein	30	20
Calories	180	140
Cost	$0.70	$0.30

Solution Let x and y denote respectively the number of units of chicken and tuna to be used. Then chicken will contribute $30x$ grams of protein and tuna will contribute $20y$ grams of protein. Chicken will contribute $\$0.70x$ to the cost and tuna will contribute $\$0.30y$ to the cost. Using the given restrictions on the protein and cost, and the fact that x and y cannot be negative, we arrive at the following constraints.

$$x \geq 0$$
$$y \geq 0$$
$$30x + 20y \geq 25$$
$$0.70x + 0.30y \leq 0.45$$

The feasible set is the solution set of this system, which is shown in Figure 53.1. Any point in the feasible set will give a combination that meets the requirements on protein and cost, but we want a point where the number of calories is as small as possible.

Chicken will contribute $180x$ calories and tuna will contribute $140y$ calories. To make the number of calories as small as possible we must minimize the function

$$f(x, y) = 180x + 140y.$$

By the Linear Programming Theorem the minimum will occur at a vertex of the feasible set. From

$$f\left(0, \frac{3}{2}\right) = 180(0) + 140\left(\frac{3}{2}\right) = 210$$

$$f\left(0, \frac{5}{4}\right) = 180(0) + 140\left(\frac{5}{4}\right) = 175$$

$$f\left(\frac{3}{10}, \frac{8}{10}\right) = 180\left(\frac{3}{10}\right) + 140\left(\frac{8}{10}\right) = 166$$

we see that the dietician should use $\frac{3}{10}$ units of chicken and $\frac{8}{10}$ units of tuna. This will give 166 calories. ∎

Example 53.3 Table 53.2 shows the nutrient contents and costs for two brands of fertilizer. A farmer wants to mix the two brands in the most economical way possible, subject to the restrictions that the mixture have at least 13 units of nitrogen, at least 8

TABLE 53.2

	A	B
Nitrogen	3 units per bag	2 units per bag
Phosphorus	3 units per bag	1 unit per bag
Potassium	1 unit per bag	3 units per bag
Cost	$15 per bag	$20 per bag

units of phosphorus, and at least 9 units of potassium. What combination should the farmer choose?

Solution Let x and y denote respectively the number of bags of brands A and B to be used. The restrictions give three inequalities.

$$3x + 2y \geq 13 \qquad \text{(nitrogen)}$$

$$3x + y \geq 8 \qquad \text{(phosphorus)}$$

$$x + 3y \geq 9 \qquad \text{(potassium)}$$

Also, $x \geq 0$ and $y \geq 0$. The feasible set that these inequalities produce is shown in Figure 53.2.

Brand A will contribute $15x$ to the total cost and brand B will contribute $20y$ to the total cost. To minimize the cost we minimize the function

$$f(x, y) = 15x + 20y.$$

By the Linear Programming Theorem, if there is a minimum it will occur at one of the four vertices. From

$$f(0, 8) = 15(0) + 20(8) = 160$$

$$f(1, 5) = 15(1) + 20(5) = 115$$

$$f(3, 2) = 15(3) + 20(2) = 85$$

$$f(9, 0) = 15(9) + 20(0) = 135$$

we see that the farmer should use 3 bags of brand A and 2 bags of brand B. Then the total cost will be $85. ∎

C. REMARKS

Notice that the Linear Programming Theorem states that *if* the linear function f attains a maximum (or a minimum) value, then that value is attained at a vertex. However, the function need not attain a maximum (or a minimum). A simple example is provided by $f(x, y) = x + y$ with the feasible set $x \geq 0$, $y \geq 0$; in this case there is clearly no maximum. For some feasible sets both a maximum and a minimum are guaranteed, whatever the linear function may be. Specifically, a subset of the plane is said to be **bounded** if there is a circle containing all of its points; otherwise the set is said to be **unbounded.** If the feasible set is bounded (as in Figures 52.5 and 53.1), then the linear function will attain a maximum and a minimum value. However, if the feasible set is unbounded (as in Figure 53.2), then the function may not attain a maximum or a minimum value.

In some cases a maximum or a minimum value is attained at two different vertices. When this happens $f(x, y)$ is constant on the segment connecting the two vertices. Exercise 5 gives an example.

The Linear Programming Theorem will not be proved here, but one way to think about the theorem is as follows. Imagine a z-axis, perpendicular to the Cartesian plane, directed out from the page, and with 0 on the z-axis at the origin of the plane. For the function f in a linear programming problem consider the equation $z = f(x, y)$. This equation will have a graph in three-dimensional space just as an equation in x and y has a graph in the Cartesian plane. It can be proved that, because f is linear, the graph of $z = f(x, y)$ is a plane. The Linear Programming Theorem is a consequence of the fact that among the points of this plane corresponding to (x, y) in a feasible set (an intersection of closed half-planes), the ones farthest from and closest to the Cartesian (x and y) plane occur for (x, y) at the vertices of the feasible set.

The theory of linear programming can handle much more general problems than those stated here. For example, many problems covered by the theory involve more than two variables.

EXERCISES FOR SECTION 53

The feasible set for Exercises 1–6 is the solution set of the system $x \geq 0$, $y \geq 0$, *and* $2x + y \leq 4$. *In each case determine the maximum and minimum values of the function f for* (x, y) *in the feasible set, and also the pairs* (x, y) *where those maximum and minimum values are attained. Remember that if a maximum or a minimum value is attained at two different vertices, then it is also attained at all points on the segment connecting those vertices.*

1. $f(x, y) = x + 3y - 1$
2. $f(x, y) = x - 5y + 5$
3. $f(x, y) = -3x - y - 2$
4. $f(x, y) = x - 7$
5. $f(x, y) = 4x + 2y$
6. $f(x, y) = -y + 1$

The feasible set for Exercises 7–12 is the solution set of the system $x \geq 0$, $y \geq 0$, $x - 5y \geq -15$, *and* $4x + y \leq 24$. *In each case determine the maximum and minimum values of the function f for* (x, y) *in the feasible set, and also the pairs* (x, y) *where those maximum and minimum values are attained.*

7. $f(x, y) = 2x - 2y + 1$
8. $f(x, y) = -x + y$
9. $f(x, y) = 5x + y - 2$
10. $f(x, y) = 5$
11. $f(x, y) = 8x + 2y$
12. $f(x, y) = -2x + 10y + 1$

13. Redo Example 53.2 assuming that the cost of chicken has decreased to $0.50 per unit and the cost of tuna has increased to $0.40 per unit.

14. A confectioner sells two variety packages of chocolate bars, Combination A with 30 plain bars and 10 almond bars each, and Combination B with 20 plain bars and 15 almond bars each. The profit on Combination A is $0.35 per package, and the profit on Combination B is $0.40 per package. If the confectioner has 30,000 plain bars and 10,500 almond bars to be used in variety packages, how many of each combination should be made in order to maximize the profit? What will this maximum be?

15. Redo Example 53.3 assuming that the cost of both brands A and B is $20 per bag.

16. Consider Example 53.3 and assume that the cost of brand A increases while the cost of brand B remains constant. At what point would it be to the farmer's advantage to use only brand B?

17. Refer to Table 53.3. How many cups each of orange juice and tomato juice would provide the most iron subject to the restrictions that the total calories not exceed 300 and the total amount of vitamin A be at least 2000 IU (international units)? How much iron would be provided?

TABLE 53.3

	Orange juice (one cup)	Tomato juice (one cup)
Calories	120	50
Vitamin A (IU)	550	2000
Vitamin C (mg)	120	49
Iron (mg)	0.2	2.2

18. Refer to Table 53.3. How many cups each of orange juice and tomato juice would provide the most vitamin C subject to the restrictions that the total calories not exceed 300 and the total amount of vitamin A be at least 2000 IU (international units)?

The feasible set for Exercises 19–21 is the solution set for the system $x \geq 0$, $y \geq 0$, $3x + y \geq 5$, and $2x + 5y \geq 12$. Notice that this set is unbounded, and remember that a linear function with an unbounded feasible set may not attain a maximum or a minimum value (Subsection C).

19. Consider the function $f(x, y) = x - y + 5$.
 (a) Verify that f does not attain a maximum value for (x, y) in the given feasible set. [What happens to $f(x, 0)$ as x increases?]
 (b) Verify that f does not attain a minimum value for (x, y) in the given feasible set. [What happens to $f(0, y)$ as y increases?]

20. Consider the function $f(x, y) = -x - y$.
 (a) Verify that f does not attain a minimum value for (x, y) in the given feasible set. [What happens to $f(x, 0)$ as x increases?]
 (b) Verify that $f(1, 2)$ is more than the output of f at any other vertex of the feasible set. [In fact, $f(1, 2)$ is the maximum value of f.]

21. Determine a and b so that $f(x, y) = ax + by$ has no maximum value but has the same value for all (x, y) on the line $3x + y = 5$. (This common value will be a minimum value for the function.)

REVIEW EXERCISES FOR CHAPTER XI

In Exercises 1–10, find all of the points of intersection of the graphs of the given equations.

1. $x - 3y = 2$
  $-2x + 6y = -4$

2. $3x - y = 5$
  $x - 2y = -10$

3. $3x + 4y = -1$
  $6x - 2y = 3$

4. $-5x + 3y = 2$
  $15x - 9y = -5$

5. $4x - y - 1 = 0$
  $-8x + 2y - 3 = 0$

6. $12x - 6y - 9 = 0$
  $-4x + 2y + 3 = 0$

7. $x^2 + y^2 = 10$
  $x - y = 5$

8. $(x + 2)^2 + y^2 = 9$
  $x + y = 0$

9. $x^2 + y^2 - 4 = 0$
  $x^2 + y^2 - 10x + 21 = 0$

10. $x^2 + y^2 + 2x + 2y - 2 = 0$
  $x^2 + y^2 + 2x - 4y + 4 = 0$

Use either substitution or elimination to determine all of the solutions of each system.

11. $x - 3y = 2$
  $-2x + 6y = -4$

12. $3x - y = 5$
  $x - 2y = -10$

13. $3x + 4y = -1$
  $6x - 2y = 3$

14. $-5x + 3y = 2$
  $15x - 9y = -5$

15. $4x - y - 1 = 0$
  $-8x + 2y - 3 = 0$

16. $12x - 6y - 9 = 0$
  $-4x + 2y + 3 = 0$

Use Gaussian elimination to determine all of the solutions of each system.

17. $x - 2y + z = 8$
  $-x \quad\quad - 2z = -8$
  $3x + 4y + 3z = 4$

18. $2x - 3y + z = 1$
  $2x + 2y + 2z = 3$
  $-2x - 4y + 4z = 1$

19. $x - y + z = 3$
  $2x + y - z = 1$
  $3y - 3z = -5$

20. $\quad 3x - 2y - z = 4$
$\quad\quad x + y + z = 18$
$\quad -2x + y + z = -5$

21. $\quad 2x - 3y + 4z - 10 = 0$
$\quad\quad 2x + y - 5z - 18 = 0$
$\quad -x + 4y + z = 0$

22. $\quad x - y - 5z - 1 = 0$
$\quad\quad 2x + 5y + 4z - 6 = 0$
$\quad\quad\quad 7y + 14z - 3 = 0$

Solve each system by reducing the augmented matrix to reduced echelon form.

23. $\quad x + y + z = 7$
$\quad\quad x - 3y - z = 5$
$\quad\quad 3x + z = 17$

24. $\quad x + y + 2z = 8$
$\quad\quad 3x - y - z = -3$
$\quad\quad 2x + 4y + z = 3$

25. $\quad\quad y - 9z = 5$
$\quad\quad 3x + 2y + 3z = 1$
$\quad\quad x - 2z = 3$

26. $\quad -x + 7y - 3z = 4$
$\quad\quad x + y - z = 2$
$\quad\quad 2x - 2y = 1$

27. $4x - 2y + 3z - 2 = 0$
$\quad\quad y - 3z - 12 = 0$
$\quad x + 3y + 10z - 3 = 0$

28. $2x - 3z - 4 = 0$
$\quad -x + y + 2z - 9 = 0$
$\quad x - 2y - z + 5 = 0$

Find the real solutions of each system.

29. $3x - y = -7$
$\quad\quad y = 3x^2 + 1$

30. $2x - y = -4$
$\quad\quad y = 2x^2$

31. $x^2 + y^2 = 25$
$\quad x^2 + 9y^2 = 36$

32. $x^2 - y^2 = 5$
$\quad x^2 + 4y^2 = 25$

Graph the solution set of each system.

33. $x + 2y \geq 10$
$\quad x^2 + y^2 < 25$

34. $x < y^2$
$\quad x^2 + y^2 \leq 36$

35. Determine the maximum and minimum values of $f(x, y) = 3x - 9y$ for (x, y) restricted to the feasible set $x \geq 0$, $y \geq 0$, $x + y \geq 1$, $x + 3y \leq 3$.

36. Determine the maximum and minimum values of $f(x, y) = x + 2y - 3$ for (x, y) restricted to the feasible set $y \leq 2x$, $3y \geq x$, $x + 2y \leq 5$.

CHAPTER XII

MATRICES AND DETERMINANTS

This chapter continues the study of matrices that was begun in Chapter XI. It includes definitions, properties, and applications of *determinants*, which are special functions of matrices.

SECTION **54**
Matrix Algebra

A. TERMINOLOGY

In Section 51 we used matrices merely as a convenience—to avoid writing variables, +'s, and ='s in solving systems of linear equations. In many other applications matrices are more than a mere convenience. The most significant applications depend on combining matrices through the operations of *addition* and *multiplication*, which will be studied in this section. We begin with some general terminology.

A matrix having m rows and n columns is said to be **m × n** (read "m by n"). We also say the matrix has **size** $m \times n$. The number in the ith row and jth column of a matrix A is called its (i, j)-entry and is denoted a_{ij}. Thus, if A is an $m \times n$ matrix, then

$$A = \begin{bmatrix} a_{11} & a_{12} & \cdots & a_{1n} \\ a_{21} & a_{22} & \cdots & a_{2n} \\ \vdots & \vdots & & \vdots \\ a_{m1} & a_{m2} & \cdots & a_{mn} \end{bmatrix}.$$

This is often abbreviated $A = [a_{ij}]$. In this book matrix entries will be real numbers.

Two matrices A and B are said to be **equal** if they have the same size (same number of rows and columns) and all of their corresponding entries are equal—that is, if $a_{ij} = b_{ij}$ for all i and j.

Example 54.1 Let

$$A = \begin{bmatrix} 3 & 7 \\ -1 & 0 \end{bmatrix}, \quad B = \begin{bmatrix} 3 & 7 & -4 \\ -1 & 0 & 5 \end{bmatrix}, \quad C = \begin{bmatrix} c_{11} & c_{12} \\ -1 & 0 \end{bmatrix}.$$

Then $A \neq B$ and $B \neq C$ because the sizes are different in each case. And

$$A = C \quad \text{iff} \quad c_{11} = 3 \quad \text{and} \quad c_{12} = 7. \qquad \blacksquare$$

B. MATRIX ADDITION

Definition The **sum** of two $m \times n$ matrices A and B, denoted $A + B$, is the $m \times n$ matrix whose (i, j)-entry is $a_{ij} + b_{ij}$.

In other words, matrices of the same size are added by adding their corresponding entries. Matrices can be added only if they have the same size.

Example 54.2
$$\begin{bmatrix} 3 & 0 \\ 6 & 2 \\ -8 & 4 \end{bmatrix} + \begin{bmatrix} 1 & 8 \\ -1 & 3 \\ 1 & -5 \end{bmatrix} = \begin{bmatrix} 3+1 & 0+8 \\ 6+(-1) & 2+3 \\ -8+1 & 4+(-5) \end{bmatrix} = \begin{bmatrix} 4 & 8 \\ 5 & 5 \\ -7 & -1 \end{bmatrix}$$

Matrix addition satisfies many of the properties of addition of real numbers. Here are the first two such properties.

Commutative law for matrix addition.

If A and B are $m \times n$ matrices, then
$$A + B = B + A.$$

Associative law for matrix addition.

If A, B, and C are $m \times n$ matrices, then
$$A + (B + C) = (A + B) + C.$$

Partial Proof The (i, j)-entry of $A + B$ is $a_{ij} + b_{ij}$. The (i, j)-entry of $B + A$ is $b_{ij} + a_{ij}$. These entries are equal because $a_{ij} + b_{ij} = b_{ij} + a_{ij}$ for all a_{ij} and b_{ij} by the commutative law for addition of real numbers (Section 1). The proof of the associative law is similar. □

For each pair (m, n) the $m \times n$ matrix whose entries are all 0's is called the $m \times n$ **zero matrix**. For example, the 2×3 zero matrix is

$$\begin{bmatrix} 0 & 0 & 0 \\ 0 & 0 & 0 \end{bmatrix}.$$

Each zero matrix will be denoted by 0; if there is a chance of confusion about which zero matrix is meant, the $m \times n$ zero matrix can be denoted $0_{m \times n}$. The following property of zero matrices is obvious.

If A is any $m \times n$ matrix and 0 is the $m \times n$ zero matrix, then
$$A + 0 = A \quad \text{and} \quad 0 + A = A.$$

The **negative** of a matrix $A = [a_{ij}]$ is the matrix whose (i, j)-entry is $-a_{ij}$ for all i and j. The negative of A is denoted $-A$.

Example 54.3 If $A = \begin{bmatrix} \pi & 0 \\ -1 & \sqrt{2} \end{bmatrix}$, then $-A = \begin{bmatrix} -\pi & 0 \\ 1 & -\sqrt{2} \end{bmatrix}$. ■

If A is any $m \times n$ matrix, then
$$A + (-A) = (-A) + A = 0_{m \times n}$$
and
$$-(-A) = A.$$

The **difference** $A - B$ of two $m \times n$ matrices is defined by
$$A - B = A + (-B).$$

We now know enough about matrix addition to be able to solve any matrix equation of the form $A + X = B$ (provided A and B have the same size; otherwise the equation is impossible). By subtracting $-A$ from both sides of $A + X = B$, we can conclude that if there is a solution, then it must be $X = (-A) + B$. Substitution shows that $(-A) + B$ is indeed a solution:

$$A + X = B$$

$$A + [(-A) + B] \stackrel{?}{=} B \qquad \text{Substitution.}$$

$$[A + (-A)] + B \stackrel{?}{=} B \qquad \text{Associative law.}$$

$$0 + B \stackrel{?}{=} B \qquad A + (-A) = 0.$$

$$B = B \qquad 0 + B = B.$$

Example 54.4 Solve $A + X = B$ if

$$A = \begin{bmatrix} 0.06 & 0.18 \\ -0.1 & 0.4 \end{bmatrix} \quad \text{and} \quad B = \begin{bmatrix} 2 & -1 \\ 3 & 2 \end{bmatrix}.$$

Solution The solution is

$$X = (-A) + B = \begin{bmatrix} -0.06 & -0.18 \\ 0.1 & -0.4 \end{bmatrix} + \begin{bmatrix} 2 & -1 \\ 3 & 2 \end{bmatrix} = \begin{bmatrix} 1.94 & -1.18 \\ 3.1 & 1.6 \end{bmatrix}. \qquad ■$$

C. MATRIX MULTIPLICATION

Two matrices can be added if they have the same size. In contrast, we shall see that the *product* of two matrices is defined iff the number of columns in the first matrix is equal to the number of rows in the second matrix. When the matrix product AB is defined, its size will satisfy the following pattern.

$$\begin{array}{ccc} A & B & = & C \\ m \times n & n \times p & & m \times p \end{array}$$

These must be equal.

Number of rows in A.

Number of columns in B.

Example 54.5 (a) If A is 3×2 and B is 2×4, then AB will be 3×4 and BA will be undefined.

(b) If A is 3×2 and B is 3×4, then neither AB nor BA will be defined.

(c) If A is 2×2 and B is 2×2, then both AB and BA will be 2×2. ■

In the definition of the product AB, which follows, we shall see that the (i, j)-entry of AB depends only on the *ith row* of A and the *jth column* of B.

*j*th column ↓

$$ith \text{ row} \rightarrow \begin{bmatrix} a_{11} & a_{12} & \cdots & a_{1n} \\ \vdots & \vdots & & \vdots \\ a_{i1} & a_{i2} & \cdots & a_{in} \\ \vdots & \vdots & & \vdots \\ a_{m1} & a_{m2} & \cdots & a_{mn} \end{bmatrix} \begin{bmatrix} b_{11} & \cdots & b_{1j} & \cdots & b_{1p} \\ b_{21} & \cdots & b_{2j} & \cdots & b_{2p} \\ \vdots & & \vdots & & \vdots \\ b_{n1} & \cdots & b_{nj} & \cdots & b_{np} \end{bmatrix}$$

Definition Assume that A is an $m \times n$ matrix and that B is an $n \times p$ matrix. The **product** of A and B, denoted AB, is the $m \times p$ matrix whose (i, j)-entry is

$$a_{i1}b_{1j} + a_{i2}b_{2j} + \cdots + a_{in}b_{nj}.$$

Example 54.6 Assume that A is 4×3, B is 3×2, and $AB = C$. Then C will be 4×2, and, for example,

$$c_{21} = a_{21}b_{11} + a_{22}b_{21} + a_{23}b_{31}.$$

That is, to compute the $(2, 1)$-entry of the product AB, use the second row of A and the first column of B: form the products of the pairs of entries that are connected in Figure 54.1, and then add the results.

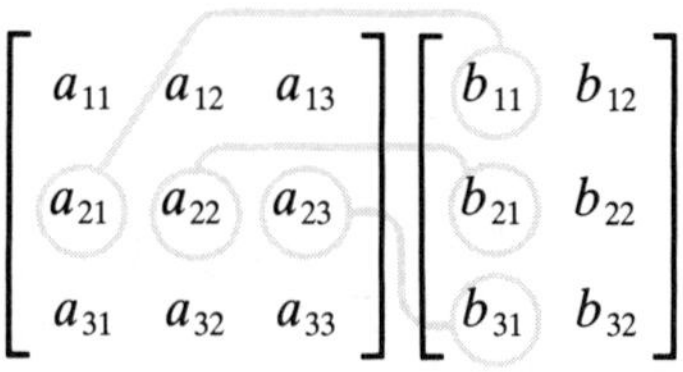

$$\begin{bmatrix} a_{11} & a_{12} & a_{13} \\ a_{21} & a_{22} & a_{23} \\ a_{31} & a_{32} & a_{33} \end{bmatrix} \begin{bmatrix} b_{11} & b_{12} \\ b_{21} & b_{22} \\ b_{31} & b_{32} \end{bmatrix}$$

FIGURE 54.1

Example 54.7 If

$$\begin{bmatrix} 2 & 0 & -5 \\ 3 & 7 & 9 \end{bmatrix} \begin{bmatrix} 4 & 8 \\ 2 & -3 \\ -1 & 5 \end{bmatrix} = \begin{bmatrix} c_{11} & c_{12} \\ c_{21} & c_{22} \end{bmatrix}$$

then

$$c_{11} = 2 \cdot 4 + 0 \cdot 2 + (-5) \cdot (-1) = 13$$

$$c_{12} = 2 \cdot 8 + 0 \cdot (-3) + (-5) \cdot (5) = -9$$

$$c_{21} = 3 \cdot 4 + 7 \cdot 2 + 9 \cdot (-1) = 17$$

$$c_{22} = 3 \cdot 8 + 7 \cdot (-3) + 9 \cdot 5 = 48.$$

Therefore,

$$\begin{bmatrix} 2 & 0 & -5 \\ 3 & 7 & 9 \end{bmatrix} \begin{bmatrix} 4 & 8 \\ 2 & -3 \\ -1 & 5 \end{bmatrix} = \begin{bmatrix} 13 & -9 \\ 17 & 48 \end{bmatrix}.$$

The commutative law for addition states that $A + B = B + A$ whenever both sums are defined (that is, whenever A and B have the same size). The corresponding law for multiplication, $AB = BA$, is not universally true, even if both AB and BA are defined. For one thing, both AB and BA may be defined and yet have different sizes: for example, if A is 2×3 and B is 3×2, then AB will be 2×2 but BA will be 3×3. But even if AB and BA are both defined and of the same size, they may be unequal, as in the next example.

Example 54.8 This example shows that the commutative law is not true for matrix multiplication:

$$\begin{bmatrix} 2 & -4 \\ 1 & -2 \end{bmatrix} \begin{bmatrix} 1 & -2 \\ -3 & 6 \end{bmatrix} = \begin{bmatrix} 14 & -28 \\ 7 & -14 \end{bmatrix}$$

but

$$\begin{bmatrix} 1 & -2 \\ -3 & 6 \end{bmatrix} \begin{bmatrix} 2 & -4 \\ 1 & -2 \end{bmatrix} = \begin{bmatrix} 0 & 0 \\ 0 & 0 \end{bmatrix}.$$

The preceding equation also shows that the product of two matrices may be the zero matrix even though neither of the two matrices is the zero matrix. This contrasts with the property of real numbers that guarantees $ab = 0$ iff $a = 0$ or $b = 0$.

The associative law for matrix multiplication, stated below, is true even though the commutative law is not. Exercise 34 asks you to prove it in the special case where A, B, and C are all 2×2 matrices; the proof for all cases can be found in books on matrices or linear algebra.

Associative law for matrix multiplication.

If A, B, and C are matrices such that both $A(BC)$ and $(AB)C$ are defined, then

$$A(BC) = (AB)C.$$

A **square matrix** is one in which the number of rows equals the number of columns. The **main diagonal** of an $n \times n$ (square) matrix consists of the elements $a_{11}, a_{22}, \ldots, a_{nn}$. For each n, the $n \times n$ matrix that contains all 1's on the main diagonal and all 0's elsewhere is called the $n \times n$ **identity matrix**. For example, the 3×3 identity matrix is

$$\begin{bmatrix} 1 & 0 & 0 \\ 0 & 1 & 0 \\ 0 & 0 & 1 \end{bmatrix}.$$

Each identity matrix will be denoted by I. If there is a chance of confusion about which identity matrix is meant, the $n \times n$ identity matrix can be denoted I_n.

The next property shows that identity matrices are to matrix multiplication what the number 1 is to multiplication of real numbers.

If A is an $m \times n$ matrix, then

$$I_m A = A \quad \text{and} \quad A I_n = A.$$

Exercise 35 asks you to prove the preceding statement for $m = n = 2$. Exercises 36 and 37 ask you to prove special cases of the following distributive laws.

Distributive laws

Each of the following equations is true for all matrices A, B, and C for which the sums and products in the equations are defined.

$$A(B + C) = AB + AC$$
$$(A + B)C = AC + BC.$$

If c is a real number and $A = [a_{ij}]$ is a matrix, then cA is defined to be the matrix whose (i, j)-entry is ca_{ij}.

Example 54.9 If $A = \begin{bmatrix} \sqrt{2} & 4 \\ -1 & 0 \end{bmatrix}$,

then

$$3A = \begin{bmatrix} 3\sqrt{2} & 12 \\ -3 & 0 \end{bmatrix}, \quad 0A = \begin{bmatrix} 0 & 0 \\ 0 & 0 \end{bmatrix}, \quad \text{and} \quad (-1)A = \begin{bmatrix} -\sqrt{2} & -4 \\ 1 & 0 \end{bmatrix}. \quad \blacksquare$$

The last equation illustrates the general property that $(-1)A = -A$ for every matrix A. Other properties of the operation cA are stated in the exercises.

Further properties of matrix multiplication, along with an application, will be discussed in the next section.

EXERCISES FOR SECTION 54

Determine the variables in each case so that the matrices are equal.

1. $\begin{bmatrix} 7 & -5 \\ -4 & 3 \end{bmatrix} = \begin{bmatrix} 7 & x \\ y & 3 \end{bmatrix}$

2. $\begin{bmatrix} 4 & a & -1 \\ 6 & -9 & b \end{bmatrix} = \begin{bmatrix} x & 2 & -1 \\ y & z & 8 \end{bmatrix}$

3.
$\begin{bmatrix} 1 & 2 \\ c & -\dfrac{1}{2} \\ \dfrac{7}{8} & d \end{bmatrix} = \begin{bmatrix} x & 2 \\ \dfrac{3}{4} & y \\ z & 7 \end{bmatrix}$

Perform the indicated operations.

4. $\begin{bmatrix} 0 & -9 \\ 3 & 8 \end{bmatrix} + \begin{bmatrix} 4 & 5 \\ -3 & 4 \end{bmatrix}$

5. $\begin{bmatrix} 5 & 27 \\ 14 & -13 \end{bmatrix} + \begin{bmatrix} -6 & 1 \\ 0 & 5 \end{bmatrix}$

6.
$\begin{bmatrix} 7 & \dfrac{4}{5} \\ \dfrac{1}{4} & -5 \\ 0 & 2 \end{bmatrix} + \begin{bmatrix} -7 & \dfrac{2}{3} \\ -\dfrac{1}{3} & 5 \\ \dfrac{1}{8} & -6 \end{bmatrix}$

7. $\begin{bmatrix} 9 & 5 & -3 \\ 2 & 0 & 17 \end{bmatrix} - \begin{bmatrix} 7 & -1 & 9 \\ 6 & 4 & 1 \end{bmatrix}$

8. $[7 \quad 12 \quad -4] - [8 \quad 9 \quad -1]$

9. $\begin{bmatrix} 6 \\ 13 \\ -7 \end{bmatrix} - \begin{bmatrix} -7 \\ 10 \\ -2 \end{bmatrix}$

Solve for X in each case.

10. $\begin{bmatrix} 9 & -6 \\ 0 & 2 \end{bmatrix} + X = \begin{bmatrix} 3 & 12 \\ -11 & 10 \end{bmatrix}$

11. $X + [4 \quad 0 \quad -7] = [14 \quad 1 \quad 8]$

12. $\begin{bmatrix} -2 & 5 & 6 \\ 9 & 7 & 11 \end{bmatrix} + X = \begin{bmatrix} 12 & 9 & -3 \\ -5 & -1 & 0 \end{bmatrix}$

13. If A is 3×5 and B is 5×1, what are the dimensions of AB?

14. If A is 3×1, B is $m \times 4$, and AB is defined, what is m and what are the dimensions of AB?

15. If A is $m \times 3$, B is $3 \times p$, and AB is 6×7, what are m and p?

Compute the matrices indicated in Exercises 16–33 with A, B, C, D, E, and F defined as follows.

$$A = \begin{bmatrix} 3 & 6 \\ 1 & -4 \end{bmatrix} \quad B = \begin{bmatrix} -8 & 1 \\ 1 & 6 \end{bmatrix}$$

$$C = \begin{bmatrix} 6 & 5 & -2 \\ 3 & 0 & -1 \\ 2 & 1 & 2 \end{bmatrix} \quad D = \begin{bmatrix} 0 & 1 & 4 \\ -7 & 1 & 2 \\ 3 & -2 & 0 \end{bmatrix}$$

$$E = \begin{bmatrix} 7 & 5 \\ 0 & 0 \\ -5 & 6 \end{bmatrix} \quad F = \begin{bmatrix} 2 & 9 & -5 \\ 0 & 3 & -6 \end{bmatrix}$$

16. AB **17.** BA

18. AF **19.** BF

20. EA **21.** EB

22. FC **23.** CD

24. DC **25.** CE

26. FE **27.** DE

28. $2A$

30. $3F$

32. $(A + B)F$

29. $-2E$

31. $E(A + B)$

33. $F(C + D)$

In Exercises 34–42, use

$$A = \begin{bmatrix} a_{11} & a_{12} \\ a_{21} & a_{22} \end{bmatrix}, \quad B = \begin{bmatrix} b_{11} & b_{12} \\ b_{21} & b_{22} \end{bmatrix},$$

$$C = \begin{bmatrix} c_{11} & c_{12} \\ c_{21} & c_{22} \end{bmatrix}.$$

34. Prove that $A(BC) = (AB)C$.

35. Prove that $AI_2 = A$ and $I_2A = A$.

36. Prove that $A(B + C) = AB + AC$.

37. Prove that $(A + B)C = AC + BC$.

38. It can be proved that if A is any matrix and 0 is an appropriately sized zero matrix in each case, then $A0 = 0$ and $0A = 0$. Write a proof for the special cases where A is 2×2.

39. Prove that $(-A)B = A(-B) = -(AB)$.

40. Prove that $(-A)(-B) = AB$.

41. Prove that $A(B - C) = AB - AC$.

42. Prove that $(A - B)C = AC - BC$.

43. For real numbers, $(a + b)^2 = a^2 + 2ab + b^2$. This exercise examines the analogous relationship for matrices.

  (a) Show that $(A + B)^2 \neq A^2 + 2AB + B^2$ for the matrices A and B preceding Exercise 16.

  (b) Prove that if A and B are $n \times n$ matrices, then $(A + B)(A - B) = A^2 - AB + BA - B^2$.
  (Suggestion: Use the distributive laws.)

44. For real numbers, $(a + b)(a - b) = a^2 - b^2$.

  (a) Show that $(A + B)(A - B) \neq A^2 - B^2$ for the matrices A and B preceding Exercise 16.

  (b) Prove that if A and B are $n \times n$ matrices, then $(A + B)(A - B) = A^2 - AB + BA - B^2$.

45. For real numbers, if $ab = ac$ and $a \neq 0$, then $b = c$. Show that the analogous fact does not hold for matrices, by finding 2×2 matrices A, B, and C such that $AB = AC$ and $A \neq 0$ but $B \neq C$. (Suggestion: Let A and B be the two matrices on the left in the second equation of Example 54.8, and let C be the zero matrix.)

SECTION **55**

The Inverse of a Matrix.
Application to Linear Systems

A. THE INVERSE OF A MATRIX

We know that identity matrices have the following similarity with the real number 1:

$$I_m A = A = A I_n \quad \text{for every } m \times n \text{ matrix } A$$

just as

$$1a = a = a1 \quad \text{for every real number } a.$$

Real numbers have the further property that if $a \neq 0$, then a has an inverse (reciprocal), a^{-1}, such that

$$aa^{-1} = 1 = a^{-1}a.$$

Some nonzero matrices do not have an inverse in this sense, however. Those matrices that do are singled out by the following definition. Notice that the definition applies only to square matrices.

Definition An $n \times n$ matrix A is said to be **invertible** if there is an $n \times n$ matrix B such that

$$AB = I_n = BA.$$

The matrix B is called an **inverse** of A in this case.

It can be proved that if a matrix is invertible, then its inverse is unique (see Exercise 28). Therefore, we can refer to *the* inverse of an invertible matrix A; this unique inverse is denoted A^{-1}.

Example 55.1 The matrix $\begin{bmatrix} 2 & -3 \\ 3 & -4 \end{bmatrix}$ is the inverse of $\begin{bmatrix} -4 & 3 \\ -3 & 2 \end{bmatrix}$, because

$$\begin{bmatrix} -4 & 3 \\ -3 & 2 \end{bmatrix} \begin{bmatrix} 2 & -3 \\ 3 & -4 \end{bmatrix} = \begin{bmatrix} 1 & 0 \\ 0 & 1 \end{bmatrix}$$

and

$$\begin{bmatrix} 2 & -3 \\ 3 & -4 \end{bmatrix} \begin{bmatrix} -4 & 3 \\ -3 & 2 \end{bmatrix} = \begin{bmatrix} 1 & 0 \\ 0 & 1 \end{bmatrix}.$$

Example 55.2 Prove that the matrix $A = \begin{bmatrix} 0 & 1 \\ 0 & 0 \end{bmatrix}$ is not invertible.

Solution If $AB = I$, then

$$\begin{bmatrix} 0 & 1 \\ 0 & 0 \end{bmatrix} \begin{bmatrix} b_{11} & b_{12} \\ b_{21} & b_{22} \end{bmatrix} = \begin{bmatrix} 1 & 0 \\ 0 & 1 \end{bmatrix}$$

and thus

$$\begin{bmatrix} b_{21} & b_{22} \\ 0 & 0 \end{bmatrix} = \begin{bmatrix} 1 & 0 \\ 0 & 1 \end{bmatrix}.$$

The last equation is impossible since the $(2, 2)$-entries, 0 and 1, are unequal. Thus there is no matrix B such that $AB = I$, and A is not invertible.

The last example points up the general fact that a matrix A is not invertible if either $AB = I$ is impossible or $BA = I$ is impossible. It can be proved that, in fact, for square matrices the two equations $AB = I$ and $BA = I$ are equivalent, in the sense that if either one is true, then the other is also true. This means that to show that B is the inverse of A we need to check only one of the two equations, $AB = I$ or $BA = I$. (Most books on matrices or linear algebra give a proof.)

B. COMPUTING INVERSES: THE 2 × 2 CASE BY DETERMINANTS

Following is an easy-to-use criterion for determining whether a 2×2 matrix is invertible, together with a formula for the inverse if the matrix *is* invertible.

A matrix $A = \begin{bmatrix} a & b \\ c & d \end{bmatrix}$ is invertible if $ad - bc \neq 0$. If $ad - bc \neq 0$, then

$$A^{-1} = \frac{1}{ad - bc} \begin{bmatrix} d & -b \\ -c & a \end{bmatrix}. \tag{55.1}$$

The number $ad - bc$ in the denominator in (55.1) is called the **determinant** of the matrix A; notice that it is the product of the two elements on the main diagonal minus the product of the two elements on the other diagonal. The matrix shown in (55.1) results from interchanging the two entries on the main diagonal of A and changing the signs of the other two entries of A.

Proof If $AB = I$ with $B = \begin{bmatrix} w & x \\ y & z \end{bmatrix}$, then

$$\begin{bmatrix} aw + by & ax + bz \\ cw + dy & cx + dz \end{bmatrix} = \begin{bmatrix} 1 & 0 \\ 0 & 1 \end{bmatrix}. \tag{55.2}$$

The equality of corresponding entries in the first columns in (55.2) dictates that

$$aw + by = 1 \quad \text{and} \quad cw + dy = 0. \tag{55.3}$$

The equality of the entries in the second columns in (55.2) dictates that

$$ax + bz = 0 \quad \text{and} \quad cx + dz = 1. \tag{55.4}$$

Since the matrix A is given, we think of w, x, y, and z as unknowns in the Systems (55.3) and (55.4).

To solve System (55.3) for y, compute a times the second equation minus c times the first equation; this eliminates w and leads to

$$(ad - bc)y = -c,$$

from which

$$y = \frac{-c}{ad - bc} \quad \text{if} \quad ad - bc \neq 0.$$

Moreover, if $ad - bc = 0$, then there is no solution for y and A has no inverse.

To solve System (55.3) for w, compute d times the first equation minus b times the second equation; this eliminates y and leads to

$$(ad - bc)w = d,$$

from which

$$w = \frac{d}{ad - bc} \quad \text{if} \quad ad - bc \neq 0.$$

The system (55.4) can be solved in the same way, giving

$$x = \frac{-b}{ad - bc} \quad \text{if} \quad ad - bc \neq 0$$

and

$$z = \frac{a}{ad - bc} \quad \text{if} \quad ad - bc \neq 0.$$

Substitution in B of these solutions for w, x, y, and z will give A^{-1} in Equation (55.1). Exercise 29 asks you to verify that $A^{-1}A = I$ and $AA^{-1} = I$ for A^{-1} as in (55.1). ☐

As pointed out in the proof, if $ad - bc = 0$, then A is *not* invertible. Thus we can make the following statement.

> A 2×2 matrix is invertible
> iff
> its determinant is nonzero.

The next two sections will consider determinants of larger square matrices and will give a generalization of the preceding statement.

Example 55.3 Determine whether $A = \begin{bmatrix} 5 & 1 \\ 6 & -3 \end{bmatrix}$ is invertible, and if it is invertible compute its inverse.

Solution The determinant of A is $(5)(-3) - (1)(6) = -21$, which is nonzero; therefore, A is invertible. To compute the inverse, apply (55.1):

$$A^{-1} = \begin{bmatrix} 5 & 1 \\ 6 & -3 \end{bmatrix}^{-1} = \frac{1}{-21}\begin{bmatrix} -3 & -1 \\ -6 & 5 \end{bmatrix} = \begin{bmatrix} \dfrac{1}{7} & \dfrac{1}{21} \\ \dfrac{2}{7} & -\dfrac{5}{21} \end{bmatrix}.$$

To check this answer, you can verify directly that $AA^{-1} = I$. ▪

C. COMPUTING INVERSES BY ELEMENTARY ROW OPERATIONS

We now consider a method that can be used to compute the inverse of an invertible matrix of any size. The method uses *elementary row operations* and *reduced echelon form*, which you may want to review from Section 51B.

It can be proved that if A is an $n \times n$ invertible matrix, then there is a sequence of elementary row operations that will transform A into I_n; moreover, the same sequence of operations will transform I_n into A^{-1}. We can apply this as follows.

> To compute the inverse of an $n \times n$ invertible matrix A, form the $n \times 2n$ matrix
>
> $$[A \mid I_n].$$
>
> Use elementary row operations on this $n \times 2n$ matrix to transform A, the left half, into I_n. Then the result in the right half, which was originally I_n, will be A^{-1}.

Example 55.4 The matrix $A = \begin{bmatrix} 1 & 0 & 3 \\ 2 & -5 & 4 \\ 1 & -2 & 2 \end{bmatrix}$ is invertible. Use elementary row operations to compute A^{-1}.

Solution Begin with $[A \mid I_3]$.

$$\begin{bmatrix} 1 & 0 & 3 & | & 1 & 0 & 0 \\ 2 & -5 & 4 & | & 0 & 1 & 0 \\ 1 & -2 & 2 & | & 0 & 0 & 1 \end{bmatrix}$$

$$\xrightarrow[\text{(Row 3)} - \text{(Row 1)}]{\text{(Row 2)} - 2\text{(Row 1)}} \begin{bmatrix} 1 & 0 & 3 & | & 1 & 0 & 0 \\ 0 & -5 & -2 & | & -2 & 1 & 0 \\ 0 & -2 & -1 & | & -1 & 0 & 1 \end{bmatrix}$$

$$\xrightarrow{-\frac{1}{5}\text{(Row 2)}} \begin{bmatrix} 1 & 0 & 3 & | & 1 & 0 & 0 \\ 0 & 1 & \frac{2}{5} & | & \frac{2}{5} & -\frac{1}{5} & 0 \\ 0 & -2 & -1 & | & -1 & 0 & 1 \end{bmatrix}$$

$$\xrightarrow{\text{(Row 3)} + 2\text{(Row 2)}} \begin{bmatrix} 1 & 0 & 3 & | & 1 & 0 & 0 \\ 0 & 1 & \frac{2}{5} & | & \frac{2}{5} & -\frac{1}{5} & 0 \\ 0 & 0 & -\frac{1}{5} & | & -\frac{1}{5} & -\frac{2}{5} & 1 \end{bmatrix}$$

$$\xrightarrow{-5\text{(Row 3)}} \begin{bmatrix} 1 & 0 & 3 & | & 1 & 0 & 0 \\ 0 & 1 & \frac{2}{5} & | & \frac{2}{5} & -\frac{1}{5} & 0 \\ 0 & 0 & 1 & | & 1 & 2 & -5 \end{bmatrix}$$

$$\xrightarrow[\text{(Row 2)} - \frac{2}{5}\text{(Row 3)}]{\text{(Row 1)} - 3\text{(Row 3)}} \begin{bmatrix} 1 & 0 & 0 & | & -2 & -6 & 15 \\ 0 & 1 & 0 & | & 0 & -1 & 2 \\ 0 & 0 & 1 & | & 1 & 2 & -5 \end{bmatrix}$$

The result is $[I_3 \mid A^{-1}]$. Check:

$$AA^{-1} = \begin{bmatrix} 1 & 0 & 3 \\ 2 & -5 & 4 \\ 1 & -2 & 2 \end{bmatrix} \begin{bmatrix} -2 & -6 & 15 \\ 0 & -1 & 2 \\ 1 & 2 & -5 \end{bmatrix} = \begin{bmatrix} 1 & 0 & 0 \\ 0 & 1 & 0 \\ 0 & 0 & 1 \end{bmatrix} = I_3.$$

     No sequence of elementary row operations will transform a *noninvertible* matrix into an identity matrix. If the method in Example 55.4 is attempted with a matrix A that is not invertible, then A will be transformed into a matrix with a row of 0's. When that happens we can simply stop.

Example 55.5 Use elementary row operations to determine whether $A = \begin{bmatrix} 1 & 1 & 2 \\ 0 & 1 & -4 \\ 3 & 1 & 14 \end{bmatrix}$ is invertible.

Solution

$$\begin{bmatrix} 1 & 1 & 2 & | & 1 & 0 & 0 \\ 0 & 1 & -4 & | & 0 & 1 & 0 \\ 3 & 1 & 14 & | & 0 & 0 & 1 \end{bmatrix}$$

$$\xrightarrow[\text{(Row 3)} - 3\text{(Row 1)}]{} \begin{bmatrix} 1 & 1 & 2 & \vdots & 1 & 0 & 0 \\ 0 & 1 & -4 & \vdots & 0 & 1 & 0 \\ 0 & -2 & 8 & \vdots & -3 & 0 & 1 \end{bmatrix}$$

$$\xrightarrow[\substack{\text{(Row 1)} - \text{(Row 2)} \\ \text{(Row 3)} + 2\text{(Row 2)}}]{} \begin{bmatrix} 1 & 0 & 6 & \vdots & 1 & -1 & 0 \\ 0 & 1 & -4 & \vdots & 0 & 1 & 0 \\ 0 & 0 & 0 & \vdots & -3 & 2 & 1 \end{bmatrix}$$

The left half of the third row contains only 0's, so A is not invertible. ∎

D. APPLICATION TO LINEAR SYSTEMS OF EQUATIONS

A linear system

$$a_1 x + b_1 y + c_1 z = d_1$$

$$a_2 x + b_2 y + c_2 z = d_2 \qquad (55.5)$$

$$a_3 x + b_3 y + c_3 z = d_3$$

can be represented in matrix form as

$$AX = B, \qquad (55.6)$$

where

$$A = \begin{bmatrix} a_1 & b_1 & c_1 \\ a_2 & b_2 & c_2 \\ a_3 & b_3 & c_3 \end{bmatrix}, \quad X = \begin{bmatrix} x \\ y \\ z \end{bmatrix}, \quad \text{and} \quad B = \begin{bmatrix} d_1 \\ d_2 \\ d_3 \end{bmatrix}.$$

The product AX in (55.6) is the matrix product, and the equality in (55.6) is matrix equality. The matrix A is called the **coefficient matrix** of the system. Systems with other variables can be represented similarly.

Example 55.6 The matrix form of the system

$$x \qquad + 3z = -2$$

$$2x - 5y + 4z = 4$$

$$x - 2y + 2z = 1$$

is

$$AX = B,$$

where

$$A = \begin{bmatrix} 1 & 0 & 3 \\ 2 & -5 & 4 \\ 1 & -2 & 2 \end{bmatrix}, \quad X = \begin{bmatrix} x \\ y \\ z \end{bmatrix}, \quad \text{and} \quad B = \begin{bmatrix} -2 \\ 4 \\ 1 \end{bmatrix}. \qquad ∎$$

If the coefficient matrix in Equation (55.6) is invertible, then we can multiply both sides of the equation on the left by A^{-1} to obtain

$$A^{-1}(AX) = A^{-1}B$$

$$(A^{-1}A)X = A^{-1}B$$

$$IX = A^{-1}B$$

$$X = A^{-1}B. \qquad (55.7)$$

Thus, by computing A^{-1} and then $A^{-1}B$ we can solve System (55.5). [*Important*: In general, $A^{-1}B \neq BA^{-1}$. Equation (55.7) requires $A^{-1}B$.]

The inverse of the coefficient matrix for Example 55.6 was computed in Example 55.4:

$$A^{-1} = \begin{bmatrix} -2 & -6 & 15 \\ 0 & -1 & 2 \\ 1 & 2 & -5 \end{bmatrix}$$

Therefore, the solution of the system in Example 55.6 is given by

$$X = A^{-1}B = \begin{bmatrix} -2 & -6 & 15 \\ 0 & -1 & 2 \\ 1 & 2 & -5 \end{bmatrix}\begin{bmatrix} -2 \\ 4 \\ 1 \end{bmatrix} = \begin{bmatrix} -5 \\ -2 \\ 1 \end{bmatrix}.$$

That is, the solution is $x = -5$, $y = -2$, $z = 1$. ■

If the coefficient matrix of System (55.5) is not invertible, then the system does not have a unique solution; it may have no solution or infinitely many solutions.

EXERCISES FOR SECTION 55

Verify that $B = A^{-1}$ *in each case by computing* AB.

1. $A = \begin{bmatrix} 2 & 3 \\ -3 & -5 \end{bmatrix}$ $B = \begin{bmatrix} 5 & 3 \\ -3 & -2 \end{bmatrix}$

2. $A = \begin{bmatrix} 2 & -4 \\ 3 & -7 \end{bmatrix}$ $B = \dfrac{1}{2}\begin{bmatrix} 7 & -4 \\ 3 & -2 \end{bmatrix}$

3. $A = \begin{bmatrix} 5 & -5 & -5 \\ -4 & 3 & 6 \\ 2 & 1 & -3 \end{bmatrix}$ $B = \dfrac{1}{5}\begin{bmatrix} 3 & 4 & 3 \\ 0 & 1 & 2 \\ 2 & 3 & 1 \end{bmatrix}$

Determine whether each matrix is invertible by computing its determinant. If the matrix is invertible, compute its inverse by using Equation (55.1). Check each inverse by computing its product with the given matrix.

4. $\begin{bmatrix} 6 & -3 \\ -3 & 2 \end{bmatrix}$

5. $\begin{bmatrix} 5 & 7 \\ 2 & 2 \end{bmatrix}$

6. $\begin{bmatrix} 15 & 2 \\ -30 & -4 \end{bmatrix}$

7. $\begin{bmatrix} 1 & 1 \\ 2 & 4 \\ \frac{1}{3} & \frac{1}{3} \end{bmatrix}$

8. $\begin{bmatrix} 3 & -\frac{1}{5} \\ 2 & \frac{2}{5} \\ -3 & \frac{2}{5} \end{bmatrix}$

9. $\begin{bmatrix} 1.7 & -0.5 \\ -10 & 3 \end{bmatrix}$

10. $\begin{bmatrix} 0.5 & -4 \\ -1.5 & 12 \end{bmatrix}$

11. $\begin{bmatrix} 14 & -2 \\ 35 & 5 \end{bmatrix}$

12. $\begin{bmatrix} \sqrt{2} & \sqrt{3} \\ 3\sqrt{3} & -2\sqrt{2} \end{bmatrix}$

Use elementary row operations to determine whether each matrix is invertible, and if the matrix is invertible give its inverse.

13. $\begin{bmatrix} 1 & 0 & 1 \\ 5 & -2 & 2 \\ 1 & 3 & 4 \end{bmatrix}$

14. $\begin{bmatrix} 1 & 3 & -5 \\ 4 & 10 & 3 \\ 2 & 0 & 2 \end{bmatrix}$

15. $\begin{bmatrix} -1 & 2 & 3 \\ 1 & 9 & 3 \\ 5 & 1 & 8 \end{bmatrix}$

16. $\begin{bmatrix} 2 & 2 & -1 \\ 6 & -1 & 7 \\ 3 & 0 & -1 \end{bmatrix}$

17. $\begin{bmatrix} 5 & -3 & 6 \\ 1 & 0 & 7 \\ -1 & -2 & 9 \end{bmatrix}$

18. $\begin{bmatrix} 1 & 5 & 2 \\ 4 & -2 & -2 \\ 1 & 3 & 8 \end{bmatrix}$

19. $\begin{bmatrix} 1 & 0 & 4 & 0 \\ 0 & -2 & -3 & -1 \\ -2 & 1 & -10 & 1 \\ 1 & 0 & 5 & 2 \end{bmatrix}$

20. $\begin{bmatrix} 1 & -2 & 4 & 1 \\ 3 & -5 & 13 & 1 \\ 1 & -2 & 5 & 1 \\ 1 & -2 & 4 & 2 \end{bmatrix}$

21.
$$\begin{bmatrix} 1 & -1 & 2 & -1 \\ 2 & 1 & -2 & 1 \\ -5 & -2 & 4 & -2 \\ 3 & 1 & -2 & 1 \end{bmatrix}$$

Solve each system by first computing the inverse of the coefficient matrix, as in Example 55.7.

22. $2x - 3y = 1$
$\quad\ x + \ y = 5$

23. $-x - \ y = 3$
$\quad\ 4x + 2y = 7$

24. $5x + \ y = 3$
$\quad\ 7x - 2y = 0$

25. $\ x + \ y + z = 1$
$\quad 2x - \ y - z = 3$
$\quad\quad\quad 5y + z = 0$

26. $2x - 2y + 3z = 0$
$\quad\ x - \ y + 2z = 5$
$\quad 2x + 2y - 3z = 1$

27. $\quad x + 2y \quad\ = 1$
$\quad -3x \quad\quad + z = 1$
$\quad\ 4x - \ y \quad\ = 3$

28. To prove that each invertible matrix has a *unique* inverse, suppose that $AB = I$ and $BA = I$ (so that B is an inverse of A) and that $AC = I$ and $CA = I$ (so that C is an inverse of A). Now give a reason for each equality that follows; together, they prove that $B = C$: $B = BI = B(AC) = (BA)C = IC = C$.

29. Verify by direct multiplication that $AA^{-1} = I$ and $A^{-1}A = I$ for A^{-1} as in Equation (55.1) and for A as in the same display.

30. If A is an invertible matrix, then A^{-1} is also

invertible. What is the inverse of A^{-1}? Justify your answer.

In Exercises 31-33, first write a linear system of equations and then solve the system by the method of Subsection D.

31. The total world gold production in 1975 was estimated to be 39 million ounces. The production in South Africa was 9 million ounces less than twice the production outside of South Africa. Find the production in South Africa and outside of South Africa. (All figures are given to the nearest million ounces.)

32. The Consumer Price Index increased 27 points from 1960 to 1970 and then increased 113 more points from 1970 to 1980. The index for 1980 was 42 points more than the sum of the indices for 1960 and 1970. Find the index in each of the three years, 1960, 1970, and 1980. (Indices have been rounded to the nearest point for simplicity.) Compute the percentage increases from 1960 to 1970 and from 1970 to 1980.

33. Energy use in the United States in 1980 was 78 quads (quadrillion BTU's). Coal and imported oil provided equal amounts of this energy, and natural gas and domestic oil also provided equal amounts. Natural gas and domestic oil together provided 10 quads more than coal and imported oil together. Nuclear and other sources provided 6 quads. Let C, G, D, and I denote the number of quads provided by coal, natural gas, domestic oil, and imported oil, respectively. Determine C, G, D, and I. (Begin by writing four linear equations in C, G, D, and I. All figures are approximate and adjusted slightly for simplicity.)

SECTION **56**
Determinants

A. SECOND- AND THIRD-ORDER DETERMINANTS

Associated with each square matrix A is a real number called the *determinant* of A, which we denote by either *det A* or $|A|$. In the language of functions, *det* is a function whose domain is the set of square matrices and whose range is the set of real numbers. This section introduces determinants of 2×2 and 3×3 matrices. Determinants of larger matrices are introduced in the section that follows.

Definition
$$\det \begin{bmatrix} a_{11} & a_{12} \\ a_{21} & a_{22} \end{bmatrix} = \begin{vmatrix} a_{11} & a_{12} \\ a_{21} & a_{22} \end{vmatrix} = a_{11}a_{22} - a_{12}a_{21}$$

Example 56.1
$$\begin{vmatrix} 12 & 4 \\ 5 & -1 \end{vmatrix} = (12)(-1) - (4)(5) = -12 - 20 = -32$$

To define determinants of 3×3 matrices we use *minors* and *cofactors*, which are defined as follows.

Definition The **(i, j)-minor** of a 3×3 matrix A, denoted M_{ij}, is the determinant of the 2×2 matrix that remains when the ith row and jth column of A have been deleted.

Example 56.2 Assume $A = \begin{bmatrix} 2 & -7 & 0 \\ 3 & -3 & -5 \\ -2 & 6 & 1 \end{bmatrix}$.

(a) To compute M_{11} we first delete the first row and first column of A:
$$\begin{bmatrix} 2 & -7 & 0 \\ 3 & -3 & -5 \\ -2 & 6 & 1 \end{bmatrix}.$$
Thus
$$M_{11} = \begin{vmatrix} -3 & -5 \\ 6 & 1 \end{vmatrix} = (-3)(1) - (-5)(6) = 27.$$

(b) To compute M_{32} we first delete the third row and second column of A:
$$\begin{bmatrix} 2 & -7 & 0 \\ 3 & -3 & -5 \\ -2 & 6 & 1 \end{bmatrix}.$$
Thus
$$M_{32} = \begin{vmatrix} 2 & 0 \\ 3 & -5 \end{vmatrix} = (2)(-5) - (0)(3) = -10.$$

Definition The **(i, j)-cofactor** of a 3×3 matrix A, denoted C_{ij}, is defined by $C_{ij} = (-1)^{i+j}M_{ij}$.

Since $(-1)^{i+j}$ is $+1$ if $i + j$ is even and -1 if $i + j$ is odd, it follows that either $C_{ij} = M_{ij}$ or $C_{ij} = -M_{ij}$, depending on whether $i + j$ is even or odd, respectively. Here is an easy way to determine the appropriate sign: the minor and cofactor that result from deleting a given row and column will have the same or opposite signs depending on whether there is a $+$ or a $-$ at the intersection for the given row and column in the following matrix.

$$\begin{bmatrix} + & - & + \\ - & + & - \\ + & - & + \end{bmatrix}$$

Example 56.3 (a) The (1, 1)-cofactor of the matrix in Example 56.2 is
$$C_{11} = +M_{11} = 27.$$
(b) The (3, 2)-cofactor of the matrix in Example 56.2 is
$$C_{32} = -M_{32} = -(-10) = 10. \qquad ■$$

Definition

$$\det \begin{bmatrix} a_{11} & a_{12} & a_{13} \\ a_{21} & a_{22} & a_{23} \\ a_{31} & a_{32} & a_{33} \end{bmatrix} = \begin{vmatrix} a_{11} & a_{12} & a_{13} \\ a_{21} & a_{22} & a_{23} \\ a_{31} & a_{32} & a_{33} \end{vmatrix}$$

$$= a_{11}C_{11} + a_{12}C_{12} + a_{13}C_{13} \qquad (56.1)$$

In words: To compute the determinant of a 3×3 matrix multiply each entry in the first row of the matrix by the corresponding cofactor and then add the results. Using minors in place of cofactors, the determinant is

$$a_{11}M_{11} - a_{12}M_{12} + a_{13}M_{13}.$$

Example 56.4

$$\begin{vmatrix} 2 & -7 & 0 \\ 3 & -3 & -5 \\ -2 & 6 & 1 \end{vmatrix} = 2 \begin{vmatrix} -3 & -5 \\ 6 & 1 \end{vmatrix} - (-7) \begin{vmatrix} 3 & -5 \\ -2 & 1 \end{vmatrix} + 0 \begin{vmatrix} 3 & -3 \\ -2 & 6 \end{vmatrix}$$

$$= 2(27) - (-7)(-7) + 0(12)$$

$$= 5 \qquad ■$$

It can be proved that the determinant of a 3×3 matrix can also be obtained by using the idea in (56.1) with any other row or column in place of the first row. Specifically:

To compute the determinant of a 3×3 matrix, multiply each entry in any row or column by the corresponding cofactor and then add the results.

Using the second column rather than the first row, for example, (56.1) becomes

$$\begin{vmatrix} a_{11} & a_{12} & a_{13} \\ a_{21} & a_{22} & a_{23} \\ a_{31} & a_{32} & a_{33} \end{vmatrix} = a_{12}C_{12} + a_{22}C_{22} + a_{32}C_{32}.$$

This is called *the expansion of the determinant by the second column.* A similar expression is used for the expansion by any other row or column.

If a 3×3 matrix has only 0's in some row or column, then we can expand the determinant by using that row or column and the final result will obviously be 0 regardless of the cofactors.

B. CRAMER'S RULE

We now look at an application of determinants in solving systems of linear equations. We begin with systems having two equations and two variables.

With the system

$$a_1 x + b_1 y = c_1 \tag{56.2}$$
$$a_2 x + b_2 y = c_2$$

we associate three determinants:

$$D = \begin{vmatrix} a_1 & b_1 \\ a_2 & b_2 \end{vmatrix}, \quad D_x = \begin{vmatrix} c_1 & b_1 \\ c_2 & b_2 \end{vmatrix}, \quad Dy = \begin{vmatrix} a_1 & c_1 \\ a_2 & c_2 \end{vmatrix}. \tag{56.3}$$

Notice that D is the determinant of the coefficient matrix, that D_x is obtained from D by replacing the coefficients of x (a_1 and a_2) by the constant terms (c_1 and c_2), and that D_y is obtained from D by replacing the coefficients of y by the constant terms. Exercise 37 indicates how to prove the following result.

Cramer's Rule (2 × 2 Case)

If $D \neq 0$ in (56.3), then the system (56.2) has the unique solution

$$x = \frac{D_x}{D}, \quad y = \frac{D_y}{D}.$$

Example 56.5   Use Cramer's Rule to solve the system

$$-2x + y = 5$$
$$x + 3y = 7.$$

Solution

$$D = \begin{vmatrix} -2 & 1 \\ 1 & 3 \end{vmatrix} = -7, \quad D_x = \begin{vmatrix} 5 & 1 \\ 7 & 3 \end{vmatrix} = 8, \quad D_y = \begin{vmatrix} -2 & 5 \\ 1 & 7 \end{vmatrix} = -19.$$

Therefore, the solution is

$$x = \frac{8}{-7} = -\frac{8}{7}, \quad y = \frac{-19}{-7} = \frac{19}{7}.$$

Cramer's Rule extends to three equations with three variables as follows. With the system

$$a_1 x + b_1 y + c_1 z = d_1$$
$$a_2 x + b_2 y + c_2 z = d_2 \tag{56.4}$$
$$a_3 x + b_3 y + c_3 z = d_3$$

we associate four determinants: the determinant of the coefficient matrix,

$$D = \begin{vmatrix} a_1 & b_1 & c_1 \\ a_2 & b_2 & c_2 \\ a_3 & b_3 & c_3 \end{vmatrix}, \tag{56.5}$$

and the three determinants D_x, D_y, and D_z, where D_x is obtained from D by replacing the coefficients of x by the constant terms, and D_y and D_z are obtained similarly.

Cramer's Rule (3 × 3 Case)

If $D \neq 0$ in (56.5), then the system (56.4) has the unique solution

$$x = \frac{D_x}{D}, \quad y = \frac{D_y}{D}, \quad z = \frac{D_z}{D}.$$

Example 56.6 Use Cramer's Rule to solve the system

$$\begin{aligned} 2x - 7y &= 1 \\ 3x - 3y - 5z &= -2 \\ -2x + 6y + z &= 0. \end{aligned}$$

Solution The determinant D was evaluated in Example 56.4. The computations for D_x, D_y, and D_z can be carried out in the same way.

$$D = \begin{vmatrix} 2 & -7 & 0 \\ 3 & -3 & -5 \\ -2 & 6 & 1 \end{vmatrix} = 5, \quad D_x = \begin{vmatrix} 1 & -7 & 0 \\ -2 & -3 & -5 \\ 0 & 6 & 1 \end{vmatrix} = 13,$$

$$D_y = \begin{vmatrix} 2 & 1 & 0 \\ 3 & -2 & -5 \\ -2 & 0 & 1 \end{vmatrix} = 3, \quad D_z = \begin{vmatrix} 2 & -7 & 1 \\ 3 & -3 & -2 \\ -2 & 6 & 0 \end{vmatrix} = 8.$$

Therefore, the solution is

$$x = \frac{13}{5}, \quad y = \frac{3}{5}, \quad z = \frac{8}{5}.$$

For a proof of Cramer's Rule in the 3 × 3 case, as well as the statements that follow, see a book on matrices or linear algebra.

In both the 2 × 2 and 3 × 3 cases Cramer's Rule covers all cases where $D \neq 0$. The cases where $D = 0$ are covered by the following statement.

Assume $D = 0$.

(a) If D_x and D_y (and D_z in the 3 × 3 case) all equal zero, then the system is dependent (has infinitely many solutions).

(b) If one or more of D_x, D_y, and D_z is nonzero, then the system is inconsistent (has no solution).

(56.6)

A linear system in which the constant terms are all 0 is said to be **homogeneous**. For a homogeneous system each of the determinants D_x and D_y (and D_z in the 3 × 3 case) will contain a column of 0's and will, therefore, have the value 0. It follows, using Cramer's Rule, that if $D \neq 0$ for a homogeneous system, then the only solution of the system is the one in which each variable equals 0. And, using (a) above, if $D = 0$ for a homogeneous system, then the system is dependent. Hence for $D = 0$ the homogeneous system has nontrivial solutions, that is, solutions in which some of the variables are not 0.

Example 56.7 For the homogeneous system

$$2x - y = 0$$
$$-7x + 2y = 0$$

$$D = \begin{vmatrix} 2 & -1 \\ -7 & 2 \end{vmatrix} = -3, \quad D_x = \begin{vmatrix} 0 & -1 \\ 0 & 2 \end{vmatrix} = 0, \quad D_y = \begin{vmatrix} 2 & 0 \\ -7 & 0 \end{vmatrix} = 0.$$

Thus $(0, 0)$ is the unique solution.

EXERCISES FOR SECTION 56

Compute the indicated minors and cofactors for the matrix

$$\begin{bmatrix} 2 & 8 & -6 \\ 4 & 1 & 5 \\ 0 & 2 & -2 \end{bmatrix}.$$

1. M_{21} **2.** M_{33}

3. M_{13} **4.** C_{21}

5. C_{33} **6.** C_{13}

7. C_{22} **8.** C_{31}

9. C_{12}

Evaluate by expanding by the first row [Equation (56.1)].

10. $\begin{vmatrix} 2 & -1 & 3 \\ 6 & 2 & -5 \\ 0 & 2 & 1 \end{vmatrix}$ **11.** $\begin{vmatrix} 2 & 1 & 0 \\ -5 & 4 & 2 \\ 1 & 0 & 6 \end{vmatrix}$

12. $\begin{vmatrix} 3 & 0 & 7 \\ -6 & -3 & -3 \\ 4 & 1 & 2 \end{vmatrix}$ **13.** $\begin{vmatrix} 1 & 4 & 9 \\ 6 & -1 & 2 \\ 2 & 3 & -1 \end{vmatrix}$

14. $\begin{vmatrix} 2 & 1 & 2 \\ 3 & -5 & -1 \\ -2 & 0 & 4 \end{vmatrix}$ **15.** $\begin{vmatrix} -2 & 8 & 3 \\ -7 & 1 & 3 \\ 0 & 3 & -4 \end{vmatrix}$

Evaluate by expanding by the second column.

16. $\begin{vmatrix} 1 & 4 & 1 \\ 3 & -1 & 5 \\ -2 & 6 & 0 \end{vmatrix}$ **17.** $\begin{vmatrix} -3 & 4 & 2 \\ 4 & -1 & 2 \\ 9 & 5 & -6 \end{vmatrix}$

18. $\begin{vmatrix} -9 & 2 & 0 \\ 7 & 1 & -5 \\ 7 & 3 & 4 \end{vmatrix}$

Evaluate by expanding by the row or column with the most 0's.

19. $\begin{vmatrix} -1 & 0 & \frac{4}{3} \\ 0 & 7 & 0 \\ 6 & \frac{1}{2} & 2 \end{vmatrix}$ **20.** $\begin{vmatrix} 4 & \frac{1}{5} & 0 \\ \frac{3}{5} & 0 & 2 \\ 3 & 0 & -1 \end{vmatrix}$

21. $\begin{vmatrix} 0 & -6 & 2 \\ -5 & \frac{1}{8} & \frac{1}{3} \\ 7 & 0 & 0 \end{vmatrix}$

Solve each system using Cramer's Rule.

22. $x - 2y = 5$ **23.** $3x + y = 0$
      $3x + y = 1$ $-x - 5y = 4$

24. $7x - y = 1$ **25.** $2u - 6v = 1$
      $-6x + y = 3$ $8u - 13v = 5$

26. $8w + z = 5$ **27.** $-r + 2s = 3$
      $2w - z = 0$ $5r - 3s = 2$

28. $x + y + z = 9$ **29.**
      $2x - y - z = 1$ $3y - z = 1$
      $3x + z = 2$ $x + z = 5$
                               $2x - y = 6$

30. $x - y + z = 3$
      $2x + y + 5z = 0$
      $x - z = 7$

Each system is either dependent or inconsistent. Use determinants and (56.6) to determine which in each case.

31. $4x - y = 5$ **32.** $6x - 4y = 10$
      $-8x + 2y = 1$ $9x - 6y = 15$

33. $12x - 8y = 3$ **34.** $x - y - z = 1$
      $-9x + 6y = 0$ $2x + 3y - z = -2$
                               $2x - 7y - 3z = 6$

35. $5x + y + z = 2$ **36.** $3x + y + z = 0$
      $x - y + z = 1$ $4x - y - 2z = 1$
      $3x + 3y - z = -1$ $-2x - 3y - 4z = 1$

37. Prove the 2×2 case of Cramer's Rule by substitution. [That is, substitute D_x/D for x and D_y/D for y in the System (56.2).]

38. Verify that

$$\begin{vmatrix} 1 & a & a^2 \\ 1 & b & b^2 \\ 1 & c & c^2 \end{vmatrix} = (b - a)(c - a)(c - b).$$

39. Prove that the equation of the line through the points (x_1, y_1) and (x_2, y_2) is

$$\begin{vmatrix} x & y & 1 \\ x_1 & y_1 & 1 \\ x_2 & y_2 & 1 \end{vmatrix} = 0.$$

40. Bach, Beethoven, and Mozart lived a total of 157 years. Bach lived 9 years more than Beethoven, and Beethoven lived 20 years more than

Mozart. How long did each live? (All ages are to the nearest year.)

41. The United States paid a total of $125 billion for imported oil in the years 1970, 1978, and 1980. The cost in 1978 was 14 times the cost in 1970, and the cost in 1980 was $4 billion less than twice the cost in 1978. Find the cost in each of the three years.

42. In the 50 years from 1927 through 1976, the Montreal Canadiens won the Stanley Cup six more times than the Toronto Maple Leafs, and the Canadiens and the Maple Leafs together won the Cup six more times than all other teams combined. How many times was the Cup won by the Canadiens, by the Maple Leafs, and by all other teams combined?

SECTION **57**

More About Determinants

A. GENERAL DEFINITION

The definition of *determinant* for 3×3 matrices can be extended in an obvious way to obtain the definition for larger matrices. First, the **(i, j)-minor** of an $n \times n$ matrix A, denoted M_{ij}, is the determinant of the $(n - 1) \times (n - 1)$ matrix that remains when the ith row and jth column of A have been deleted. The **(i, j)-cofactor** is $C_{ij} = (-1)^{i+j}M_{ij}$. The pattern for the signs relating M_{ij} and C_{ij} is given by simply extending the pattern from the 3×3 case.

$$\begin{bmatrix} + & - & + & - & \cdots \\ - & + & - & + & \cdots \\ + & - & + & - & \cdots \\ - & + & - & + & \cdots \\ \vdots & \vdots & \vdots & \vdots & \vdots \end{bmatrix}$$

Example 57.1 If

$$A = \begin{bmatrix} 8 & -2 & 1 & -4 \\ -1 & 0 & 0 & 2 \\ 3 & -3 & 7 & -6 \\ 2 & 1 & -1 & 8 \end{bmatrix}$$

then

$$M_{14} = \begin{vmatrix} -1 & 0 & 0 \\ 3 & -3 & 7 \\ 2 & 1 & -1 \end{vmatrix} = -1 \begin{vmatrix} -3 & 7 \\ 1 & -1 \end{vmatrix} - 0 \begin{vmatrix} 3 & 7 \\ 2 & -1 \end{vmatrix} + 0 \begin{vmatrix} 3 & -3 \\ 2 & 1 \end{vmatrix}$$

$$= -1(-4) = 4.$$

And

$$C_{14} = -M_{14} = -4.$$

■

Definition If A is an $n \times n$ matrix ($n \geq 2$), then

$$\det A = |A| = a_{11}C_{11} + a_{12}C_{12} + \cdots + a_{1n}C_{1n}. \qquad (57.1)$$

In terms of minors,

$$\det A = |A| = a_{11}M_{11} - a_{12}M_{12} + \cdots + (-1)^{1+n}a_{1n}M_{1n}.$$

As in the 3×3 case, we can expand a determinant by using any other row or column in place of the first row.

> To compute the determinant of an $n \times n$ matrix ($n \geq 2$), multiply each entry in any row or column by the corresponding cofactor and then add the results.

Example 57.2 To compute $|A|$ for A as in Example 57.1, we can take advantage of the two zeros in the second row; by expanding by the second row in this example we need to compute only two 3×3 cofactors.

$$\begin{vmatrix} 8 & -2 & 1 & -4 \\ -1 & 0 & 0 & 2 \\ 3 & -3 & 7 & -6 \\ 2 & 1 & -1 & 8 \end{vmatrix} = -(-1) \begin{vmatrix} -2 & 1 & -4 \\ -3 & 7 & -6 \\ 1 & -1 & 8 \end{vmatrix} + 0 \begin{vmatrix} 8 & 1 & -4 \\ 3 & 7 & -6 \\ 2 & -1 & 8 \end{vmatrix}$$

$$- 0 \begin{vmatrix} 8 & -2 & -4 \\ 3 & -3 & -6 \\ 2 & 1 & 8 \end{vmatrix} + 2 \begin{vmatrix} 8 & -2 & 1 \\ 3 & -3 & 7 \\ 2 & 1 & -1 \end{vmatrix}$$

$$= 1(-66) + 0 - 0 + 2(-57)$$

$$= -180$$

■

Exercise 26 asks you to verify that in the 2×2 case the definition in (57.1) agrees with the definition in Section 56.

B. PROPERTIES OF DETERMINANTS

In each of the following properties A is assumed to be an $n \times n$ matrix for $n \geq 2$. The examples will indicate some of the proofs for 3×3 matrices. For general proofs consult a book on matrices on linear algebra.

Property I If any row or column of A contains only 0's, then $|A| = 0$.

Example 57.3 Suppose the second row of a 3×3 matrix A contains only 0's. If we expand

$|A|$ by the second row, we get

$$\begin{vmatrix} a_{11} & a_{12} & a_{13} \\ 0 & 0 & 0 \\ a_{31} & a_{32} & a_{33} \end{vmatrix} = 0C_{21} + 0C_{22} + 0C_{23} = 0.$$

The operations in Properties II, III, and V are similar to the elementary row operations on matrices (Section 51B). Here, however, the operations are also applied to columns, and we must pay careful attention to the effect on the determinant.

Property II If B is obtained from A by multiplying each element of any row or column of A by the real number k, then $|B| = k|A|$.

Example 57.4 Suppose that each element of the third column of a 3×3 matrix A is multiplied by k to produce B. Using expansion by the third column, we have

$$|B| = \begin{vmatrix} a_{11} & a_{12} & ka_{13} \\ a_{21} & a_{22} & ka_{23} \\ a_{31} & a_{32} & ka_{33} \end{vmatrix}$$

$$= ka_{13}C_{13} + ka_{23}C_{23} + ka_{33}C_{33}$$

$$= k(a_{13}C_{13} + a_{23}C_{23} + a_{33}C_{33})$$

$$= k|A|.$$

Example 57.5 Property II implies that a common factor of the elements of any row or column can be removed by placing it in front of the determinant. For example,

$$\begin{vmatrix} 0 & 4 & -3 \\ 2 & -6 & 1 \\ -7 & 2 & 5 \end{vmatrix} = \begin{vmatrix} 0 & 2(2) & -3 \\ 2 & 2(-3) & 1 \\ -7 & 2(1) & 5 \end{vmatrix} = 2 \begin{vmatrix} 0 & 2 & -3 \\ 2 & -3 & 1 \\ -7 & 1 & 5 \end{vmatrix}.$$

Property III If B is obtained from A by interchanging two rows (or two columns) of A, then $|B| = -|A|$.

Example 57.6 Exercises 27 and 28 ask you to verify these special cases.

$$\begin{vmatrix} a_1 & b_1 & c_1 \\ a_2 & b_2 & c_2 \\ a_3 & b_3 & c_3 \end{vmatrix} = - \begin{vmatrix} a_3 & b_3 & c_3 \\ a_2 & b_2 & c_2 \\ a_1 & b_1 & c_1 \end{vmatrix} \quad \text{Interchange rows 1 and 3.}$$

$$\begin{vmatrix} a_1 & b_1 & c_1 \\ a_2 & b_2 & c_2 \\ a_3 & b_3 & c_3 \end{vmatrix} = - \begin{vmatrix} b_1 & a_1 & c_1 \\ b_2 & a_2 & c_2 \\ b_3 & a_3 & c_3 \end{vmatrix} \quad \text{Interchange columns 1 and 2.}$$

Property IV If two rows (or two columns) of A are identical, then $|A| = 0$.

Example 57.7 Exercise 29 asks you to verify that

$$\begin{vmatrix} a_1 & b_1 & c_1 \\ a_1 & b_1 & c_1 \\ a_3 & b_3 & c_3 \end{vmatrix} = 0.$$

Property V If B is obtained from A by adding k times one row (column) of A to another row (column) of A, then $|A| = |B|$.

Example 57.8 Exercise 30 asks you to use Properties II and IV to verify the following special case.

$$\begin{vmatrix} a_1 & b_1 & c_1 \\ a_2 & b_2 & c_2 \\ a_3 & b_3 & c_3 \end{vmatrix} = \begin{vmatrix} a_1 & b_1 + ka_1 & c_1 \\ a_2 & b_2 + ka_2 & c_2 \\ a_3 & b_3 + ka_3 & c_3 \end{vmatrix}$$

C. CRAMER'S RULE

Cramer's Rule, stated in Section 56 for systems with two or three variables, can also be extended to systems with more than three variables. For example, if D, the determinant of the coefficient matrix, is nonzero for the system

$$a_1 w + b_1 x + c_1 y + d_1 z = k_1$$

$$a_2 w + b_2 x + c_2 y + d_2 z = k_2$$

$$a_3 w + b_3 x + c_3 y + d_3 z = k_3$$

$$a_4 w + b_4 x + c_4 y + d_4 z = k_4,$$

then the system has a unique solution which is given by

$$w = \frac{D_w}{D}, \quad x = \frac{D_x}{D}, \quad y = \frac{D_y}{D}, \quad z = \frac{D_z}{D},$$

where D_w is obtained from D by replacing the coefficients of w by the constant terms, and D_x, D_y, and D_z are obtained similarly. In general this method of solving a system is not as efficient as the method using augmented matrices and elementary row operations (Section 51).

EXERCISES FOR SECTION 57

Compute the indicated minors and cofactors for the matrix

$$\begin{bmatrix} 8 & -2 & 3 & 7 \\ 1 & 6 & 1 & 0 \\ 5 & 0 & -1 & 0 \\ 4 & 1 & 9 & 5 \end{bmatrix}$$

1. M_{41} **2.** M_{33}

3. M_{12} **4.** C_{41}

5. C_{33} **6.** C_{12}

7. C_{22} **8.** C_{34}

9. C_{13}

Evaluate by expanding by a row or column with the maximum number of 0's.

10. $\begin{vmatrix} 4 & 0 & -1 & 2 \\ 0 & 3 & 1 & 0 \\ 5 & -2 & 6 & 7 \\ 1 & 2 & -1 & 1 \end{vmatrix}$

11. $\begin{vmatrix} 0 & -4 & 3 & 2 \\ 2 & -3 & 0 & 1 \\ 0 & 4 & 0 & 0 \\ 5 & 7 & 6 & -2 \end{vmatrix}$

12. $\begin{vmatrix} 4 & -1 & 7 & -3 \\ 0 & -1 & 6 & 4 \\ 0 & 3 & 7 & 2 \\ 1 & 2 & -2 & 5 \end{vmatrix}$

13. $\begin{vmatrix} 0 & -4 & 2 & 2 & 6 \\ 1 & 3 & -2 & 0 & 1 \\ 0 & 1 & 7 & 0 & 2 \\ 4 & -5 & -3 & 1 & 0 \\ 0 & 1 & 3 & -1 & 0 \end{vmatrix}$

14. $\begin{vmatrix} 1 & 0 & 0 & 2 & 3 \\ 4 & 5 & 0 & 0 & 6 \\ 7 & 8 & 9 & 0 & 0 \\ 0 & 0 & -6 & -5 & -4 \\ -3 & 0 & 0 & -2 & -1 \end{vmatrix}$

15. $\begin{vmatrix} 1 & 0 & -1 & 0 & 1 \\ 0 & -1 & 0 & -2 & 0 \\ -1 & 0 & 2 & 0 & -3 \\ 0 & 1 & 0 & 1 & 0 \\ 1 & 0 & 1 & 0 & 1 \end{vmatrix}$

Each equation in Exercises 16–21 is true because of one of the Properties I–V in Subsection B. Provide the reason in each case.

16. $\begin{vmatrix} 2 & 0 & 10 \\ 1 & -3 & 7 \\ 5 & 4 & -6 \end{vmatrix} = -\begin{vmatrix} 2 & 0 & 10 \\ 5 & 4 & -6 \\ 1 & -3 & 7 \end{vmatrix}$

17. $\begin{vmatrix} 12 & 0 & 6 \\ -7 & 4 & 11 \\ 12 & 0 & 6 \end{vmatrix} = 0$

18. $\begin{vmatrix} 6 & -1 & 0 \\ 3 & 2 & 0 \\ 4 & 1 & 1 \end{vmatrix} = \begin{vmatrix} 6 & -1 & 0 \\ 11 & 4 & 2 \\ 4 & 1 & 1 \end{vmatrix}$

19. $\begin{vmatrix} 19 & 7 & 4 \\ -1 & 1 & 1 \\ 0 & 0 & 0 \end{vmatrix} = 0$

20. $\begin{vmatrix} 4 & 7 & 0 \\ -6 & 2 & 5 \\ 10 & 8 & -3 \end{vmatrix} = 2\begin{vmatrix} 2 & 7 & 0 \\ -3 & 2 & 5 \\ 5 & 8 & -3 \end{vmatrix}$

21. $-3\begin{vmatrix} 1 & 0 & 2 \\ -1 & 4 & 2 \\ 7 & 8 & -3 \end{vmatrix} = \begin{vmatrix} 1 & 0 & 2 \\ 3 & -12 & -6 \\ 7 & 8 & -3 \end{vmatrix}$

Solve each system using Cramer's Rule.

22.
$$\begin{aligned}
w - 3x + y & = 2 \\
w \quad\; + 2y - 4z & = 2 \\
6x \quad\; + 2z & = 3 \\
3w \quad - 2y & = 4
\end{aligned}$$

23.
$$\begin{aligned}
w + x - 3y + 4z & = -5 \\
3x \quad - 2z & = 1 \\
2w + 2x + 3y & = 0 \\
w - x \quad\; 2z & = -2
\end{aligned}$$

24.
$$\begin{aligned}
-s + 4t - 5u & = 4 \\
r + s - 8t & = -6 \\
-r \quad\; + 10u & = 0 \\
2s + 4t & = -1
\end{aligned}$$

25. Show that
$$\begin{vmatrix} a_{11} & a_{12} & a_{13} & a_{14} \\ a_{21} & a_{22} & a_{23} & a_{24} \\ 0 & 0 & a_{33} & a_{34} \\ 0 & 0 & a_{43} & a_{44} \end{vmatrix} = \begin{vmatrix} a_{11} & a_{12} \\ a_{21} & a_{22} \end{vmatrix} \cdot \begin{vmatrix} a_{33} & a_{34} \\ a_{43} & a_{44} \end{vmatrix}$$

26. Show that with $n = 2$ the definition in Equation (57.1) agrees with the definition of the determinant of a 2×2 matrix at the beginning of Section 56, if the determinant of a 1×1 matrix is equal to its entry.

27. Verify the first equation in Example 57.6 by expanding both sides.

28. Verify the second equation in Example 57.6 by expanding both sides.

29. Verify the equation in Example 57.7 by expanding the left side.

30. Verify the equation in Example 57.8 by expanding both sides. (Properties II and IV will help.)

REVIEW EXERCISES FOR CHAPTER XII

Perform the indicated operations.

1. $\begin{bmatrix} -3 & 14 \\ 0 & 5 \end{bmatrix} + \begin{bmatrix} 5 & 9 \\ -2 & -2 \end{bmatrix}$

2. $\begin{bmatrix} 0 & 6 \\ -5 & -2 \\ 9 & 11 \end{bmatrix} + \begin{bmatrix} 0 & -3 \\ 2 & 1 \\ 4 & 7 \end{bmatrix}$

3. $\begin{bmatrix} 4 & 1 & 0 \\ -1 & 3 & 5 \end{bmatrix} \begin{bmatrix} -1 & -2 \\ 5 & 6 \\ 8 & 7 \end{bmatrix}$

4. $\begin{bmatrix} 1 & -1 & 0 \\ 4 & 6 & 2 \\ -1 & -2 & 5 \end{bmatrix} \begin{bmatrix} 2 & -1 & 5 \\ 2 & 4 & -3 \\ 1 & -2 & 0 \end{bmatrix}$

Compute the inverse of each matrix by using the determinant [Equation (55.1)].

5. $\begin{bmatrix} 4 & 1 \\ -5 & 2 \end{bmatrix}$

6. $\begin{bmatrix} 1 & 1 \\ 2 & 3 \\ 1 & 1 \\ 4 & 5 \end{bmatrix}$

Use elementary row operations to determine whether each matrix is invertible, and if the matrix is invertible give its inverse.

7. $\begin{bmatrix} 1 & 0 & 2 \\ 3 & 2 & 7 \\ -1 & 2 & 6 \end{bmatrix}$

8. $\begin{bmatrix} 4 & 1 & 0 \\ -2 & 3 & 1 \\ 2 & -10 & 3 \end{bmatrix}$

9. $\begin{bmatrix} 3 & 1 & 3 \\ 4 & 0 & -2 \\ 1 & 3 & 13 \end{bmatrix}$ **10.** $\begin{bmatrix} 1 & 2 & 4 \\ 5 & 0 & -3 \\ -6 & 0 & -1 \end{bmatrix}$

Solve each system by first computing the inverse of the coefficient matrix (that is, use that the solution of $AX = B$ *is* $X = A^{-1}B$*).*

11. $\begin{aligned} 0.5x + y &= 2 \\ -0.1x + 3y &= 1.5 \end{aligned}$ **12.** $\begin{aligned} 2x - y - z &= 0 \\ x + y + z &= 4 \\ x - y \quad\quad &= 1 \end{aligned}$

Evaluate each determinant.

13. $\begin{vmatrix} 2 & -5 & 1 \\ 4 & 0 & 2 \\ -7 & 1 & 5 \end{vmatrix}$ **14.** $\begin{vmatrix} -\dfrac{1}{2} & 1 & \dfrac{1}{2} \\[2mm] 2 & \dfrac{1}{4} & -2 \\[2mm] \dfrac{1}{3} & 1 & -\dfrac{1}{3} \end{vmatrix}$

15. $\begin{vmatrix} 0 & -1 & 1 & 0 \\ 2 & 5 & 0 & -3 \\ 1 & 0 & -2 & 1 \\ 0 & 2 & -1 & 1 \end{vmatrix}$ **16.** $\begin{vmatrix} a & 0 & b & 0 \\ 0 & b & 0 & c \\ b & 0 & c & 0 \\ 0 & c & 0 & d \end{vmatrix}$

Use determinants to determine whether each system has a unique solution, is dependent, or is inconsistent. If the system has a unique solution, find it using Cramer's Rule.

17. $\begin{aligned} 4x + y &= 3 \\ -x + y &= 1 \end{aligned}$ **18.** $\begin{aligned} x - 2y &= 3 \\ -2x + 4y &= 6 \end{aligned}$

19. $\begin{aligned} x - y - z &= 1 \\ 2x - 2y - z &= 3 \\ 4x - 4y - 3z &= 5 \end{aligned}$ **20.** $\begin{aligned} 2x \quad\quad + z &= 0 \\ x - y - z &= 1 \\ 4x + y + z &= 5 \end{aligned}$

CHAPTER XIII

MATHEMATICAL INDUCTION AND SEQUENCES

The first section of this chapter is devoted to an important method of proof for many statements that involve positive integers. The other sections treat ideas about sequences and sums that arise often throughout mathematics and its applications. Section 61, on the Binomial Theorem, can be covered before the other sections if you are willing to delay the proof of the theorem.

SECTION **58**
Mathematical Induction

A. SEQUENCES AND SERIES

A collection of real numbers arranged so that there is a first, a second, a third, and so on, is called a **sequence** (or, if we need to be more precise, a **sequence of real numbers**). The successive numbers in the sequence are called its **terms:** the **first term,** the **second term,** the **third term,** and so on. A sequence with a last term is said to be **finite;** if there is no last term the sequence is said to be **infinite.**

Example 58.1 (a) $5, 10, 15, \ldots, 100$

(b) $\dfrac{1}{1}, \dfrac{1}{3}, \dfrac{1}{5}, \dfrac{1}{7}, \ldots$

(c) $1, -1, 1, -1, \ldots$

(d) $4, 4, 4, 4, \ldots$

Sequence (a) is finite. The other sequences are infinite. Sequences (c) and (d) make the point that different terms in the same sequence can be equal. ◼

To represent general sequences we use notation like

$$a_1, a_2, \ldots, a_m \tag{58.1}$$

(for finite sequences) and

$$a_1, a_2, \ldots, a_n, \ldots \tag{58.2}$$

(for infinite sequences).

A sequence is actually a function. The sequence in (58.1), for example, can be thought of as a function whose domain is the set $\{1, 2, \ldots, m\}$; the output of the function for the input k is a_k. The sequence in (58.2) can be thought of as a function whose domain is the set of all positive integers; and again, the output is a_k for the input k.

For many sequences a rule or formula can be given for a_n. It is best to give such a rule or formula whenever possible; that eliminates the possibility of uncertainty or ambiguity about the terms that are not listed.

Example 58.2 Here are the sequences from Example 58.1 with the nth term specified in each case. If you replace n by 1 in the nth term, you will get the first term, and so on for $n = 2, 3, \ldots$.

(a) $5, 10, 15, \ldots, 5n, \ldots, 100$

(b) $\dfrac{1}{1}, \dfrac{1}{3}, \dfrac{1}{5}, \ldots, \dfrac{1}{2n-1}, \ldots$

(c) $1, -1, 1, \ldots, (-1)^{n+1}, \ldots$

(d) $4, 4, 4, \ldots, 4, \ldots$

Part (b) illustrates that the nth positive odd integer is $2n - 1$. Part (c) illustrates that to introduce alternating signs we can use $(-1)^{n+1}$. [To alternate signs starting with -1, use $(-1)^n$.]

Example 58.3 Write the first seven terms of the sequence defined by

$$a_n = \begin{cases} n^2 & \text{if } n \text{ is odd} \\ 1 & \text{if } n \text{ is even.} \end{cases}$$

Solution 1, 1, 9, 1, 25, 1, 49.

Expressions like

$$a_1 + a_2 + a_3 + \cdots + a_n \tag{58.3}$$

and

$$a_1 + a_2 + a_3 + \cdots + a_n + \cdots, \tag{58.4}$$

obtained from adding the terms of a sequence, are called **series.** The series in this section will be **finite,** like (58.3). In Section 60 we shall also consider **infinite series,** like (58.4).

B. PRINCIPLE OF MATHEMATICAL INDUCTION

In Example 58.4 we shall prove that if n is a positive integer, then

$$1 + 2 + 3 + \cdots + n = \frac{n(n + 1)}{2}. \tag{58.5}$$

We can verify (58.5) for small values of n by direct computation. Here are two examples.

$n = 5$ case:

$$1 + 2 + 3 + 4 + 5 = 15 \quad \text{and} \quad \frac{5 \cdot 6}{2} = 15.$$

$n = 10$ case:

$$1 + 2 + 3 + \cdots + 10 = 55 \quad \text{and} \quad \frac{10 \cdot 11}{2} = 55.$$

To prove (58.5) once and for all, for *every* value of n, we need a different idea. The Principle of Mathematical Induction provides such an idea; it applies not only to (58.5) but to a very broad class of similar statements.

Principle of Mathematical Induction

For each positive integer n, let P_n represent a statement depending on n. If

(a) P_1 is true, and

(b) the truth of P_k implies the truth of P_{k+1} for each positive integer k,

then P_n is true for every positive integer n.

The Principle of Mathematical Induction can be visualized in terms of an infinite sequence of dominoes, originally standing on edge (Figure 58.1): the fall of the nth domino corresponds to the truth of P_n. Part (a) guarantees that the first domino will fall, and part (b) guarantees that if any particular domino [the kth] falls, then the next one [the $(k + 1)$st] will also fall; it follows that every domino will fall.

To apply the Principle of Mathematical Induction we must verify both parts, (a) and (b). Notice that to verify (b) we must prove that *if* P_k is true, *then* P_{k+1} is true; that is, we must establish P_{k+1} based on the assumption of P_k.

1   2   3

FIGURE 58.1

k   k + 1   k + 2   k + 3

C. EXAMPLES

Example 58.4 Prove that Equation (58.5) is true for every positive integer n.

Proof Let P_n represent Equation (58.5).

(a) P_1 is true because

$$1 = \frac{1 \cdot 2}{2}.$$

(b) Assume P_k, that is,

$$1 + 2 + 3 + \cdots + k = \frac{k(k + 1)}{2}. \tag{58.6}$$

We must prove P_{k+1}, that is, (58.5) with $k + 1$ in place of n:

$$1 + 2 + 3 + \cdots + (k + 1) = \frac{(k + 1)(k + 2)}{2}. \tag{58.7}$$

To emphasize the strategy involved, notice that we are to *prove* (58.7), and we may use (58.6) in the process. If we begin with (58.6) and add $k + 1$ to both sides, we obtain

$$1 + 2 + 3 + \cdots + k + (k + 1) = \frac{k(k + 1)}{2} + (k + 1)$$

$$1 + 2 + 3 + \cdots + k + (k + 1) = \frac{k(k + 1) + 2(k + 1)}{2}$$

$$1 + 2 + 3 + \cdots + k + (k + 1) = \frac{(k + 1)(k + 2)}{2}.$$

The last equation is the same as (58.7), which is what we were to prove. ☐

Example 58.5 Prove that

$$\frac{1}{1 \cdot 3} + \frac{1}{3 \cdot 5} + \frac{1}{5 \cdot 7} + \cdots + \frac{1}{(2n-1)(2n+1)} = \frac{n}{2n+1} \quad (58.8)$$

for every positive integer n.

Remark To help clarify the meaning of (58.8), here is the special case $n = 5$.

$$\frac{1}{1 \cdot 3} + \frac{1}{3 \cdot 5} + \frac{1}{5 \cdot 7} + \frac{1}{7 \cdot 9} + \frac{1}{9 \cdot 11} = \frac{5}{11}$$

(Even with a calculator the direct computation of the left-hand side would be tedious, and the answer would only be a decimal approximation of the exact sum $\frac{5}{11}$. More important, a calculator could not provide the general formula $n/(2n+1)$ for the sum.)

Proof Let P_n represent Equation (58.8).

(a) P_1 is true because

$$\frac{1}{1 \cdot 3} = \frac{1}{2 \cdot 1 + 1}.$$

(b) Assume P_k, that is,

$$\frac{1}{1 \cdot 3} + \frac{1}{3 \cdot 5} + \frac{1}{5 \cdot 7} + \cdots + \frac{1}{(2k-1)(2k+1)} = \frac{k}{2k+1}. \quad (58.9)$$

To find P_{k+1}, which is what we must prove from P_k, substitute $k + 1$ for n in (58.8) and simplify; the result is

$$\frac{1}{1 \cdot 3} + \frac{1}{3 \cdot 5} + \frac{1}{5 \cdot 7} + \cdots + \frac{1}{(2k+1)(2k+3)} = \frac{k+1}{2k+3}. \quad (58.10)$$

To repeat, (58.10) is what we are to prove, and we may use (58.9) in the process. We add

$$\frac{1}{(2k+1)(2k+3)}$$

to both sides of (58.9) and simplify the right-hand side.

$$\frac{1}{1 \cdot 3} + \frac{1}{3 \cdot 5} + \frac{1}{5 \cdot 7} + \cdots + \frac{1}{(2k+1)(2k+3)}$$

$$= \frac{k}{2k+1} + \frac{1}{(2k+1)(2k+3)}$$

$$= \frac{k(2k+3) + 1}{(2k+1)(2k+3)}$$

$$= \frac{2k^2 + 3k + 1}{(2k+1)(2k+3)}$$

$$= \frac{(2k+1)(k+1)}{(2k+1)(2k+3)}$$

$$= \frac{k+1}{2k+3} \qquad \qquad \square$$

The next example illustrates a slight variation on the Principle of Mathematical Induction. In place of beginning with P_1 we begin with P_5. That is, we show that

(a) P_5 is true, and

(b) the truth of P_k implies the truth of P_{k+1} for each integer $k \geq 5$;

it will follow that P_n is true for every integer $n \geq 5$. (This corresponds to starting with the fifth domino rather than the first domino in Figure 58.1.) In other problems the same idea can be applied with 5 replaced by another appropriate starting number.

Example 58.6 Prove that $2^n > n^2$ for every integer n such that $n \geq 5$.

Proof Let P_n represent $2^n > n^2$.

(a) P_5 is true because $2^5 > 5^2$, that is, $32 > 25$.

(b) Assume that $k \geq 5$ and that P_k is true, that is,

$$2^k > k^2. \tag{58.11}$$

We must prove P_{k+1}, that is,

$$2^{k+1} > (k + 1)^2. \tag{58.12}$$

If we multiply both sides of (58.11) by 2, we have

$$2^{k+1} > 2k^2.$$

To get (58.12), it will suffice to prove that

$$2k^2 > (k + 1)^2, \tag{58.13}$$

for then we shall have $2^{k+1} > 2k^2 > (k + 1)^2$, from which $2^{k+1} > (k + 1)^2$. Inequality (58.13) is equivalent to

$$2k^2 > k^2 + 2k + 1$$

$$k^2 - 2k > 1$$

$$k(k - 2) > 1. \tag{58.14}$$

The latter inequality is satisfied because $k \geq 5$. This completes the proof. □

Remark Notice that (58.14) is true for $k \geq 3$. Thus in part (b) we have actually shown that P_k implies P_{k+1} for $k \geq 3$. However, that does not prove P_n for $n \geq 3$, because P_3 is not true (P_3 would say $2^3 > 3^2$, which is false). What about P_1? P_2? P_4?

The other sections of this chapter will give more proofs by mathematical induction.

EXERCISES FOR SECTION 58

Write the first five terms and the $(n + 1)$st term of the sequence whose nth term is a_n.

1. $a_n = 5n$

2. $a_n = n^3$

3. $a_n = n/(n + 1)$

4. $a_n = -2$

5. $a_n = (-2)^n$

6. $a_n = (-1)^{n+1}[1 + (1/n)]$

Use mathematical induction to prove each statement for every positive integer n *or for those* n *satisfying the stated restriction.*

7. $1 + 2 + 2^2 + \cdots + 2^n = 2^{n+1} - 1$

8. $1^2 + 2^2 + 3^2 + \cdots + n^2$
$= \dfrac{n(n+1)(2n+1)}{6}$

9. $5 + 10 + 15 + \cdots + 5n = \dfrac{5n(n+1)}{2}$

10. $1 + 3 + 5 + \cdots + (2n+1) = n^2$

11. $1^3 + 2^3 + \cdots + n^3 = \left(\dfrac{n(n+1)}{2}\right)^2$

12. $1^3 + 3^3 + \cdots + (2n-1)^3 = n^2(2n^2 - 1)$

13. $(1 \cdot 2) + (2 \cdot 3) + (3 \cdot 4) + \cdots + n(n+1)$
$= \dfrac{n(n+1)(n+2)}{3}$

14. $\dfrac{1}{1 \cdot 2} + \dfrac{1}{2 \cdot 3} + \dfrac{1}{3 \cdot 4} + \cdots + \dfrac{1}{n(n+1)}$
$= \dfrac{n}{n+1}$

15. $\dfrac{1}{1 \cdot 2 \cdot 3} + \dfrac{1}{2 \cdot 3 \cdot 4} + \dfrac{1}{3 \cdot 4 \cdot 5} + \cdots$
$+ \dfrac{1}{n(n+1)(n+2)} = \dfrac{n(n+3)}{4(n+1)(n+2)}$

16. $\left(1 + \dfrac{1}{1}\right)\left(1 + \dfrac{1}{2}\right)\left(1 + \dfrac{1}{3}\right) \cdots \left(1 + \dfrac{1}{n}\right)$
$= n + 1$

17. $\left(1 - \dfrac{1}{4}\right)\left(1 - \dfrac{1}{9}\right)\left(1 - \dfrac{1}{16}\right) \cdots \left(1 - \dfrac{1}{n^2}\right)$
$= \dfrac{n+1}{2n}$

18. $\dfrac{1}{2} + \dfrac{2}{2^2} + \dfrac{3}{2^3} + \cdots + \dfrac{n}{2^n} = 2 - \dfrac{n+2}{2^n}$

19. $2n \le 2^n$

20. $n^3 < 3^n$ for $n \ge 4$.

21. $\left(1 + \dfrac{1}{2}\right)^n \ge 1 + (1/2^n)$

22. If $a > 1$, then $a^n > 1$.

23. If $0 < a < 1$, then $a^n < 1$.

24. $\log(a_1 a_2 \cdots a_n) = \log a_1 + \log a_2 + \cdots + \log a_n$

25. The sum of the interior angles of an n-sided convex polygon is $(n - 2) \cdot 180°$ ($n \ge 3$). (A polygon is *convex* if all of the points on a line segment are inside the polygon whenever the endpoints of the segment are inside the polygon. You may assume that the sum of the interior angles of a triangle is 180°.)

26. The number of lines determined by n points, no three of which are collinear, equals $\dfrac{1}{2} n(n - 1)$ ($n \ge 2$).

27. The number of diagonals of an n-sided convex polygon is $n(n - 3)/2$ for $n \ge 3$. (See Exercise 25 for the definition of a convex polygon.)

SECTION 59

Arithmetic Sequences and Series

A. SUMS OF ARITHMETIC SERIES

Definition A finite or infinite sequence $a_1, a_2, a_3, \ldots$ is called an **arithmetic sequence** if there is a constant number d (the **common difference**) such that

$$a_{k+1} - a_k = d \qquad (59.1)$$

for every k. (Arithmetic sequences are sometimes called **arithmetic progressions.**)

By writing (59.1) in the form $a_{k+1} = a_k + d$, we see that each term after the first can be obtained by adding d to the preceding term. Once the first term, the common difference, and the number of terms have been specified, the sequence is completely determined.

Example 59.1 A six-term arithmetic sequence has first term 4 and common difference 5. Write the sequence.

Solution 4, 9, 14, 19, 24, 29.

If a_n denotes the nth term of an arithmetic sequence with first term a_1 and common difference d, then

$$a_n = a_1 + (n - 1)d. \qquad (59.2)$$

Proof We let P_n represent (59.2) and use mathematical induction.

(a) P_1 is true because

$$a_1 = a_1 + (1 - 1)d.$$

(b) Assume P_k, that is,

$$a_k = a_1 + (k - 1)d.$$

Then

$$a_{k+1} = a_k + d \qquad \text{by (59.1)}$$

$$= a_1 + (k - 1)d + d \qquad \text{by } P_k$$

$$= a_1 + kd.$$

The equation $a_{k+1} = a_1 + kd$ is P_{k+1}, so the proof is complete. $\square$

Example 59.2 Write the first three terms and the 20th term of the arithmetic sequence with first term 5 and common difference -3.

Solution Using $a_{k+1} = a_k + d$, we have

$$a_1 = 5$$

$$a_2 = 5 + (-3) = 2$$

$$a_3 = 2 + (-3) = -1.$$

From (59.2), we have

$$a_{20} = a_1 + 19d = 5 + 19(-3) = -52.$$

The next formulas give the sum of an **arithmetic series,** that is, a series whose terms form an arithmetic sequence.

If S_n denotes the sum of the first n terms of an arithmetic sequence with first term a_1 and common difference d, then

$$S_n = \frac{n}{2} [2a_1 + (n - 1)d] \tag{59.3}$$

and

$$S_n = \frac{n}{2} (a_1 + a_n). \tag{59.4}$$

Proof We shall prove (59.3) by mathematical induction. Then (59.4) will follow from (59.2) and (59.3) (Exercise 31). Let P_n denote (59.3).

(a) P_1 is true because

$$S_1 = a_1 \quad \text{and} \quad \frac{1}{2} [2a_1 + (1 - 1)d] = a_1.$$

(b) Assume P_k, that is,

$$S_k = \frac{k}{2} [2a_1 + (k - 1)d]. \tag{59.5}$$

Because the sum of the first $k + 1$ terms equals the sum of the first k terms added to the $(k + 1)$st term, we can use (59.5) and $a_{k+1} = a_1 + kd$ [from Equation (59.2)] to write

$$S_{k+1} = S_k + a_{k+1}$$

$$S_{k+1} = \frac{k}{2} [2a_1 + (k - 1)d] + a_1 + kd$$

$$S_{k+1} = \frac{k + 1}{2} (2a_1 + kd). \tag{59.6}$$

(Exercise 32 asks you to verify that the right-hand sides of the last two equations are equal.) Equation (59.6) is P_{k+1}, so the proof is complete. □

Example 59.3 Find the sum of the first 20 terms of the arithmetic sequence with first term 5 and common difference -3.

First solution Use (59.3) with $a_1 = 5$, $d = -3$, and $n = 20$:

$$S_{20} = \frac{20}{2} [2 \cdot 5 + 19(-3)] = -470.$$

Second solution Use (59.4) with $a_1 = 5$, $n = 20$, and $a_{20} = -52$ (from Example 59.2):

$$S_{20} = \frac{20}{2} (5 - 52) = -470. \qquad ■$$

Example 59.4 Find the sum of the positive integers that are multiples of 3 and less than 100.

Solution We must sum the series

$$3 + 6 + 9 + \cdots + 99.$$

Use (59.4) with $a_1 = 3$, $n = 33$, and $a_{33} = 99$:

$$S_{33} = \frac{33}{2}(3 + 99) = 1683.$$

B. SUMMATION NOTATION

The notations

$$\sum_{k=1}^{n} a_k \quad \text{and} \quad \sum_{k=1}^{\infty} a_k \tag{59.7}$$

are used to represent the series

$$a_1 + a_2 + a_3 + \cdots + a_n$$

and

$$a_1 + a_2 + a_3 + \cdots + a_n + \cdots,$$

respectively. (The symbol Σ is a capital Greek *sigma*.) The variable k in (59.7) is called the **index of summation.** The equation $k = 1$ below Σ indicates that the first term is to be a_1. To begin with a_2 rather than a_1, write $k = 2$ below Σ; similarly for other possibilities. An n above Σ indicates that the last term is to be a_n. The symbol ∞ (read *infinity*) is used above Σ to indicate that there is to be no last term. Always begin with the smallest indicated value for the index of summation and repeatedly increase by 1 until reaching the largest indicated value (if there is one). If there is a formula for the kth term, we can write it in place of a_k.

Example 59.5 (a) $\displaystyle\sum_{k=1}^{5} a_k = a_1 + a_2 + a_3 + a_4 + a_5$

(b) $\displaystyle\sum_{k=4}^{\infty} a_k = a_4 + a_5 + a_6 + \cdots + a_k + \cdots$

(c) $\displaystyle\sum_{k=0}^{10} (-1)^k(2k + 1) = 1 - 3 + 5 - 7 + \cdots - 19 + 21$

Example 59.6 Changing the index of summation does not change the series.

(a) $\displaystyle\sum_{k=1}^{n} k, \quad \sum_{j=1}^{n} j, \quad \text{and} \quad \sum_{t=1}^{n} t$ all represent the series

$$1 + 2 + 3 + \cdots + n.$$

(b) $\displaystyle\sum_{k=1}^{5} 3, \quad \sum_{j=1}^{5} 3, \quad \text{and} \quad \sum_{t=1}^{5} 3$ all represent the series

$$3 + 3 + 3 + 3 + 3.$$

(There are five terms, and each term is 3.)

EXERCISES FOR SECTION 59

Write the arithmetic sequence having the given first term a_1, common difference d, and number of terms n.

1. $a_1 = 5, d = 3, n = 5$

2. $a_1 = -10, d = 15, n = 6$

3. $a_1 = 40, d = -12, n = 5$

Write the first three terms and the 10th term of the arithmetic sequence having the given first term a_1 and common difference d.

4. $a_1 = 12, d = 0.5$

5. $a_1 = \dfrac{3}{2}, d = \dfrac{1}{4}$

6. $a_1 = 7, d = -1.5$

Find the first three terms and the sum of the arithmetic series having the given first term a_1, common difference d, and the number of terms n.

7. $a_1 = 1, d = 7, n = 11$

8. $a_1 = 0, d = -2.1, n = 10$

9. $a_1 = \dfrac{1}{4}, d = \dfrac{1}{8}, n = 20$

10. $a_1 = \dfrac{7}{3}, d = -\dfrac{1}{3}, n = 15$

11. $a_1 = 120, d = 5, n = 60$

12. $a_1 = 100, d = -\dfrac{5}{2}, n = 200$

13. (a) Solve Equation (59.2) for a_1.

   (b) Use part (a) to find the first term of a 25-term arithmetic sequence having last term 20 and common difference $-\frac{1}{3}$.

14. (a) Solve Equation (59.2) for d.

   (b) Use part (a) to find the common difference for a 15-term arithmetic sequence having first term 4 and last term 25.

15. (a) Solve Equation (59.2) for n.

   (b) Use part (a) to find the number of terms in an arithmetic sequence having first term 10, last term -60, and common difference $-\frac{1}{5}$.

Find the sum of each series. You may want to use Equation (59.4) and Exercise 15(a).

16. $6 + 10 + 14 + \cdots + 102$

17. $1 - 5 - 11 - 17 - \cdots - 71$

18. $1 + 1.05 + 1.10 + \cdots + 1.95$

Write each sum without using summation notation. (Write all of the terms; do not compute their sum.)

19. $\displaystyle\sum_{k=1}^{5} b_k$

20. $\displaystyle\sum_{k=1}^{4} 2^k$

21. $\displaystyle\sum_{k=1}^{6} (k + 3)$

22. $\displaystyle\sum_{j=3}^{8} (-1)^j 2j$

23. $\displaystyle\sum_{j=5}^{10} \dfrac{1}{j+1}$

24. $\displaystyle\sum_{j=2}^{6} j^{-1}$

Represent each series using summation (Σ) notation.

25. $3 + 3^2 + 3^3 + \cdots + 3^{10}$

26. $11 - 12 + 13 - 14 + \cdots - 30 + 31$

27. $1 \cdot 5 + 2 \cdot 5^2 + 3 \cdot 5^3 + \cdots + 20 \cdot 5^{20}$

Each series in Exercises 28–30 is an arithmetic series. Find its sum.

28. $\displaystyle\sum_{k=1}^{n} (2k + 3)$

29. $\displaystyle\sum_{k=1}^{n} 5(3k - 1)$

30. $\displaystyle\sum_{k=1}^{n} (3k + k\sqrt{2})$

31. Use Equations (59.2) and (59.3) to prove (59.4).

32. Verify that Equation (59.6) follows from the equation that precedes it.

33. Use an equation in this section to help justify the following statement: The sum of an arithmetic series with n terms is n times the average of the first and last terms.

34. To raise money for a charity, each of the 50 members of a club agrees to draw a ticket at random from a box containing tickets numbered from 1 through 50, and then contribute an amount of dollars equal to the number on the ticket. (No two members can get the same number.) Compute the total amount the club will contribute.

35. A small auditorium has 20 rows of seats. The first row contains 15 seats and each successive row contains one seat more than the previous row. How many seats are there in all?

36. If you save $0.50 today, $1.00 tomorrow, $1.50 the next day, and so on through 30 days, what will your total savings be?

37. A tank originally contains 1000 gallons of water.

Ten gallons are withdrawn one day, 10.5 gallons the next, 11 gallons the next, and so on, so that each day one-half gallon more is withdrawn than the previous day. (a) How much water will remain in the tank after 30 days? (b) On which day will the tank be emptied?

38. You want to put $50 in an account this month, and then add an additional amount each month, increasing the amount that you add each month by a constant amount in such a way that the account will have $3250 at the end of 26 months. How much must the payments increase from month to month? (Ignore interest.)

39. You plan to make monthly withdrawals from an account that does not earn interest and that originally contains $20,100, and you plan to increase the amount of the withdrawal by $10 each month. What should the first withdrawal be so that the account will be exhausted at the end of exactly 5 years?

SECTION **60**
Geometric Sequences and Series

A. GEOMETRIC SEQUENCES

Definition A finite or infinite sequence $a_1, a_2, a_3, \ldots$ is called a **geometric sequence** if there is a constant number r (the **common ratio**) such that

$$a_{k+1} = a_k r \qquad (60.1)$$

for $k \geq 1$. (Geometric sequences are sometimes called **geometric progressions.**)

By applying (60.1) repeatedly, we see that a geometric sequence has the form

$$a_1, a_1 r, a_1 r^2, \ldots, a_1 r^{n-1}, \ldots$$

Exercise 31 asks you to use mathematical induction to prove that

$$a_n = a_1 r^{n-1} \text{ for } n \geq 1. \qquad (60.2)$$

Example 60.1 Write the first 4 terms, the 15th term, and the nth term of the geometric sequence with first term 5 and common ratio $\frac{2}{3}$.

Solution
$$a_1 = 5$$

$$a_2 = a_1 \left(\frac{2}{3}\right) = 5 \cdot \frac{2}{3} = \frac{10}{3}$$

$$a_3 = a_2 \left(\frac{2}{3}\right) = \frac{10}{3} \cdot \frac{2}{3} = \frac{20}{9}$$

$$a_4 = a_3 \left(\frac{2}{3}\right) = \frac{20}{9} \cdot \frac{2}{3} = \frac{40}{27}$$

$$a_{15} = 5 \left(\frac{2}{3}\right)^{14}$$

$$a_n = 5 \left(\frac{2}{3}\right)^{n-1}$$

This sequence can be written as follows.

$$5, 5\left(\frac{2}{3}\right), 5\left(\frac{2}{3}\right)^2, 5\left(\frac{2}{3}\right)^3, \ldots$$

Example 60.2 The first term of a geometric sequence is 4 and the sixth term is 8. What is the common ratio?

Solution Use Equation (60.2) with $n = 6$, $a_1 = 4$, and $a_6 = 8$. This gives

$$8 = 4r^5, \quad r^5 = 2, \quad \text{and} \quad r = \sqrt[5]{2}.$$

B. SUMS OF FINITE GEOMETRIC SERIES

A **geometric series** is a series whose terms form a geometric sequence. Thus a geometric series will have the form

$$a_1 + a_1 r + a_1 r^2 + \cdots + a_1 r^{n-1} \qquad \text{(if finite)}$$

or

$$a_1 + a_1 r + a_1 r^2 + \cdots + a_1 r^{n-1} + \cdots \qquad \text{(if infinite)}.$$

If S_n denotes the sum of the first n terms of a geometric series with first term a_1 and common ratio $r \neq 1$, then

$$S_n = \frac{a_1 - a_1 r^n}{1 - r} \tag{60.3}$$

and

$$S_n = \frac{a_1 - r a_n}{1 - r}. \tag{60.4}$$

Proof We prove (60.3) by mathematical induction. Then (60.4) will follow from (60.3) and (60.2) (Exercise 32). Let P_n denote Equation (60.3).

(a) P_1 is true because $S_1 = a_1$ and

$$\frac{a_1 - a_1 r}{1 - r} = \frac{a_1(1 - r)}{1 - r} = a_1.$$

(b) Assume P_k. Then

$$S_k = \frac{a_1 - a_1 r^k}{1 - r}.$$

Therefore,

$$S_{k+1} = a_1 + a_1 r + a_1 r^2 + \cdots + a_1 r^{k-1} + a_1 r^k$$

$$= S_k + a_1 r^k$$

$$= \frac{a_1 - a_1 r^k}{1 - r} + a_1 r^k$$

$$= \frac{a_1 - a_1 r^k + (1 - r)a_1 r^k}{1 - r}$$

$$= \frac{a_1 - a_1 r^{k+1}}{1 - r},$$

which is P_{k+1}. □

Example 60.3 Find the sum of the first 8 terms of the geometric sequence with first term 5 and common ratio 2.

Solution Use (60.3) with $a_1 = 5$, $r = 2$, and $n = 8$:

$$S_8 = \frac{5 - 5 \cdot 2^8}{1 - 2} = 5(2^8 - 1) = 5 \cdot 255 = 1275.$$

C. APPLICATION: ANNUITIES

A common way to save for some future event such as college or retirement is by the investment of equal amounts of money at equal intervals of time. Any such savings plan is an example of an **annuity,** which is simply a sequence of payments at periodic intervals. By using the formulas from the preceding subsection we now compute the amount accumulated under a **simple annuity,** which is one in which the payment intervals and compounding intervals coincide.

Example 60.4 At the end of each year $1000 is added to an account that pays 10% compounded annually. How much will be in the account at the end of the fifth year?

Solution

	Added Excluding Interest	Interest Earned	Accumulated Amount
End of year 1:	$1000	$ 0	$1000
End of year 2:	$1000	$100	$2100
End of year 3:	$1000	$210	$3310
End of year 4:	$1000	$331	$4641
End of year 5:	$1000	$464.10	$6105.10

We see that the accumulated amount is $6105.10. Notice that the total amount added to the account, excluding interest, is $5000. Interest adds an additional $1105.10.

Remark The following observations about Example 60.4 will help us discover a general formula.

The $1000 added at the end of year 5 earns interest for 0 years.

The $1000 added at the end of year 4 earns interest for 1 year.

The $1000 added at the end of year 3 earns interest for 2 years.

The $1000 added at the end of year 2 earns interest for 3 years.

The $1000 added at the end of year 1 earns interest for 4 years.

We now consider a general problem of which Example 60.4 is a special case. Suppose that an amount R is added to an account at the end of each of n equal time intervals. Also suppose that the account earns interest at a rate of i percent per time interval (compounding interval).* If we think in terms of the remark at the end of Example 60.4, we find that:

The $\$R$ added at the end of period n earns interest for 0 periods.

The $\$R$ added at the end of period $n - 1$ earns interest for 1 period.

The $\$R$ added at the end of period $n - 2$ earns interest for 2 periods.

$$\cdot \ \cdot \ \cdot$$

The $\$R$ added at the end of period 2 earns interest for $n - 2$ periods.

The $\$R$ added at the end of period 1 earns interest for $n - 1$ periods.

By results from Section 2C, an amount R invested for k compounding periods at an interest rate of i per period accumulates to

$$R(1 + i)^k.$$

Therefore, the accumulated amount for the annuity is

$$R + R(1 + i) + R(1 + i)^2 + \cdots + R(1 + i)^{n-1}. \tag{60.5}$$

To sum (60.5), we apply (60.3) with $a_1 = R$ and $r = 1 + i$. Let S denote the sum. Then

$$S = \frac{R - R(1 + i)^n}{1 - (1 + i)} = R \cdot \frac{(1 + i)^n - 1}{i}.$$

As in previous formulas of this type, the interest rate i must be in decimal form: for example, use 0.06 rather than 6%.

If an amount R is added to an account at the end of each of n equal compounding intervals, and the account earns interest at an interest rate of i per compounding interval, then the total amount accumulated from the n payments will be

$$S = R \cdot \frac{(1 + i)^n - 1}{i}. \tag{60.6}$$

Example 60.5 Suppose that at the end of each month $50 is added to an account that pays interest at an annual rate of 6% compounded monthly. How much will be in the account at the end of the tenth year?

Solution The annual rate of 6% converts to a monthly rate of $6/12\% = 1/2\%$, so we use $i = 0.005$ in (60.6). Also, $R = 50$ and $n = 120$ (10 years times 12 months per year).

$$S = 50 \cdot \frac{(1 + 0.005)^{120} - 1}{0.005}$$

$$S = \$8193.97. \qquad \boxed{\text{C}} \ \blacksquare$$

*The letter i here denotes an interest rate, not an imaginary number as in Chapter X.

The fraction on the right in (60.6) arises so often that it has been given a special symbol, $s_{\overline{n}|i}$. Thus

$$s_{\overline{n}|i} = \frac{(1 + i)^n - 1}{i}.$$

This quantity has been tabulated extensively for different combinations of i and n. Table 60.1 gives the values needed for the exercises and for the following example.

TABLE 60.1

n	$s_{\overline{n}|0.005}$	$s_{\overline{n}|0.01}$	n	$s_{\overline{n}|0.05}$	$s_{\overline{n}|0.10}$
10	10.2280	10.4622	1	1.0000	1.0000
20	20.9791	22.0190	2	2.0500	2.1000
30	32.2800	34.7849	3	3.1525	3.3100
40	44.1588	48.8864	4	4.3101	4.6410
50	56.6452	64.4632	5	5.5256	6.1051
60	69.7700	81.6697	6	6.8019	7.7156
70	83.5661	100.6763	7	8.1420	9.4872
80	98.0677	121.6715	8	9.5491	11.4359
90	113.3109	144.8633	9	11.0266	13.5795
100	129.3337	170.4814	10	12.5779	15.9374
110	146.1759	198.7797	11	14.2068	18.5312
120	163.8793	230.0387	12	15.9171	21.3843

Example 60.6 Suppose that at the end of each year $500 is added to an account that pays 5% compounded annually. How much will be in the account at the end of 10 years?

Solution To the nearest dollar,

$$S = 500\, s_{\overline{10}|0.05} = 500(12.5779) = \$6,289.$$

Here we have considered only simple annuities used for saving. Other uses of annuities include repayment of loans by equal installments and equal monthly payments from retirement accounts. These other kinds of annuities are discussed in books on the mathematics of finance.

D. SUMS OF INFINITE GEOMETRIC SERIES

We shall now see that it can make sense to talk about the sum of an *infinite* series of numbers.

Example 60.7 Consider the following geometric series

$$\frac{1}{2} + \frac{1}{4} + \frac{1}{8} + \cdots + \frac{1}{2^n} + \cdots. \tag{60.7}$$

Since $a_1 = \frac{1}{2}$ and $r = \frac{1}{2}$, we can apply (60.3) to get

$$S_n = \frac{\dfrac{1}{2} - \dfrac{1}{2}\left(\dfrac{1}{2}\right)^n}{1 - \dfrac{1}{2}} = 1 - \left(\frac{1}{2}\right)^n = 1 - \frac{1}{2^n}.$$

Thus

$$S_1 = \frac{1}{2}, \ S_2 = \frac{3}{4}, \ S_3 = \frac{7}{8}, \ldots, \ S_n = 1 - \frac{1}{2^n}, \ldots. \tag{60.8}$$

The sequence in (60.8) is called the *sequence of partial sums* of the series in (60.7). Because $1/2^n$ approaches closer and closer to 0 as n increases, we see that the successive terms in (60.8) approach closer and closer to 1 as n increases. We abbreviate this by writing

$$\lim_{n \to \infty} S_n = \lim_{n \to \infty} \left(1 - \frac{1}{2^n} \right) = 1.$$

We say in this case that the series *converges* to 1, and also that 1 is the *sum* of the series.

Now consider an arbitrary geometric series

$$a_1 + a_1 r + a_1 r^2 + \cdots + a_1 r^{n-1} + \cdots. \tag{60.9}$$

By factoring a_1 from the numerator $a_1 - a_1 r^n$ of (60.3), we can write the sum of the first n terms of (60.9) as

$$S_n = \frac{a_1}{1 - r} (1 - r^n). \tag{60.10}$$

If $|r| < 1$, then r^n approaches 0 as n increases. (Try some specific examples, like $r = 0.1$, or $r = -\frac{1}{2}$.) Therefore, we can write

$$\lim_{n \to \infty} S_n = \frac{a_1}{1 - r} \text{ if } |r| < 1.$$

As in Example 60.7, we say that the series **converges** to $a_1/(1 - r)$, and that $a_1/(1 - r)$ is the **sum** of the series (60.9). Denoting the sum by S_∞, we can summarize as follows.

> If an infinite geometric series has first term a_1 and common ratio r, and $|r| < 1$, then the series converges with sum
>
> $$S_\infty = \frac{a_1}{1 - r}. \tag{60.11}$$

If $a_1 \neq 0$ and $|r| \geq 1$, then the series (60.9) does not converge; Exercises 34 and 35 give examples. Infinite series are treated more fully in calculus.

Example 60.8 Find the sum of the infinite geometric series

$$1 - \frac{2}{3} + \frac{4}{9} - \frac{8}{27} + \cdots + \left(-\frac{2}{3} \right)^{n-1} + \cdots.$$

Solution Use (60.11) with $a_1 = 1$ and $r = -\frac{2}{3}$.

$$S_\infty = \frac{1}{1 - \left(-\frac{2}{3} \right)} = \frac{1}{\frac{5}{3}} = \frac{3}{5}.$$

We know from Section 1 that every repeating decimal represents a rational number; that is, every repeating decimal number can be expressed in the form a/b, where a and b are integers and $b \neq 0$. The next example illustrates how to find a fraction a/b representing a given repeating decimal number.

Example 60.9 Express $5.\overline{12}$ in the form a/b, where a and b are integers and $b \neq 0$.

Solution

$$5.\overline{12} = 5.121212 \cdots$$

$$= 5 + 0.12 + 0.0012 + 0.000012 + \cdots$$

$$= 5 + \frac{12}{100} + \frac{12}{10000} + \frac{12}{1000000} + \cdots.$$

The terms following 5 form an infinite geometric series with $a_1 = \frac{12}{100}$ and $r = \frac{1}{100}$; thus they can be summed by Equation (60.11). Here are the computations.

$$5.\overline{12} = 5 + \frac{\dfrac{12}{100}}{1 - \dfrac{1}{100}} = 5 + \frac{\dfrac{12}{100}}{\dfrac{99}{100}} = 5 + \frac{12}{99}$$

$$= 5 + \frac{4}{33} = \frac{169}{33}$$

EXERCISES FOR SECTION 60

Write the geometric sequence with the given first term a_1, common ratio r, and number of terms n.

1. $a_1 = \dfrac{1}{2}, r = 4, n = 5$

2. $a_1 = 3, r = -\dfrac{1}{2}, n = 6$

3. $a_1 = 10^{-1}, r = 10^{-3}, n = 5$

Write the first three terms and the tenth term of the geometric sequence with the given first term a_1 and common ratio r.

4. $a_1 = -1, r = -\dfrac{1}{2}$

5. $a_1 = 100, r = 0.1$

6. $a_1 = 8, r = \sqrt{2}$

Find the kth term and the sum of the geometric series with the given first term a_1, common ratio r, and number of terms n.

7. $a_1 = 5, r = \dfrac{2}{5}, n = 5$

8. $a_1 = -1, r = -1, n = 11$

9. $a_1 = 3, r = -\sqrt{3}, n = 5$

10. $a_1 = 0.3, r = 3, n = 6$

11. $a_1 = 1 - \sqrt{2}, r = 1 + \sqrt{2}, n = 3$

12. $a_1 = u, r = v^2, n = m$

Find the sum of each series.

13. $\dfrac{1}{4} + \dfrac{3}{8} + \dfrac{9}{16} + \cdots + \dfrac{243}{128}$

14. $2 + 2\sqrt{2} + 4 + \cdots + 32$

15. $\displaystyle\sum_{k=1}^{10} 0.6^k$

16. $\dfrac{1}{5} + \dfrac{1}{25} + \dfrac{1}{125} + \dfrac{1}{625} + \cdots$

17. $\displaystyle\sum_{k=1}^{\infty} 20(0.1)^{k-1}$

18. $5 - \dfrac{10}{3} + \dfrac{20}{3} - \dfrac{40}{27} + \cdots$

19. (a) Solve Equation (60.2) for a_1.

   (b) Use part (a) to find the first term of the five-term geometric sequence with last term 6 and common ratio $\frac{1}{3}$.

20. (a) Solve Equation (60.2) for r.

(b) Use part (a) to find the common ratio for the six-term geometric sequence with first term 2 and last term $-\frac{243}{16}$.

21. (a) Solve Equation (60.3) for a_1.

(b) Use part (a) to find the first term of the five-term geometric sequence with common ratio $-\sqrt{2}$ and sum $1 + 4\sqrt{2}$.

Use Table 60.1 to find the amount accumulated under each annuity.

22. $100 per month for 30 months at 12% (annual rate) compounded monthly.

23. $2000 per year for 6 years at 10% (annual rate) compounded annually.

24. $25 per month for 10 years at 6% (annual rate) compounded monthly.

25. $500 per year for 12 years at 5% (annual rate) compounded annually.

26. $50 per month for $7\frac{1}{2}$ years at 6% (annual rate) compounded monthly.

27. $1200 per year for 10 years for 10% (annual rate) compounded annually.

28. A parent wants $10,000 available 10 years from now for a daughter's college expenses. To carry this out, how much should the parent invest at the end of each month in an account that pays 12% (annual rate) compounded monthly?

29. Suppose you want to save $3000 for use $2\frac{1}{2}$ years from now. To accomplish this, how much should you invest at the end of each month in an account that pays 6% (annual rate) compounded monthly?

30. I plan to invest $1000 at the end of each year for the next 10 years in an account that pays 10% (annual rate) compounded annually. You plan to invest $75 at the end of each month for the next 10 years in an account that pays 12% (annual rate) compounded monthly. Which of us will accumulate the most, and by how much?

31. Use mathematical induction to prove (60.2).

32. Use (60.2) and (60.3) to prove (60.4).

33. The text proves (60.3) by mathematical induction. The following statements provide the steps in an alternate proof. Justify each step.

(a) $rS_n = a_1r + a_1r^2 + \cdots + a_1r^{n-1} + a_1r^n.$

(b) $S_n - rS_n = a_1 - a_1r^n.$

(c) $S_n = (a_1 - a_1r^n)/(1 - r).$

34. Determine S_n for a geometric series with first term a_1 and common ratio 1. What happens to S_n as $n \to \infty$?

35. Determine S_n for a geometric series with first term a_1 and common ratio -1. What happens to S_n as $n \to \infty$?

36. Show that (60.2) implies that

$$n = \frac{\log(a_n r/a_1)}{\log r}.$$

(This exercise assumes acquaintance with Section 27.)

Express each number in the form a/b *where* a *and* b *are integers and* b $\neq$ 0.

37. $3.\overline{4}$

38. $0.\overline{18}$

39. $21.\overline{12}$

40. $1.\overline{02}$

41. $0.\overline{102}$

42. $5.1\overline{23}$

43. Prove that if a, b, c is a geometric sequence, then $b = \sqrt{ac}$.

44. Prove that if $a_1, a_2, a_3, \ldots$ is an arithmetic sequence, then $10^{a_1}, 10^{a_2}, 10^{a_3}, \ldots$ is a geometric sequence.

45. Prove that if $a_1, a_2, a_3, \ldots$ is a geometric sequence, then $\log a_1, \log a_2, \log a_3, \ldots$ is an arithmetic sequence.

46. For the geometric sequence 2, 6, 18, 54, ..., the sequence of the differences of successive terms, 4, 12, 36, ..., also forms a geometric sequence. Prove that the sequence of the differences of successive terms of every geometric sequence forms a geometric sequence.

47. The frequencies of the semitones of a well-tempered scale in music form a geometric sequence with common ratio $\sqrt[12]{2}$. If the frequency of concert A is 440 cycles per second, what is the frequency of middle C, which is 9 semitones lower than concert A? (The higher the semitone, the higher the frequency.)

48. A sequence of squares is constructed by beginning with a square one unit on each side and then, at each step thereafter, connecting the midpoints of the adjacent sides of the square immediately preceding. Prove that the sequence of perimeters forms a geometric sequence. Also prove that the sequence of areas forms a geometric sequence.

SECTION **61**
The Binomial Theorem

A. INTRODUCTION. FACTORIAL NOTATION

In this section we show how to extend the following list to include all higher integral powers of $a + b$.

$$(a + b)^0 = 1$$
$$(a + b)^1 = a + b$$
$$(a + b)^2 = a^2 + 2ab + b^2$$
$$(a + b)^3 = a^3 + 3a^2b + 3ab^2 + b^3$$

The theorem that gives $(a + b)^n$ in general is called the *Binomial Theorem*. Before stating and proving this theorem, however, we need some preliminary facts about *factorial notation* and *binomial coefficients*.

The notation $n!$ (read **n factorial**) is defined for positive integers as follows:

$$1! = 1$$
$$2! = 2 \cdot 1 = 2$$
$$3! = 3 \cdot 2 \cdot 1 = 6$$
$$4! = 4 \cdot 3 \cdot 2 \cdot 1 = 24$$

and, in general,

$$n! = n(n - 1)(n - 2) \cdots \cdots 2 \cdot 1.$$

You can easily verify that $5! = 120$. Fortunately, we will not need larger specific values like

$$20! = 2,432,902,008,176,640,000.$$

Notice that if n is any integer greater than 1, then $n! = n \cdot (n - 1)!$. With $n = 4$, for example, we have $4! = 4 \cdot 3! = 4 \cdot 3 \cdot 2 \cdot 1$. If we use $n = 1$ in $n! = n \cdot (n - 1)!$, we get $1! = 1 \cdot 0!$. Partly for this reason, and partly to make many formulas simpler, we *define* $0!$ by $0! = 1$.

B. BINOMIAL COEFFICIENTS

The coefficients needed to expand $(a + b)^n$ are called **binomial coefficients** (remember from Section 4 that $a + b$ is called a *binomial*). For each pair of integers n and k such that $n \geq k \geq 0$, there is a binomial coefficient denoted by $\binom{n}{k}$ and defined by

$$\binom{n}{k} = \frac{n!}{k!(n - k)!}. \tag{61.1}$$

Example 61.1 (a) $\dbinom{5}{2} = \dfrac{5!}{2!3!} = \dfrac{5 \cdot 4 \cdot \cancel{3 \cdot 2 \cdot 1}}{2 \cdot 1 \cdot \cancel{3 \cdot 2 \cdot 1}} = 10$

(b) $\dbinom{10}{2} = \dfrac{10!}{2!8!} = \dfrac{10 \cdot 9 \cdot \cancel{8!}}{2 \cdot 1 \cdot \cancel{8!}} = 45$

(c) $\dbinom{12}{9} = \dfrac{12!}{9!3!} = \dfrac{12 \cdot 11 \cdot 10 \cdot \cancel{9!}}{\cancel{9!} \cdot 3 \cdot 2 \cdot 1} = 220$

(d) $\dbinom{n}{0} = \dfrac{n!}{0!n!} = \dfrac{n!}{1 \cdot n!} = 1 \quad (n \geq 0)$

(e) $\dbinom{n}{n} = \dfrac{n!}{n!0!} = 1 \quad (n \geq 0)$

(f) $\dbinom{n}{1} = \dfrac{n!}{1!(n-1)!} = \dfrac{n \cdot (n-1)!}{1 \cdot (n-1)!} = n \quad (n \geq 1)$

Following are two fundamental facts about binomial coefficients. They are valid whenever $n \geq k \geq 0$.

$$\dbinom{n}{k} = \dbinom{n}{n-k} \tag{61.2}$$

Proof
$$\dbinom{n}{k} = \dfrac{n!}{k!(n-k)!}$$

$$= \dfrac{n!}{(n-k)!k!}$$

$$= \dfrac{n!}{(n-k)!(n-(n-k))!} = \dbinom{n}{n-k} \qquad \square$$

Example 61.2 (a) $\dbinom{20}{3} = \dbinom{20}{17}$

(b) $\dbinom{n}{1} = \dbinom{n}{n-1}$

(c) $\dbinom{n}{0} = \dbinom{n}{n}$

$$\dbinom{n+1}{k} = \dbinom{n}{k-1} + \dbinom{n}{k} \tag{61.3}$$

Proof We add the two terms on the right in (61.3) using a common denominator of $k!(n-k+1)!$. The result will turn out to be the term on the left.

$$\dbinom{n}{k-1} + \dbinom{n}{k} = \dfrac{n!}{(k-1)!(n-(k-1))!} + \dfrac{n!}{k!(n-k)!}$$

$$= \dfrac{k \cdot n!}{k \cdot (k-1)!(n-k+1)!} + \dfrac{n!(n-k+1)}{k!(n-k)!(n-k+1)}$$

$$= \frac{k \cdot n!}{k!(n - k + 1)!} + \frac{(n - k + 1) \cdot n!}{k!(n - k + 1)!}$$

$$= \frac{(k + n - k + 1) \cdot n!}{k!(n - k + 1)!}$$

$$= \frac{(n + 1)!}{k!(n - k + 1)!}$$

$$= \binom{n + 1}{k} \qquad\qquad \square$$

Equation (61.3) provides a convenient scheme for computing binomial coefficients for reasonably small n. This scheme is shown by the triangle in Figure 61.1, named for the French mathematician Blaise Pascal (1623–1662) but known to the Chinese at least 300 years before his time. In Pascal's triangle $\binom{n}{k}$ stands

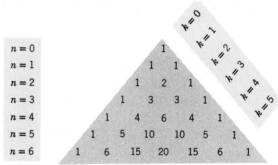

FIGURE 61.1

at the intersection of the nth row and the kth diagonal. Equation (61.3) tells us that each entry can be obtained by adding the numbers immediately to the left and right in the row above it. Equation (61.2) is reflected in the left-right symmetry of the table. For example, the pair of 15's in the sixth row represent $\binom{6}{2}$ and $\binom{6}{4}$, which must be equal by (61.2).

Other notation used for the binomial coefficient $\binom{n}{k}$ includes $C(n, k)$, $_nC_k$, and C_k^n.

C. THE BINOMIAL THEOREM

The Binomial Theorem

If n is any nonnegative integer then

$$(a + b)^n = \binom{n}{0} a^n + \binom{n}{1} a^{n-1}b + \binom{n}{2} a^{n-2}b^2 + \cdots$$

$$+ \binom{n}{k} a^{n-k}b^k + \cdots + \binom{n}{n} b^n. \tag{61.4}$$

The proof of the theorem will follow some observations and examples.

Observations

- The top entry in each binomial coefficient is n.
- The bottom entries in the binomial coefficients increase successively from 0 to n as we move to the right.
- In each term, the exponent on b is the same as the bottom entry of the term's binomial coefficient.
- The exponents on a decrease successively from n to 0 as we move to the right, and the exponents on b increase successively from 0 to n.
- In each term, the sum of the exponents on a and b is n.
- There are $n + 1$ terms in the expansion of $(a + b)^n$.

Example 61.3

$$(a + b)^4 = \binom{4}{0} a^4 + \binom{4}{1} a^3b + \binom{4}{2} a^2b^2 + \binom{4}{3} ab^3 + \binom{4}{4} b^4$$

The binomial coefficients needed here can be read from the fourth row of Pascal's triangle: 1, 4, 6, 4, 1. Thus

$$(a + b)^4 = a^4 + 4a^3b + 6a^2b^2 + 4ab^3 + b^4.$$

■

Example 61.4 Expand $(x^2 - 2y)^5$.

Solution Use (61.4) with $a = x^2$, $b = -2y$, and $n = 5$. The binomial coefficients can be read from the fifth row of Pascal's triangle. [You may want to write out $(a + b)^5$ first and then substitute x^2 for a and $-2y$ for b.]

$$(x^2 - 2y)^5 = (x^2)^5 + 5(x^2)^4(-2y) + 10(x^2)^3(-2y)^2$$
$$+ 10(x^2)^2(-2y)^3 + 5(x^2)(-2y)^4 + (-2y)^5$$
$$= x^{10} - 10x^8y + 40x^6y^2 - 80x^4y^3$$
$$+ 80x^2y^4 - 32y^5$$

■

Example 61.5 Find the coefficient of a^4b^{16} in the expansion of $(a + b)^{20}$.

Solution Use (61.4) with $n = 20$. The coefficient of a^4b^{16} is

$$\binom{20}{16} = \frac{20!}{16!4!} = \frac{20 \cdot 19 \cdot 18 \cdot 17 \cdot 16!}{16! \cdot 4 \cdot 3 \cdot 2 \cdot 1} = 4845.$$

■

Example 61.6 Find the coefficient of u^6v^{12} in the expansion of $(2u^2 - v^3)^7$.

Solution Use (61.4) with $n = 7$ and with k chosen so that $(2u^2)^{7-k}(-v^3)^k$ involves v^{12}. This requires $k = 4$, and thus the term involving u^6v^{12} is

$$\binom{7}{4} (2u^2)^3(-v^3)^4 = \frac{7!}{4!3!} 8u^6(v^{12})$$
$$= 35 \cdot 8u^6v^{12}$$
$$= 280u^6v^{12}.$$

Therefore, the answer is 280.

■

Proof of the Binomial Theorem

Let P_n represent Equation (61.4) and use induction on n.

(a) P_1 is true because

$$(a + b)^1 = a + b = \binom{1}{0} a^1 + \binom{1}{1} b^1.$$

(b) We now show that P_m implies P_{m+1} for each positive integer m. (We are using m in place of k in the statement of the Principle of Mathematical Induction in Section 58.) Assume P_m, that is,

$$(a + b)^m = \binom{m}{0} a^m + \binom{m}{1} a^{m-1}b + \binom{m}{2} a^{m-2}b^2 + \cdots \qquad (61.5)$$
$$+ \binom{m}{k} a^{m-k}b^k + \cdots + \binom{m}{m} b^m.$$

To prove P_{m+1} from (61.5), we multiply both sides of (61.5) by $a + b$. The result on the left will be $(a + b)^{m+1}$, as required. The result on the right will be

$$\left[\binom{m}{0} a^m + \binom{m}{1} a^{m-1}b + \cdots + \binom{m}{k} a^{m-k}b^k + \cdots + \binom{m}{m} b^m \right] (a + b)$$

$$= \binom{m}{0} a^{m+1} + \binom{m}{1} a^m b + \cdots + \binom{m}{k} a^{m-k+1}b^k + \cdots$$

$$+ \binom{m}{m} ab^m + \binom{m}{0} a^m b + \binom{m}{1} a^{m-1}b^2 + \cdots$$

$$+ \binom{m}{k} a^{m-k}b^{k+1} + \cdots + \binom{m}{m} b^{m+1}$$

The term involving b^k in the last expansion is seen to be

$$\binom{m}{k} a^{m-k+1}b^k + \binom{m}{k-1} a^{m-k+1}b^k = \binom{m+1}{k} a^{m-k+1}b^k,$$

where we have used Equation (61.3). But $\binom{m+1}{k} a^{m-k+1}b^k$ is also the term involving b^k in P_{m+1}. Since the terms involving b^k are equal for all k ($0 \le k \le m + 1$), the proof is complete. $\square$

EXERCISES FOR SECTION 61

Compute

1. $\binom{10}{3}$

2. $\binom{12}{1}$

3. $\binom{9}{4}$

4. $\binom{50}{50}$

5. $\binom{20}{3}$

6. $\binom{14}{0}$

Simplify as much as possible.

7. $\dfrac{(n + 2)!}{n!}$

8. $\dfrac{(n + 1)!}{(n - 1)!}$

9. $\dfrac{(n + 1)!(n - 1)!}{(n!)^2}$

10. $\binom{n}{0} + \binom{n}{n}$

11. $\binom{n}{1} + \binom{n}{n - 1}$

12. $\binom{n}{2} + \binom{n}{n - 2}$

13. Write rows 7 and 8 of Pascal's triangle.

14. Prove that

$$\binom{n}{k} = \frac{n}{k}\binom{n-1}{k-1}.$$

15. Prove that

$$\binom{n+2}{k} = \binom{n}{k-2} + 2\binom{n}{k-1} + \binom{n}{k}.$$

Expand and simplify.

16. $(x + y)^5$

17. $(u + v)^6$

18. $(a + 2)^4$

19. $(a - b)^6$

20. $\left(x - \dfrac{2}{y}\right)^5$

21. $(2u - v^2)^4$

22. Find and simplify the coefficient of u^7v^2 in the expansion of $(u + v)^9$.

23. Find and simplify the coefficient of $x^{23}y^2$ in the expansion of $(x + y)^{25}$.

24. Find and simplify the coefficient of a^3b^9 in the expansion of $(a + b)^{12}$.

25. Find and simplify the first four terms in the expansion of $(a^2 + 2b^3)^{10}$.

26. Find and simplify the first three terms in the expansion of $(2u - 5v)^{20}$.

27. Find and simplify the first four terms in the expansion of $(x + y)^{100}$.

28. Find and simplify the coefficient of u^6v^{15} in the expansion of $(2u^2 - v^3)^8$.

29. Find and simplify the coefficient of x^8 in $(x - \sqrt{x})^{10}$.

30. There is a constant term in the expansion of $[x - (1/x^2)]^{30}$. What is it?

31. The sum of the entries in the nth row of Pascal's triangle is 2^n. You can verify this for $1 \le n \le 6$ simply by addition and Figure 61.1. Prove it for every row. [Suggestion: Look at what (61.4) becomes when $a = b = 1$.]

32. The product of the first n positive even integers is $n!$ times a function of n. What is the function of n?

33. Show that if n is a positive integer, then

$$1 \cdot 3 \cdot 5 \cdots \cdots (2n - 1) = \frac{(2n)!}{2^n(n!)}.$$

REVIEW EXERCISES FOR CHAPTER XIII

1. Find and simplify the first four terms and the $(n + 1)$st term of the series

$$\sum_{k=2}^{\infty} 3^{k-1}\binom{k}{2}.$$

2. Represent the infinite series

$$1 - 20 + 300 - 4000 + \cdots$$

using summation (Σ) notation.

3. Use mathematical induction to prove that

$$3 + 7 + 11 + \cdots + (4n - 1)$$
$$= n(2n + 1)$$

for every positive integer n.

4. Use mathematical induction to prove that

$$1^2 + 3^2 + 5^2 + \cdots + (2n - 1)^2$$
$$= \frac{n(2n - 1)(2n + 1)}{3}$$

for every positive integer n.

5. Find the sum of the first ten terms of the sequence $1, 1 - \pi, 1 - 2\pi, 1 - 3\pi, \ldots$.

6. Verify the equation in Review Exercise 3 by using a formula for the sum of an arithmetic sequence.

7. Write the first three terms and the tenth term of the geometric sequence having first term $2\sqrt{3}$ and common ratio $-\sqrt{3}$.

8. A geometric sequence has first term x and fifth term $0.0001xy^8$. What is the common ratio?

9. Find the sum of the first ten terms of the geometric series with first term 5 and common ratio -2.

10. Find the sum of the series

$$a + a^2b^2 + a^3b^4 + \cdots + a^nb^{2n-2}.$$

11. An infinite series has first term 1 and common ratio $5t$. Determine the values of t for which the series converges. Also find the sum of the series when it converges.

12. Find the sum of the series

$$4 + \frac{4}{3} + \frac{4}{9} + \frac{4}{27} + \cdots.$$

13. At the end of each month you place $25 in an account that pays interest at an annual rate of 6% compounded monthly. How much will be in the account at the end of the 40th month?

14. I plan to deposit a constant amount in an account at the end of each month, and I want $5000 in the account at the end of 5 years. The account will pay 6% (annual rate) compounded monthly. What is the minimum amount, rounded up to the nearest $5, that I should deposit each month?

15. Simplify as much as possible:

$$\binom{n+1}{k} \Big/ \binom{n}{k-1}.$$

16. Prove that

$$\binom{2n}{n} + \binom{2n}{n-1} = \frac{1}{2}\binom{2n+2}{n+1}.$$

Expand and simplify.

17. $(a^2 - b)^5$

18. $(10x + 0.1y)^6$

CHAPTER XIV

COMBINATORICS AND PROBABILITY

This chapter introduces the most fundamental ideas of combinatorics and probability: combinatorics is concerned with systematic enumeration and counting; probability is concerned with the mathematics of random behavior or chance. Both subjects form rich branches of mathematics with applications throughout business, technology, and the natural and social sciences.

SECTION **62**
Permutations

A. BASIC COUNTING PRINCIPLE

The list of all possibilities in many enumeration problems can be determined most easily with the aid of a *tree*. Here are some examples.

Example 62.1 In how many ways can the three letters *A*, *B*, and *C* be arranged in a row if we require that no letter be repeated?

Solution Although it isn't necessary in an example as simple as this one, all of the different ways can be shown as in Figure 62.1, which is an example of a *tree*. The letters in the first column (reading from the left) indicate the possibilities for the first letter. The letters in the second column indicate the possibilities for the second letter corresponding to each choice for the first letter. The letters in the third column in each case represent the (only) possibility for the third letter. By following the different *paths* through the tree we get the six different ways to arrange the letters, which are indicated to the right.

$$
A \begin{cases} B - C \\ C - B \end{cases} \qquad \begin{matrix} ABC \\ ACB \end{matrix}
$$

$$
B \begin{cases} A - C \\ C - A \end{cases} \qquad \begin{matrix} BAC \\ BCA \end{matrix}
$$

$$
C \begin{cases} A - B \\ B - A \end{cases} \qquad \begin{matrix} CAB \\ CBA \end{matrix} \qquad \text{FIGURE 62.1}
$$

Example 62.2 For a toss of a coin, let the outcomes heads and tails be denoted by *H* and *T*, respectively. How many different possible outcomes are there if a coin is tossed three times?

Solution The tree is shown in Figure 62.2, with the eight possible outcomes listed to the right.

$$
H \begin{cases} H \begin{cases} H \\ T \end{cases} \\ T \begin{cases} H \\ T \end{cases} \end{cases} \qquad \begin{matrix} HHH \\ HHT \\ HTH \\ HTT \end{matrix}
$$

$$
T \begin{cases} H \begin{cases} H \\ T \end{cases} \\ T \begin{cases} H \\ T \end{cases} \end{cases} \qquad \begin{matrix} THH \\ THT \\ \\ TTH \\ TTT \end{matrix} \qquad \text{FIGURE 62.2}
$$

Example 62.3 Assume that students in three departments—English, History, and Philosophy— are to be classified according to sex (F, M), year (1, 2, 3, 4), and department (E, H, P). How many categories are possible?

Solution See Figure 62.3. For each sex there are four possible years, and then there are three possible departments for each combination of sex and year. Thus altogether there are $2 \cdot 4 \cdot 3 = 24$ different categories. ■

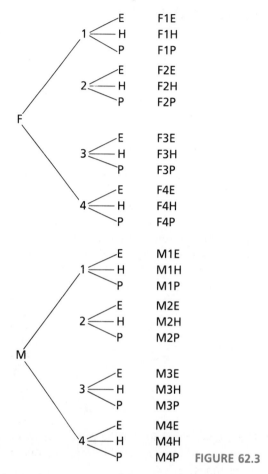

FIGURE 62.3

Thinking about the appropriate tree can help even when the size of the problem makes it impractical to write out the tree in full.

Example 62.4 In how many ways can ten persons $(P_1, P_2, \ldots, P_{10})$ be seated in a row?

Solution Applying the ideas from the previous examples, we can see that the relevant tree would have each of P_1 through P_{10} in the first column, and for each of these there would be nine possibilities in the second column, and so on, until altogether there would be $10 \cdot 9 \cdot 8 \cdot 7 \cdot 6 \cdot 5 \cdot 4 \cdot 3 \cdot 2 \cdot 1 = 10!$ different paths through the tree. Thus $10! \ (= 3,628,800)$ is the number of possible seating arrangements. ■

These examples lead to the following counting principle.

Basic Counting Principle

If one thing can be done in any of n_1 ways, and after it has been done in any of those ways a second thing can be done in any of n_2 ways, and after any of those a third thing can be done in any of n_3 ways, and so on until we get to a kth thing that can be done in any of n_k ways, then the k things can be done together in $n_1 \cdot n_2 \cdots \cdots n_k$ different ways.

A tree representing the Basic Counting Principle would have k columns, with n_1 places in the first column, n_2 places corresponding to each of those in the second column, and so on.

B. PERMUTATIONS

Example 62.5 In how many ways can three officers—President, Vice President, and Secretary—be selected from a club with ten members?

Solution

First select a President:	10 possibilities
Then select a Vice President:	9 possibilities
Finally, select a Secretary:	8 possibilities

Therefore, by the Basic Counting Principle, there are $10 \cdot 9 \cdot 8 \, (= 720)$ different ways of selecting the officers.

If we think of the officers as being arranged in order (President, Vice President, Secretary), we see that this example is a special case of the idea in the following definition. ▧

Definition An arrangement of k different elements, selected from a set of n elements, is called a **permutation** of the n elements taken k at a time. We denote the number of such permutations by $P(n, k)$.

Example 62.5 shows that $P(10, 3) = 10 \cdot 9 \cdot 8 = 720$. To obtain a general formula for $P(n, k)$, we first record the number of possibilities for each of the k choices to be made.

1st element:	n possibilities
2nd element:	$n - 1$ possibilities
3rd element:	$n - 2$ possibilities
	$\cdots$
kth element:	$n - (k - 1) = n - k + 1$ possibilities

Applying the Basic Counting Principle, we see that

$$P(n, k) = n(n - 1)(n - 2) \cdots (n - k + 1). \tag{62.1}$$

The product on the right begins with n and continues till there are k terms. We can write this using factorial notation by multiplying and dividing by $(n - k)!$, as follows.

$$P(n, k) = n(n - 1)(n - 2) \cdots (n - k + 1)$$

$$P(n, k) = \frac{n(n - 1)(n - 2) \cdots (n - k + 1)(n - k)(n - k - 1) \cdots 1}{(n - k)(n - k - 1) \cdots 1}$$

$$P(n, k) = \frac{n!}{(n - k)!} \qquad (62.2)$$

Example 62.6 (a) $P(6, 2) = 6 \cdot 5 = 30$

(b) $P(n, 1) = n \quad (n \geq 1)$

(c) $P(n, n) = n! \quad (n \geq 1)$

Example 62.7 How many three-digit numbers can be formed with the digits 1, 2, 3, 4, 5 (a) if no digit can be repeated? (b) if digits can be repeated?

Solution (a) $P(5, 3) = 5 \cdot 4 \cdot 3 = 60$

(b) Apply the Basic Counting Principle. We must choose three digits, with five possibilities for each choice. Thus the answer is 5^3 or 125.

EXERCISES FOR SECTION 62

Compute.

1. $P(10, 4)$ **2.** $P(6, 4)$

3. $P(50, 3)$ **4.** $P(6, 5)$

5. $P(20, 2)$ **6.** $P(10, 5)$

Simplify as much as possible.

7. $P(n, n - 1)/P(n, n)$

8. $P(n, k)P(n - k, n - k)$

9. $P(n, k)/P(n - 1, k - 1)$

10. How many possible outcomes are there if a coin is tossed four times? five times? n times?

11. How many possible outcomes are there if a die (one of a pair of dice) is tossed two times? (An outcome will consist of two numbers.) What about three times? n times?

12. There are three routes connecting points A and B.

(a) In how many ways can one travel from A to B and back to A?

(b) In how many ways can one travel from A to B and back to A if the route from B to A is required to be different from the route from A to B?

13. A man has three sport coats and five pairs of slacks. He can wear any of the coats with any of the pairs of slacks. How many combinations are possible?

14. In how many ways can ten marching bands be arranged to make up a parade?

15. In how many ways (orders) can the teams in a ten-team soccer league finish if there are no ties?

16. Ten riders are to be assigned randomly to ten horses. In how many ways can this be done?

17. How many even three-digit numbers can be formed with the digits 1, 2, 3, 4, and 5

(a) if no digit can be repeated?

(b) if digits can be repeated?

18. In how many ways can three officers—President, Vice President, and Secretary—be selected from a club with 10 male and 15 female members so that the President and Secretary are female and the Vice President is male?

19. In how many ways can 5 men and 4 women be seated in a row so that no men are in adjacent seats?

20. In how many ways can $n + 1$ men and n women be seated in a row under each of the following conditions?

(a) No restrictions by sex.

(b) No men are in adjacent seats.

(c) At least two men are in adjacent seats.

21. In an eight-team basketball league, in how many ways can the top three positions in the final standings be filled?

22. For a given set of pairings in an eight-team basketball tournament, in how many ways can the top three positions in the final standings be filled? (The top two teams must be from different brackets. Compare Exercise 21.)

23. How many positive integers are factors of 4? (Don't forget 1 and 4.) 8? 16? 2^n?

24. How many positive integers are factors of $2^2 \cdot 3^2$? $2^2 \cdot 3^2 \cdot 5^2$? $2^r \cdot 3^s \cdot 5^t$ (r, s, and t positive integers)? (Compare Exercise 23.)

SECTION **63**
Combinations

A. BASIC IDEAS

In any enumeration problem the following question is critical: Does the order in which elements are chosen (or things are arranged, or events happen) matter?

Example 63.1 Consider the problem of selecting two letters from A, B, and C.

(a) If order matters, there are six possibilities:

$$AB, AC, BA, BC, CA, CB.$$

Both AB and BA are listed, for instance, because they give A and B in different orders, and order matters.

(b) If order does not matter, there are only three possibilities:

$$AB, AC, BC.$$

The number of possibilities in part (a) is $P(3, 2)$, the number of *permutations* of three elements taken two at a time (Section 62). In contrast, the number of possibilities in part (b) is $C(3, 2)$, the number of *combinations* of three elements taken two at a time, as described by the following definition. ◼

Definition A selection of k different elements from a set of n elements, without regard to order, is called a **combination** of the n elements taken k at a time. We denote the number of such combinations by $C(n, k)$.

Example 63.1(b) shows that $C(3, 2) = 3$. The discussion that follows will lead to a formula for $C(n, k)$ for all n and k.

In Example 62.5 we answered the following question: In how many ways can three officers—President, Vice-President, and Secretary—be selected from a club with ten members? The answer was $P(10, 3)$ ($= 10 \cdot 9 \cdot 8 = 720$). We now consider a related but different question for which the answer is $C(10, 3)$.

Example 63.2 In how many ways can a three-member committee be selected from a club with ten members?

Solution The critical difference between this question and that in Example 62.5 concerns order. Order mattered in Example 62.5: for example, having Smith as President,

Jones as Vice-President, and Brown as Secretary is not the same as having Brown as President, Jones as Vice-President, and Smith as Secretary. But order does not matter in the present example: the committee {Smith, Jones, Brown} is the same as the committee {Brown, Jones, Smith}.

The answer in this example is $C(10, 3)$. To obtain a numerical value for this answer we'll redo Example 62.5 and get $C(10, 3)$ as a by-product.

Our first solution for Example 62.5 gave the answer $P(10, 3)$. Here is a second solution. Instead of electing the three officers directly, suppose we elect a slate of three, not specifying who will be President, and so on, but simply agreeing that the persons on the slate will be the officers, in a way to be agreed on among themselves. This method still allows for the same number of possibilities as the first method, but it is conceptually different for counting purposes. To compute the number of possibilities by the second method, use the Basic Counting Principle with two steps: first count the number of ways of selecting three members from ten [this is $C(10, 3)$], then count the number of ways the three chosen members can be given the three different offices, and then multiply. The number of ways three members can be given the three different offices is $3 \cdot 2 \cdot 1$ (any of the three for President, then any of the remaining two for Vice-President, then the remaining one for Secretary). Thus the second solution of Example 62.5 gives the answer $C(10, 3) \cdot 3 \cdot 2 \cdot 1$. Since this solution and the original solution must agree, we must have

$$C(10, 3) \cdot 3 \cdot 2 \cdot 1 = P(10, 3).$$

Therefore,

$$C(10, 3) = \frac{10 \cdot 9 \cdot 8}{3 \cdot 2 \cdot 1} = 120.$$

To obtain a general formula for $C(n, k)$ we proceed just as in the example. From Equation (62.2),

$$P(n, k) = \frac{n!}{(n - k)!}. \tag{63.1}$$

But we can also get $P(n, k)$ as follows. First, think of selecting k elements from n elements without regard to order; this can be done in $C(n, k)$ different ways. Second, each of these k-element subsets can be arranged in $k(k - 1) \cdot (k - 2) \cdots 1$ different ways. Therefore, as Example 63.2 showed in a special case,

$$P(n, k) = C(n, k) \cdot k(k - 1)(k - 1) \cdots 1. \tag{63.2}$$

Thus

$$\boxed{C(n, k) = \frac{P(n, k)}{k!}.}$$

If we use $P(n, k) = n!/(n - k)!$ [Equation (63.1)], we find that

$$C(n, k) = \frac{n!}{k!(n - k)!}. \tag{63.3}$$

If you have studied Section 61 you will recognize the expression on the right as a binomial coefficient. Specifically,

$$C(n, k) = \binom{n}{k}. \tag{63.4}$$

B. MORE EXAMPLES

Example 63.3   How many different 10-member committees can be formed from the 100 members of the U.S. Senate?

Solution   The answer is the number of combinations of 100 elements taken 10 at a time, which is

$$C(100, 10) = \frac{100!}{10!90!}.$$

This is approximately 1.7×10^{13}.   ⓒ ▪

Example 63.4   Assume that 52 members of the Senate are Republicans and that 48 are Democrats. How many 10-member committees can be formed subject to the condition that 6 are Republicans and 4 are Democrats?

Solution   The Republicans can be chosen in $C(52, 6)$ different ways. The Democrats can be chosen in $C(48, 4)$ different ways. Therefore, by the Basic Counting Principle, the total number of committees is

$$C(52, 6) \cdot C(48, 4) = \frac{52!}{6!46!} \cdot \frac{48!}{4!44!}.$$

This is approximately 4.0×10^{12}.   ⓒ ▪

A *bridge deck* contains 52 cards, 13 in each of four suits: spades, hearts, diamonds, and clubs. Each suit contains an ace, 2, 3, 4, 5, 6, 7, 8, 9, 10, jack, queen, and king.

Example 63.5   (a) The number of different 13-card bridge hands is the number of combinations of 52 elements taken 13 at a time, which is $C(52, 13)$. With a calculator you can show that this is approximately 6.4×10^{11}.   ⓒ

(b) The number of different 5-card poker hands that can be dealt from a bridge deck is $C(52, 5)$. This is

$$\frac{52!}{5!47!} = \frac{52 \cdot 51 \cdot 50 \cdot 49 \cdot 48}{5 \cdot 4 \cdot 3 \cdot 2 \cdot 1} = 2,598,960.$$ ▪

Example 63.6   In the notation of this section, Equation (61.3) states that

$$C(n + 1, k) = C(n, k - 1) + C(n, k). \tag{63.5}$$

This equation was proved algebraically in Section 61. Following is a combinatorial proof.

Let S be a set with $n + 1$ elements, $x_1, \cdots, x_{n+1}$. Then $C(n + 1, k)$ is the number of k-element subsets of S, and these can be divided into two classes, those that do contain x_1 and those that do not. There are $C(n, k - 1)$ subsets of the first kind (the number of ways we can choose $k - 1$ elements from x_2, $\cdots, x_{n+1}$); there are $C(n, k)$ subsets of the second kind (the number of ways we can choose k elements from $x_2, \cdots, x_n$). Adding these we get Equation (63.5). ■

EXERCISES FOR SECTION 63

Compute.

1. $C(5, 3)$ **2.** $C(7, 4)$

3. $C(10, 9)$ **4.** $C(n, n - 1)$

5. $C(n, 3)$ **6.** $C(n, 2)$

Simplify as much as possible.

7. $P(n, k)/C(n, k)$

8. $C(n, k)/C(n, n - k)$

9. $C(n + 1, n - 1)/(n + 1)$

10. An organization contains ten men and ten women. How many six-member committees can be chosen under the following conditions?

    (a) With three members of each sex.

    (b) With four men and two women.

    (c) With four of one sex and two of the other.

11. Ten basketball players want to divide themselves into two teams of five players each. In how many ways can this be done?

12. Ten basketball players want to divide themselves into two teams of five players each, in such a way that the two best players are on opposite teams. In how many ways can this be done?

13. How many hands of 13 cards, none of which is a spade, can be dealt from a 52-card bridge deck?

14. How many hands of 13 cards containing all 4 aces can be dealt from a 52-card bridge deck?

15. How many hands of 5 cards containing no aces can be dealt from a 52-card bridge deck?

16. How many hands of 13 cards containing 5 spaces, 4 hearts, 3 diamonds, and 1 club can be dealt from a 52-card bridge deck?

17. How many hands of 13 cards containing 4 spades, 3 hearts, 3 diamonds, and 3 clubs can be dealt from a 52-card bridge deck?

18. How many other hands of 13 cards containing 4 of any one suit and 3 of each of the other suits can be dealt from a 52-card bridge deck?

19. (a) A baseball league has eight teams, and each team plays six games against each of the other teams. How many games are there altogether?
(b) Repeat part (a) with n in place of eight and k in place of six.

20. In how many ways can a ten-question true-false examination be answered so that there are five true and five false responses? (Any way of answering amounts to a choice of which questions to mark *true*.)

21. A drawer contains seven good light bulbs and three defective light bulbs. In how many ways can four bulbs be chosen under the following conditions?

    (a) None is defective.

    (b) Two are good and two are defective.

    (c) At least two are good.

22. How many lines are determined by n points if no three of the points are collinear?

23. How many planes are determined by n points if no four of the points are coplanar? (Three non-collinear points determine a plane.)

24. How many triangles can be formed if the vertices are chosen from a set of n points all lying on a circle?

25. How many quadrilaterals can be formed if the vertices are chosen from a set of n points all lying on a circle?

26. Ten parallel lines are all perpendicular to ten other parallel lines. How many rectangles are formed?

27. How many diagonals does a regular n-sided polygon have? (A polygon is *regular* if all of its angles are equal and all of its sides are equal.

A *diagonal* is a segment joining two nonadjacent vertices.)

28. How many different sums of money can be formed by choosing two coins from a penny, a nickel, a dime, a quarter, and a half dollar?

29. How many different sums of money can be formed by choosing one or more coins from a penny, a nickel, a dime, a quarter, and a half dollar? (Compare Exercise 28.)

30. Explain why the number of subsets of an *n*-element set is

$$C(n, 0) + C(n, 1) \\ + C(n, 2) + \cdots + C(n, n).$$

[The empty set, which is accounted for by $C(n, 0)$, is included.] Use the Binomial Theorem (Section 61) to show that the indicated sum equals 2^n, so that an *n*-element set has 2^n subsets. [Suggestion: Use $a = b = 1$ in Equation (61.4). Compare Exercise 31 in Section 61.]

SECTION **64**
Probability

A. INTRODUCTION

Mathematical probability is concerned with events whose occurrence is subject to chance. The goal is to represent the likelihood of such an event by a real number p, called the *probability* of the event, with $0 \le p \le 1$. This is done in such a way that the more likely the event, the larger the number p. If an event is *certain* to occur, then $p = 1$; if an event is certain *not* to occur, then $p = 0$.

Not all events involving uncertainty fall within the province of mathematical probability. A question such as "What is the probability good will win out over evil?", however important, is not one for mathematical probability. Still, mathematical probability is extremely useful, with applications throughout business, science, technology, and elsewhere. Unfortunately, all that can be given here is a brief introduction to the subject. To illustrate the basic ideas we begin with three sample questions, which will be answered later in the section.

Example 64.1 If a coin is tossed three times, what is the probability a head will occur exactly twice?

Example 64.2 If the six persons making up three married couples are seated randomly in a row, what is the probability a particular husband and wife will be in adjacent seats?

Example 64.3 In a room of 25 randomly chosen persons, what is the probability they all have different birthdays?

The underlying process in any such example is called an **experiment.** The set of all possible outcomes in an experiment is called the **sample space** of the experiment. Any subset of the sample space is called an **event.**

Example 64.1 In this example the *experiment* consists of tossing three coins and observing
(Continued) which land heads (H) and which land tails (T). The *sample space* is

$$\{HHH, HHT, HTH, THH, HTT, THT, TTH, TTT\}. \qquad (64.1)$$

Thus there are eight possible outcomes. The subset of those outcomes in which H occurs exactly twice is

$$\{HHT, HTH, THH\}. \tag{64.2}$$

This subset represents the *event* with which Example 64.1 is concerned; thus there are three possible outcomes corresponding to this event. ■

In this book we consider only experiments for which the sample space is finite.

Definition The **probability** of an event E, in an experiment in which all possible outcomes are equally likely, is defined by

$$P(E) = \frac{n(E)}{n(S)}, \tag{64.3}$$

where $n(S)$ denotes the number of elements in the sample space (the total number of possible outcomes) and $n(E)$ denotes the number of possible outcomes corresponding to the event E.

Example 64.1 (*Concluded*) For the question in Example 64.1, $n(S) = 8$ [see (64.1)] and $n(E) = 3$ [see (64.2)]. Therefore, $P(E) = \frac{3}{8}$. ■

The number $P(E)$ in (64.3) corresponds to the number p in the opening paragraph of this section. Our definition of probability applies only to experiments having finitely many possible outcomes, and those outcomes must all be equally likely. The computation of a probability in such cases reduces to the computation of the two numbers $n(E)$ and $n(S)$, and this will very often involve permutations and combinations (Sections 62 and 63). Before we do those computations for Examples 64.2 and 64.3, let us look at the following simpler example. When we say that we *select a number at random* in this example we mean that all of the outcomes are equally likely.

Example 64.4 Suppose a number is selected at random from the list 1, 2, 3, 4, 5, 6, 7, 8, 9. Compute the probability of each of the following events.

(a) An even number is selected.

(b) An odd number is selected.

(c) A negative number is selected.

(d) A positive number is selected.

Solution In each part of this example the experiment consists of selecting a number at random from the list 1, 2, 3, 4, 5, 6, 7, 8, 9. The sample space is $S = \{1, 2, 3, 4, 5, 6, 7, 8, 9\}$. Thus $n(S) = 9$.

(a) $E = \{2, 4, 6, 8\}$, so $P(E) = \dfrac{4}{9}$.

(b) $E = \{1, 3, 5, 7, 9\}$, so $P(E) = \dfrac{5}{9}$.

(c) E is the empty set, so $P(E) = \dfrac{0}{9} = 0$.

(d) $E = S$, so $P(E) = \dfrac{9}{9} = 1$. ▪

Solution for The experiment consists of arranging the six persons randomly in a row. The
Example 64.2 sample space S consists of all of the ways this can be done. Thus $n(S)$ is the
number of permutations (arrangements) of six elements taken six at a time, so
that $n(S) = 6!$ (Section 62).

The event E corresponds to all of the arrangements in which the particular
husband and wife are in adjacent seats. Let's compute $n(E)$ in two steps; first
we'll compute the number of arrangements in which they are adjacent with the
husband to the right of the wife. In this case we have five elements to arrange
in a row (the given couple, considered as one, and the other four persons); this
can be done in 5! ways. There will be the same number of ways with the husband
to the left of the wife. Thus $n(E) = 2(5!)$.

Since $n(S) = 6!$ and $n(E) = 2(5!)$, we arrive at $P(E) = 2(5!)/6! = \frac{2}{6} = \frac{1}{3}$. ▪

Solution for The experiment consists of selecting 25 persons at random and recording their
Example 64.3 birthdays. For simplicity we ignore leap years, so that there are 365 possibilities
for the birthday of each person. Since these 25 birthdays are mutually independent, the Basic Counting Principle (Section 62) shows that $n(S) = 365^{25}$,
where S denotes the sample space.

The event E corresponds to all of the possibilities in which no two birthdays
are the same. To compute $n(E)$ we again apply the Basic Counting Principle.
If we record the birthdays one at a time, then there are 365 possibilities for the
first birthday. Then there are 364 possibilities for the second birthday, which
must be different from the first. Then there are 363 possibilities for the third
birthday, which must be different from both of the first two. If we continue in
this way and apply the Basic Counting Principle we will arrive at $n(E) = 365 \cdot 364 \cdot \cdots \cdot 341$ (25 factors). In the notation of Section 62, $n(E) = P(365, 25)$.
Since $n(S) = 365^{25}$ and $n(E) = P(365, 25)$, we arrive at

$$P(E) = \frac{n(E)}{n(S)} = \frac{365 \cdot 364 \cdot \cdots \cdot 341}{365^{25}} \approx 0.4313. \qquad \boxed{\text{C}}$$

Notice carefully what this implies: Among 25 randomly chosen persons, the
probability is less than $\frac{1}{2}$ that their birthdays will all be different. Put another
way, the probability is more than $\frac{1}{2}$ that at least two of the persons will have the
same birthday. ▪

Continuation of The reasoning in the solution for Example 64.3 shows that in a room of n
Example 64.3 randomly chosen persons the probability they all have different birthdays is

$$P(E) = \frac{n(E)}{n(S)} = \frac{P(365, n)}{365^n} = \frac{365 \cdot 364 \cdot \cdots \cdot (365 - n + 1)}{365^n}. \qquad (64.4)$$

Table 64.1 shows these probabilities (to four decimal places) for selected values
of n. The last row shows that more than 97% of the time in a room of 50
randomly chosen persons at least two of the persons will have the same birthday

TABLE 64.1

Number of persons	Probability all birthdays are different
5	0.9729
10	0.8831
15	0.7471
20	0.5886
22	0.5243
23	0.4927
25	0.4313
50	0.0296

$(1 - 0.0296 = 0.9704 > 0.97)$. The table also shows that more than 50% of the time in a room of 23 (or more) randomly chosen persons at least two of the persons will have the same birthday.

B. PROPERTIES AND MORE EXAMPLES

The opening paragraph of this section stated that if p represents the probability of an event, then it should be true that $0 \leq p \leq 1$, with $p = 1$ for an event that is certain to occur and $p = 0$ for an event that is certain not to occur. We can now prove these properties. We use Ø (Greek letter *phi*) to denote the empty set, which corresponds to an event that is certain not to occur.

If S denotes a sample space and E an event, then

$$0 \leq P(E) \leq 1.$$

In particular,

$$P(\emptyset) = 0 \quad \text{and} \quad P(S) = 1.$$

Proof If E denotes any subset of S, then $0 \leq n(E) \leq n(S)$, so that

$$\frac{0}{n(S)} \leq \frac{n(E)}{n(S)} \leq \frac{n(S)}{n(S)}$$

$$0 \leq P(E) \leq 1.$$

Also, $P(\emptyset) = n(\emptyset)/n(S) = 0/n(S) = 0$ and $P(S) = n(S)/n(S) = 1$. □

Recall that if A and B are sets, then $A \cup B$ denotes the union of A and B, and $A \cap B$ denotes the intersection of A and B (Section 10). If A and B are finite sets, then

$$n(A \cup B) = n(A) + n(B) - n(A \cap B). \tag{64.5}$$

We must subtract $n(A \cap B)$ here because otherwise each element of $A \cap B$ would be counted twice on the right, once as an element of A and again as an element of B.

If A and B represent events, then $A \cup B$ occurs iff either A occurs or B occurs (or both A and B occur); $A \cap B$ occurs iff both A and B occur.

> If A and B denote events from a sample space S, then
>
> $$P(A \cup B) = P(A) + P(B) - P(A \cap B). \tag{64.6}$$

Proof Apply Equation (64.5):

$$n(A \cup B) = n(A) + n(B) - n(A \cap B)$$

$$\frac{n(A \cup B)}{n(S)} = \frac{n(A)}{n(S)} + \frac{n(B)}{n(S)} - \frac{n(A \cap B)}{n(S)}$$

$$P(A \cup B) = P(A) + P(B) - P(A \cap B).$$

□

Example 64.5 Assume that a die (one of a pair of dice) is rolled twice. Find the probability that either the first number is 1 or the sum of the two numbers is at least 6.

Solution Figure 64.1 shows the sample space. Let A denote the event that the first number is 1, and let B denote the event that the sum of the two numbers is at least 6. The outcomes in A and B are indicated in the figure. We see that

$$n(A) = 6 \text{ so } P(A) = \frac{n(A)}{n(S)} = \frac{6}{36},$$

$$n(B) = 26 \text{ so } P(A) = \frac{n(B)}{n(S)} = \frac{26}{36}, \quad \text{and}$$

$$n(A \cap B) = 2 \text{ so } P(A \cap B) = \frac{n(A \cap B)}{n(S)} = \frac{2}{36}.$$

A {	1,1	1,2	1,3	1,4	1,5	1,6
	2,1	2,2	2,3	2,4	2,5	2,6
	3,1	3,2	3,3	3,4	3,5	3,6
	4,1	4,2	4,3	4,4	4,5	4,6
B {	5,1	5,2	5,3	5,4	5,5	5,6
	6,1	6,2	6,3	6,4	6,5	6,6

FIGURE 64.1

Therefore, by (64.6),

$$P(A \cup B) = \frac{6}{36} + \frac{26}{36} - \frac{2}{36} = \frac{30}{36} = \frac{5}{6}.$$

■

Events A and B are said to be **mutually exclusive** if $A \cap B = \emptyset$. Since $P(\emptyset) = 0$, we have the following special case of (64.6).

> If the events A and B are mutually exclusive, then
>
> $$P(A \cup B) = P(A) + P(B). \tag{64.7}$$

If E denotes an event from a sample space S, then the **complement** of E, denoted $\overline{E}$, represents all of the outcomes that are in S but not in E. For example, in rolling a die, the complement of the event "the outcome is at least 3" is the event "the outcome is less than 3." We have the following convenient result between the probability of an event and the probability of its complement.

> If E is any event with complement $\overline{E}$, then
>
> $$P(\overline{E}) = 1 - P(E). \qquad (64.8)$$

Proof From the definition of $\overline{E}$ it follows that $E \cup \overline{E} = S$ and $E \cap \overline{E} = \emptyset$. Therefore, by (64.7) with $A = E$ and $B = \overline{E}$, we have $P(S) = P(E) + P(\overline{E})$, so that $1 = P(E) + P(\overline{E})$, which gives (64.8). □

One convenient use of (64.8) comes by realizing that the complement of "at least one" is "none." Here is an example.

Example 64.6 A hand of 13 cards is chosen randomly from a bridge deck of 52 cards. What is the probability the hand contains at least 1 ace? (The makeup of a bridge deck is explained preceding Example 63.5.)

Solution The sample space S consists of all possible 13-card hands from a 52-card bridge deck. Therefore, from Example 63.5, $n(S) = C(52, 13)$.

Let E represent the event "the hand contains at least one ace." Then $\overline{E}$ represents the event "the hand contains no ace." The number $n(\overline{E})$ is the number of 13-card hands from the 48-card deck that remains when the 4 aces have been removed. Thus $n(\overline{E}) = C(48, 13)$. Therefore,

$$P(E) = 1 - P(\overline{E})$$

$$= 1 - \frac{C(48, 13)}{C(52, 13)}$$

$$= 1 - \left[\left(\frac{48!}{13!35!}\right) \bigg/ \left(\frac{52!}{13!39!}\right)\right]$$

$$= 1 - \frac{39 \cdot 38 \cdot 37 \cdot 36}{52 \cdot 51 \cdot 50 \cdot 49}$$

$$\approx 0.70.$$

EXERCISES FOR SECTION 64

1. If a coin is tossed three times, what is the probability a head will occur exactly once? Begin by listing all of the elements of the sample space.

2. If a coin is tossed four times, what is the probability a tail will occur at least twice? Begin by listing all of the elements of the sample space.

3. If a coin is tossed three times, what is the prob-

ability of an odd number of heads? Begin by listing all of the elements of the sample space.

4. A number is formed by arranging the digits 1, 2, and 3 in random order (with no digit repeated). Find the probability of each of the following events.

   (a) The number is greater than 100.

(b) The number is greater than 200.

(c) The number is greater than 220.

(d) The number is greater than 330.

5. A number is chosen randomly from the set {17, 18, 19, 20, 21, 22}. Find the probability of each of the following events.

(a) The number is even.

(b) The sum of the digits is even.

(c) Either the number is even or the sum of its digits is even.

(d) The number is a perfect square (that is, the square of an integer).

6. A number is selected randomly from the set of eight primes less than 20. Find the probability of each of the following events.

(a) The number is even.

(b) The number is less than 10.

(c) The number is two more than another prime.

(d) The number is a sum of three different primes.

7. Assume that a pair of dice is rolled. Find the probability of each of the following events. (An array such as in Figure 64.1 should help.)

(a) Double 4 is obtained.

(b) Neither die falls 3.

(c) Neither 3 nor 4 appears.

(d) Each die shows more than 3.

(e) At least one die shows more than 3.

(f) Exactly one die shows more than 3.

8. A card is chosen randomly from a bridge deck of 52 cards. Find the probability of each of the following events.

(a) The card is a spade.

(b) The card is not a spade.

(c) The card is an ace.

(d) The card is not an ace.

(e) The card is either a spade or an ace.

(f) The card is a jack, queen, or king.

9. Assume that in families with three children, all of the eight arrangements GGG, GGB, GBG, . . . , BBB are equally likely, where, for example, GGB denotes that the oldest is a girl, the next is also a girl, and the youngest is a boy. Compute the probability of each of the following events.

(a) There is exactly one boy.

(b) There is at least one boy.

(c) There is at most one boy.

(d) All are girls.

(e) The oldest is a boy.

(f) There is no boy older than a girl.

10. Six dice are rolled. What is the probability that no number occurs twice?

11. Four friends enter an ice cream parlor, and each one chooses a flavor randomly from among the 20 that are offered. What is the probability that no two choose the same flavor?

12. Four cards are drawn at random from a bridge deck of 52 cards. What is the probability no two of the cards are from the same suit?

13. Two shoes are chosen randomly from the twelve shoes making up six distinguishable pairs. What is the probability the two shoes belong to the same pair?

14. A committee contains six men and six women. What is the probability that a randomly chosen three-member subcommittee will contain only women?

15. Ten basketball players, three of whom are brothers, are divided randomly into two five-member teams. What is the probability the three brothers are on the same team?

16. A 25-member university committee contains 7 female faculty members, 8 male faculty members, 6 female student members, and 4 male student members. A 5-member subcommittee is chosen at random. Find the probability of each of the following events. How does Equation (64.6) apply?

(a) The committee contains only males.

(b) The committee contains only faculty members.

(c) The committee is made up entirely of male faculty members.

(d) The committee contains only males or only faculty members.

17. A die is rolled twice. Find the probability of each of the following events. (An array such as that in Figure 64.1 will help.)

(a) The first number is odd.

(b) The sum of the numbers is odd.

(c) The first number is odd and the sum of the numbers is odd.

(d) The first number is odd or the sum of the numbers is odd.

18. A subset is chosen randomly from the 15 non-empty subsets of $\{A, B, C, D\}$. Find the probability of each of the following events. How does Equation (64.6) apply?

(a) The subset contains D.

(b) The subset contains exactly three elements.

(c) The subset contains exactly three elements, one of which is D.

(d) The subset contains exactly three elements or the subset contains D.

19. A hand of 13 cards is chosen randomly from a bridge deck of 52 cards. Compute the probability of each of the following events.

(a) The hand contains no spades.

(b) The hand contains only spades.

(c) The hand contains no aces.

(d) The hand contains all four aces.

20. Ninety lottery tickets are numbered from 1 to 90. Assume that two tickets are chosen at random from the 90. Compute the probability of each of the following events.

(a) Each of the tickets has a number less than 10.

(b) Each of the tickets drawn has an even number.

(c) The first ticket has an even number and the second ticket has an odd number.

(d) One of the tickets has an even number and one of the tickets has an odd number.

21. Imagine a game in which a player either wins, in which case he receives a certain prize, or loses, in which case he receives nothing. The player's **mathematical expectation** for such a game can be defined to be his probability of winning times the value of the prize. This can be thought of as the reasonable amount to pay for the privilege of playing. For example, in 100 rounds of coin tossing with prize $1 for each tail, a player can reasonably expect to win $50, or, on the average, $0.50 (the mathematical expectation) per round.

(a) If a die is to be thrown once and a prize of $3 is to be awarded if it turns up 6, what is the mathematical expectation?

(b) How much should a gambling house charge for each turn at a certain game if the player has probability $\frac{1}{8}$ of winning, the prize is $10, and the house wants a profit of $0.50 per turn on the average?

(c) Would you pay $1 for a chance to win $1,000,000 if your chances of winning were only 1 in 5 million? Compute the relevant mathematical expectation. (Be honest—lotteries thrive because commercial value can exceed mathematical expectation.)

(d) Would you pay $10,000 for a chance to win $100,000 if your chance of winning were 1 in 5? Compute the relevant mathematical expectation. (The point here is the opposite of that in (c)—mathematical expectation can exceed commercial value. Of course, your answer here may depend on how much $10,000 means to you.)

REVIEW EXERCISES FOR CHAPTER XIV

Compute.

1. $P(12, 3)$ **2.** $P(8, 4)$

3. $C(7, 5)$ **4.** $C(11, 5)$

5. Five flutists audition for the first, second, and third chairs in the flute section of a school orchestra. In how many ways can the chairs be filled (ignoring ability)?

6. The four faces on Mount Rushmore are in the order Washington, Jefferson, Roosevelt, Lincoln. How many other orders would have been possible?

7. A menu lists three appetizers, four main courses, and five desserts. How many different ways are there to choose one of each? (Assume all combinations are acceptable.)

8. A school play will have five roles: a king, a queen, a prince, a princess, and a (male) court jester. At the auditions, 10 boys try for the 3 male roles, and 12 girls try for the 2 female roles. In how many different ways can the 5 roles be filled?

9. A sergeant must choose four "volunteers" for a clean-up crew. He has ten recruits from which to choose. How many different crews are possible?

10. To fulfill her graduation requirements, a student must choose four courses: two electives from a group of six courses in the humanities, and two other electives from a group of five courses in the sciences. How many different combinations are possible?

11. The three numbered volumes of a three-volume set of books are placed on a shelf in random order. What is the probability they are in the correct order?

12. You are one of 20 students in a class in which the teacher asks questions of three randomly chosen students each day. What is the probability that you will be chosen on a particular day?

13. A box contains 20 blue cards numbered 1–20, and 15 white cards numbered 1–15. One card is chosen randomly from the box.

(a) What is the probability the card is white?

(b) What is the probability the number on the card is odd?

(c) What is the probability the card is either white or has an odd number?

14. A hand of 5 cards is chosen randomly from a bridge deck of 52 cards. What is the probability the cards are all diamonds? What is the probability they all belong to the same suit?

15. An investor randomly chooses two stocks from a set of five that look equally attractive. Over the next year, two of the original five increase in value while the other three decrease. What is the probability that the two stocks that increase are those the investor chose? If three of the five increase and the other two decrease, what is the probability that both of the investor's choices increase?

16. A three-member subcommittee is chosen randomly from a ten-member committee of five males and five females. What is the probability the committee contains at least one male?

CHAPTER XV

CONIC SECTIONS

Earlier chapters have covered a number of topics from analytic (coordinate) geometry, including lines, circles, graphs of general classes of functions, vectors, and polar coordinates. This chapter continues the coverage of analytic geometry by giving a full treatment of conic sections.

SECTION **65**
Parabolas

A. INTRODUCTION

A **conic section** is a curve formed by the intersection of a plane and a double-napped right circular cone (Figure 65.1). The most important conic sections are *parabolas, ellipses,* and *hyperbolas,* which will be studied in turn in this and the next two sections. In each case the conic section will be defined in a way that is equivalent to, but more convenient than, that given in Figure 65.1. With a few minor exceptions, the conic sections are the same as the graphs of the quadratic equations with variables x and y; in Section 67D this will be explained in more detail.

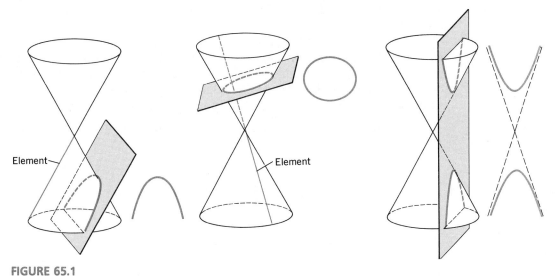

FIGURE 65.1

(a) Parabola. The intersecting plane is parallel to an element.

(b) Ellipse. The intersecting plane cuts all of the elements of the cone.

(c) Hyperbola. The intersecting plane cuts both nappes (parts) of the cone.

B. DEFINITION

Definition A **parabola** is the set of all points in a plane that are equidistant from a fixed line (the **directrix**) in the plane and a fixed point (the **focus**) in the plane but not on the line.

In Figure 65.2, P is a typical point on the parabola with directrix D and focus F. Thus the distance between P and D, denoted d_1, is equal to the distance between P and F, denoted d_2. The line through F perpendicular to D is called

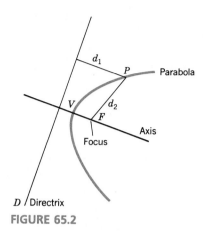

FIGURE 65.2

the **axis** of the parabola. The point V where the parabola intersects its axis is called the **vertex** of the parabola. The vertex is the midpoint of the segment cut off on the axis by the focus and the directrix. A parabola is symmetric with respect to its axis.

To derive equations for parabolas in a Cartesian plane, we first consider the special case in which the directrix is horizontal with equation $y = -p$ and the focus is $F(0, p)$, where p is a nonzero constant. Whether $p > 0$ or $p < 0$, the vertex in this case is at the origin and the y-axis is the axis of the parabola. Figure 65.3 shows a case with $p > 0$. A point $P(x, y)$ is on the parabola iff

$$\text{distance between } P \text{ and } Q = \text{distance between } P \text{ and } F \qquad (65.1)$$

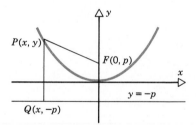

FIGURE 65.3

where $Q(x, -p)$ is the foot of the perpendicular to the directrix from $P(x, y)$. Equation (65.1) is equivalent to

$$\sqrt{(x - x)^2 + (y + p)^2} = \sqrt{(x - 0)^2 + (y - p)^2}$$

$$(y + p)^2 = x^2 + (y - p)^2$$

$$y^2 + 2py + p^2 = x^2 + y^2 - 2py + p^2$$

$$x^2 = 4py. \qquad (65.2)$$

If $p > 0$ the parabola opens upward, as in Figure 65.3. If $p < 0$ the parabola opens downward. Summarizing:

> The graph of $x^2 = 4py$ is a parabola opening upward if $p > 0$ and downward if $p < 0$. The directrix is $y = -p$, the focus is $(0, p)$, and the vertex is at the origin. $\qquad (65.3)$

If $p = \frac{1}{4}$, then Equation (65.2) becomes $y = x^2$, whose graph is that of the quadratic function $f(x) = x^2$ (Figure 16.1). More generally, the graph of Equation (65.2) is that of the quadratic function $f(x) = (1/4p)x^2$.

If the directrix is vertical with equation $x = -p$ and the focus is at $(p, 0)$, then a derivation like that for Equation (65.2) shows that the equation of the parabola is

$$y^2 = 4px. \tag{65.4}$$

Figure 65.4 shows a case with $p > 0$, for which the parabola opens to the right. If $p < 0$ the parabola opens to the left. In either case, the vertex is at the origin and the x-axis is the axis of the parabola.

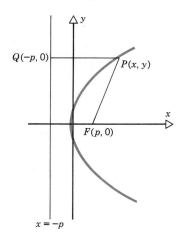

FIGURE 65.4

The graph of $y^2 = 4px$ is a parabola opening to the right if $p > 0$ and to the left if $p < 0$. The directrix is $x = -p$, the focus is $(p, 0)$, and the vertex is at the origin. $\tag{65.5}$

A parabola with vertex at the origin is uniquely determined by its directrix, or by its focus, or by any (nonvertex) point on the parabola together with the direction the parabola opens.

Example 65.1 Find an equation of the parabola with vertex at the origin and that opens to the left and passes through $(-2, 5)$ (Figure 65.5).

Solution The equation will have the form $y^2 = 4px$, given by Equation (65.4). Since $(-2, 5)$ is on the parabola, we must have

$$5^2 = 4p(-2)$$

$$p = -\frac{25}{8}.$$

Thus the equation is

$$y^2 = 4\left(-\frac{25}{8}\right)x$$

$$y^2 = -\frac{25}{2}x.$$

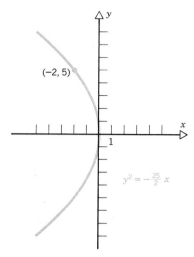

FIGURE 65.5

Example 65.2 Find the directrix and focus of the parabola whose equation is $y = 2x^2$.

Solution Rewrite $y = 2x^2$ in the form of Equation (65.2).

$$x^2 = \frac{1}{2} y$$

$$x^2 = 4 \cdot \frac{1}{8} y$$

This shows that $p = \frac{1}{8}$, so the parabola opens upward $(p > 0)$, the equation of the directrix is $y = -\frac{1}{8}$, and the focus is at $(0, \frac{1}{8})$. Figure 65.6 shows the graph.

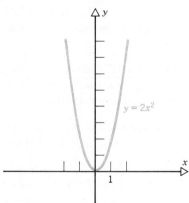

FIGURE 65.6

To draw the graph of a parabola, always plot the vertex and several points on one side of the axis, and then take advantage of the symmetry with respect to the axis.

C. TRANSLATIONS

For a parabola whose directrix is parallel to a coordinate axis but whose vertex is not at the origin, we use the results from Subsection B together with the ideas about translations that were presented in Section 20. The reasoning used to justify Statements (20.2) and (20.3) will also yield the following statements (where the term *Cartesian equation* refers to an equation with variables x and y).

If x is replaced by $x - h$ in a Cartesian equation, the resulting graph will be the graph of the original equation translated to the right by $|h|$ units if $h > 0$, and translated to the left by $|h|$ units if $h < 0$. If y is replaced by $y - k$ in a Cartesian equation, the resulting graph will be the graph of the original equation translated upward by $|k|$ units if $k > 0$, and translated downward by $|k|$ units if $k < 0$. $\qquad$ (65.6)

Example 65.3 Draw the graph of $y + 3 = 2(x - 1)^2$.

Solution The given equation is $y = 2x^2$ with x replaced by $x - 1$ and y replaced by $y + 3$, or $y - (-3)$. Therefore, by (65.6), the graph of $y + 3 = 2(x - 1)^2$ is the graph of $y = 2x^2$ translated to the right by 1 unit and downward by 3 units. The graph of $y = 2x^2$ is shown in Figure 65.6. The graph of $y + 3 = 2(x - 1)^2$ is shown in Figure 65.7. ■

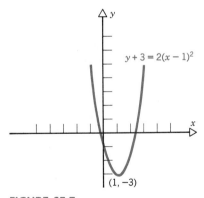

FIGURE 65.7

By applying Statement (65.6) with Statements (65.3) and (65.5), we obtain the following general results.

The graph of
$$(x - h)^2 = 4p(y - k)$$
is a parabola with directrix $y = k - p$, focus $(h, k + p)$, vertex (h, k), and axis $x = h$.
The graph of $\qquad$ (65.7)
$$(y - k)^2 = 4p(x - h)$$
is a parabola with directrix $x = h - p$, focus $(h + p, k)$, vertex (h, k), and axis $y = k$.

The equations in (65.7) are called the **standard forms** for the equations of the parabolas they represent. The graph of any equation that can be written in the form

$$y = ax^2 + bx + c, \quad \text{with } a \neq 0,$$

is a parabola with directrix parallel to the x-axis. The graph of any equation that can be written in the form

$$x = ay^2 + by + c, \quad \text{with } a \neq 0,$$

is a parabola with directrix parallel to the y-axis. By completing a square (Section 8), an equation of either of the latter types can be converted to the appropriate standard form. Here is an example.

Example 65.4 Rewrite $x = \frac{1}{6}y^2 - \frac{2}{3}y - \frac{4}{3}$ in standard form and draw the graph of the parabola it represents.

Solution Multiply both sides by 6 to clear fractions, and isolate the y terms on the right.

$$6x = y^2 - 4y - 8$$

$$6x + 8 = y^2 - 4y$$

Now complete the square on the y terms by adding 4 to both sides. Then put the equation in the second standard form from (65.7).

$$6x + 8 + 4 = y^2 - 4y + 4$$

$$6(x + 2) = (y - 2)^2$$

$$(y - 2)^2 = 4 \cdot \frac{3}{2}[x - (-2)]$$

This has the second form in (65.7) with $h = -2$, $k = 2$, and $p = \frac{3}{2}$. Thus the graph is a parabola with directrix $x = -2 - \frac{3}{2}$, or $x = -\frac{7}{2}$, focus $(-2 + \frac{3}{2}, 2)$ $= (-\frac{1}{2}, 2)$, vertex $(-2, 2)$, and axis $y = 2$. Figure 65.8 shows the graph. ■

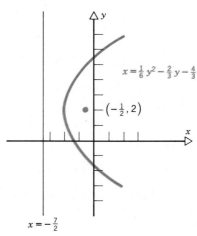

$$x = \tfrac{1}{6}y^2 - \tfrac{2}{3}y - \tfrac{4}{3}$$

$$\left(-\tfrac{1}{2}, 2\right)$$

$$x = -\tfrac{7}{2}$$

FIGURE 65.8

Example 65.5 Find an equation for the parabola with focus $(1, -2)$ and vertex $(4, -2)$. Also find the directrix and the axis of the parabola.

Solution The axis is the line through the focus and the vertex, so in this case the axis has equation $y = -2$. The distance between the focus and the vertex is $|p|$, so in this case $|p| = 3$. The focus is to the left of the vertex so the parabola opens to the left and $p = -3$. Using $h = 4$ and $k = -2$, since the vertex is $(4, -2)$, we get the standard form

$$(y + 2)^2 = -12(x - 4).$$

The directrix is $x = 7$. Figure 65.9 shows the result.

Applications of parabolas are given in Section 16.

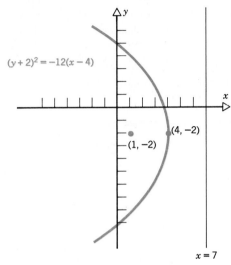

$(y + 2)^2 = -12(x - 4)$

$(4, -2)$

$(1, -2)$

$x = 7$

FIGURE 65.9

EXERCISES FOR SECTION 65

The graph of each equation in Exercises 1–15 is a parabola. In each case, find the directrix, the focus, the vertex, and the axis, and draw the graph.

1. $2x^2 + y = 0$ **2.** $4x^2 - y = 0$

3. $12x + y^2 = 0$ **4.** $6x = y^2$

5. $y^2 = -x$ **6.** $x^2 = 4y$

7. $(x + 2) + (y - 1)^2 = 0$

8. $8x - (y + 4)^2 = 0$

9. $(x - 3)^2 = -10(y - 3)$

10. $x = y^2 + 2y - 2$

11. $2x^2 + y + 3 = 0$

12. $x - 9 = y^2 - 6y$

13. $4y = 4x^2 + 12x + 9$

14. $18x = 9y^2 - 24y + 16$

15. $8x^2 - 8x + 2 = 8y$

In each of Exercises 16–24 find an equation of the parabola satisfying the given conditions.

16. Vertex $(0, 0)$, focus $(-4, 0)$.

17. Vertex $(0, 0)$, directrix $y = 5$.

18. Vertex $(0, 0)$, contains $(-4, -1)$, axis the y-axis.

19. Vertex $(-1, 4)$, directrix $y = 0$.

20. Vertex $(2, 3)$, contains $(0, 0)$, axis $x = 2$.

21. Vertex $(0, -4)$, focus $(0, 0)$.

22. Focus $(-3, 3)$, directrix $x = -5$.

23. Vertex $(5, 1)$, contains $(0, 2)$, symmetric with respect to the line $y = 1$.

24. Focus $(5, 5)$, directrix $y = 10$.

25. A parabolic arch on a bridge has span 30 feet, and 10 feet from each end the rise is 5 feet. (See

Exercise 51 in Section 16.) What is the rise at the center?

26. Derive Equation (65.4) for a parabola with focus $(p, 0)$ and directrix $x = -p$. The derivation of (65.3) is similar.

27. The segment cut off by a parabola on the line that is parallel to the directrix and that passes through the focus is called the **latus rectum** of the parabola. Prove that if $p > 0$, then the length of the latus rectum of $x^2 = 4py$ is $4p$.

28. The **focal radius** of a point on a parabola is the distance between the point and the focus of the parabola. Prove that the focal radius of the point $P(x, y)$ on the parabola $x^2 = 4py$ is $|y + p|$.

29. Prove that the line $y = mx + b$ intersects the parabola $y^2 = 4px$ in two points if $p^2 > mbp$, in one point if $p^2 = mbp$, and in no point if $p^2 < mbp$.

30. Suppose $p > 0$ and $k > 0$. By (65.6), the graph of $x^2 = 4p(y - k)$ is the graph of $x^2 = 4py$ translated upward by k units, and therefore the two graphs do not intersect. Prove that if $p > 0$, $q > 0$, and $k > 0$, then the graphs of $x^2 = 4q(y - k)$ and $x^2 = 4py$ intersect iff $q > p$. [Interpretation: If two different parabolas have the same axis and open the same direction, then they intersect iff the one whose vertex is "inside" the other is "flatter" (by any amount).]

SECTION **66**
Ellipses

A. DEFINITION

Definition An **ellipse** is the set of all points in a plane such that the sum of the distances from two fixed points (the **foci**) in the plane is a constant. (*Foci* is the plural of **focus**.)

Figure 66.1 shows an example with foci $F'(-c, 0)$ and $F(c, 0)$. The sum $d_1 + d_2$ of the distances to P from F' and F is the same for all points P on the ellipse.

To derive equations for ellipses in a Cartesian plane, we first consider the special case shown in Figure 66.1. Assume $d_1 + d_2 = 2a$ (a constant) for all

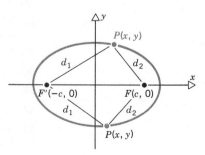

FIGURE 66.1 $d_1 + d_2$ is the same for all points P on the ellipse.

$P(x, y)$ on the ellipse. Using the distance formula, we have

$$d_1 + d_2 = 2a$$

$$\sqrt{(x + c)^2 + y^2} + \sqrt{(x - c)^2 + y^2} = 2a$$

$$\sqrt{(x + c)^2 + y^2} = 2a - \sqrt{(x - c)^2 + y^2}.$$

Now square both sides and simplify.

$$x^2 + 2cx + c^2 + y^2 = 4a^2 - 4a\sqrt{(x - c)^2 + y^2} + x^2 - 2cx + c^2 + y^2$$

$$a^2 - cx = a\sqrt{(x - c)^2 + y^2}.$$

Now square again and simplify.

$$a^4 - 2a^2cx + c^2x^2 = a^2(x^2 - 2cx + c^2 + y^2)$$

$$(a^2 - c^2)x^2 + a^2y^2 = a^2(a^2 - c^2)$$

To simplify our result, we introduce the constant b defined by $b > 0$ and

$$b^2 = a^2 - c^2. \tag{66.1}$$

(Since $d_1 + d_2 = 2a$, we see from Figure 66.1 that $2a > 2c$, or $a > c$, so $a^2 > c^2$.) We now have

$$b^2x^2 + a^2y^2 = a^2b^2$$

$$\frac{x^2}{a^2} + \frac{y^2}{b^2} = 1. \tag{66.2}$$

The line through the foci intersects the ellipse in two points called the **vertices** of the ellipse. In Figure 66.1, the vertices correspond to the x-intercepts; these are at $(-a, 0)$ and $(a, 0)$, as can be seen by using $y = 0$ in Equation (66.2). The segment connecting the vertices is called the **major axis** of the ellipse. See Figure 66.2. The midpoint of the major axis is called the **center** of the ellipse. In Figure 66.1 the center is at the origin. Each segment connecting the center to a vertex is a **semimajor axis.**

Consider the line through the center perpendicular to the major axis. The segment cut off on this line by the ellipse is called the **minor axis** of the ellipse. In Figure 66.1, the minor axis is the segment joining $(0, -b)$ and $(0, b)$. Each segment connecting the center to an endpoint of the minor axis is called a **semiminor axis.**

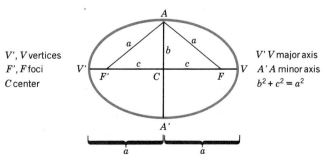

FIGURE 66.2

The length of the minor axis is less than the length of the major axis. For Figure 66.1, this follows from Equation (66.1), which shows $b^2 < a^2$ so that $b < a$ and $2b < 2a$. An ellipse is symmetric with respect to both its major and minor axes. For the ellipse represented by Equation (66.2), this follows because the equation remains unchanged if x is replaced by $-x$ and y is replaced by $-y$.

If the foci of the ellipse are $F'(0, -c)$ and $F(0, c)$ on the y-axis, rather than on the x-axis, then the equation of the ellipse is

$$\frac{x^2}{b^2} + \frac{y^2}{a^2} = 1. \tag{66.3}$$

As in the derivation of Equation (66.2), the sum of the distances from the foci to any point on the ellipse is $2a$, $b^2 = a^2 - c^2$, and $b < a$. Thus, with the foci on the y-axis, the larger number, a^2, is under y^2 when the equation is written in the form (66.3). The major axis lies along with y-axis in this case. See Figure 66.3.

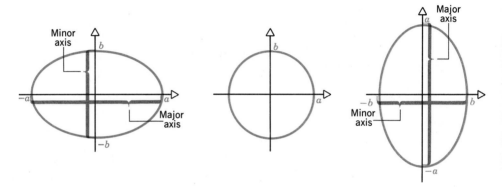

FIGURE 66.3

If $a > b$, then the graph of

$$\frac{x^2}{a^2} + \frac{y^2}{b^2} = 1$$

in an ellipse with center at the origin and foci $(-c, 0)$ and $(c, 0)$, where $c^2 = a^2 - b^2$. The major axis has endpoints $(-a, 0)$ and $(a, 0)$, and the minor axis has endpoints $(0, -b)$ and $(0, b)$. If $a > b$, then the graph of

$$\frac{x^2}{b^2} + \frac{y^2}{a^2} = 1$$

is an ellipse with center at the origin and foci $(0, -c)$ and $(0, c)$, where $c^2 = a^2 - b^2$. The major axis has endpoints $(0, -a)$ and $(0, a)$, and the minor axis has endpoints $(-b, 0)$ and $(b, 0)$.

(66.4)

Example 66.1 The graph of $9x^2 + 4y^2 - 36 = 0$ is an ellipse. Find the foci and vertices and draw the graph.

Solution The equation can be rewritten as

$$\frac{x^2}{4} + \frac{y^2}{9} = 1 \quad \text{or} \quad \frac{x^2}{2^2} + \frac{y^2}{3^2} = 1.$$

The larger of 2^2 and 3^2 is under y^2, so the foci and vertices are on the y-axis. The vertices are $(0, -3)$ and $(0, 3)$. And from $a = 3$ and $b = 2$ we have $c^2 = 3^2 - 2^2 = 5$, so $c = \sqrt{5}$. Thus the foci are $(0, -\sqrt{5})$ and $(0, \sqrt{5})$. Figure 66.4

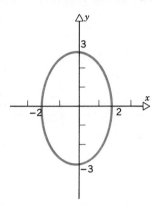

FIGURE 66.4

shows the graph. In addition to the endpoints of the major and minor axes, some other coordinate pairs used to draw the graph are shown in Table 66.1.

TABLE 66.1

x	0	$\pm\dfrac{1}{2}$	± 1	± 2
y (exact)	± 3	$\pm 3\sqrt{15}/4$	$\pm 3\sqrt{3}/2$	0
y (approximate)	± 3	± 2.90	± 2.60	0

B. TRANSLATIONS

For an ellipse whose axes are parallel to the coordinate axes but whose center is not at the origin, we use the results from (66.4) together with translations. Reasoning as we did for parabolas in Section 65C, we obtain the following results.

If $a > b$, then the graph of

$$\frac{(x - h)^2}{a^2} + \frac{(y - k)^2}{b^2} = 1$$

is an ellipse with center (h, k) and foci $(h - c, k)$ and $(h + c, k)$, where $c^2 = a^2 - b^2$. The major axis has endpoints $(h - a, k)$ and $(h + a, k)$, and the minor axis has endpoints $(h, k - b)$ and $(h, k + b)$.

If $a > b$, then the graph of

$$\frac{(x - h)^2}{b^2} + \frac{(y - k)^2}{a^2} = 1$$

(66.5)

is an ellipse with center (h, k) and foci $(h, k - c)$ and $(h, k + c)$, where $c^2 = a^2 - b^2$. The major axis has endpoints $(h, k - a)$ and $(h, k + a)$, and the minor axis endpoints $(h - b, k)$ and $(h + b, k)$.

The equations in (66.5) are called the **standard forms** for the equations of the ellipses they represent. Figure 66.5 shows a typical case of the first possibility in (66.5).

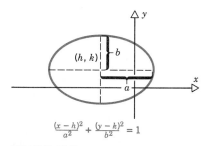

$$\frac{(x - h)^2}{a^2} + \frac{(y - k)^2}{b^2} = 1$$

FIGURE 66.5

Example 66.2 Draw the graph of $9x^2 + 16y^2 - 18x + 64y - 71 = 0$.

Solution We first rewrite the equation in standard form.

Step I. Factor the coefficient of x^2 from the x terms, factor the coefficient of y^2 from the y terms, and write the equation with the constant term on the right of the equality sign.

$$9(x^2 - 2x \quad) + 16(y^2 + 4y \quad) = 71.$$

Step II. Complete the squares on the x and y terms, and add appropriate constants to the right side to compensate.

$$9(x^2 - 2x + 1) + 16(y^2 + 4y + 4) = 71 + 9 + 64$$

$$9(x - 1)^2 + 16(y + 2)^2 \qquad = 144.$$

Step III. Divide both sides by the constant on the right, to put the equation in the appropriate form from (66.5).

$$\frac{(x - 1)^2}{16} + \frac{(y + 2)^2}{9} = 1$$

$$\frac{(x - 1)^2}{4^2} + \frac{[y - (-2)]^2}{3^2} = 1. \qquad (66.6)$$

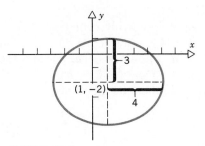

FIGURE 66.6

Step IV. Draw the graph of (66.6) by following the pattern in Figure 66.5. See Figure 66.6 for the result.

C. APPLICATION

One application of ellipses is given by Kepler's three laws of planetary motion (Figure 66.7):

I. The planets move in elliptical orbits with the sun at one focus.

II. The imaginary line connecting the sun to a planet sweeps out equal areas in equal intervals of time.

III. The ratio of the cube of the semimajor axis to the square of the period is a constant, which is the same for all planets.

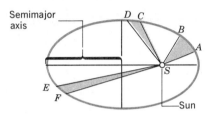

FIGURE 66.7

While the orbits described by the first law are ellipses, they do not differ greatly from circular orbits. To make this more precise, we consider the *eccentricity* of the orbits: Using the notation of (66.4), the **eccentricity** of an ellipse is the number e defined by $e = c/a$. (This number e is not the constant introduced in Section 28, which is the base for the natural logarithms.) From $c^2 = a^2 - b^2$ and $b > 0$, we have $0 \le c^2 < a^2$ and $0 \le c < a$, so

$$0 \le e < 1,$$

with $e = 0$ if $b = a$, and e approaching 1 if b approaches 0. If $e = 0$, then the foci coincide (see Figure 66.2) and the ellipse becomes a circle. If b approaches 0, so that e approaches 1, then F' approaches V' and F approaches V (Figure 66.2), and the ellipse becomes elongated, approaching the segment joining F' and F.

For planetary orbits the eccentricities vary from 0.250 (for Pluto) down to 0.007 (for Venus), with $e = 0.017$ for the Earth. In contrast, some comets travel in elliptical orbits with very large eccentricities.

Kepler's second law implies that if areas such as ASB, CSD, and ESF in Figure 66.7 are all equal, then the planet will move from A to B, from C to D, and from E to F in equal intervals of time. In particular, a planet moves fastest when it is closest to the sun.

Kepler's third law establishes a uniformity among the motions of the different planets. Let R denote the length of the semimajor axis for a planet, measured in millions of miles. Let T denote the *period* of the planet—that is, the time required for one complete orbital revolution about the sun; measure T in Earth years, so that $T = 1$ for the Earth. Then Kepler's third law states that the ratio R^3/T^2 is the same for all planets. If this constant ratio is denoted by K, then

$$R^3/T^2 = K. \tag{66.7}$$

If R and T are known for a single planet, then K can be computed from Equation (66.7). Then, for example, if T can be determined for a second planet, the value of R for the second planet can be computed by solving Equation (66.7) for R:

$$R^3 = KT^2$$
$$R = \sqrt[3]{KT^2}. \tag{66.8}$$

It can be proved that the length of the semimajor axis for an ellipse is also the average distance between a focus and the points on the ellipse. Thus the number R for a planet is also the average distance of the planet from the sun.

Example 66.3 The average distance between the Earth and the sun is approximately 93 million miles. Determine K in Equation (66.7).

Solution With $R = 93$ and $T = 1$, equation (66.7) becomes

$$K = 93^3/1^2 \approx 8.04 \times 10^5. \qquad \boxed{\text{C}} \tag{66.9}$$

Example 66.4 The period of Mars is approximately 1.88 Earth years. What is the average distance between Mars and the sun?

Solution Use Equation (66.8) with $T = 1.88$ and $K = 8.04 \times 10^5$ (from the previous example). Then

$$R = \sqrt[3]{8.04 \times 10^5 \times 1.88^2} \approx 142. \qquad \boxed{\text{C}}$$

That is, Mars is approximately 142 million miles from the sun.

EXERCISES FOR SECTION 66

The graph of each equation in Exercises 1–12 is an ellipse. In each case, find the center, the foci, and the endpoints of the major and minor axes, and draw the graph.

1. $16x^2 + 9y^2 - 144 = 0$

2. $9x^2 + 4y^2 - 36 = 0$

3. $25x^2 + 4y^2 - 100 = 0$

4. $(x - 1)^2 + 25y^2 - 25 = 0$

5. $3x^2 + 9(y + 1)^2 - 27 = 0$

6. $3(x - 2)^2 + 4(y - 3)^2 - 12 = 0$

7. $4x^2 + y^2 + 4y = 0$

8. $x^2 + 16y^2 - 32y = 0$

9. $16x^2 + 4y^2 + 160x + 256 = 0$

10. $3x^2 + y^2 - 12x - 6y + 12 = 0$

11. $25x^2 + 16y^2 + 50x + 96y - 231 = 0$

12. $3x^2 + 2y^2 - 24x + 4y + 44 = 0$

In each of Exercises 13–21 find an equation of the ellipse satisfying the given conditions.

13. Endpoints of major axis (2, 0) and (2, 8), endpoints of minor axis (0, 4) and (4, 4).

14. Endpoints of minor axis (0, −1) and (0, 1), semimajor axis of length 5.

15. Center (−3, 2), one vertex (−3, −2), minor axis of length 6.

16. Foci (0, 0) and (6, 0), endpoints of minor axis (3, −2) and (3, 2).

17. Endpoints of major axis (−5, 0) and (5, 0), contains $\left(3, \dfrac{16}{5}\right)$.

18. Center (2, 4), one vertex (2, 0), semiminor axis of length 1.

19. Center (0, 0), one vertex (0, −6), contains (1, 1).

20. Foci $(-2, 3)$ and $(2, 3)$, the sum of the distances from the foci to each point on the ellipse equal to 6.

21. Endpoints of minor axis $(-5, 0)$ and $(5, 0)$, contains $(1, 5)$.

22. Find all of the points of intersection of the graphs of $y = x + 1$ and $2x^2 + y^2 - 2 = 0$.

23. Draw the graphs of $3x^2 + 4y^2 - 16 = 0$ and $y = \frac{1}{2}x^2 - 1$ on a single Cartesian coordinate system and find all of the points where the two graphs intersect. (Do the second part of the problem algebraically; do not just estimate the answer from the graph.)

24. Find all of the points of intersection of the graphs of $2x^2 + y^2 - 1 = 0$ and $x^2 + 2y^2 - 1 = 0$.

25. The London Bridge designed by John Rennie and completed in 1831 had five semi-elliptical arches. The central arch had a span of 152 feet 6 inches and a rise at its center of 37 feet 6 inches. Find the rise of this arch 20 feet from either end (Figure 66.8). ⓒ

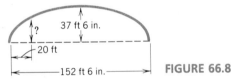

— 152 ft 6 in. — **FIGURE 66.8**

26. The lines $y = x$ and $y = -x$ intersect the ellipse $\frac{x^2}{a^2} + \frac{y^2}{b^2} = 1$ in four points that determine a square. What is the area of the square?

27. Draw the graph of
$(4x^2 + 9y^2 - 36)(x^2 + y^2 - 25) \le 0$.

28. Derive Equation (66.3) for an ellipse with foci $F'(0, -c)$ and $F(0, c)$ on the y-axis. The conditions determining a, b, and c are given following (66.3). The derivation of (66.2) is similar.

29. The segment cut off by an ellipse on a line that is perpendicular to the major axis and that passes through a focus is called a **latus rectum** of the ellipse. Prove that the length of a latus rectum of the ellipse determined by Equation (66.2) is $2b^2/a$.

30. Find an equation of the ellipse with center at $(0, 0)$, one focus at $(0, -3)$ and eccentricity $\frac{1}{4}$.

31. The orbit of Pluto, with an eccentricity of 0.250, is the least circular of all the planetary orbits. Find b for an ellipse whose eccentricity is 0.250

and whose equation is $\frac{x^2}{a^2} + \frac{y^2}{b^2} = 1$, with $a = 10$. Now draw the graph of this equation. This graph will have the shape of Pluto's orbit and will give an idea of how close to being circular the planetary orbits are. (The value $a = 10$ was chosen for convenience; it is the relative size of a and b that is important here.)

32. The points in the (elliptical) orbit of a satellite of the earth that are farthest from and nearest to the earth are called the **apogee** and **perigee**, respectively. Let A and P denote the distance from the earth's center to the apogee and perigee, respectively. Prove that the eccentricity of the orbit is $e = (A - P)/(A + P)$.

33. The first artificial satellite was *Sputnik I*, launched October 4, 1957. The greatest and least distances of its orbit from the earth's surface were 584 miles and 143.5 miles, respectively. Use the result in Exercise 32 to compute the eccentricity of its orbit. Use 3960 miles for the radius of the earth.

The next six exercises relate to Subsection C and blanks in Table 66.2. ⓒ

34. What is the average distance between Mercury and the sun?

35. What is the average distance between Venus and the sun?

36. What is the average distance between Jupiter and the sun?

37. What is the period for Saturn?

38. What is the period for Uranus?

39. What is the period for Neptune?

TABLE 66.2

Planet	Average distance from sun[a]	Period[b]
Mercury		0.24
Venus		0.62
Earth	93	1.00
Mars	142	1.88
Jupiter		11.86
Saturn	886	
Uranus	1782	
Neptune	2793	
Pluto	3672	248.43

[a]In millions of miles.
[b]As multiples of the Earth's period.

SECTION **67**

Hyperbolas

A. DEFINITION

Definition A **hyperbola** is the set of all points in a plane such that the difference of the distances from two fixed points (called **foci**) in the plane is a constant.

Figure 67.1 shows an example with foci $F'(-c, 0)$ and $F(c, 0)$: $|d_1 - d_2|$ is the same for all points $P(x, y)$ on the hyperbola. The line through the foci intersects the hyperbola in two points called the **vertices** of the hyperbola. The segment connecting the vertices is called the **transverse axis.** The midpoint of the transverse axis is called the **center.**

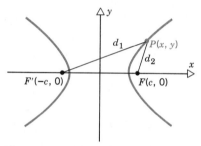

FIGURE 67.1 $|d_1 - d_2|$ is the same for all points P on the hyperbola.

Suppose $a > 0$ and $c > 0$, and consider the hyperbola shown in Figure 67.1 with $|d_1 - d_2| = 2a$. The method used to derive Equation (66.2) for an ellipse can be used to show that if $b^2 = c^2 - a^2$, then the equation of the hyperbola in Figure 67.1 is

$$\frac{x^2}{a^2} - \frac{y^2}{b^2} = 1. \tag{67.1}$$

The segment connecting $(0, -b)$ and $(0, b)$ is called the **conjugate axis.**

Now suppose $a > 0$ and $c > 0$, and consider the hyperbola in Figure 67.2 with the foci $F'(0, -c)$ and $F(0, c)$ on the y-axis, and, again, $|d_1 - d_2| = 2a$. It can be shown that if $b^2 = c^2 - a^2$, then the equation of this hyperbola is

$$\frac{y^2}{a^2} - \frac{x^2}{b^2} = 1. \tag{67.2}$$

In this case, the segment connecting $(-b, 0)$ and $(b, 0)$ is the conjugate axis.

Figure 67.3 shows examples representing both of Equations (67.1) and (67.2). The graph of (67.1) has x-intercepts $x = \pm a$, but it has no y-intercept. The x-intercepts are also the vertices. The graph of (67.2) has y-intercepts $y = \pm a$, but it has no x-intercept. The y-intercepts are the vertices in this case. The

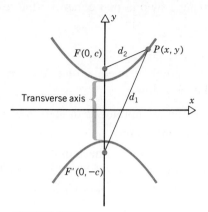

FIGURE 67.2

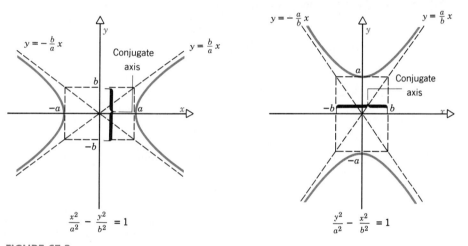

FIGURE 67.3

two parts of the graph of a hyperbola are called **branches.** The graphs of both (67.1) and (67.2) are symmetric with respect to both the x- and y-axes.

Each dashed line in Figure 67.3 is called an **asymptote** of its hyperbola. This means that the distance to the dashed line from a variable point (x, y) on a branch of the hyperbola approaches zero as the point moves outward along the hyperbola. (The asymptotes are an aid in graphing; they are not part of the hyperbola.) It can be shown that the equations of the asymptotes for Equation (67.1) arise from replacing 1 by 0 on the right side of the equation:

$$\frac{x^2}{a^2} - \frac{y^2}{b^2} = 0$$

$$y^2 = \frac{b^2}{a^2} x^2$$

$$y = \pm \frac{b}{a} x. \qquad (67.3)$$

(See Exercise 30.) Similarly, the asymptotes for Equation (67.2) arise from replacing 1 by 0 in that equation; the result in this case is

$$y = \pm \frac{a}{b} x. \qquad (67.4)$$

The graph of an equation of the form in either (67.1) or (67.2) can be drawn quickly as follows:

- Draw the asymptotes as dashed lines.
- Plot the intercepts.
- Determine several other points by direct calculation. Also plot the additional points that arise from these by symmetry.
- Complete the graph so that it has the general shape of the appropriate form from Figure 67.3.

Example 67.1 Draw the graph of

$$25x^2 - 4y^2 + 100 = 0. \qquad (67.5)$$

Solution Divide both sides by 100 and then rearrange the terms as shown.

$$\frac{x^2}{4} - \frac{y^2}{25} + 1 = 0$$

$$\frac{y^2}{5^2} - \frac{x^2}{2^2} = 1$$

This is Equation (67.2) with $a = 5$ and $b = 2$. The equations of the asymptotes are

$$\frac{y^2}{5^2} - \frac{x^2}{2^2} = 0 \quad \text{or} \quad y = \pm \frac{5}{2} x.$$

These asymptotes are shown in Figure 67.4, along with the intercepts, $(0, 5)$ and $(0, -5)$. If $x = 1$ in Equation (67.5), then $y^2 = \frac{125}{4}$ so $y = \pm 5\sqrt{5}/2 \approx \pm 5.6$. This gives the points $(1, 5.6)$ and $(1, -5.6)$. Symmetry gives the additional points $(-1, 5.6)$ and $(-1, -5.6)$. Other points can be determined in the same way. ■

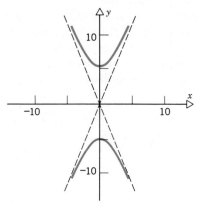

FIGURE 67.4

B. TRANSLATIONS

For a hyperbola whose transverse axis is parallel to a coordinate axis but whose center is not at the origin, the statements in Subsection A together with translations will yield the following results.

If $a > 0$ and $b > 0$, then the graph of

$$\frac{(x - h)^2}{a^2} - \frac{(y - k)^2}{b^2} = 1$$

is a hyperbola with center (h, k) and foci $(h - c, k)$ and $(h + c, k)$, where $c^2 = a^2 + b^2$. The vertices and endpoints of the transverse axis are $(h - a, k)$ and $(h + a, k)$. The conjugate axis has endpoints $(h, k - b)$ and $(h, k + b)$. The asymptotes have equations $y - k = \pm\dfrac{b}{a}(x - h)$.

If $a > 0$ and $b > 0$, then the graph of

$$\frac{(y - k)^2}{a^2} - \frac{(x - h)^2}{b^2} = 1$$

is a hyperbola with center (h, k) and foci $(h, k - c)$ and $(h, k + c)$, where $c^2 = a^2 + b^2$. The vertices and endpoints of the transverse axis are $(h, k - a)$ and $(h, k + a)$. The conjugate axis has endpoints $(h - b, k)$ and $(h + b, k)$. The asymptotes have equations $y - k = \pm\dfrac{a}{b}(x - h)$.

(67.6)

The equations in (67.6) are called the **standard forms** for the equations of the hyperbolas they represent. Figure 67.5 shows a typical case of the first possibility in (67.6).

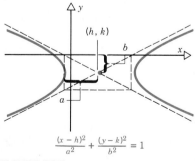

$$\frac{(x - h)^2}{a^2} + \frac{(y - k)^2}{b^2} = 1$$

FIGURE 67.5

Example 67.2 Draw the graph of $9x^2 - 16y^2 + 36x + 32y + 164 = 0$.

Solution We first rewrite the equation in standard form. The details are similar to those in Example 66.2.

$$9(x^2 + 4x \quad\quad) - 16(y^2 - 2y \quad\quad) = -164$$

$$9(x^2 + 4x + 4) - 16(y^2 - 2y + 1) = -164 + 36 - 16$$

$$9(x + 2)^2 - 16(y - 1)^2 = -144$$

$$\frac{(y - 1)^2}{9} - \frac{(x + 2)^2}{16} = 1$$

$$\frac{(y - 1)^2}{3^2} - \frac{[x - (-2)]^2}{4^2} = 1$$

Thus we have the second possibility in (67.6). The center is at $(-2, 1)$. The vertices and endpoints of the transverse axis are at $(-2, -2)$ and $(-2, 4)$. The conjugate axis has endpoints $(-6, 1)$ and $(2, 1)$.

To draw the asymptotes, we first draw the rectangle having sides parallel to the coordinate axes and passing through the endpoints of the transverse and conjugate axes (Figure 67.6). The asymptotes are the lines through the pairs of opposite corners of this rectangle.

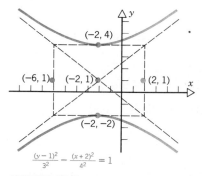

FIGURE 67.6

In addition to the vertices, we can obtain other points by substituting values for x and solving for y. For example, $x = -1$ gives $y = 1 \pm \sqrt{153}/4 \approx 4.09$, -2.09, that is, the points $(-1, 4.09)$ and $(-1, -2.09)$. These two points are located symmetrically with respect to the conjugate axis. And each one gives another point because of the symmetry of the hyperbola with respect to its transverse axis; $(-1, 4.09)$ corresponds to $(-3, 4.09)$, and $(-1, -2.09)$ corresponds to $(-3, -2.09)$. Figure 67.6 shows the graph.

C. APPLICATION

The defining property of a hyperbola is fundamental to the *loran* system of navigation. (*Loran* is an acronym for *long range navigation*.) Imagine a pair of stations that can transmit radio signals, such as those labelled master and slave A in Figure 67.7. Suppose pulse signals are transmitted from master and slave A at a predetermined time interval. If these signals are received at a ship at different times, then, since the velocity of radio waves is known, this time difference can be converted into a difference in distance of the ship from master

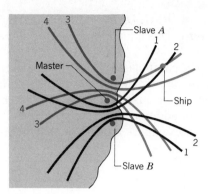

FIGURE 67.7

and slave A. This locates the ship on a particular one of the hyperbolas with foci at master and slave A. Applying the same idea to master and slave B, the ship can be located on a particular one of the hyperbolas with foci at master and slave B. The intersection of the hyperbolas gives the location of the ship.

D. CONIC SECTIONS AND QUADRATIC EQUATIONS

In Section 65 we defined a conic section as a curve formed by the intersection of a plane and a double-napped right circular cone. In Sections 65–67 we studied three general types of conic sections: parabolas, ellipses, and hyperbolas. We shall now see that the other possibilities for conic sections are circles and points (special cases of ellipses), lines, and two intersecting lines. Then we shall consider the relationship between conic sections and the graphs of quadratic equations.

In Figure 65.1*b*, the ellipse is a circle if the intersecting plane is perpendicular to the axis of the cone, and the ellipse is a point if the intersecting plane contains the vertex of the cone. The definition of ellipse in Section 66 produces a circle if the foci coincide, and it produces a point if, in addition, the sum of the distances to the foci is allowed to be zero.

Next consider Figure 65.1*a*. If the intersecting plane contains an element of the cone, then the intersection degenerates into a single line. The parabola in Figure 65.2 approaches a straight line as the focus moves away from the directrix, but a straight line is not a parabola (by the definition of parabola in Section 65B).

Now consider Figure 65.1*c*. If the intersecting plane contains the axis of the cone, then the hyperbola degenerates into a pair of intersecting lines. The hyperbola in Figure 67.1 approaches two parallel lines as the distance between the foci increases, but the definition of hyperbola (Section 67A) does not include two parallel lines. Summarizing:

> A conic section (as defined in Section 65) is either a parabola, an ellipse (with a circle and a point as special cases), a hyperbola, a line, or two intersecting lines.   (67.7)

The **general quadratic equation** with variables x and y has the form

$$Ax^2 + Bxy + Cy^2 + Dx + Ey + F = 0, \qquad (67.8)$$

where A, B, C, D, E, and F are real numbers with at least one of A, B, and C not zero. This equation allows for all possible terms of degree two or less in x and y.

The parabolas, ellipses, and hyperbolas in Sections 65–67 all have equations that can be written in the form (67.8). Figure 67.8 shows three other possibilities arising from quadratic equations: intersecting lines, parallel lines, and single lines. In addition, the empty set is the graph of a quadratic equation ($x^2 + 1 = 0$, for example). It turns out that these examples illustrate all of the possibilities

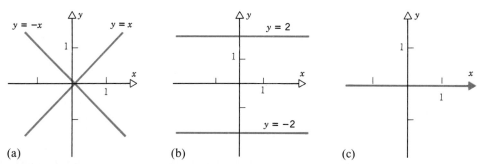

FIGURE 67.8 (a) **Intersecting lines:** $x^2 - y^2 = 0$ or $y = \pm x$. (b) **Parallel lines:** $y^2 - 4 = 0$ or $y = \pm 2$. (c) **One line:** $y^2 = 0$ or $y = 0$.

arising as graphs of quadratic equations in x and y. If we recall from Statement (67.7) that intersecting lines and single lines are conic sections, we can summarize this as follows.

> The graph of a quadratic equation in x and y is either a conic section, two parallel lines, or the empty set.　　　　(67.9)

We shall not prove Statement (67.9). However, recall that the parabolas, ellipses, and hyperbolas in Sections 65–67 all have directrices and axes parallel to the coordinate axes. The significant difference between these cases and the most general possibilities from Equation (67.8) involve the cross product term Bxy; by allowing an xy term we can write equations for all conic sections appearing as graphs in a Cartesian plane, whatever their orientations. This is done by rotation of axes, which is discussed in the author's *Precalculus* (John Wiley & Sons, New York, 1986) or *Trigonometry* (John Wiley & Sons, New York, 1987).

EXERCISES FOR SECTION 67

The graph of each equation in Exercises 1–12 is a hyperbola. In each case, find the center, the foci, the endpoints of the transverse and conjugate axes, and draw the graph. Draw the asymptotes as dashed lines.

1. $4x^2 - 9y^2 - 36 = 0$

2. $9x^2 - 16y^2 - 144 = 0$

3. $16x^2 - 4y^2 + 64 = 0$

4. $4(x + 1)^2 - y^2 + 4 = 0$

5. $9(x - 2)^2 - 25(y + 1)^2 + 225 = 0$

6. $x^2 - 25(y + 3)^2 - 25 = 0$

7. $x^2 - 3y^2 - 12y - 21 = 0$

8. $25x^2 - 4y^2 - 200x + 300 = 0$

9. $9x^2 - 16y^2 - 18x - 64y - 199 = 0$

10. $9x^2 - 16y^2 - 54x - 64y + 161 = 0$

11. $9x^2 - 5y^2 - 36x - 60y - 99 = 0$

12. $2x^2 - 3y^2 + 30y - 69 = 0$

In each of Exercises 13–21 find an equation of the hyperbola satisfying the given conditions.

13. Endpoints of transverse axis $(0, -3)$ and $(0, 3)$, endpoints of conjugate axis $(-2, 0)$ and $(2, 0)$.

14. Foci $(-4, 0)$ and $(4, 0)$, vertices $(-2, 0)$ and $(2, 0)$.

15. Foci $(0, -5)$ and $(0, 5)$, endpoints of conjugate axis $(-2, 0)$ and $(2, 0)$.

16. Foci $(0, 1)$ and $(4, 1)$, vertices $(1, 1)$ and $(3, 1)$.

17. Vertices $(-4, 0)$ and $(4, 0)$, contains $(5, \frac{3}{4} \sqrt{5})$.

18. Center $(-2, 1)$, vertex $(-2, 2)$, focus $(-2, 3)$.

19. Vertex $(0, 5)$, asymptotes $y = \pm\frac{5}{3} x$.

20. Center $(1, -1)$, vertex $(1, 1)$, one endpoint of conjugate axis $(0, -1)$.

21. Foci $(-5, 3)$ and $(5, 3)$, the difference of the distances from the foci to each point on the hyperbola equal to 6.

22. Find all of the points of intersection of the graphs of $x - y - 1 = 0$ and $4x^2 - y^2 = 16$.

23. Find all of the points of intersection of the graphs of $y = x^2 - 1$ and $y^2 - x^2 = 1$.

24. Find all of the points of intersection of the graphs of $4x^2 - 9y^2 = 36$ and $x^2 + y^2 = 16$.

25. Two hyperbolas are said to be **conjugate** if the transverse axis of either is the conjugate axis of

the other. What is the conjugate of the hyperbola whose equation is $\dfrac{x^2}{a^2} - \dfrac{y^2}{b^2} = 1$?

26. If a circle with its center at the origin intersects the graph of $\dfrac{y^2}{a^2} - \dfrac{x^2}{b^2} = 1$ in exactly two points, what is the area of the circle?

27. A hyperbola is said to be **equilateral** if the length of its transverse and conjugate axes are equal. Find the equations of the asymptotes of an equilateral hyperbola of the form (67.1).

28. An equilateral hyperbola (see Exercise 27) with foci on the x-axis intersects the circle with equation $x^2 + y^2 = 100$ in exactly two points. What is the equation of the hyperbola?

29. Derive Equation (67.1) for a hyperbola with foci $F'(-c, 0)$ and $F(c, 0)$ and with a and b as defined just preceding (67.1).

30. Verify that Equation (67.3) gives asymptotes for the hyperbola defined by Equation (67.1) by carrying out all of the following steps.

(a) Show that Equation (67.1) can be rewritten as
$$y^2 = \frac{b^2}{a^2} (x^2 - a^2).$$

(b) Show that the equation in (a) can be rewritten as
$$y^2 = \left(\frac{bx}{a}\right)^2 \left(1 - \frac{a^2}{x^2}\right).$$

(c) Explain why $1 - (a^2/x^2)$ approaches 1 as $|x|$ becomes larger and larger.

(d) Use the preceding steps to explain why a variable point on the graph of Equation (67.1) approaches the graph of $y = \frac{b}{a} x$ or $y = -\frac{b}{a} x$ as $|x|$ becomes larger and larger.

REVIEW EXERCISES FOR CHAPTER XV

Draw the graph of each equation in Exercises 1–14.

1. $y^2 = 2x$

2. $x^2 + 4y^2 - 36 = 0$

3. $9x^2 - y^2 + 1 = 0$

4. $x^2 + 3y = 0$

5. $x^2 - 4y^2 = 0$

6. $x^2 + y^2 - 2x + 8y - 8 = 0$

7. $25x^2 + 9y^2 + 100x - 116 = 0$

8. $3x^2 + 2y^2 - 6x + 4y - 1 = 0$

9. $2(x - 5)^2 = y$

10. $x^2 + y^2 + 1 = 0$

11. $4x^2 - y^2 + 4y - 8 = 0$

12. $x^2 + 4x - y + 6 = 0$

13. $x^2 - y^2 - 2x + 2 = 0$

14. $4x^2 - 9 = 0$

In each of Exercises 15–20 find an equation of the conic section satisfying the given conditions.

15. Parabola with focus $(0, -5)$ and directrix $y = 5$.

16. Ellipse with endpoints of minor axis $(-2, 0)$ and $(2, 0)$ and major axis of length 10.

17. Hyperbola with foci $(-2, 0)$ and $(-2, 6)$, and vertices $(-2, 1)$ and $(-2, 5)$.

18. Parabola with vertex $(2, 0)$ and focus $(-1, 0)$.

19. Ellipse with foci $(-1, -2)$ and $(-1, 2)$, and the sum of the distances from the foci to each point on the ellipse equal to 8.

20. Hyperbola with vertex $(3, 0)$ and asymptotes $y = \pm 2x$.

REAL NUMBERS

ORDER PROPERTIES

In Chapter I, *positive* real numbers were defined as those to the right of 0 on a real line directed to the right. To define *positive* without appealing to a real line, we amend the axioms for a field (Section 1B) by adding the following *order axiom.*

Order Axiom

> A field F is said to be an **ordered field** if there is a subset F^p of F satisfying the following three conditions.
>
> | | |
> |---|---|
> | If $a \in F^p$ and $b \in F^p$, then $a + b \in F^p$. | **Closure under addition** |
> | If $a \in F^p$ and $b \in F^p$, then $ab \in F^p$. | **Closure under multiplication** |
> | If $a \in F$, then exactly one of the·following is true: $a = 0$, $a \in F^p$, or $-a \in F^p$. | **Law of trichotomy** |
>
> The elements of F^p are called the **positive elements** of F. Elements that are neither zero nor positive are said to be **negative.**

The system of real numbers is an ordered field. So is the system of rational numbers. It can be proved, however, that the system of complex numbers is a field that is *not* ordered (regardless of what one tries for the set of positive elements). For a more extensive discussion of fields and related algebraic systems, see any introductory book on modern algebra, such as the author's *Modern Algebra: An Introduction,* second edition (John Wiley & Sons, New York, 1985).

DECIMAL NUMBERS

Section 1 states that a decimal number represents a rational number iff it either terminates or repeats. To understand why, first consider a rational number a/b, where a and b are integers and $b > 0$. In computing the decimal representation of a/b, the remainders in the division process will be either $0, 1, 2, \ldots,$ or $b - 1$. If one of these remainders is 0 after we start bringing down zeros in the division process, the decimal terminates. If not, one of the numbers $1, 2, \ldots,$ or $b - 1$ must repeat as a remainder (in fewer than b steps after we start bringing down zeros in the division process); this means that the decimal has started to repeat. Thus, if a decimal number represents a rational number, then it either terminates or repeats.

Now consider the converse, namely, if a decimal number either terminates or repeats, then it represents a rational number. For a terminating decimal this can be verified by multiplying and dividing by an appropriate power of 10. For example, $7.321 = 7.321(10^3/10^3) = 7321/1000$, which is a quotient of two integers. For repeating decimals see Example 60.9, which shows how to convert each repeating decimal to a quotient of two integers.

IRRATIONALITY OF $\sqrt{2}$

Section 1 states that $\sqrt{2}$ *is irrational*. We can prove this by showing that the alternative—$\sqrt{2}$ is rational—leads to a contradiction. Thus assume that $\sqrt{2}$ is rational, so that $\sqrt{2} = a/b$ for some integers a and b with $b \neq 0$. We can also assume that a/b is reduced to lowest terms, so that a and b have no common factor except ± 1. Since $\sqrt{2} = a/b$, we have $2 = a^2/b^2$ and $2b^2 = a^2$. The left side of this equation, $2b^2$, is even; therefore the right side, a^2, must also be even. But if a^2 is even, then a must be even, and hence $a = 2k$ for some integer k. Substituting $a = 2k$ in $2b^2 = a^2$, we obtain $2b^2 = (2k)^2$, $2b^2 = 4k^2$, and $b^2 = 2k^2$. The last equation shows that since $2k^2$ is even, b^2 must also be even, and thus b must be even. Thus we have deduced that a and b are even, and they therefore have 2 as a common factor. This contradicts the assumption that a/b was reduced to lowest terms, and completes the proof.

SETS

The text uses a number of ideas about sets. This appendix gives more examples of these ideas.

SET-BUILDER NOTATION

Section 7A introduces the *set-builder notation*

$$\{x: \cdots\},$$

which means

the set of all x such that $\cdots$.

The notation $\{a, b, \cdots\}$ denotes the set whose members are $a, b, \cdots$.

Example B.1 (a) $\{x: \quad x$ is a positive integer$\}$ denotes the set of all positive integers.

(b) $\{x: \quad x$ is a positive even integer$\}$ denotes the set $\{2, 4, 6, \cdots\}$.

(c) $\{a/b: \quad a$ and b are integers with $b \neq 0\}$ denotes the set of all rational numbers (Section 1). ▪

When the context implies that a variable represents a number, and no other restriction is specified (such as "x is an integer"), then (in this book) x is assumed to represent a real number.

Example B.2 (a) $\{x: \quad x < 0\}$ denotes the set of all negative real numbers.

(b) $\{x: \quad x^2 > 2\}$ denotes the set of all real numbers whose squares are greater than 2. ▪

OTHER SET NOTATION

The following notation was used in Section 10C.

$x \in A$ denotes that x *is an element of* (or is a member of, or belongs to) the set A.

$\emptyset$ denotes the *empty-set*—that is, the set that contains no elements.

$A \cup B$ denotes $\{x: \quad x \in A$ or $x \in B\}$, which is called the *union* of A and B.

$A \cap B$ denotes $\{x: \quad x \in A$ and $x \in B\}$, which is called the *intersection* of A and B.

Example B.3 Let $S = \{a, b, c\}$, $T = \{c, d, e\}$, and $U = \{d, e\}$. Then

$$S \cup T = \{a, b, c, d, e\} \qquad S \cap T = \{c\}$$
$$S \cup U = \{a, b, c, d, e\} \qquad S \cap U = \emptyset$$
$$T \cup U = T \qquad\qquad\qquad T \cap U = U.$$

Example B.4 If A is any set, then $A \cup \emptyset = A$ and $A \cap \emptyset = \emptyset$.

The following ideas are not used elsewhere in the book, but they are so basic that they are included here for completeness.

To indicate that x is *not* an element of a set A, we write $x \notin A$. Thus, for example,

$x \notin \emptyset$ is always true, and

$x \in \emptyset$ is always false.

If A and B are sets and each element of A is an element of B, then A is called a **subset** of B; this is denoted by

$$A \subseteq B \quad \text{or} \quad B \supseteq A.$$

Notice that $A \subseteq B$ does not preclude the possibility that $A = B$. In fact,

$$A = B \text{ iff } A \subseteq B \text{ and } A \supseteq B.$$

The following three statements are equivalent:

(i) $A \subseteq B$.

(ii) If $x \in A$, then $x \in B$.

(iii) If $x \notin B$, then $x \notin A$.

If A is not a subset of B, we write $A \not\subseteq B$. This is true, of course, iff A contains at least one element that is not in B. Set inclusion ($\subseteq$) has the following properties:

$\emptyset \subseteq A$ for every set A.

$A \subseteq A$ for every set A.

If $A \subseteq B$ and $B \subseteq C$, then $A \subseteq C$.

Example B.5 If S, T, and U are as in Example B.3, then $S \not\subseteq T$ and $T \supseteq U$.

CARTESIAN PRODUCTS, FUNCTIONS, AND RELATIONS

The **Cartesian product** of sets S and T is denoted by $S \times T$, and is defined by

$$S \times T = \{(x, y): \ x \in S \text{ and } y \in T\}.$$

Here each (x, y) is an *ordered pair* (see the footnote on page 95).

Example B.6 If $S = \{1, 2\}$ and $T = \{u, v, w\}$, then

$$S \times T = \{1, u), (1, v), (1, w), (2, u), (2, v), (2, w)\}.$$

Notice that in this case $S \times T \neq T \times S$. In general, $S \times T = T \times S$ iff $S = T$. ▪

Coordinate geometry (Section 12) begins with a *Cartesian plane,* which is a one-to-one correspondence between the points of a plane and the set of ordered pairs of real numbers—that is, the elements of the Cartesian product $R \times R$, where R denotes the set of all real numbers. This example explains the choice of the name "Cartesian product."

The notion of Cartesian product can be used to give a definition of *function* that is preferred by some to that given in Section 14: A *function* from a set S to a set T is a subset of $S \times T$ such that each $x \in S$ is a first member of precisely one pair in the subset. The connection between this definition and that given in Section 14 is that if f is a function from S to T (in the sense of Section 14), then each $x \in S$ contributes the pair $(x, f(x))$ to the subset of $S \times T$ in the definition of function as a subset of $S \times T$.

Example B.7 If f denotes the function such that $f(x) = x^2$ for each real number x, then in the sense introduced above f is

$$\{(x, x^2): \quad x \text{ is a real number}\}. \qquad ▪$$

A more general idea than that of a function is that of a *relation:* A **relation** is any subset of a Cartesian product. Although relations occur often in mathematics, we have no need to dwell on them here.

APPENDIX C

TABLES

TABLE I Squares and Square Roots

n	n^2	$\sqrt{n}$	n	n^2	$\sqrt{n}$
1	1	1.000	51	2,601	7.141
2	4	1.414	52	2,704	7.211
3	9	1.732	53	2,809	7.280
4	16	2.000	54	2,916	7.348
5	25	2.236	55	3,025	7.416
6	36	2.449	56	3,136	7.483
7	49	2.646	57	3,249	7.550
8	64	2.828	58	3,364	7.616
9	81	3.000	59	3,481	7.681
10	100	3.162	60	3,600	7.746
11	121	3.317	61	3,721	7.810
12	144	3.464	62	3,844	7.874
13	169	3.606	63	3,969	7.937
14	196	3.742	64	4,096	8.000
15	225	3.873	65	4,225	8.062
16	256	4.000	66	4,356	8.124
17	289	4.123	67	4,489	8.185
18	324	4.243	68	4,624	8.246
19	361	4.359	69	4,761	8.307
20	400	4.472	70	4,900	8.367
21	441	4.583	71	5,041	8.426
22	484	4.690	72	5,184	8.485
23	529	4.796	73	5,329	8.544
24	576	4.899	74	5,476	8.602
25	625	5.000	75	5,625	8.600
26	676	5.099	76	5,776	8.718
27	729	5.196	77	5,929	8.775
28	784	5.292	78	6,084	8.832
29	841	5.385	79	6,241	8.888
30	900	5.477	80	6,400	8.944
31	961	5.568	81	6,561	9.000
32	1,024	5.657	82	6,724	9.055
33	1,089	5.745	83	6,889	9.100
34	1,156	5.831	84	7,056	9.165
35	1,225	5.916	85	7,225	9.220
36	1,296	6.000	86	7,396	9.274
37	1,369	6.083	87	7,569	9.327
38	1,444	6.164	88	7,744	9.381
39	1,521	6.245	89	7,921	9.434
40	1,600	6.325	90	8,100	9.487
41	1,681	6.403	91	8,281	9.539
42	1,764	6.481	92	8,464	9.592
43	1,849	6.557	93	8,649	9.644
44	1,936	6.633	94	8,836	9.695
45	2,025	6.708	95	9,025	9.747
46	2,116	6.782	96	9,216	9.798
47	2,209	6.856	97	9,409	9.849
48	2,304	6.928	98	9,604	9.899
49	2,401	7.000	99	9,801	9.950
50	2,500	7.071	100	10,000	10.000

TABLE II Common Logarithms

n	0	1	2	3	4	5	6	7	8	9
1.0	0.0000	0043	0086	0128	0170	0212	0253	0294	0334	0374
1.1	.0414	0453	0492	0531	0569	0607	0645	0682	0719	0755
1.2	.0792	0828	0864	0899	0934	0969	1004	1038	1072	1106
1.3	.1139	1173	1206	1239	1271	1303	1335	1367	1399	1430
1.4	.1461	1492	1523	1553	1584	1614	1644	1673	1703	1732
1.5	.1761	1790	1818	1847	1875	1903	1931	1959	1987	2014
1.6	.2041	2068	2095	2122	2148	2175	2201	2227	2253	2279
1.7	.2304	2330	2355	2380	2405	2430	2455	2480	2504	2529
1.8	.2553	2577	2601	2625	2648	2672	2695	2718	2742	2765
1.9	.2788	2810	2833	2856	2878	2900	2923	2945	2967	2989
2.0	.3010	3032	3054	3075	3096	3118	3139	3160	3181	3201
2.1	.3222	3243	3263	3284	3304	3324	3345	3365	3385	3404
2.2	.3424	3444	3464	3483	3502	3522	3541	3560	3579	3598
2.3	.3617	3636	3655	3674	3692	3711	3729	3747	3766	3784
2.4	.3802	3820	3838	3856	3874	3892	3909	3927	3945	3962
2.5	.3979	3997	4014	4031	4048	4065	4082	4099	4116	4133
2.6	.4150	4166	4183	4200	4216	4232	4249	4265	4281	4298
2.7	.4314	4330	4346	4362	4378	4393	4409	4425	4440	4456
2.8	.4472	4487	4502	4518	4533	4548	4564	4579	4594	4609
2.9	.4624	4639	4654	4669	4683	4698	4713	4728	4742	4757
3.0	.4771	4786	4800	4814	4829	4843	4857	4871	4886	4900
3.1	.4914	4928	4942	4955	4969	4983	4997	5011	5024	5038
3.2	.5051	5065	5079	5092	5105	5119	5132	5145	5159	5172
3.3	.5185	5198	5211	5224	5237	5250	5263	5276	5289	5302
3.4	.5315	5328	5340	5353	5366	5378	5391	5403	5416	5428
3.5	.5441	5453	5465	5478	5490	5502	5514	5527	5539	5551
3.6	.5563	5575	5587	5599	5611	5623	5635	5647	5658	5670
3.7	.5682	5694	5705	5717	5729	5740	5752	5763	5775	5786
3.8	.5798	5809	5821	5832	5843	5855	5866	5877	5888	5899
3.9	.5911	5922	5933	5944	5955	5966	5977	5988	5999	6010
4.0	.6021	6031	6042	6053	6064	6075	6085	6096	6107	6177
4.1	.6128	6138	6149	6160	6170	6180	6191	6201	6212	6222
4.2	.6232	6243	6253	6263	6274	6284	6294	6304	6314	6325
4.3	.6335	6345	6355	6365	6375	6385	6395	6405	6415	6425
4.4	.6435	6444	6454	6464	6474	6484	6493	6503	6513	6522
4.5	.6532	6542	6551	6561	6571	6580	6590	6599	6609	6618
4.6	.6628	6637	6646	6656	6665	6675	6684	6693	6702	6712
4.7	.6721	6730	6739	6749	6758	6767	6776	6785	6794	6803
4.8	.6812	6821	6830	6839	6848	6857	6866	6875	6884	6893
4.9	.6902	6911	6920	6928	5937	6946	6955	6964	6972	6981

TABLE II (continued)

n	0	1	2	3	4	5	6	7	8	9
5.0	.6990	6998	7007	7016	7024	7033	7042	7050	7059	7067
5.1	.7076	7084	7093	7101	7110	7118	7126	7135	7143	7152
5.2	.7160	7168	7177	7185	7193	7202	7210	7218	7226	7235
5.3	.7243	7251	7259	7267	7275	7284	7292	7300	7308	7316
5.4	.7324	7332	7340	7348	7356	7364	7372	7380	7388	7396
5.5	.7404	7412	7419	7427	7435	7443	7451	7459	7466	7474
5.6	.7482	7490	7497	7505	7513	7520	7528	7536	7543	7551
5.7	.7559	7566	7574	7582	7589	7597	7604	7612	7619	7627
5.8	.7634	7642	7649	7657	7664	7672	7679	7686	7694	7701
5.9	.7709	7716	7723	7731	7738	7745	7752	7760	7767	7774
6.0	.7782	7789	7796	7803	7810	7818	7825	7832	7839	7846
6.1	.7853	7860	7868	7875	7882	7889	7896	7903	7910	7917
6.2	.7924	7931	7938	7945	7952	7959	7966	7973	7980	7987
6.3	.7993	8000	8007	8014	8021	8028	8035	8041	8048	8055
6.4	.8062	8069	8075	8082	8089	8096	8102	8109	8116	8122
6.5	.8129	8136	8142	8149	8156	8162	8169	8176	8182	8189
6.6	.8195	8202	8209	8215	8222	8228	8235	8241	8248	8254
6.7	.8261	8267	8274	8280	8287	8293	8299	8306	8312	8319
6.8	.8325	8331	8338	8344	8351	8357	8363	8370	8376	8382
6.9	.8388	8395	8401	8407	8414	8420	8426	8432	8439	8445
7.0	.8451	8457	8463	8470	8476	8482	8488	8494	8500	8506
7.1	.8513	8519	8525	8531	8537	8543	8549	8555	8561	8567
7.2	.8573	8579	8585	8591	8597	8603	8609	8615	8621	8627
7.3	.8633	8639	8645	8651	8657	8663	8669	8675	8681	8686
7.4	.8692	8698	8704	8710	8716	8722	8727	8733	8739	8745
7.5	.8751	8756	8762	8768	8774	8779	8785	8791	8797	8802
7.6	.8808	8814	8820	8825	8831	8837	8842	8848	8854	8859
7.7	.8865	8871	8876	8882	8887	8893	8899	8904	8910	8915
7.8	.8921	8927	8932	8938	8943	8949	8954	8960	8965	8971
7.9	.8976	8982	8987	8993	8998	9004	9009	9015	9020	9025
8.0	.9031	9036	9042	9047	9053	9058	9063	9069	9074	9079
8.1	.9085	9090	9096	9101	9106	9112	9117	9122	9128	9133
8.2	.9138	9143	9149	9154	9195	9165	9170	9175	9180	9186
8.3	.9191	9196	9201	9206	9212	9217	9222	9227	9232	9238
8.4	.9243	9248	9253	9258	9263	9269	9274	9279	9284	9289
8.5	.9294	9299	9304	9309	9315	9320	9325	9330	9335	9340
8.6	.9345	9350	9355	9360	9365	9370	9375	9380	9385	9390
8.7	.9395	9400	9405	9410	9415	9420	9425	9430	9435	9440
8.8	.9445	9450	9455	9460	9465	9469	9474	9479	9484	9489
8.9	.9494	9499	9504	9509	9513	9518	9523	9528	9533	9538
9.0	.9542	9547	9552	9557	9562	9566	9571	9576	9581	9586
9.1	.9590	9595	9600	9605	9609	9614	9619	9624	9628	9633
9.2	.9638	9643	9647	9652	9657	9661	9666	9671	9675	9680
9.3	.9685	9689	9694	9699	9703	9708	9713	9717	9722	9727
9.4	.9731	9736	9741	9745	9750	9754	9759	9763	9768	9773
9.5	.9777	9782	9786	9791	9795	9800	9805	9809	9814	9818
9.6	.9823	9827	9832	9836	9841	9845	9850	9854	9859	9863
9.7	.9868	9872	9877	9881	9886	9890	9894	9899	9903	9908
9.8	.9912	9917	9921	9926	9930	9934	9939	9943	9948	9952
9.9	.9956	9961	9965	9969	9974	9978	9983	9987	9991	9996

TABLE III Powers of e

x	e^x	e^{-x}
0.00	1.0000	1.00000
0.01	1.0101	0.99005
0.02	1.0202	.98020
0.03	1.0305	.97045
0.04	1.0408	.96079
0.05	1.0513	.95123
0.06	1.0618	.94176
0.07	1.0725	.93239
0.08	1.0833	.92312
0.09	1.0942	.91393
0.10	1.1052	.90484
0.11	1.1163	.89583
0.12	1.1275	.88692
0.13	1.1388	.87809
0.14	1.1503	.86936
0.15	1.1618	.86071
0.16	1.1735	.85214
0.17	1.1853	.84366
0.18	1.1972	.83527
0.19	1.2092	.82696
0.20	1.2214	.81873
0.21	1.2337	.81058
0.22	1.2461	.80252
0.23	1.2586	.79453
0.24	1.2712	.78663
0.25	1.2840	.77880
0.26	1.2969	.77105
0.27	1.3100	.76338
0.28	1.3231	.75578
0.29	1.3364	.74826
0.30	1.3499	.74082
0.31	1.3634	.73345
0.32	1.3771	.72615
0.33	1.3910	.71892
0.34	1.4049	.71177
0.35	1.4191	.70469
0.36	1.4333	.69768
0.37	1.4477	.69073
0.38	1.4623	.68386
0.39	1.4770	.67706
0.40	1.4918	.67032
0.41	1.5068	.66365
0.42	1.5220	.65705
0.43	1.5373	.65051
0.44	1.5527	.64404
0.45	1.5683	.63763
0.46	1.5841	.63128
0.47	1.6000	.62500
0.48	1.6161	.61878
0.49	1.6323	.61263
0.50	1.6487	.60653

TABLE III (continued)

x	e^x	e^{-x}
0.51	1.6653	.60060
0.52	1.6820	.59452
0.53	1.6989	.58860
0.54	1.7160	.58275
0.55	1.7333	.57695
0.56	1.7507	.57121
0.57	1.7683	.56553
0.58	1.7860	.55990
0.59	1.8040	.55433
0.60	1.8221	.54881
0.61	1.8404	.54335
0.62	1.8589	.53794
0.63	1.8776	.53259
0.64	1.8965	.52729
0.65	1.9155	.52205
0.66	1.9348	.51685
0.67	1.9542	.51171
0.68	1.9739	.50662
0.69	1.9937	.50158
0.70	2.0138	.49659
0.71	2.0340	.49164
0.72	2.0544	.48675
0.73	2.0751	.48191
0.74	2.0959	.47711
0.75	2.1170	.47237
0.76	2.1383	.46767
0.77	2.1598	.46301
0.78	2.1815	.45841
0.79	2.2034	.45384
0.80	2.2255	.44933
0.81	2.2479	.44486
0.82	2.2705	.44043
0.83	2.2933	.43605
0.84	2.3164	.43171
0.85	2.3396	.42741
0.86	2.3632	.42316
0.87	2.3869	.41895
0.88	2.4109	.41478
0.89	2.4351	.41066
0.90	2.4596	.40657
0.91	2.4843	.40252
0.92	2.5093	.39852
0.93	2.5345	.39455
0.94	2.5600	.39063
0.95	2.5857	.38674
0.96	2.6117	.38289
0.97	2.6379	.37908
0.98	2.6645	.37531
0.99	2.6912	.37158
1.00	2.7183	.36788

TABLE IV Natural Logarithms

x	$\ln x$	x	$\ln x$
0.1	−2.3026	5.1	1.6292
0.2	−1.6094	5.2	1.6487
0.3	−1.2040	5.3	1.6677
0.4	−0.9163	5.4	1.6864
0.5	−0.6931	5.5	1.7047
0.6	−0.5108	5.6	1.7228
0.7	−0.3567	5.7	1.7405
0.8	−0.2231	5.8	1.7579
0.9	−0.1054	5.9	1.7750
1.0	0.0000	6.0	1.7918
1.1	0.0953	6.1	1.8083
1.2	0.1823	6.2	1.8245
1.3	0.2624	6.3	1.8405
1.4	0.3365	6.4	1.8563
1.5	0.4055	6.5	1.8718
1.6	0.4700	6.6	1.8871
1.7	0.5306	6.7	1.9021
1.8	0.5878	6.8	1.9169
1.9	0.6419	6.9	1.9315
2.0	0.6931	7.0	1.9459
2.1	0.7419	7.1	1.9601
2.2	0.7885	7.2	1.9741
2.3	0.8329	7.3	1.9879
2.4	0.8755	7.4	2.0015
2.5	0.9163	7.5	2.0149
2.6	0.9555	7.6	2.0281
2.7	0.9933	7.7	2.0412
2.8	1.0296	7.8	2.0541
2.9	1.0647	7.9	2.0669
3.0	1.0986	8.0	2.0794
3.1	1.1314	8.1	2.0919
3.2	1.1632	8.2	2.1041
3.3	1.1939	8.3	2.1163
3.4	1.2238	8.4	2.1282
3.5	1.2528	8.5	2.1401
3.6	1.2809	8.6	2.1518
3.7	1.3083	8.7	2.1633
3.8	1.3350	8.8	2.1748
3.9	1.3610	8.9	2.1861
4.0	1.3863	9.0	2.1972
4.1	1.4110	9.1	2.2083
4.2	1.4351	9.2	2.2192
4.3	1.4586	9.3	2.2300
4.4	1.4816	9.4	2.2407
4.5	1.5041	9.5	2.2513
4.6	1.5261	9.6	2.2618
4.7	1.5476	9.7	2.2721
4.8	1.5686	9.8	2.2824
4.9	1.5892	9.9	2.2925
5.0	1.6094	10	2.3026

TABLE V Trigonometric Functions of Degrees

θ deg	$\sin \theta$	$\cos \theta$	$\tan \theta$	$\cot \theta$	$\sec \theta$	$\csc \theta$	
0.0°	0.0000	1.0000	0.0000	undefined	1.0000	undefined	90.0°
0.1	0.0017	1.0000	0.0017	572.96	1.0000	572.96	89.9
0.2	0.0035	1.0000	0.0035	286.48	1.0000	286.48	89.8
0.3	0.0052	1.0000	0.0052	190.98	1.0000	190.99	89.7
0.4	0.0070	1.0000	0.0070	143.24	1.0000	143.24	89.6
0.5	0.0087	1.0000	0.0087	114.59	1.0000	114.59	89.5
0.6	0.0105	0.9999	0.0105	95.490	1.0001	95.495	89.4
0.7	0.0122	0.9999	0.0122	81.847	1.0001	81.853	89.3
0.8	0.0140	0.9999	0.0140	71.615	1.0001	71.622	89.2
0.9	0.0157	0.9999	0.0157	63.657	1.0001	63.665	89.1
1.0°	0.0175	0.9998	0.0175	57.290	1.0002	57.299	89.0°
1.1	0.0192	0.9998	0.0192	52.081	1.0002	52.090	88.9
1.2	0.0209	0.9998	0.0209	47.740	1.0002	47.750	88.8
1.3	0.0227	0.9997	0.0227	44.066	1.0003	44.077	88.7
1.4	0.0244	0.9997	0.0244	40.917	1.0003	40.930	88.6
1.5	0.0262	0.9997	0.0262	38.188	1.0003	38.202	88.5
1.6	0.0279	0.9996	0.0279	35.801	1.0004	35.815	88.4
1.7	0.0297	0.9996	0.0297	33.694	1.0004	33.708	88.3
1.8	0.0314	0.9995	0.0314	31.821	1.0005	31.836	88.2
1.9	0.0332	0.9995	0.0332	30.145	1.0005	30.161	88.1
2.0°	0.0349	0.9994	0.0349	28.636	1.0006	28.654	88.0°
2.1	0.0366	0.9993	0.0367	27.271	1.0007	27.290	87.9
2.2	0.0384	0.9993	0.0384	26.031	1.0007	26.050	87.8
2.3	0.0401	0.9992	0.0402	24.898	1.0008	24.918	87.7
2.4	0.0419	0.9991	0.0419	23.859	1.0009	23.880	87.6
2.5	0.0436	0.9990	0.0437	22.904	1.0010	22.926	87.5
2.6	0.0454	0.9990	0.0454	22.022	1.0010	22.044	87.4
2.7	0.0471	0.9989	0.0472	21.205	1.0011	21.229	87.3
2.8	0.0488	0.9988	0.0489	20.446	1.0012	20.471	87.2
2.9	0.0506	0.9987	0.0507	19.740	1.0013	19.766	87.1
3.0°	0.0523	0.9986	0.0524	19.081	1.0014	19.107	87.0°
3.1	0.0541	0.9985	0.0542	18.464	1.0015	18.492	86.9
3.2	0.0558	0.9984	0.0559	17.886	1.0016	17.914	86.8
3.3	0.0576	0.9983	0.0577	17.343	1.0017	17.372	86.7
3.4	0.0593	0.9982	0.0594	16.832	1.0018	16.862	86.6
3.5	0.0610	0.9981	0.0612	16.350	1.0019	16.380	86.5
3.6	0.0628	0.9980	0.0629	15.895	1.0020	15.926	86.4
3.7	0.0645	0.9979	0.0647	15.464	1.0021	15.496	86.3
3.8	0.0663	0.9978	0.0664	15.056	1.0022	15.089	86.2
3.9	0.0680	0.9977	0.0682	14.669	1.0023	14.703	86.1
4.0°	0.0698	0.9976	0.0699	14.301	1.0024	14.336	86.0°
4.1	0.0715	0.9974	0.0717	13.951	1.0026	13.987	85.9
4.2	0.0732	0.9973	0.0734	13.617	1.0027	13.654	85.8
4.3	0.0750	0.9972	0.0752	13.300	1.0028	13.337	85.7
4.4	0.0767	0.9971	0.0769	12.996	1.0030	13.035	85.6
4.5	0.0785	0.9969	0.0787	12.706	1.0031	12.746	85.5
4.6	0.0802	0.9968	0.0805	12.429	1.0032	12.469	85.4
4.7	0.0819	0.9966	0.0822	12.163	1.0034	12.204	85.3
4.8	0.0837	0.9965	0.0840	11.909	1.0035	11.951	85.2
4.9	0.0854	0.9963	0.0857	11.665	1.0037	11.707	85.1°
	$\cos \theta$	$\sin \theta$	$\cot \theta$	$\tan \theta$	$\csc \theta$	$\sec \theta$	θ deg

TABLE V (continued)

θ deg	sin θ	cos θ	tan θ	cot θ	sec θ	csc θ	
5.0°	0.0872	0.9962	0.0875	11.430	1.0038	11.474	85.0°
5.1	0.0889	0.9960	0.0892	11.205	1.0040	11.249	84.9
5.2	0.0906	0.9959	0.0910	10.988	1.0041	11.034	84.8
5.3	0.0924	0.9957	0.0928	10.780	1.0043	10.826	84.7
5.4	0.0941	0.9956	0.0945	10.579	1.0045	10.626	84.6
5.5	0.0958	0.9954	0.0963	10.385	1.0046	10.433	84.5
5.6	0.0976	0.9952	0.0981	10.199	1.0048	10.248	84.4
5.7	0.0993	0.9951	0.0998	10.019	1.0050	10.069	84.3
5.8	0.1011	0.9949	0.1016	9.8448	1.0051	9.8955	84.2
5.9	0.1028	0.9947	0.1033	9.6768	1.0053	9.7283	84.1
6.0°	0.1045	0.9945	0.1051	9.5144	1.0055	9.5668	84.0°
6.1°	0.1063	0.9943	0.1069	9.3573	1.0057	9.4105	83.9
6.2	0.1080	0.9942	0.1086	9.2052	1.0059	9.2593	83.8
6.3	0.1097	0.9940	0.1104	9.0579	1.0061	9.1129	83.7
6.4	0.1115	0.9938	0.1122	8.9152	1.0063	8.9711	83.6
6.5	0.1132	0.9936	0.1139	8.7769	1.0065	8.8337	83.5
6.6	0.1149	0.9934	0.1157	8.6428	1.0067	8.7004	83.4
6.7	0.1167	0.9932	0.1175	8.5126	1.0069	8.5711	83.3
6.8	0.1184	0.9930	0.1192	8.3863	1.0071	8.4457	83.2
6.9	0.1201	0.9928	0.1210	8.2636	1.0073	8.3238	83.1
7.0°	0.1219	0.9925	0.1228	8.1444	1.0075	8.2055	83.0°
7.1	0.1236	0.9923	0.1246	8.0285	1.0077	8.0905	82.9
7.2	0.1253	0.9921	0.1263	7.9158	1.0079	7.9787	82.8
7.3	0.1271	0.9919	0.1281	7.8062	1.0082	7.8700	82.7
7.4	0.1288	0.9917	0.1299	7.6996	1.0084	7.7642	82.6
7.5	0.1305	0.9914	0.1317	7.5958	1.0086	7.6613	82.5
7.6	0.1323	0.9912	0.1334	7.4947	1.0089	7.5611	82.4
7.7	0.1340	0.9910	0.1352	7.3962	1.0091	7.4635	82.3
7.8	0.1357	0.9907	0.1370	7.3002	1.0093	7.3684	82.2
7.9	0.1374	0.9905	0.1388	7.2066	1.0096	7.2757	82.1
8.0°	0.1392	0.9903	0.1405	7.1154	1.0098	7.1853	82.0°
8.1	0.1409	0.9900	0.1423	7.0264	1.0101	7.0972	81.9
8.2	0.1426	0.9898	0.1441	6.9395	1.0103	7.0112	81.8
8.3	0.1444	0.9895	0.1459	6.8548	1.0106	6.9273	81.7
8.4	0.1461	0.9893	0.1477	6.7720	1.0108	6.8454	81.6
8.5	0.1478	0.9890	0.1495	6.6912	1.0111	6.7655	81.5
8.6	0.1495	0.9888	0.1512	6.6122	1.0114	6.6874	81.4
8.7	0.1513	0.9885	0.1530	6.5350	1.0116	6.6111	81.3
8.8	0.1530	0.9882	0.1548	6.4596	1.0119	6.5366	81.2
8.9	0.1547	0.9880	0.1566	6.3859	1.0122	6.4637	81.1
9.0°	0.1564	0.9877	0.1584	6.3138	1.0125	6.3925	81.0°
9.1	0.1582	0.9874	0.1602	6.2432	1.0127	6.3228	80.9
9.2	0.1599	0.9871	0.1620	6.1742	1.0130	6.2547	80.8
9.3	0.1616	0.9869	0.1638	6.1880	1.0133	6.1066	80.7
9.4	0.1633	0.9866	0.1655	6.1227	1.0136	6.0405	80.6
9.5	0.1650	0.9863	0.1673	5.9758	1.0139	6.0589	80.5
9.6	0.1668	0.9860	0.1691	5.9124	1.0142	5.9963	80.4
9.7	0.1685	0.9857	0.1709	5.8502	1.0145	5.9351	80.3
9.8	0.1702	0.9854	0.1727	5.7894	1.0148	5.8751	80.2
9.9	0.1719	0.9851	0.1745	5.7297	1.0151	5.8164	80.1°
	cos θ	sin θ	cot θ	tan θ	csc θ	sec θ	θ deg

TABLE V (continued)

θ deg	sin θ	cos θ	tan θ	cot θ	sec θ	csc θ	
10.0°	0.1736	0.9848	0.1763	5.6713	1.0154	5.7588	80.0°
10.1	0.1754	0.9845	0.1781	5.6140	1.0157	5.7023	79.9
10.2	0.1771	0.9842	0.1799	5.5578	1.0161	5.6470	79.8
10.3	0.1788	0.9839	0.1817	5.5027	1.0164	5.5928	79.7
10.4	0.1805	0.9836	0.1835	5.4486	1.0167	5.5396	79.6
10.5	0.1822	0.9833	0.1853	5.3955	1.0170	5.4874	79.5
10.6	0.1840	0.9829	0.1871	5.3435	1.0174	5.4362	79.4
10.7	0.1857	0.9826	0.1890	5.2924	1.0177	5.3860	79.3
10.8	0.1874	0.9823	0.1908	5.2422	1.0180	5.3367	79.2
10.9	0.1891	0.9820	0.1926	5.1929	1.0184	5.2883	79.1
11.0°	0.1908	0.9816	0.1944	5.1446	1.0187	5.2408	79.0°
11.1	0.1925	0.9813	0.1962	5.0970	1.0191	5.1942	78.9
11.2	0.1942	0.9810	0.1980	5.0504	1.0194	5.1484	78.8
11.3	0.1959	0.9806	0.1998	5.0045	1.0198	5.1043	78.7
11.4	0.1977	0.9803	0.2016	4.9595	1.0201	5.0593	78.6
11.5	0.1994	0.9799	0.2035	4.9152	1.0205	5.0159	78.5
11.6	0.2011	0.9796	0.2053	4.8716	1.0209	4.9732	78.4
11.7	0.2028	0.9792	0.2071	4.8288	1.0212	4.9313	78.3
11.8	0.2045	0.9789	0.2089	4.7867	1.0216	4.8901	78.2
11.9	0.2062	0.9785	0.2107	4.7453	1.0220	4.8496	78.1
12.0°	0.2079	0.9781	0.2126	4.7046	1.0223	4.8097	78.0°
12.1	0.2096	0.9778	0.2144	4.6646	1.0227	4.7706	77.9
12.2	0.2113	0.9774	0.2162	4.6252	1.0231	4.7321	77.8
12.3	0.2130	0.9770	0.2180	4.5864	1.0235	4.6942	77.7
12.4	0.2147	0.9767	0.2199	4.5483	1.0239	4.6569	77.6
12.5	0.2164	0.9763	0.2217	4.5107	1.0243	4.6202	77.5
12.6	0.2181	0.9759	0.2235	4.4737	1.0247	4.5841	77.4
12.7	0.2198	0.9755	0.2254	4.4374	1.0251	4.5486	77.3
12.8	0.2215	0.9751	0.2272	4.4015	1.0255	4.5137	77.2
12.9	0.2232	0.9748	0.2290	4.3662	1.0259	4.4793	77.1
13.0°	0.2250	0.9744	0.2309	4.3315	1.0263	4.4454	77.0°
13.1	0.2267	0.9740	0.2327	4.2972	1.0267	4.4121	76.9
13.2	0.2284	0.9736	0.2345	4.2635	1.0271	4.3792	76.8
13.3	0.2300	0.9732	0.2364	4.2303	1.0276	4.3469	76.7
13.4	0.2317	0.9728	0.2382	4.1976	1.0280	4.3150	76.6
13.5	0.2334	0.9724	0.2401	4.1653	1.0284	4.2837	76.5
13.6	0.2351	0.9720	0.2419	4.1335	1.0288	4.2528	76.4
13.7	0.2368	0.9715	0.2438	4.1022	1.0293	4.2223	76.3
13.8	0.2385	0.9711	0.2456	4.0713	1.0297	4.1923	76.2
13.9	0.2402	0.9707	0.2475	4.0408	1.0302	4.1627	76.1
14.0°	0.2419	0.9703	0.2493	4.0108	1.0306	4.1336	76.0°
14.1	0.2436	0.9699	0.2512	3.9812	1.0311	4.1048	75.9
14.2	0.2453	0.9694	0.2530	3.9520	1.0315	4.0765	75.8
14.3	0.2470	0.9690	0.2549	3.9232	1.0320	4.0486	75.7
14.4	0.2487	0.9686	0.2568	3.8947	1.0324	4.0211	75.6
14.5	0.2504	0.9681	0.2586	3.8667	1.0329	3.9939	75.5
14.6	0.2521	0.9677	0.2605	3.8391	1.0334	3.9672	75.4
14.7	0.2538	0.9673	0.2623	3.8118	1.0338	3.9408	75.3
14.8	0.2554	0.9668	0.2642	3.7849	1.0343	3.9147	75.2
14.9	0.2571	0.9664	0.2661	3.7583	1.0348	3.8890	75.1°
	cos θ	sin θ	cot θ	tan θ	csc θ	sec θ	θ deg

TABLE V (continued)

θ deg	$\sin \theta$	$\cos \theta$	$\tan \theta$	$\cot \theta$	$\sec \theta$	$\csc \theta$	
15.0°	0.2588	0.9659	0.2679	3.7321	1.0353	3.8637	75.0°
15.1	0.2605	0.9655	0.2698	3.7062	1.0358	3.8387	74.9
15.2	0.2622	0.9650	0.2717	3.6806	1.0363	3.8140	74.8
15.3	0.2639	0.9646	0.2736	3.6554	1.0367	3.7897	74.7
15.4	0.2656	0.9641	0.2754	3.6305	1.0372	3.7657	74.6
15.5	0.2672	0.9636	0.2773	3.6059	1.0377	3.7420	74.5
15.6	0.2689	0.9632	0.2792	3.5816	1.0382	3.7186	74.4
15.7	0.2706	0.9627	0.2811	3.5576	1.0388	3.6955	74.3
15.8	0.2723	0.9622	0.2830	3.5339	1.0393	3.6727	74.2
15.9	0.2740	0.9617	0.2849	3.5105	1.0398	3.6502	74.1
16.0°	0.2756	0.9613	0.2867	3.4874	1.0403	3.6280	74.0°
16.1	0.2773	0.9608	0.2886	3.4646	1.0408	3.6060	73.9
16.2	0.2790	0.9603	0.2905	3.4420	1.0413	3.5843	73.8
16.3	0.2807	0.9598	0.2924	3.4197	1.0419	3.5629	73.7
16.4	0.2823	0.9593	0.2943	3.3977	1.0424	3.5418	73.6
16.5	0.2840	0.9588	0.2962	3.3759	1.0429	3.5209	73.5
16.6	0.2857	0.9583	0.2981	3.3544	1.0435	3.5003	73.4
16.7	0.2874	0.9578	0.3000	3.3332	1.0440	3.4800	73.3
16.8	0.2890	0.9573	0.3019	3.3122	1.0446	3.4598	73.2
16.9	0.2907	0.9568	0.3038	3.2914	1.0451	3.4399	73.1
17.0°	0.2924	0.9563	0.3057	3.2709	1.0457	3.4203	73.0°
17.1	0.2940	0.9558	0.3076	3.2506	1.0463	3.4009	72.9
17.2	0.2957	0.9553	0.3096	3.2305	1.0468	3.3817	72.8
17.3	0.2974	0.9548	0.3115	3.2106	1.0474	3.3628	72.7
17.4	0.2990	0.9542	0.3134	3.1910	1.0480	3.3440	72.6
17.5	0.3007	0.9537	0.3153	3.1716	1.0485	3.3255	72.5
17.6	0.3024	0.9532	0.3172	3.1524	1.0491	3.3072	72.4
17.7	0.3040	0.9527	0.3191	3.1334	1.0497	3.2891	72.3
17.8	0.3057	0.9521	0.3211	3.1146	1.0503	3.2712	72.2
17.9	0.3074	0.9516	0.3230	3.0961	1.0509	3.2536	72.1
18.0°	0.3090	0.9511	0.3249	3.0777	1.0515	3.2361	72.0°
18.1	0.3107	0.9505	0.3268	3.0595	1.0521	3.2188	71.9
18.2	0.3123	0.9500	0.3288	3.0415	1.0527	3.2017	71.8
18.3	0.3140	0.9494	0.3307	3.0237	1.0533	3.1848	71.7
18.4	0.3156	0.9489	0.3327	3.0061	1.0539	3.1681	71.6
18.5	0.3173	0.9483	0.3346	2.9887	1.0545	3.1515	71.5
18.6	0.3190	0.9478	0.3365	2.9714	1.0551	3.1352	71.4
18.7	0.3206	0.9472	0.3385	2.9544	1.0557	3.1190	71.3
18.8	0.3223	0.9466	0.3404	2.9375	1.0564	3.1030	71.2
18.9	0.3239	0.9461	0.3424	2.9208	1.0570	3.0872	71.1
19.0°	0.3256	0.9455	0.3443	2.9042	1.0576	3.0716	71.0°
19.1	0.3272	0.9449	0.3463	2.8878	1.0583	3.0561	70.9
19.2	0.3289	0.9444	0.3482	2.8716	1.0589	3.0407	70.8
19.3	0.3305	0.9438	0.3502	2.8556	1.0595	3.0256	70.7
19.4	0.3322	0.9432	0.3522	2.8397	1.0602	3.0106	70.6
19.5	0.3338	0.9426	0.3541	2.8239	1.0608	2.9957	70.5
19.6	0.3355	0.9421	0.3561	2.8083	1.0615	2.9811	70.4
19.7	0.3371	0.9415	0.3581	2.7929	1.0622	2.9665	70.3
19.8	0.3387	0.9409	0.3600	2.7776	1.0628	2.9521	70.2
19.9	0.3404	0.9403	0.3620	2.7625	1.0635	2.9379	70.1°
	$\cos \theta$	$\sin \theta$	$\cot \theta$	$\tan \theta$	$\csc \theta$	$\sec \theta$	θ deg

TABLE V (continued)

θ deg	sin θ	cos θ	tan θ	cot θ	sec θ	csc θ	
20.0°	0.3420	0.9397	0.3640	2.7475	1.0642	2.9238	70.0°
20.1	0.3437	0.9391	0.3659	2.7326	1.0649	2.9099	69.9
20.2	0.3453	0.9385	0.3679	2.7179	1.0655	2.8960	69.8
20.3	0.3469	0.9379	0.3699	2.7034	1.0662	2.8824	69.7
20.4	0.3486	0.9373	0.3719	2.6889	1.0669	2.8688	69.6
20.5	0.3502	0.9367	0.3739	2.6746	1.0676	2.8555	69.5
20.6	0.3518	0.9361	0.3759	2.6605	1.0683	2.8422	69.4
20.7	0.3535	0.9354	0.3779	2.6464	1.0690	2.8291	69.3
20.8	0.3551	0.9348	0.3799	2.6325	1.0697	2.8161	69.2
20.9	0.3567	0.9342	0.3819	2.6187	1.0704	2.8032	69.1
21.0°	0.3584	0.9336	0.3839	2.6051	1.0711	2.7904	69.0°
21.1	0.3600	0.9330	0.3859	2.5916	1.0719	2.7778	68.9
21.2	0.3616	0.9323	0.3879	2.5782	1.0726	2.7653	68.8
21.3	0.3633	0.9317	0.3899	2.5649	1.0733	2.7529	68.7
21.4	0.3649	0.9311	0.3919	2.5517	1.0740	2.7407	68.6
21.5	0.3665	0.9304	0.3939	2.5386	1.0748	2.7285	68.5
21.6	0.3681	0.9298	0.3959	2.5257	1.0755	2.7165	68.4
21.7	0.3697	0.9291	0.3979	2.5129	1.0763	2.7046	68.3
21.8	0.3714	0.9285	0.4000	2.5002	1.0770	2.6927	68.2
21.9	0.3730	0.9278	0.4020	2.4876	1.0778	2.6811	68.1
22.0°	0.3746	0.9272	0.4040	2.4751	1.0785	2.6695	68.0°
22.1	0.3762	0.9265	0.4061	2.4627	1.0793	2.6580	67.9
22.2	0.3778	0.9259	0.4081	2.4504	1.0801	2.6466	67.8
22.3	0.3795	0.9252	0.4101	2.4383	1.0808	2.6354	67.7
22.4	0.3811	0.9245	0.4122	2.4262	1.0816	2.6242	67.6
22.5	0.3827	0.9239	0.4142	2.4142	1.0824	2.6131	67.5
22.6	0.3843	0.9232	0.4163	2.4023	1.0832	2.6022	67.4
22.7	0.3859	0.9225	0.4183	2.3906	1.0848	2.5913	67.3
22.8	0.3875	0.9219	0.4204	2.3789	1.0848	2.5805	67.2
22.9	0.3891	0.9212	0.4224	2.3673	1.0856	2.5699	67.1
23.0°	0.3907	0.9205	0.4245	2.3559	1.0864	2.5593	67.0°
23.1	0.3923	0.9198	0.4265	2.3445	1.0872	2.5488	66.9
23.2	0.3939	0.9191	0.4286	2.3332	1.0880	2.5384	66.8
23.3	0.3955	0.9184	0.4307	2.3220	1.0888	2.5282	66.7
23.4	0.3971	0.9178	0.4327	2.3109	1.0896	2.5180	66.6
23.5	0.3987	0.9171	0.4348	2.2998	1.0904	2.5078	66.5
23.6	0.4003	0.9164	0.4369	2.2889	1.0913	2.4978	66.4
23.7	0.4019	0.9157	0.4390	2.2781	1.0921	2.4879	66.3
23.8	0.4035	0.9150	0.4411	2.2673	1.0929	2.4780	66.2
23.9	0.4051	0.9143	0.4431	2.2566	1.0938	2.4683	66.1
24.0°	0.4067	0.9135	0.4452	2.2460	1.0946	2.4586	66.0°
24.1	0.4083	0.9128	0.4473	2.2355	1.0955	2.4490	65.9
24.2	0.4099	0.9121	0.4494	2.2251	1.0963	2.4395	65.8
24.3	0.4115	0.9114	0.4515	2.2148	1.0972	2.4301	65.7
24.4	0.4131	0.9107	0.4536	2.2045	1.0981	2.4207	65.6
24.5	0.4147	0.9100	0.4557	2.1943	1.0989	2.4114	65.5
24.6	0.4163	0.9092	0.4578	2.1842	1.0998	2.4022	65.4
24.7	0.4179	0.9085	0.4599	2.1742	1.1007	2.3931	65.3
24.8	0.4195	0.9078	0.4621	2.1642	1.1016	2.3841	65.2
24.9	0.4210	0.9070	0.4642	2.1543	1.1025	2.3751	65.1°
	cos θ	sin θ	cot θ	tan θ	csc θ	sec θ	θ deg

TABLE V (continued)

θ deg	sin θ	cos θ	tan θ	cot θ	sec θ	csc θ	
25.0°	0.4226	0.9063	0.4663	2.1445	1.1034	2.3662	65.0°
25.1	0.4242	0.9056	0.4684	2.1348	1.1043	2.3574	64.9
25.2	0.4258	0.9048	0.4706	2.1251	1.1052	2.3486	64.8
25.3	0.4274	0.9041	0.4727	2.1155	1.1061	2.3400	64.7
25.4	0.4289	0.9033	0.4748	2.1060	1.1070	2.3314	64.6
25.5	0.4305	0.9026	0.4770	2.0965	1.1079	2.3228	64.5
25.6	0.4321	0.9018	0.4791	2.0872	1.1089	2.3144	64.4
25.7	0.4337	0.9011	0.4813	2.0778	1.1098	2.3060	64.3
25.8	0.4352	0.9003	0.4834	2.0686	1.1107	2.2976	64.2
25.9	0.4368	0.8996	0.4856	2.0594	1.1117	2.2894	64.1
26.0°	0.4384	0.8988	0.4877	2.0503	1.1126	2.2812	64.0°
26.1	0.4399	0.8980	0.4899	2.0413	1.1136	2.2730	63.9
26.2	0.4415	0.8973	0.4921	2.0323	1.1145	2.2650	63.8
26.3	0.4431	0.8965	0.4942	2.0233	1.1155	2.2570	63.7
26.4	0.4446	0.8957	0.4964	2.0145	1.1164	2.2490	63.6
26.5	0.4462	0.8949	0.4986	2.0057	1.1174	2.2412	63.5
26.6	0.4478	0.8942	0.5008	1.9970	1.1184	2.2333	63.4
26.7	0.4493	0.8934	0.5029	1.9883	1.1194	2.2256	63.3
26.8	0.4509	0.8926	0.5051	1.9797	1.1203	2.2179	63.2
26.9	0.4524	0.8918	0.5073	1.9711	1.1213	2.2103	63.1
27.0°	0.4540	0.8910	0.5095	1.9626	1.1223	2.2027	63.0°
27.1	0.4555	0.8902	0.5117	1.9542	1.1233	2.1952	62.9
27.2	0.4571	0.8894	0.5139	1.9458	1.1243	2.1877	62.8
27.3	0.4586	0.8886	0.5161	1.9375	1.1253	2.1803	62.7
27.4	0.4602	0.8878	0.5184	1.9292	1.1264	2.1730	62.6
27.5	0.4617	0.8870	0.5206	1.9210	1.1274	2.1657	62.5
27.6	0.4633	0.8862	0.5228	1.9128	1.1284	2.1584	62.4
27.7	0.4648	0.8854	0.5250	1.9047	1.1294	2.1513	62.3
27.8	0.4664	0.8846	0.5272	1.8967	1.1305	2.1441	62.2
27.9	0.4679	0.8838	0.5295	1.8887	1.1315	2.1371	62.1
28.0°	0.4695	0.8829	0.5317	1.8807	1.1326	2.1301	62.0°
28.1	0.4710	0.8821	0.5339	1.8728	1.1336	2.1231	61.9
28.2	0.4726	0.8813	0.5362	1.8650	1.1347	2.1162	61.8
28.3	0.4741	0.8805	0.5384	1.8572	1.1357	2.1093	61.7
28.4	0.4756	0.8796	0.5407	1.8495	1.1368	2.1025	61.6
28.5	0.4772	0.8788	0.5430	1.8418	1.1379	2.0957	61.5
28.6	0.4787	0.8780	0.5452	1.8341	1.1390	2.0890	61.4
28.7	0.4802	0.8771	0.5475	1.8265	1.1401	2.0824	61.3
28.8	0.4818	0.8763	0.5498	1.8190	1.1412	2.0758	61.2
28.9	0.4833	0.8755	0.5520	1.8115	1.1423	2.0692	61.1
29.0°	0.4848	0.8746	0.5543	1.8040	1.1434	2.0627	61.0°
29.1	0.4863	0.8738	0.5566	1.7966	1.1445	2.0562	60.9
29.2	0.4879	0.8729	0.5589	1.7893	1.1456	2.0598	60.8
29.3	0.4894	0.8721	0.5612	1.7820	1.1467	2.0434	60.7
29.4	0.4909	0.8712	0.5635	1.7747	1.1478	2.0371	60.6
29.5	0.4924	0.8704	0.5658	1.7675	1.1490	2.0308	60.5
29.6	0.4939	0.8695	0.5681	1.7603	1.1501	2.0245	60.4
29.7	0.4955	0.8686	0.5704	1.7532	1.1512	2.0183	60.3
29.8	0.4970	0.8678	0.5727	1.7461	1.1524	2.0122	60.2
29.9	0.4985	0.8669	0.5750	1.7391	1.1535	2.0061	60.1°
	cos θ	sin θ	cot θ	tan θ	csc θ	sec θ	θ deg

TABLE V (continued)

θ deg	sin θ	cos θ	tan θ	cot θ	sec θ	csc θ	
30.0°	0.5000	0.8660	0.5774	1.7321	1.1547	2.0000	60.0°
30.1	0.5015	0.8652	0.5797	1.7251	1.1559	1.9940	59.9
30.2	0.5030	0.8643	0.5820	1.7182	1.1570	1.9880	59.8
30.3	0.5045	0.8634	0.5844	1.7113	1.1582	1.9821	59.7
30.4	0.5060	0.8625	0.5867	1.7045	1.1594	1.9762	59.6
30.5	0.5075	0.8616	0.5890	1.6977	1.1606	1.9703	59.5
30.6	0.5090	0.8607	0.5914	1.6909	1.1618	1.9645	59.4
30.7	0.5105	0.8599	0.5938	1.6842	1.1630	1.9587	59.3
30.8	0.5120	0.8590	0.5961	1.6775	1.1642	1.9530	59.2
30.9	0.5135	0.8581	0.5985	1.6709	1.1654	1.9473	59.1
31.0°	0.5150	0.8572	0.6009	1.6643	1.1666	1.9416	59.0°
31.1	0.5165	0.8563	0.6032	1.6577	1.1679	1.9360	58.9
31.2	0.5180	0.8554	0.6056	1.6512	1.1691	1.9304	58.8
31.3	0.5195	0.8545	0.6080	1.6447	1.1703	1.9249	58.7
31.4	0.5210	0.8536	0.6104	1.6383	1.1716	1.9194	58.6
31.5	0.5225	0.8526	0.6128	1.6319	1.1728	1.9139	58.5
31.6	0.5240	0.8517	0.6152	1.6255	1.1741	1.9084	58.4
31.7	0.5255	0.8508	0.6176	1.6191	1.1753	1.9031	58.3
31.8	0.5270	0.8499	0.6200	1.6128	1.1766	1.8977	58.2
31.9	0.5284	0.8490	0.6224	1.6066	1.1779	1.8924	58.1
32.0°	0.5299	0.8480	0.6249	1.6003	1.1792	1.8871	58.0°
32.1	0.5314	0.8471	0.6273	1.5941	1.1805	1.8818	57.9
32.2	0.5329	0.8462	0.6297	1.5880	1.1818	1.8766	57.8
32.3	0.5344	0.8453	0.6322	1.5818	1.1831	1.8714	57.7
32.4	0.5358	0.8443	0.6346	1.5757	1.1844	1.8663	57.6
32.5	0.5373	0.8434	0.6371	1.5697	1.1857	1.8612	57.5
32.6	0.5388	0.8425	0.6395	1.5637	1.1870	1.8561	57.4
32.7	0.5402	0.8415	0.6420	1.5577	1.1883	1.8510	57.3
32.8	0.5417	0.8406	0.6445	1.5517	1.1897	1.8460	57.2
32.9	0.5432	0.8396	0.6469	1.5458	1.1910	1.8410	57.1
33.0°	0.5446	0.8387	0.6494	1.5399	1.1924	1.8361	57.0°
33.1	0.5461	0.8377	0.6519	1.5340	1.1937	1.8312	56.9
33.2	0.5476	0.8368	0.6544	1.5282	1.1951	1.8263	56.8
33.3	0.5490	0.8358	0.6569	1.5224	1.1964	1.8214	56.7
33.4	0.5505	0.8348	0.6594	1.5166	1.1978	1.8166	56.6
33.5	0.5519	0.8339	0.6619	1.5108	1.1992	1.8118	56.5
33.6	0.5534	0.8329	0.6644	1.5051	1.2006	1.8070	56.4
33.7	0.5548	0.8320	0.6669	1.4994	1.2020	1.8023	56.3
33.8	0.5563	0.8310	0.6694	1.4938	1.2034	1.7976	56.2
33.9	0.5577	0.8300	0.6720	1.4882	1.2048	1.7929	56.1
34.0°	0.5592	0.8290	0.6745	1.4826	1.2062	1.7883	56.0°
34.1	0.5606	0.8281	0.6771	1.4770	1.2076	1.7837	55.9
34.2	0.5621	0.8271	0.6796	1.4715	1.2091	1.7791	55.8
34.3	0.5635	0.8261	0.6822	1.4659	1.2105	1.7745	55.7
34.4	0.5650	0.8251	0.6847	1.4605	1.2120	1.7700	55.6
34.5	0.5664	0.8241	0.6873	1.4550	1.2134	1.7655	55.5
34.6	0.5678	0.8231	0.6899	1.4496	1.2149	1.7610	55.4
34.7	0.5693	0.8221	0.6924	1.4442	1.2163	1.7566	55.3
34.8	0.5707	0.8211	0.6950	1.4388	1.2178	1.7522	55.2
34.9	0.5721	0.8202	0.6976	1.4335	1.2193	1.7478	55.1°
	cos θ	sin θ	cot θ	tan θ	csc θ	sec θ	θ deg

TABLE V (continued)

θ deg	sin θ	cos θ	tan θ	cot θ	sec θ	csc θ	
35.0°	0.5736	0.8192	0.7002	1.4281	1.2208	1.7434	55.0°
35.1	0.5750	0.8181	0.7028	1.4229	1.2223	1.7391	54.9
35.2	0.5764	0.8171	0.7054	1.4176	1.2238	1.7348	54.8
35.3	0.5779	0.8161	0.7080	1.4124	1.2253	1.7305	54.7
35.4	0.5793	0.8151	0.7107	1.4071	1.2268	1.7263	54.6
35.5	0.5807	0.8141	0.7133	1.4019	1.2283	1.7221	54.5
35.6	0.5821	0.8131	0.7159	1.3968	1.2299	1.7179	54.4
35.7	0.5835	0.8121	0.7186	1.3916	1.2314	1.7137	54.3
35.8	0.5850	0.8111	0.7212	1.3865	1.2329	1.7095	54.2
35.9	0.5864	0.8100	0.7239	1.3814	1.2345	1.7054	54.1
36.0°	0.5878	0.8090	0.7265	1.3764	1.2361	1.7013	54.0°
36.1	0.5892	0.8080	0.7292	1.3713	1.2376	1.6972	53.9
36.2	0.5906	0.8070	0.7319	1.3663	1.2392	1.6932	53.8
36.3	0.5920	0.8059	0.7346	1.3613	1.2408	1.6892	53.7
36.4	0.5934	0.8049	0.7373	1.3564	1.2424	1.6852	53.6
36.5	0.5948	0.8039	0.7400	1.3514	1.2440	1.6812	53.5
36.6	0.5962	0.8028	0.7427	1.3465	1.2456	1.6772	53.4
36.7	0.5976	0.8018	0.7454	1.3416	1.2472	1.6733	53.3
36.8	0.5990	0.8007	0.7481	1.3367	1.2489	1.6694	53.2
36.9	0.6004	0.7997	0.7508	1.3319	1.2505	1.6655	53.1
37.0°	0.6018	0.7986	0.7536	1.3270	1.2521	1.6616	53.0°
37.1	0.6032	0.7976	0.7563	1.3222	1.2538	1.6578	52.9
37.2	0.6046	0.7965	0.7590	1.3175	1.2554	1.6540	52.8
37.3	0.6060	0.7955	0.7618	1.3127	1.2571	1.6502	52.7
37.4	0.6074	0.7944	0.7646	1.3079	1.2588	1.6464	52.6
37.5	0.6088	0.7934	0.7673	1.3032	1.2605	1.6427	52.5
37.6	0.6101	0.7923	0.7701	1.2985	1.2622	1.6390	52.4
37.7	0.6115	0.7912	0.7729	1.2938	1.2639	1.6353	52.3
37.8	0.6129	0.7902	0.7757	1.2892	1.2656	1.6316	52.2
37.9	0.6143	0.7891	0.7785	1.2846	1.2673	1.6279	52.1
38.0°	0.6157	0.7880	0.7813	1.2799	1.2690	1.6243	52.0°
38.1	0.6170	0.7869	0.7841	1.2753	1.2708	1.6207	51.9
38.2	0.6184	0.7859	0.7869	1.2708	1.2725	1.6171	51.8
38.3	0.6198	0.7848	0.7898	1.2662	1.2742	1.6135	51.7
38.4	0.6211	0.7837	0.7926	1.2617	1.2760	1.6099	51.6
38.5	0.6225	0.7826	0.7954	1.2572	1.2778	1.6064	51.5
38.6	0.6239	0.7815	0.7983	1.2527	1.2796	1.6029	51.4
38.7	0.6252	0.7804	0.8012	1.2482	1.2813	1.5994	51.3
38.8	0.6266	0.7793	0.8040	1.2437	1.2831	1.5959	51.2
38.9	0.6280	0.7782	0.8069	1.2393	1.2849	1.5925	51.1
39.0°	0.6293	0.7771	0.8098	1.2349	1.2868	1.5890	51.0°
39.1	0.6307	0.7760	0.8127	1.2305	1.2886	1.5856	50.9
39.2	0.6320	0.7749	0.8156	1.2261	1.2904	1.5822	50.8
39.3	0.6334	0.7738	0.8185	1.2218	1.2923	1.5788	50.7
39.4	0.6347	0.7727	0.8214	1.2174	1.2941	1.5755	50.6
39.5	0.6361	0.7716	0.8243	1.2131	1.2960	1.5721	50.5
39.6	0.6374	0.7705	0.8273	1.2088	1.2978	1.5688	50.4
39.7	0.6388	0.7694	0.8302	1.2045	1.2997	1.5655	50.3
39.8	0.6401	0.7683	0.8332	1.2002	1.3016	1.5622	50.2
39.9	0.6414	0.7672	0.8361	1.1960	1.3035	1.5590	50.1°
	cos θ	sin θ	cot θ	tan θ	csc θ	sec θ	θ deg

TABLE V (continued)

θ deg	sin θ	cos θ	tan θ	cot θ	sec θ	csc θ	
40.0°	0.6428	0.7660	0.8391	1.1918	1.3054	1.5557	50.0°
40.1	0.6441	0.7649	0.8421	1.1875	1.3073	1.5525	49.9
40.2	0.6455	0.7638	0.8451	1.1833	1.3092	1.5493	49.8
40.3	0.6468	0.7627	0.8481	1.1792	1.3112	1.5461	49.7
40.4	0.6481	0.7615	0.8511	1.1750	1.3131	1.5429	49.6
40.5	0.6494	0.7604	0.8541	1.1708	1.3151	1.5398	49.5
40.6	0.6508	0.7593	0.8571	1.1667	1.3171	1.5366	49.4
40.7	0.6521	0.7581	0.8601	1.1626	1.3190	1.5335	49.3
40.8	0.6534	0.7570	0.8632	1.1585	1.3210	1.5304	49.2
40.9	0.6547	0.7559	0.8662	1.1544	1.3230	1.5273	49.1
41.0°	0.6561	0.7547	0.8693	1.1504	1.3250	1.5243	49.0°
41.1	0.6574	0.7536	0.8724	1.1463	1.3270	1.5212	48.9
41.2	0.6587	0.7524	0.8754	1.1423	1.3291	1.5182	48.8
41.3	0.6600	0.7513	0.8785	1.1383	1.3311	1.5151	48.7
41.4	0.6613	0.7501	0.8816	1.1343	1.3331	1.5121	48.6
41.5	0.6626	0.7490	0.8847	1.1303	1.3352	1.5092	48.5
41.6	0.6639	0.7478	0.8878	1.1263	1.3373	1.5062	48.4
41.7	0.6652	0.7466	0.8910	1.1224	1.3393	1.5032	48.3
41.8	0.6665	0.7455	0.8941	1.1184	1.3414	1.5003	48.2
41.9	0.6678	0.7443	0.8972	1.1145	1.3435	1.4974	48.1
42.0°	0.6691	0.7431	0.9004	1.1106	1.3456	1.4945	48.0°
42.1	0.6704	0.7420	0.9036	1.1067	1.3478	1.4916	47.9
42.2	0.6717	0.7408	0.9067	1.1028	1.3499	1.4887	47.8
42.3	0.6730	0.7396	0.9099	1.0990	1.3520	1.4859	47.7
42.4	0.6743	0.7385	0.9131	1.0951	1.3542	1.4830	47.6
42.5	0.6756	0.7373	0.9163	1.0913	1.3563	1.4802	47.5
42.6	0.6769	0.7361	0.9195	1.0875	1.3585	1.4774	47.4
42.7	0.6782	0.7349	0.9228	1.0837	1.3607	1.4746	47.3
42.8	0.6794	0.7337	0.9260	1.0799	1.3629	1.4718	47.2
42.9	0.6807	0.7325	0.9293	1.0761	1.3651	1.4690	47.1
43.0°	0.6820	0.7314	0.9325	1.0724	1.3673	1.4663	47.0°
43.1	0.6833	0.7302	0.9358	1.0686	1.3696	1.4635	46.9
43.2	0.6845	0.7290	0.9391	1.0649	1.3718	1.4608	46.8
43.3	0.6858	0.7278	0.9424	1.0612	1.3741	1.4581	46.7
43.4	0.6871	0.7266	0.9457	1.0575	1.3763	1.4554	46.6
43.5	0.6884	0.7254	0.9490	1.0538	1.3786	1.4527	46.5
43.6	0.6896	0.7242	0.9523	1.0501	1.3809	1.4501	46.4
43.7	0.6909	0.7230	0.9556	1.0464	1.3832	1.4474	46.3
43.8	0.6921	0.7218	0.9590	1.0428	1.3855	1.4448	46.2
43.9	0.6934	0.7206	0.9623	1.0392	1.3878	1.4422	46.1
44.0°	0.6947	0.7193	0.9657	1.0355	1.3902	1.4396	46.0°
44.1	0.6959	0.7181	0.9691	1.0319	1.3925	1.4370	45.9
44.2	0.6972	0.7169	0.9725	1.0283	1.3949	1.4344	45.8
44.3	0.6984	0.7157	0.9759	1.0247	1.3972	1.4318	45.7
44.4	0.6997	0.7145	0.9793	1.0212	1.3996	1.4293	45.6
44.5	0.7009	0.7133	0.9827	1.0176	1.4020	1.4267	45.5
44.6	0.7022	0.7120	0.9861	1.0141	1.4044	1.4242	45.4
44.7	0.7034	0.7108	0.9896	1.0105	1.4069	1.4217	45.3
44.8	0.7046	0.7096	0.9930	1.0070	1.4093	1.4192	45.2
44.9	0.7059	0.7083	0.9965	1.0035	1.4118	1.4167	45.1
45.0°	0.7071	0.7071	1.0000	1.0000	1.4142	1.4142	45.0°
	cos θ	sin θ	cot θ	tan θ	csc θ	sec θ	θ deg

TABLE VI Trigonometric Functions of Radians and Real Numbers

t	$\sin t$	$\cos t$	$\tan t$	$\cot t$	$\sec t$	$\csc t$
0.00	.0000	1.0000	.0000	undefined	1.000	undefined
.01	.0100	1.0000	.0100	99.997	1.000	100.00
.02	.0200	.9998	.0200	49.993	1.000	50.00
.03	.0300	.9996	.0300	33.323	1.000	33.34
.04	.0400	.9992	.0400	24.987	1.001	25.01
.05	.0500	.9988	.0500	19.983	1.001	20.01
.06	.0600	.9982	.0601	16.647	1.002	16.68
.07	.0699	.9976	.0701	14.262	1.002	14.30
.08	.0799	.9968	.0802	12.473	1.003	12.51
.09	.0899	.9960	.0902	11.081	1.004	11.13
.10	.0998	.9950	.1003	9.967	1.005	10.02
.11	.1098	.9940	.1104	9.054	1.006	9.109
.12	.1197	.9928	.1206	8.293	1.007	8.353
.13	.1296	.9916	.1307	7.649	1.009	7.714
.14	.1395	.9902	.1409	7.096	1.010	7.166
.15	.1494	.9888	.1511	6.617	1.011	6.692
.16	.1593	.9872	.1614	6.197	1.013	6.277
.17	.1692	.9856	.1717	5.826	1.015	5.911
.18	.1790	.9838	.1820	5.495	1.016	5.586
.19	.1889	.9820	.1923	5.200	1.018	5.295
.20	.1987	.9801	.2027	4.933	1.020	5.033
.21	.2085	.9780	.2131	4.692	1.022	4.797
.22	.2182	.9759	.2236	4.472	1.025	4.582
.23	.2280	.9737	.2341	4.271	1.027	4.386
.24	.2377	.9713	.2447	4.086	1.030	4.207
.25	.2474	.9689	.2553	3.916	1.032	4.042
.26	.2571	.9664	.2660	3.759	1.035	3.890
.27	.2667	.9638	.2768	3.613	1.038	3.749
.28	.2764	.9611	.2876	3.478	1.041	3.619
.29	.2860	.9582	.2984	3.351	1.044	3.497
.30	.2955	.9553	.3093	3.233	1.047	3.384
.31	.3051	.9523	.3203	3.122	1.050	3.278
.32	.3146	.9492	.3314	3.018	1.053	3.179
.33	.3240	.9460	.3425	2.920	1.057	3.086
.34	.3335	.9428	.3537	2.827	1.061	2.999
.35	.3429	.9394	.3650	2.740	1.065	2.916
.36	.3523	.9359	.3764	2.657	1.068	2.839
.37	.3616	.9323	.3879	2.578	1.073	2.765
.38	.3709	.9287	.3994	2.504	1.077	2.696
.39	.3802	.9249	.4111	2.433	1.081	2.630

TABLE VI (continued)

t	sin t	cos t	tan t	cot t	sec t	csc t
.40	.3894	.9211	.4228	2.365	1.086	2.568
.41	.3986	.9171	.4346	2.301	1.090	2.509
.42	.4078	.9131	.4466	2.239	1.095	2.452
.43	.4169	.9090	.4586	2.180	1.100	2.399
.44	.4259	.9048	.4708	2.124	1.105	2.348
.45	.4350	.9004	.4831	2.070	1.111	2.299
.46	.4439	.8961	.4954	2.018	1.116	2.253
.47	.4529	.8916	.5080	1.969	1.122	2.208
.48	.4618	.8870	.5206	1.921	1.127	2.166
.49	.4706	.8823	.5334	1.875	1.133	2.125
.50	.4794	.8776	.5463	1.830	1.139	2.086
.51	.4882	.8727	.5594	1.788	1.146	2.048
.52	.4969	.8678	.5726	1.747	1.152	2.013
.53	.5055	.8628	.5859	1.707	1.159	1.978
.54	.5141	.8577	.5994	1.668	1.166	1.945
.55	.5227	.8525	.6131	1.631	1.173	1.913
.56	.5312	.8473	.6269	1.595	1.180	1.883
.57	.5396	.8419	.6410	1.560	1.188	1.853
.58	.5480	.8365	.6552	1.526	1.196	1.825
.59	.5564	.8309	.6696	1.494	1.203	1.797
.60	.5646	.8253	.6841	1.462	1.212	1.771
.61	.5729	.8196	.6989	1.431	1.220	1.746
.62	.5810	.8139	.7139	1.401	1.229	1.721
.63	.5891	.8080	.7291	1.372	1.238	1.697
.64	.5972	.8021	.7445	1.343	1.247	1.674
.65	.6052	.7961	.7602	1.315	1.256	1.652
.66	.6131	.7900	.7761	1.288	1.266	1.631
.67	.6210	.7838	.7923	1.262	1.276	1.610
.68	.6288	.7776	.8087	1.237	1.286	1.590
.69	.6365	.7712	.8253	1.212	1.297	1.571
.70	.6442	.7648	.8423	1.187	1.307	1.552
.71	.6518	.7584	.8595	1.163	1.319	1.534
.72	.6594	.7518	.8771	1.140	1.330	1.517
.73	.6669	.7452	.8949	1.117	1.342	1.500
.74	.6743	.7385	.9131	1.095	1.354	1.483
.75	.6816	.7317	.9316	1.073	1.367	1.467
.76	.6889	.7248	.9505	1.052	1.380	1.452
.77	.6961	.7179	.9697	1.031	1.393	1.437
.78	.7033	.7109	.9893	1.011	1.407	1.422
.79	.7104	.7038	1.009	.9908	1.421	1.408

TABLE VI (continued)

t	sin t	cos t	tan t	cot t	sec t	csc t
.80	.7174	.6967	1.030	.9712	1.435	1.394
.81	.7243	.6895	1.050	.9520	1.450	1.381
.82	.7311	.6822	1.072	.9331	1.466	1.368
.83	.7379	.6749	1.093	.9146	1.482	1.355
.84	.7446	.6675	1.116	.8964	1.498	1.343
.85	.7513	.6600	1.138	.8785	1.515	1.331
.86	.7578	.6524	1.162	.8609	1.533	1.320
.87	.7643	.6448	1.185	.8437	1.551	1.308
.88	.7707	.6372	1.210	.8267	1.569	1.297
.89	.7771	.6294	1.235	.8100	1.589	1.287
.90	.7833	.6216	1.260	.7936	1.609	1.277
.91	.7895	.6137	1.286	.7774	1.629	1.267
.92	.7956	.6058	1.313	.7615	1.651	1.257
.93	.8016	.5978	1.341	.7458	1.673	1.247
.94	.8076	.5898	1.369	.7303	1.696	1.238
.95	.8134	.5817	1.398	.7151	1.719	1.229
.96	.8192	.5735	1.428	.7001	1.744	1.221
.97	.8249	.5653	1.459	.6853	1.769	1.212
.98	.8305	.5570	1.491	.6707	1.795	1.204
.99	.8360	.5487	1.524	.6563	1.823	1.196
1.00	.8415	.5403	1.557	.6421	1.851	1.188
1.01	.8468	.5319	1.592	.6281	1.880	1.181
1.02	.8521	.5234	1.628	.6142	1.911	1.174
1.03	.8573	.5148	1.665	.6005	1.942	1.166
1.04	.8624	.5062	1.704	.5870	1.975	1.160
1.05	.8674	.4976	1.743	.5736	2.010	1.153
1.06	.8724	.4889	1.784	.5604	2.046	1.146
1.07	.8772	.4801	1.827	.5473	2.083	1.140
1.08	.8820	.4713	1.871	.5344	2.122	1.134
1.09	.8866	.4625	1.917	.5216	2.162	1.128
1.10	.8912	.4536	1.965	.5090	2.205	1.122
1.11	.8957	.4447	2.014	.4964	2.249	1.116
1.12	.9001	.4357	2.066	.4840	2.295	1.111
1.13	.9044	.4267	2.120	.4718	2.344	1.106
1.14	.9086	.4176	2.176	.4596	2.395	1.101
1.15	.9128	.4085	2.234	.4475	2.448	1.096
1.16	.9168	.3993	2.296	.4356	2.504	1.091
1.17	.9208	.3902	2.360	.4237	2.563	1.086
1.18	.9246	.3809	2.427	.4120	2.625	1.082
1.19	.9284	.3717	2.498	.4003	2.691	1.077

TABLE VI (continued)

t	sin t	cos t	tan t	cot t	sec t	csc t
1.20	.9320	.3624	2.572	.3888	2.760	1.073
1.21	.9356	.3530	2.650	.3773	2.833	1.069
1.22	.9391	.3436	2.733	.3659	2.910	1.065
1.23	.9425	.3342	2.820	.3546	2.992	1.061
1.24	.9458	.3248	2.912	.3434	3.079	1.057
1.25	.9490	.3153	3.010	.3323	3.171	1.054
1.26	.9521	.3058	3.113	.3212	3.270	1.050
1.27	.9551	.2963	3.224	.3102	3.375	1.047
1.28	.9580	.2867	3.341	.2993	3.488	1.044
1.29	.9608	.2771	3.467	.2884	3.609	1.041
1.30	.9636	.2675	3.602	.2776	3.738	1.038
1.31	.9662	.2579	3.747	.2669	3.878	1.035
1.32	.9687	.2482	3.903	.2562	4.029	1.032
1.33	.9711	.2385	4.072	.2456	4.193	1.030
1.34	.9735	.2288	4.256	.2350	4.372	1.027
1.35	.9757	.2190	4.455	.2245	4.566	1.025
1.36	.9779	.2092	4.673	.2140	4.779	1.023
1.37	.9799	.1994	4.913	.2035	5.014	1.021
1.38	.9819	.1896	5.177	.1931	5.273	1.018
1.39	.9837	.1798	5.471	.1828	5.561	1.017
1.40	.9854	.1700	5.798	.1725	5.883	1.015
1.41	.9871	.1601	6.165	.1622	6.246	1.013
1.42	.9887	.1502	6.581	.1519	6.657	1.011
1.43	.9901	.1403	7.055	.1417	7.126	1.010
1.44	.9915	.1304	7.602	.1315	7.667	1.009
1.45	.9927	.1205	8.238	.1214	8.299	1.007
1.46	.9939	.1106	8.989	.1113	9.044	1.006
1.47	.9949	.1006	9.887	.1011	9.938	1.005
1.48	.9959	.0907	10.983	.0910	11.029	1.004
1.49	.9967	.0807	12.350	.0810	12.390	1.003
1.50	.9975	.0707	14.101	.0709	14.137	1.003
1.51	.9982	.0608	16.428	.0609	16.458	1.002
1.52	.9987	.0508	19.670	.0508	19.695	1.001
1.53	.9992	.0408	24.498	.0408	24.519	1.001
1.54	.9995	.0308	32.461	.0308	32.476	1.000
1.55	.9998	.0208	48.078	.0208	48.089	1.000
1.56	.9999	.0108	92.620	.0108	92.626	1.000
1.57	1.0000	.0008	1255.8	.0008	1255.8	1.000

ANSWERS TO SELECTED EXERCISES

SECTION 1

1. *Natural numbers*: 17, 6. *Integers*: 17, $-6/\sqrt{4}\,(=-3)$, 6. *Rational numbers*: 17, -3.14, $-16\frac{2}{3}$, $4.\overline{07}$, $-6/\sqrt{4}$, 6. *Irrational numbers*: π, $\sqrt{10}$.

3. *Natural numbers*: $\sqrt{64}\,(=8)$. *Integers*: $0\sqrt{2}\,(=0)$, $\sqrt{64}$, $-15/\sqrt{9}\,(=-5)$. *Rational numbers*: $0\sqrt{2}$, $4\frac{1}{7}$, $-13.\overline{5}$, $\sqrt{64}$, -2.81, $-15/\sqrt{9}$. *Irrational numbers*: $\sqrt{3}$, 2π.

5.

7. $<$ **9.** $>$ **11.** $>$ **13.** $>$ **15.** $<$

17. $99\sqrt{10}/100$, $3\frac{10}{17}$, π, $22/7$

19. 4	**21.** 15	**23.** 0	**25.** 27
27. 4	**29.** -11	**31.** -0.8	**33.** 0
35. -1	**37.** -1.1	**39.** -56	**41.** 70
43. -1	**45.** 537.9	**47.** -10	**49.** 2/3
51. 2	**53.** 3/8	**55.** 7/12	**57.** 7/12
59. 7/12	**61.** $-16/75$	**63.** $-7/29$	**65.** $-3/7$
67. $-6/5$	**69.** 6/5	**71.** $-14/55$	**73.** $-2a + b$
75. $a + b$	**77.** $15a - 5b$	**79.** $-7a + 6b$	**81.** $39a - 48b$
83. 1.9	**85.** $\pi/2$	**87.** $5/(2\pi)$	**89.** 6000
91. 12.5%	**93.** \$15,061	**95.** \$270	**97.** \$444.45
99. \$500	**101.** Approximately 35%.		

SECTION 2

1. $-\dfrac{8}{81}$	**3.** $-\dfrac{49}{125}$	**5.** 32	**7.** -50
9. -1000	**11.** t^8	**13.** $2.25u^6$	**15.** $-0.125w^2$
17. $2y/x^2$	**19.** $-b^5/a^8$	**21.** $-u^{12}/(81t^8)$	**23.** $2x^2$

25. $25/(a^2b^4)$ **27.** $1/(4a^4b)$ **29.** $16b^{13}/a^7$

31. b^{3n}/a^n **33.** y^{3n}/x^{2n} **35.** $4^n x^n y^n/z^{2n}$

37. 1×10^{-8} **39.** 3.212×10^3 **41.** 3.6×10^4

43. 3.3×10^{-9} **45.** 2.9×10^{-3} **47.** 640 **49.** 1

51. 300 **53.** $12.6100, 12.0060, 12.0001$

55. $126.68   **57.** $670.05   **59.** $443.73

61. (a) \$200 at 4%   (b) \$100 at 8%

63. 150.9

SECTION 3

1. 2 **3.** 4 **5.** 4 **7.** $\dfrac{1}{3}$ **9.** $\dfrac{1}{125}$

11. 0.03125 **13.** $-\dfrac{1}{10}$ **15.** $\dfrac{1}{16}$ **17.** -3125

19. 3125 **21.** -128 **23.** $\dfrac{1}{100}$ **25.** x^2

27. z^3 **29.** $4y^3$ **31.** $1/(a^{16}b^4)$ **33.** m

35. -2 **37.** -5 **39.** -10 **41.** -0.2

43. $-\dfrac{1}{2}$ **45.** $-\dfrac{2}{3}$ **47.** 20 **49.** -13

51. $-\dfrac{1}{12}$ **53.** -1.7 **55.** 0.027 **57.** 0.729

59. $\sqrt{3}$ **61.** 0 **63.** $-7\sqrt{3}$ **65.** 30 **67.** a^2

69. c^2 **71.** $y^3\sqrt[3]{y}$ **73.** $\sqrt{u}$ **75.** $8\sqrt{w}$

77. $3s\sqrt{2s}$ **79.** $\sqrt[12]{a}$ **81.** $1/\sqrt[20]{c}$ **83.** 81

85. $1/r^2$ **87.** t^3

89. uv/w **91.** a^{m+2} **93.** c^n/c^2 or c^{n-2}

95. 1.620656597 **97.** 0.009180550

99. 12.69300465 **101.** $\sqrt{10}$ **103.** $(3 - \sqrt{2})/7$

105. $-1 - \sqrt{6}$ **107.** $\sqrt{x + 1} + \sqrt{x}$ **109.** $\sqrt[3]{4}/2$

111. $3\sqrt[5]{4}$ **113.** $1/(7 + 4\sqrt{3})$

115. $-1/(\sqrt{x - h} + \sqrt{x})$ **117.** $1/(\sqrt{1 - x + h} + \sqrt{1 - x})$

119. $3.4366, 3.3549, 3.3540$ **121.** 8.04×10^5

123. 142 million miles

SECTION 4

1. $4x^2 - 2x - 2$ **3.** $t^2 + 3$ **5.** $2x^3 + 2x^2 - 2x - 1$

7. $-u^3v^4$ **9.** $-4x^2yz^3$ **11.** $u^4 - \dfrac{3}{2}u^3$

13. $3x^3 + 2x^2 - 7x + 2$ **15.** $4y^3 - 5y^2 - 11y + 3$

17. $0.1x - 0.4$ **19.** $2xy - 5y$ **21.** $mn + m - 2n$

23. $-dm$ **25.** $x^3 + 6x^2 + 11x + 6$ **27.** $z^4 - 1$

29. $2m^2 - mn - n^2$ **31.** $0.03y^2 + 0.02yz - 0.08z^2$

33. $0.001x^2 + 0.0202xy + 0.004y^2$ **35.** $y^2 - 2\sqrt{2}y + 2$

37. $x^2 + 2\sqrt{5}x + 5$ **39.** $z^2 - 3$ **41.** $u^2 + u + \dfrac{1}{4}$

43. $x - y$ **45.** 2 **47.** $x^{2p} + 2x^{p+q} + x^{2q}$

49. $(x + 5)(x - 5)$ **51.** $(x - 1)^2$

53. $(x - 5)^2$ **55.** $(y + 4)^2$ **57.** $(5w + 1)^2$

59. $\left(v + \dfrac{1}{2}\right)^2$ **61.** $(u + \sqrt{3})^2$ **63.** $\left(x - \dfrac{1}{3}\right)^2$

65. $(0.2a - 0.5b)^2$ **67.** $(2y - 1)(y + 1)$ **69.** $(3w + 1)(w + 1)$

71. $2(y + 1)(2y - 3)$ **73.** $(3v - 2)(2v + 3)$ **75.** $(4n - 1)(n - 3)$

77. Verify $(m^2 - n^2)^2 + (2mn)^2 = (m^2 + n^2)^2$. Also, a, b, and c are positive.

79. $(360, 38, 362)$

81. $(9, 12, 112, 113)$

SECTION 5

1. $(x + 6)^2$ **3.** $(z + 1)(z - 1)$ **5.** $z^2(z + 1)^2$

7. $(2z - 1)^2$ **9.** $(3y + 1)^2$ **11.** $(v - 1)(v^2 + v + 1)$

13. $v^2(v - 1)(v - 3)$ **15.** $u^3(u + 1)(u + 6)$

17. $(u - 10v)(u^2 + 10v + 100v^2)$ **19.** $r(r - 5)(r^2 + 5r + 25)$

21. $t(4 - t)(16 + 4t + t^2)$ **23.** $(t + 1)^2(t - 1)^2$

25. $(t - 0.1)(t^2 + 0.1t + 0.01)$ **27.** $(s + 0.2)(s^2 - 0.2s + 0.04)$

29. $(b^n + 1)(b^n - 1)$ **31.** $(b + 1)^3$ **33.** $(a - 2)^3$

35. $(a + 3b)^3$ **37.** $xy(x - 5y)^2$ **39.** $rs(r - 3s)^2$

41. $\left(\dfrac{1}{3} - n\right)^3$ **43.** $(a - b)(x + y)$ **45.** $(a + b)(x + y)$

47. $(c + d)(u + v)(u - v)$

49. $(a + b)(a - b)(a^4 + a^2b^2 + b^4)$ or
 $(a + b)(a - b)(a^2 + ab + b^2)(a^2 - ab + b^2)$

51. $(x^4 + y^4)(x^2 + y^2)(x + y)(x - y)$

53. $(2u + v - 3)(2u - v + 3)$ **55.** $x^2(4x^n - 1)(x^n - 1)$

57. $(x^n + y^n)(w^{2n} + z^{2n})$

59. $x^2y^2(x^p + y^q)^2$

61. $t^3 + 3t^2 + 3t + 1$ **63.** $v^3 + 15v^2 + 75v + 125$

65. $x^3 + \dfrac{3}{2}x^2 + \dfrac{3}{4}x + \dfrac{1}{8}$ **67.** $\dfrac{1}{8}x^6 + \dfrac{3}{4}x^4 + \dfrac{3}{2}x^2 + 1$

69. $y^3 - 2y^2 + \dfrac{4}{3}y - \dfrac{8}{27}$ **71.** $c + d$

73. $2.86a^3 - 8.17a^2b + 7.76ab^2 - 2.46b^3$

75. $333{,}000u^3 + 212{,}000u^2v^2 + 44{,}900uv^4 + 3{,}180v^6$

83. $(x + 1)(x - 1)(x^2 + x + 1)$ **85.** $(a + b)(a - b)^2$

87. $(a + b)(a - b)(a^2 + b^2)$

SECTION 6

1. $\dfrac{1}{5}$ **3.** 4   **5.** $2/(y - 3)$ **7.** $(z^2 + z + 1)/(z + 1)$

9. $(z + 1)^2/(z^2 - z + 1)$ **11.** $2/(y^2 + y)$ **13.** $a^2 - a$

15. $c - 1$ **17.** $s^4 + s^2 + 1$ **19.** $v/(4u^3 - 2u^2v + uv^2)$

21. 5   **23.** $3/z$ **25.** $a^2 - a + 1$

27. $(100 + 10c + c^2)/(10 + c)$ **29.** $(-2hx - h^2)/[x^2(x + h)^2]$

31. $a/(a^2 - b^2)$ **33.** 1   **35.** $1/[2(c - d)]$

37. $x/(x - 2)$ **39.** $-z - 2$ **41.** $(z^2 + 2z + 4)/(z^2 + 2z)$

43. $(z^2 + z + 1)/(z + 1)$ **45.** $2y/(y + 1)$ **47.** $-2/(cd)$

49. $-1/[x(x + h)]$ **51.** $-(3x^2 + 3hx + h^2)/[(x + h)^3x^3]$

53. $(x^2 + 4x)/[2(x + 1)^{5/2}]$ **55.** a^n **57.** c^{n-3}

59. $x^{2n} - 1$ **61.** $\dfrac{5}{16}$ sq mi   **63.** $1\dfrac{3}{16}$ miles

65. 400 yd/hr   **67.** 1.76 cm^3 **69.** 271 g   **71.** 33

CHAPTER I REVIEW

1. *Natural numbers*: $\sqrt{36}$. *Integers*: $0/\sqrt{2}$, $-12/\sqrt{9}$, $\sqrt{36}$, $-\sqrt{2}\sqrt{2}$. *Rational numbers*: $20.\overline{1}$, $0/\sqrt{2}$, $-12/\sqrt{9}$, $\sqrt{36}$, -3.14, $-\sqrt{2}\sqrt{2}$. *Irrational numbers*: 3π, $\sqrt{5}$.

2. See Section 1.   **3.** 0.2   **4.** -0.88 **5.** -0.4

6. 0.4   **7.** 0.66   **8.** -0.43 **9.** -46

10. -21 **11.** $2a - 14b$ **12.** $3a - 4b$ **13.** $\dfrac{8}{15}$

14. $-\dfrac{2}{21}$ **15.** 4   **16.** $\dfrac{1}{4}$ **17.** $(x + 1)/(x - 1)$

18. $(3y + 2)/(y + 2)$ **19.** $-1/a$ **20.** $2bc$

21. 4   **22.** $-\dfrac{4}{3}$ **23.** 1.99   **24.** 10

25. 49   **26.** $-\dfrac{1}{1000}$ **27.** $-\dfrac{100}{9}$ **28.** -125

29. $x^3/(27y^3)$ **30.** $32a^{10}/b^5$ **31.** $1/(36a^2)$

32. $y^8/256$ **33.** $1000a^6b^3$ **34.** $125x^9y^{15}$

35. a^{6n}/b^{9n} **36.** $2^nu^{2n}v^{4n}/9^n$

37. 7.72×10^8 **38.** 4.8×10^{-6} **39.** $2a^2$

40. $2a^2b^3$ **41.** $3x^2\sqrt{y}$ **42.** $x - 4y$

43. 64   **44.** $\dfrac{9}{4}$ **45.** $125b^6/(8a^9)$

46. c^{10} **47.** a^{2m+1} **48.** b^{m+1}

49. $2x^3 + x^2 + x - 1$ **50.** $y^3 - y^2 - 8y + 6$

51. $a^2 - \dfrac{2}{3}a + \dfrac{1}{9}$ **52.** $b^2 + 0.4b + 0.04$

53. $a^3 + 6a^2b + 12ab^2 + 8b^3$ **54.** $c^6 - 3c^4d^2 + 3c^2d^4 - d^6$

55. $(t + 3)(t - 3)$ **56.** $(u - 11)^2$

57. $(v + 6)^2$ **58.** $(2w + 1)(2w - 1)$

59. $(5x + 2)(2x - 1)$ **60.** $(3y + 2)(3y - 1)$

61. $x(3x - 1)^2$ **62.** $y^2(y + 1)(y^2 - y + 1)$

63. $(5a - 2b)(3a + 2b)$

64. $(c^2 + d^2)(c + d)(c - d)$ **65.** $(m - n)^3$

66. $(u + v + 2)(u + v - 2)$

67. $(x + y)(x^n + y^n)(x^n - y^n)$

68. $(x^n - y^n)(x^{2n} + x^ny^n + y^{2n})$

69. $x(x + 1)(x - 1)(x^2 - x + 1)$

70. $y(y - 2)^2(y^2 + 2y + 4)$

71. $2(m + 1)$ **72.** $v/[u(u + 2v)]$

73. $(x + 1)/4$ **74.** $4x/(x - 1)$

75. $1/(x^2 - 1)$ **76.** $2/[x(x + 1)(x - 2)]$

77. $-ab$ **78.** $(d + c)/d$

79. $5\sqrt{3}/3$ **80.** $(a^{1/2} + b^{1/2})/(a - b)$

81. 1.35   **82.** 147   **83.** 4.865 billion

84. 275 million   **85.** \$210   **86.** \$150

87. $\$2000(1.02)^{12} = \$2536.48$ **88.** $\$5000(1.005)^{30} = \$5807.00$

89. $\$10,000(1.02)^{-40} = \$4528.90$ **90.** $\$2000(1.025)^{-16} = \$1347.25$

91. 50 ft/min   **92.** $315.\overline{3}$ l/min

93. 77.2 g   **94.** 3.69 cm³   **95.** 0.2776, 0.2886

95. (a) 3.2751   (b) 3.0741

SECTION 7

1. $\{x: \ x \neq 0\}$ **3.** All real numbers.   **5.** $\{y: \ y \neq -2\}$

7. $\{x: \ x \neq 0 \text{ and } x \geq -1\}$ **9.** $\{x: \ x \neq \frac{1}{2} \text{ and } x \geq 0\}$

11. No   **13.** No   **15.** No   **17.** Yes

19. 6   **21.** -2 **23.** $-7/3$ **25.** -1.13 **27.** 1.377

29. $-1/2$ **31.** 4   **33.** No solution ($z = 2$ is extraneous).

35. No solution ($x = 1$ is extraneous).

37. (a) $y = (1 - 2x)/3$ (b) $x = (1 - 3y)/2$

39. (a) $y = (5x - 17)/6$ (b) $x = (6y + 17)/5$

41. (a) $y = (2x - 1)/2$ (b) $x = (2y + 1)/2$

43. $r = I/(Pt)$ **45.** $V = k/P$ **47.** $L = (P - 2W)/2$

49. $V = Bn/(n - x)$ **51.** $r = (A - P)/Pt$

53. $a = bR/(b - R)$ **55.** $S = 4\pi r^2$ **57.** $V = \dfrac{1}{3} Bh$

59. $L < \dfrac{3}{10} S$ **61.** 1.5 **63.** 0.2 **65.** 40

67. 94 ft and 50 ft **69.** 28, 29, 30 **71.** 18

73. 27.5°, 55°, 97.5° **75.** $17,300

77. 2.5 liters from A and 1.5 liters from B.

79. 360 mph **81.** 8.5 m

SECTION 8

1. ± 6 **3.** ± 10 **5.** ± 1.5 **7.** No real solution.

9. No real solution. **11.** $\pm 3\sqrt{3}/A$ **13.** 0 **15.** $\pm \dfrac{2}{7}$

17. 0, 3 **19.** 1 **21.** 3 **23.** $-1, 4$

25. 0, 3 **27.** 0.5 **29.** $-2/3, 1$ **31.** ± 3.51

33. ± 6.94 **35.** $-3 \pm \sqrt{10}$ **37.** $-1, 1/3$ **39.** $(1 \pm \sqrt{6})/5$

41. $(-1 \pm \sqrt{37})/6$ **43.** Two. **45.** Two. **47.** None.

49. One. **51.** None. **53.** $(1 \pm \sqrt{13})/6$ **55.** $-\dfrac{3}{2}, \dfrac{1}{4}$

57. $1 \pm \sqrt{5}$ **59.** $(-q \pm \sqrt{q^2 + 4q})/2$ **61.** 0, 7 **63.** $-\dfrac{1}{4}, 0$

65. $(-3 \pm \sqrt{5})/2$ **67.** ± 5 **69.** ± 0.1 **71.** $-3/2, 0$

73. $1.10, -0.561$ **75.** $-0.94, 1.65$ **77.** $\pm\sqrt{R^2 - (V/k)}$

79. $2\sqrt{2}$ in. **81.** 5 in. **83.** $\sqrt{S/P} - 1 \ (r > 0)$

85. (a) ft/sec² (b) m/sec² **87.** 400 ft, 3.75 sec

89. $\sqrt{630}/4 \approx 6.3$ sec later. **91.** 36 ft/sec

93. $\sqrt{4R^2 - t^2}/2$ **95.** $(-v_0 + \sqrt{v_0^2 + 2gs})/g$ (positive solution)

97. $\pm 2\sqrt{2}$ **99.** $b^2 - 4ac = 0$ iff $b = \pm 2\sqrt{ac}$

101. $\dfrac{-b + \sqrt{b^2 - 4ac}}{2a} \cdot \dfrac{-b - \sqrt{b^2 - 4ac}}{2a} = \dfrac{b^2 - (b^2 - 4ac)}{4a^2} = \dfrac{c}{a}$

SECTION 9

1. 7 **3.** $-\dfrac{2}{3}$ **5.** 1, 4 **7.** 5 **9.** 1, 9

11. $\dfrac{1}{4}$ **13.** 2   **15.** $-\dfrac{3}{4}$ **17.** 2   **19.** -3

21. -5 **23.** No solution.   **25.** $-\dfrac{1}{5}, -\dfrac{1}{2}$

27. $-\dfrac{6}{5}, \dfrac{24}{5}$ **29.** $\pm\sqrt{3}$ **31.** $1 + 2\sqrt{2}$

33. $-4 \pm 4\sqrt[4]{2}$ **35.** $\dfrac{1}{12}, \dfrac{5}{12}$ **37.** $\dfrac{1}{2}$ **39.** $\dfrac{18}{5}, 6$

41. -0.39 **43.** $A = \pi r^2$

45. $g = 4\pi^2 l / T^2$ **47.** $h = \sqrt{S^2 - \pi^2 r^4} / \pi r$

49. $R_1 = RR_2 / (R_2 - R)$ **51.** $r_N = n[(r_E + 1)^{1/n} - 1]$

53. $C = 2\sqrt{\pi A}$

55. (a) 168 cm³/sec   (b) $h = v^2/(0.72A^2 g)$ (c) 7 cm

57. 55 mph

SECTION 10

1. $>$ **3.** $>$ **5.** $<$ **7.** $>$ **9.** $>$ **11.** $>$

13. $\{x: \ x > -2\}$ **15.** $\{z: \ z < \frac{1}{6}\}$ **17.** $\{u: \ u > -7\}$

19. $\{v: \ v \ge -0.75\}$ **21.** $\{y: \ y \ge 3.5\}$

23.   [2, 3]

25. $\varnothing$

27.   Not an interval.

29. $\varnothing$

31. $[-4, 4]$ **33.** $[-5, 5]$ **35.** $[-60, 60]$ **37.** $(-3, 4)$

39. $[-1, 5]$ **41.** $(0, 0.4)$ **43.** $(-9, -4) \cup (4, 9)$

45. $\left(-\dfrac{4}{3}, -\dfrac{1}{3}\right) \cup (1, 2)$ **47.** $[-2, \infty)$

49. $\{x: \ -4/3 \le x < 2\}$

51. $\varnothing$ (The expression is undefined for every real number x.)

53. (a) $|x - 3| \ge 6$ (b) $[-3, 9]$

55. The discriminant, $(-3)^2 - 4(k)(2)$, must be positive. Thus the answer is $\{k: \ k < 9/8\}$.

57. If $a \ge 0$ then $|a| = a$; if $a < 0$ then $|a| = -a$ so $-|a| = a$. Thus, in either case, $-|a| \le a \le |a|$. Similarly, $-|b| \le b \le |b|$. Add these inequalities to get

$$-|a| - |b| \le a + b \le |a| + |b|.$$

From this, Equation (10.7) yields $|a + b| \le |a| + |b|$.

59. 50°F to 68°F

61. $\dfrac{1}{5} < y < 1$

63. Solve $4[x + (x + 1)] > 7(x + 2)$. This yields $x > 10$. Therefore, the answer is those sets in which the smallest of the three integers is at least 11.

SECTION 11

1. $(-1, 5)$ **3.** $(-2, 4)$ **5.** $\left[-1, \dfrac{2}{3}\right]$

7. $(-\infty, -\tfrac{1}{2}) \cup (2, \infty)$ **9.** $(-\infty, -0.2) \cup (0.1, \infty)$

11. $[0, 1] \cup [2, \infty)$ **13.** $[-2, 0] \cup [\tfrac{1}{2}, \infty)$ **15.** $(0, 1)$

17. $(-\infty, -4) \cup \{1\}$ **19.** $(-\infty, -1] \cup (0, 1]$

21. $[-12, -5) \cup (2, \infty)$ **23.** $(-\infty, -1) \cup (0, 1)$

25. $[-1, 1]$ **27.** $[-0.25, 0.25]$

29. $[-2, -1) \cup (1, 2]$ **31.** $(-\infty, -2) \cup (2, \infty)$

33. The discriminant, $-3q^2$, is negative for all $q \neq 0$.

35. $\{0, 1, 2\}, \{1, 2, 3\}, \{2, 3, 4\}$

37. $15/4 \leq R \leq 60/11$ **39.** $(-\infty, 0) \cup (1, \infty)$

CHAPTER II REVIEW

1. $\dfrac{13}{11}$ **2.** $-\dfrac{9}{2}$ **3.** $(3y + 4)/4$ **4.** $(y - 9)/4$

5. -4 **6.** No solution.   **7.** $-5, 2$ **8.** $-3/2, -1$

9. $(-3 \pm \sqrt{17})/4$ **10.** $(1 \pm \sqrt{13})/6$ **11.** $\pm\dfrac{11}{3}$

12. $(-p \pm \sqrt{p^2 - 4})/2$ **13.** $(3 \pm \sqrt{2})/7$ **14.** $\pm\dfrac{7}{5}$

15. Two.   **16.** Two.   **17.** None.   **18.** One.

19. $-\dfrac{3}{2}, -\dfrac{1}{2}$ **20.** -3 **21.** $\dfrac{1}{2}$ **22.** 2, 6

23. No real solution.   **24.** No real solution.   **25.** $4df^2L^2$

26. $-1 \pm \sqrt{r + 1}$ **27.** $(-9, \infty)$ **28.** $[\tfrac{1}{15}, \infty)$

29. $[-1, 2)$ **30.** $(-\pi, \sqrt{3}]$ **31.** $(-2, 7]$ **32.** $(-\infty, 6]$

33. $(-5, 7)$ **34.** $(-\infty, -2] \cup \left[\dfrac{1}{2}, \infty\right)$ **35.** $(-\infty, -2] \cup \left[\dfrac{6}{5}, \infty\right)$

36. $\left(-\dfrac{4}{3}, 2\right)$ **37.** $(-\infty, -3) \cup (0, 1)$ **38.** $\left(-1, -\dfrac{1}{3}\right] \cup (0, \infty)$

39. $\dfrac{4}{5}$ **40.** 20, 21, 22   **41.** $1\dfrac{1}{3}l$

42. \$350 at 8%, \$850 at 10%   **43.** $3 + \sqrt{11}$ **44.** 10

45. $\frac{1}{2}\sqrt{4h^2 + x^2}$ **46.** First, show that $\overline{OD} = \sqrt{r^2 - (s/2)^2}$.

47. $\sqrt{10}/4 \approx 0.79$ sec   **48.** 3 sec   **49.** $\sqrt{a/(4 - \pi)}$

50. $13\frac{7}{11}$ sec

SECTION 12

1.

3.

5. $(2.5, -4)$ **7.** $(0, -4)$ **9.** $(0, 3.2)$

11. $(0, 2)$ and $(4, 2)$ are. **13.** $(1, 1)$ and $(-2, 4)$ are. **15.** None is.

17. Straight line parallel to and 1 unit below the x-axis.

19. Straight line parallel to and 1.5 units below the x-axis.

21. Straight line parallel to and 2 units to the left of the y-axis.

23. Circle centered at the origin with radius 4.

25. Circle centered at $(1, 0)$ with radius $\sqrt{10}$.

27. Circle centered at $(-2, -1)$ with radius $\sqrt{12}$.

29. $y = -3$ **31.** $x = 5$ **33.** $y = 5$

35. $(x + 2)^2 + (y + 3)^2 = 2$ **37.** $x^2 + y^2 = 25$

39. $(x + 3)^2 + (y - 1)^2 = 9$ **41.** $\sqrt{34}$

43. $\sqrt{389}/20$ **45.** $\sqrt{29}/2$

47. 5.238

49. From $(x - 1)^2 + (y + 2)^2 = 9$, the graph is a circle with center at $(1, -2)$ and radius 3.

51. From $(x - 1)^2 + (y - 4)^2 = 25$, the graph is a circle with center at $(1, 4)$ and radius 5.

53. From $x^2 + (y - 1)^2 = 4$, the graph is a circle with center at $(0, 1)$ and radius 2.

55. A single point, the origin.

57. The empty set.

59. Circle with center at $(1, -2/3)$ and radius $\sqrt{2}$.

61. The points to the left of the y-axis.

63. The points in the second and fourth quadrants.

65. The points outside the circle centered at the origin with radius $\sqrt{5}$.

67.

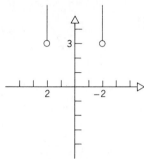

69. The points inside and on the square with vertices at $(0, 0)$, $(0, 2)$, $(2, 2)$, and $(2, 0)$.

71. $(0, -1 + \sqrt{5})$, $(0, -1 - \sqrt{5})$, $(2 + 2\sqrt{2}, 0)$, and $(2 - 2\sqrt{2}, 0)$.

73. $(5 + \sqrt{7})/3$ or $(5 - \sqrt{7})/3$.

75.

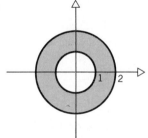

SECTION 13

1. -3 **3.** $\frac{1}{9}$ **5.** 0 **7.** $x - 2y + 5 = 0$

9. $x - 4y = 0$ **11.** $10x - 4y - 15 = 0$ **13.** $2x - 5y + 8 = 0$

15. $x - y + 3 = 0$ **17.** $6x - y - 5 = 0$ **19.** $y = -x + 2$

21. $y = 2$ **23.** $y = 0$ **25.** $m = -2, b = 3$

27. $m = \dfrac{1}{2}, b = -\dfrac{3}{2}$ **29.** $m = 1, b = 0$

31.

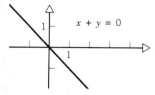

33.

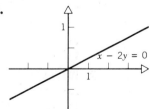

35.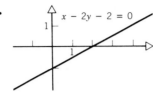

37.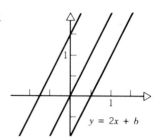

39.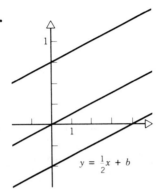

41.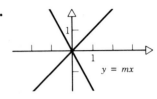

43. $y = \dfrac{2}{3} x$ **45.** $x - y + 2 = 0$

47. $4x + 3y + 9 = 0$

49. Substitute m from (13.2) into (13.4).

51. $\left(\dfrac{1}{2}, \dfrac{9}{2}\right)$

53. Substitute $x = 1$ and $y = 1$ into Equation (13.5), and then solve for m.

55. $2x - y - 7 = 0$

57. $3x + 4y - 25 = 0$

SECTION 14

1. $1, -3, -2\sqrt{5} - 3, 2t - 3, a - 3, 2b - 1, 2x^2 - 3$

3. $3, 5, \sqrt{5} + 5, -t + 5, (10 - a)/2, -b + 4, -x^2 + 5$

5. $5, 1, 6, t^2 + 1, (a^2 + 4)/4, b^2 + 2b + 2, x^4 + 1$

7. $\{x: \ x \neq 1\}$ **9.** $\{x: \ x \neq -1\}$ **11.** $\{x: \ x \leq 1\}$

13. $\{x: \ -1 \leq x \leq 1 \ \text{and} \ x \neq 0\}$ **15.** $\{x: \ x < 4 \ \text{and} \ x \neq -1\}$

17. (a) $x^2 + x$ (b) $-x^2 + x + 4$ (c) $x^3 + 2x^2 - 2x - 4$
   (d) $(x + 2)/(x^2 - 2)$ (e) 1 (f) 5
   (g) $a + 2$ (h) $-(t + 2)/2$

19. (a) $2x + a + b$ (b) $b - a$ (c) $x^2 + (a + b)x + ab$
   (d) $(x + b)/(x + a)$ (e) $a + b + 1$ (f) $3 - a + b$
   (g) $(a + b)(\sqrt{3} + a)$ (h) $(t + b)/a$

21. (a) $ax + x + 2b$ (b) $ax - x$ (c) $ax^2 + abx + bx + b^2$
    (d) $(ax + b)/(x + b)$ (e) $2b + 1$ (f) $2a + 1$
    (g) $(a^2 + b)(\sqrt{3} + b)$ (h) $(at + b)/b$

23. (a) $-4x + 2$ (b) $-4x - 8$ (c) -6 (d) 4

25. (a) $-27x^3$ (b) $-3x^3$ (c) -216 (d) 81

27. (a) $-2(x + 1)^3$ (b) $-2x^3 + 1$ (c) -54 (d) 55

29. (a) x^2 (b) $-x^2$ (c) 4 (d) -9

31. $g(x) = x - 4,\ h(x) = x^2$

33. $g(x) = 2x - 1,\ h(x) = \sqrt{x}$

35. $g(x) = \sqrt{x},\ h(x) = x - 1$

37. 4 **39.** -1 **41.** $4x + h$

43. $g(x) = 3x - 10,\ h(x) = 9x^2 + 30x + 100$

45. $g(x) = x^{1/3} + 2,\ h(x) = x^{1/3} - 2$

47. $\{x:\ x \le -12\ \text{ or }\ x \ge 8\}$

49. 24.76 **51.** 46.83

53. $\{y:\ y \ge 1\}$ **55.** $-9/8$ **57.** 1

59. $h(C) = \dfrac{9}{5} C + 32$. Domain $= \{C:\ C \ge -273.15\}$.
    Range $= \{F:\ F \ge -459.67\}$.

61. $g(t) = 2000\,(1.03)^{4t}$ **63.** $f(S) = S\sqrt{S}/6\sqrt{\pi}$

SECTION 15

1.

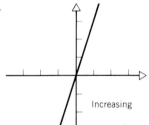

Increasing

3.

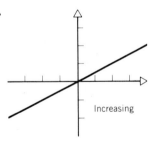

Increasing

5.

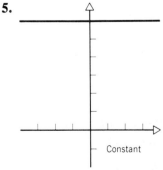

Constant

7.

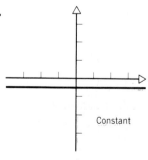

Constant

9.

(graph showing a horizontal line labeled "Constant")

11. Not a function.

13. Not a function.

15. Function. Increasing for $x \le -2.5$ and $0 \le x \le 2.5$.
Decreasing for $-2.5 \le x \le 0$ and $x \ge 2.5$.

17. Function. Increasing for $x \le 0$. Decreasing for $x \ge 0$.

19. 1   **21.** 1   **23.** 2   **25.** 4   **27.** 3

29. 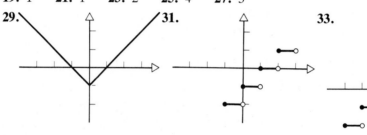   **31.   33.**

35.   **37.** $f(x) = 2x + 3$ **39.** $f(x) = x + 3$

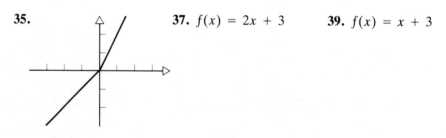

41. $f(x) = 12$ **43.** $-\sqrt{2}, 1$ **45.** $3, \pi$

47. (a) $f(W, r) = \dfrac{3W}{4\pi r^3}$ (b) $\sqrt[3]{\dfrac{3W}{4\pi d}}$

49. (a) $f(x) = \dfrac{1}{4} x$ (b) $\dfrac{13}{4}$ (c) $g(y) = 4y$

51. (a) $f(x) = 0.4x^2$ (b) 19.6   (c) $g(y) = \sqrt{2.5y}$

53. 0.1125 watts/m²

55. (a) 0.90 sec   (b) An increase from L to $4L$.

57. The new force will be 4/9 of the original force.

59. 1600 lb

SECTION 16

1.

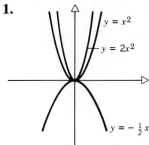

3.

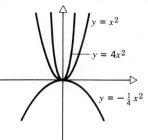

5.

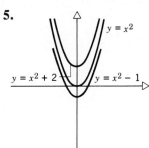

7. $f\left(-\dfrac{1}{4}\right) = \dfrac{7}{8}$, minimum.   **9.** $h(3) = -9$, minimum.

11. $g\left(\dfrac{1}{6}\right) = \dfrac{13}{12}$, maximum.

13.

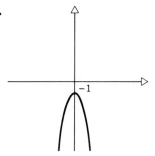

Vertex $(0, -1)$; axis $x = 0$; no x-intercepts.

15.

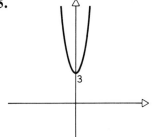

Vertex $(0, 3)$; axis $x = 0$, no x-intercepts.

17.

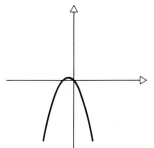

Vertex $\left(-\dfrac{1}{2}, \dfrac{1}{4}\right)$; axis $= -\dfrac{1}{2}$; x-intercepts -1 and 0.

19.

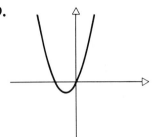

Vertex $(-1, -1)$; axis $x = -1$, x-intercepts -2 and 0.

21.

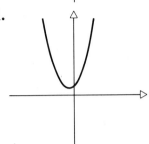

Vertex $\left(-\dfrac{1}{2}, \dfrac{3}{4}\right)$; axis $x = -\dfrac{1}{2}$; no x-intercepts.

23.

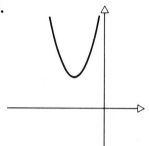

Vertex $(-3, 3)$; axis $x = -3$, no x-intercepts.

25. $f(x) = 2(x - 1)^2 + 3$, vertex $(1, 3)$.

27. $f(x) = 4(x + 1)^2 + 1$, vertex $(-1, 1)$.

29. $f(x) = 3(x - 3)^2 - 2$, vertex $(3, -2)$

31. $(-\infty, -4) \cup (\frac{1}{2}, \infty)$ **33.** $(-\infty, -\sqrt{3}) \cup (\sqrt{3}, \infty)$

35. Nowhere.   **37.** -7 **39.** $5/2$ **41.** $-2 < b < 2$

43. Maximum altitude: 100 feet above the ground. It will reach the ground 4 seconds after it is fired.

45. (a) $v_0^2/64$ (b) Increases by a factor of four.

47. 3.5, 3.5

49. $y = s - x$, the product is $x(s - x)$, and the latter is maximum when $x = s/2$, which implies $y = s/2$ also.

51. Verify that $f(x)$, as given, represents a parabola with x-intercepts at $(0, 0)$ and $(l, 0)$, with vertex at $(l/2, h)$.

53. The maximum area, 31,250/3 square meters, is obtained when the length is 125 meters and the width is 250/3 meters.

55. $r = h = 10/(4 + \pi)$.

SECTION 17

1. One-to-one.   **3.** One-to-one.   **5.** One-to-one.

7. One-to-one.   **9.** One-to-one.

11. $f^{-1}(x) = \dfrac{1}{4}x$

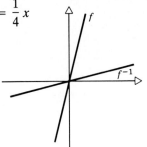

13. $f^{-1}(x) = x - 1$

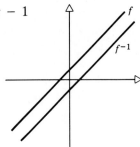

15. $f^{-1}(x) = x + 3$

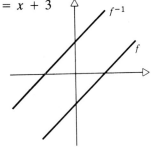

17. $f^{-1}(x) = (x - 4)/3$

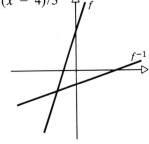

19.

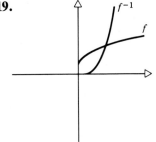

$f^{-1}(x) = (x - 1)^2, \; x \geq 1$

21.

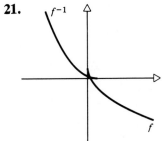

$f^{-1}(x) = (1 - x)^2/4, \; x \leq 1$

23.

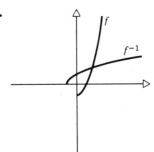

$f^{-1}(x) = \sqrt{x+1}, x \geq -1$

25. One-to-one function.

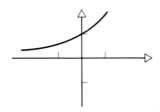

27. Not a function.
29. One-to-one function.

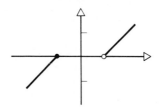

31. Not a function.
33. One-to-one function.

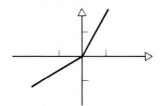

35. 0.0548
37. (a) $(f \circ f)(x) = \sqrt{1 - \sqrt{1 - x^2}^2} = \sqrt{x^2} = x$ for $0 \leq x \leq 1$.
(b)

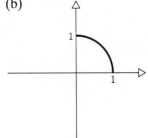

39. (a) Each vertical line intersects the graph at most one.
   (b) $f^{-1}(x) = -\sqrt{4-x}$ for $0 \le x \le 4$.
   (c) Domain f = range f^{-1} = $(-2, 0)$ (in interval notation).
       Domain f^{-1} = range f = $(0, 4)$.

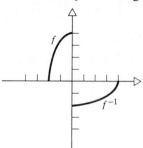

41. $f^{-1}(x) = \dfrac{3x+1}{2-x}$ for $x \neq 2$.

   Domain $f^{-1} = \{x: \ x \neq 2\}$.
   Range $f^{-1} = \{x: \ x \neq -3\}$.

43. $f^{-1}(x) = (x - b)/a$

45. $((g^{-1} \circ f^{-1}) \circ (f \circ g))(x) = g^{-1}(f^{-1}(f(g(x))))$
$$= g^{-1}(g(x))$$
$$= x.$$

CHAPTER III REVIEW

1. $2\sqrt{5}$ **2.** $\sqrt{29}$ **3.** $(x + 5)^2 + (y - 2)^2 = 16$

4. $(x - 2)^2 + (y + 4)^2 = 25$

5. Circle with center at $(6, -2)$ and radius 1.

6. The empty set.

7. $7x - 5y - 3 = 0$ **8.** $2x - 3y + 9 = 0$

9. $y = -4x + 2$ **10.** $4x + 3y + 5 = 0$

11. $x - y - 3 = 0$ **12.** $x - 2y + 1 = 0$

13. Circle with center at $(0, 0)$ and radius $\sqrt{5}$.

14. Parabola opening downward, vertex at $(0, 1)$, passing through $(2, -1)$.

15. Line through $(0, 1)$ and $(1/3, 0)$.

16. Circle with center at $(0, 0)$ and radius $2\sqrt{3}$.

17. Parabola opening downward, vertex at $(0, 1)$, passing through $(1, -2)$.

18. Vertical line through $(2/3, 0)$.

19. Horizontal line through $(0, 5/2)$.

20. Line through $(0, -3)$ and $(3/4, 0)$.

21. Circle with center at $(-3, 0)$ and radius 3.

22. Circle with center at $(0, 1)$ and radius 3.

23. Parabola opening upward, vertex at $(0, -2)$, passing through $(2, 0)$.

24. Parabola opening upward, vertex at $(0, 1)$, passing through $(1, 2)$.

25. Circle with center at $(-5, 2)$ and radius 4.

26. Circle with center at $(2, -1/2)$ and radius 2.

27. $\{x: \ x \geq -1 \ \text{ and } \ x \neq 2\}$ **28.** $\{x: \ x \geq 1/2\}$

29. (a) $x^3 + 3x + 4$ (b) $3x^3 + 5x^2 - 3x - 5$
(c) 2   (d) 0
(e) $3x^2 + 2$ (f) 195

30. (a) $x^2 - 2x + 4$ (b) $-2x^3 + 3x^2 - 2x + 3$
(c) 2   (d) 1/2
(e) $4x^2 - 12x + 10$ (f) -17

31. $g(x) = 4x + 1, h(x) = \sqrt{x}$

32. $g(x) = x^3, h(x) = x - 1$

33. -3 **34.** $2x + h$

35.
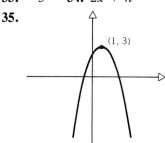

36.
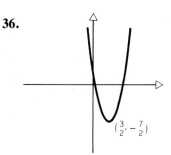

37.
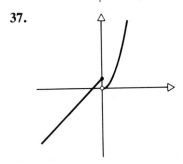

38.
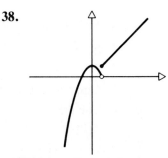

39. For $-5 < x < 2/3$ **40.** For $-1 < x < 2$.

41. 1/4   **42.** $-1/4$ **43.** 5 and -5 **44.** 5 and 2.5

45. $b = 4, c = 7$ **46.** 17.5

47. Not one-to-one. For example, $f(1) = f(-1)$.

48. Not one-to-one. For example, $g(1) = g(-1)$.

49. $f^{-1}(x) = (5 - x)/3$ **50.** $f^{-1}(x) = \sqrt{x - 2}, x \geq 2$

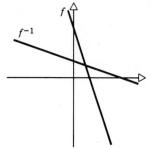
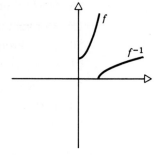

51. $f(x) = \dfrac{5}{4}x^2$ **52.** $g(v) = 30/v$

53. $f(n) = 1000 + 50n$. The variable cost is the slope.

54. $f(I) = 0.03I$ for $0 \le I \le 2000$, $f(I) = 0.04I - 20$ for $2000 < I \le 4000$, $f(I) = 0.05I - 60$ for $4000 < I \le 10{,}000$.

55. (a) $f(m_1, m_2, r, F) = Fr^2/(m_1 m_2)$ (b) $h(m_1, m_2, G, F) = \sqrt{Gm_1 m_2/F}$

56. (a) $g(d_o, d_i) = d_o d_i/(d_o + d_i)$ (b) $h(f, d_i) = d_i f/(d_i - f)$

57. $f^{-1}(x) = (x - 1)/2$, $g^{-1}(x) = 5 - x$, $h^{-1}(x) = (11 - x)/2$.

58. $f(S, r) = \sqrt{S^2 - \pi^2 r^4}/\pi r$

59. (a) $h(x) = x^3$, $g(x) = x + 0.125$
  (b) $r(x) = x + 0.5$, $s(x) = x^2 - 0.5x + 0.25$

60. $(-\infty, -2) \cup (-2, -\tfrac{3}{2}] \cup [\tfrac{3}{2}, \infty)$

61. $f(x) = 3 + \sqrt{25 - x^2}$, $-5 \le x \le 5$

62. $(-1, 1)$ and $(6, 0)$.

63. $(-\infty, -2) \cup (\tfrac{4}{7}, \infty)$

64. $f^{-1}(x) = b + \pi\sqrt{x}/a$, $x \ge 0$

65. $\dfrac{3}{2}$

SECTION 18

For Exercises 1–23, the first polynomial is the quotient and the second is the remainder.

1. $x^2 - x + 2$, 5

3. $x^2 - 2x - 3$, 4

5. $3x - 6$, $x + 2$

7. 2, $3x^2 + x - 2$

9. $\dfrac{1}{5}x^2 + \dfrac{6}{25}x + \dfrac{6}{125}, \dfrac{131}{125}x + 1$

11. 0, $x^3 + x^2 - 4$

13. $2x^2 + x + 6$, 7

15. $x^2 + 4x + 8$, 15

17. $x^2 - 5x + 24$, -120

19. $x^3 + \dfrac{1}{2}x^2 - \dfrac{1}{4}x + \dfrac{1}{8}, -\dfrac{1}{16}$

21. $x^2 - 0.5x + 1.25$, -0.625

23. $x^4 + \dfrac{3}{2}x^3 + \dfrac{7}{4}x^2 + \dfrac{15}{8}x + \dfrac{31}{16}, \dfrac{63}{32}$

25. 7, -119 **27.** 3, -45

29. $-\dfrac{94}{81}$, -1.0195

31. -0.30, 0.47

33. 0.089, 0.033

SECTION 19

1. Factor. $q(x) = 6x^2 + x - 2$

3. Factor. $q(x) = x^3 - 2x^2 + 1$

5. Factor. $q(x) = x^3 - x + 4$

7. Not a factor.

9. Not a factor.

11. 0 (one), -1 (one), 5 (two)

13. 0 (three), 10 (two), -0.1 (five)

15. 0 (one), -2 (one), -0.4 (two)

17. $x \left(x - \dfrac{1}{2} \right)^2 (x + 5)^3$ **19.** $x^7 + 69x^6 + 615x^5 - 21{,}853x^4$

21. $x^4 + 3.0x^3 - 4.2x^2 - 9.7x + 10.5$

23. 1 (one), -2 (one), 2 (one); $(x - 1)(x + 2)(x - 2)$

25. $\dfrac{1}{2}$ (one), 1 (two); $(2x - 1)(x - 1)^2$

27. -0.4 (one), -1 (two); $(5x + 2)(x + 1)^2$

29. 5 (one), $\sqrt{5}$ (one), $-\sqrt{5}$ (one); $(x - 5)(x - \sqrt{5})(x + \sqrt{5})$

31. -2 (one); $(x + 2)(x^2 + 1)$ **33.** $-\dfrac{1}{3}$ (one); $(3x + 1)(x^2 - 3x + 3)$

35. $2^6 + 64 \neq 0$ **37.** None.   **39.** n odd.

41. $a = -4$

43. $(x + 3)^2(x - 2)^2(x - 4)^2$

SECTION 20

1.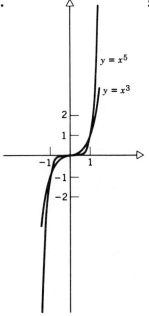

3.

5.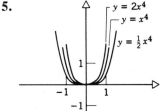

7. Odd. Symmetric about the origin.

9. Even. Symmetric about the y-axis.

11. Odd. Symmetric about the origin.

13. Even. Symmetric about the y-axis.

15. Neither.

17. Neither.

19.

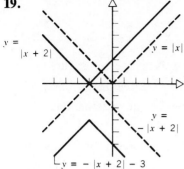

$y = |x + 2|$

$y = |x|$

$y = -|x + 2|$

$y = -|x + 2| - 3$

21.

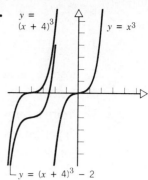

$y = (x + 4)^3$

$y = x^3$

$y = (x + 4)^3 - 2$

23.

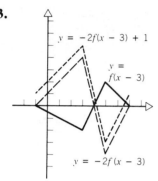

$y = -2f(x - 3) + 1$

$y = f(x - 3)$

$y = -2f(x - 3)$

25.

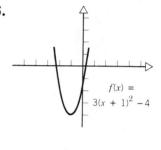

$f(x) = 3(x + 1)^2 - 4$

27. Parabola opening upward, vertex at $(1, 2)$, passing through $(0, 5/2)$.

29.

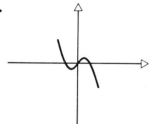

31.

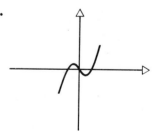

33.

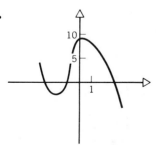

35.

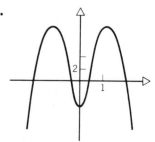

37. $x^3 - 5x^2 - 2x + 24$

39. $3x^3 - 3x^2 - 6x$

41. $(f \circ g)(-x) = f(g(-x)) = f(-g(x)) = f(g(x)) = (f \circ g)(x)$

SECTION 21

1. $\pm 1, \pm 2, \pm 3, \pm 6, \pm \frac{1}{5}, \pm \frac{2}{5}, \pm \frac{3}{5}, \pm \frac{6}{5}$ **3.** $\pm 1, \pm 2, \pm \frac{1}{2}, \pm \frac{1}{7}, \pm \frac{2}{7}, \pm \frac{1}{14}$

5. Positive zeros: two or none. Negative zeros: three or one.

7. Positive zeros: two or none. Negative zeros: one.

9. Positive zeros: three or one. Negative zeros: one.

11. Upper bound 4. Lower bound -3.

13. Between -3 and -2, between 0 and 1, and between 2 and 3.

15. Between -2 and -1, between -1 and 0, and between 2 and 3.

17. $-1, -\sqrt{3}, \sqrt{3}$ **19.** $-1, -\frac{1}{2}, \frac{3}{2}$ **21.** $-\frac{1}{2}, 2, 2$

23. $\frac{1}{2}$ **25.** 0   **27.** $0, 5 + \sqrt{3}, 5 - \sqrt{3}$

29. $-2, \frac{1}{2}, \frac{1}{2}, 2$ **31.** $0, 5$ **33.** $-\frac{4}{3}, 0, -\sqrt{3}, \sqrt{3}$

35. 3.5   **37.** -3 **39.** -6 **41.** 1.5 m

SECTION 22

1. Horizontal: $y = 2$. Vertical: $x = -5$

3. Horizontal: $y = 0$. Vertical: $x = 3$ and $x = -3$.

5. Horizontal: $y = \frac{1}{8}$. Vertical: $x = \frac{1}{2}$.

7.   **9.**   **11.**

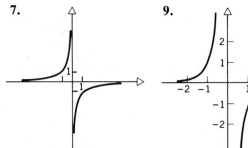
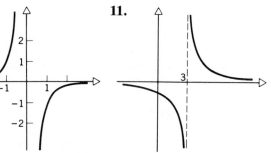

13.   **15.**   **17.**

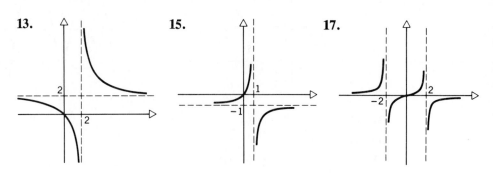

19. 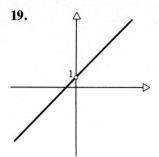 **21.** 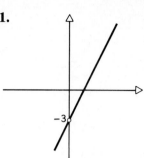 **23.**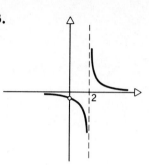

25. 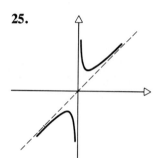 **27.** 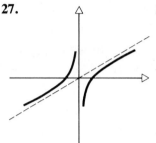 **29.**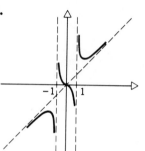

31. (a) $f(-x) = \dfrac{a(-x)}{b(-x)} = \dfrac{a(x)}{b(x)} = f(x)$

(b) Similar to (a).   (c) Similar to (a).

33. 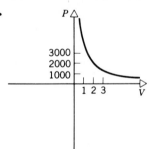 **35.**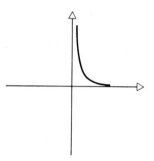

SECTION 23

1. $\dfrac{3}{x + 1} - \dfrac{2}{x - 2}$

3. $\dfrac{5}{x + 2} + \dfrac{1}{x - 4}$

5. $\dfrac{1}{3} \cdot \dfrac{1}{x + 1} - \dfrac{2}{3} \cdot \dfrac{1}{2x - 1}$

7. $\dfrac{3}{x - 1} - \dfrac{1}{(x - 1)^2}$

9. $\dfrac{-2}{2x - 3} + \dfrac{5}{(2x - 3)^2}$

11. $\dfrac{3}{x} - \dfrac{1}{x - 1} + \dfrac{3}{(x - 1)^2}$

13. $\dfrac{4}{x} - \dfrac{3}{x^2 + x + 1}$

15. $\dfrac{-3}{x} + \dfrac{4}{x^2 + 2}$

17. $\dfrac{-2}{x^2} + \dfrac{3}{x^2 + 1}$

19. $\dfrac{x}{x^2 + 1} - \dfrac{2x - 1}{(x^2 + 1)^2}$

21. $\dfrac{2x - 1}{x^2 + x + 1} + \dfrac{3x}{(x^2 + x + 1)^2}$

23. $\dfrac{4}{x - 1} - \dfrac{3}{(x - 1)^2} + \dfrac{1}{(x - 1)^3}$

25. $3x - 1 + \dfrac{1}{x} - \dfrac{2}{x + 1}$

27. $2x - 1 + \dfrac{x - 1}{x^2 + 1} - \dfrac{2x}{(x^2 + 1)^2}$

CHAPTER IV REVIEW

1. $q(x) = x - 3$, $r(x) = x + 2$

2. $q(x) = x^2 - x + 1$, $r(x) = x$

3. $q(x) = x^3 + 2x^2 + x + 3$, $r(x) = 10$

4. $q(x) = 3x^2 - 8x + 8$, $r(x) = -6$

5. $f(3) = 58$, $f(-1) = -6$ **6.** $f(2) = 26$, $f(-2) = 30$

7. $q(x) = x^2 + x + 2$ **8.** $q(x) = 4x^3 - 2$

9. $-1, 2, 3; (x + 1)(x - 2)(x - 3)$

10. $-\sqrt{3}, \sqrt{3}, 2; (x + \sqrt{3})(x - \sqrt{3})(x - 2)$

11. $-\sqrt{2}, \sqrt{2}, \dfrac{1}{2}, 1; (x + \sqrt{2})(x - \sqrt{2})(2x - 1)(x - 1)$

12. $-\sqrt{3}, \sqrt{3}, -1, \dfrac{2}{3}; (x + \sqrt{3})(x - \sqrt{3})(x + 1)(3x - 2)$

13. Between -4 and -3, between -3 and -2, and between 2 and 3. (The zeros are $-\dfrac{7}{2}, -\sqrt{5}$, and $\sqrt{5}$.)

14. Between -4 and -3, between -2 and -1, and between 3 and 4. (The zeros are $-\dfrac{5}{3}, -\sqrt{10}$, and $\sqrt{10}$.)

15. Positive zeros: four, two, or none. Negative zeros: none.

16. Positive zeros: three or one. Negative zeros: one.

17. Neither. Not symmetric about either the y-axis or the origin.

18. Even. Symmetric about the y-axis.

19. Even. Symmetric about the y-axis.

20. Odd. Symmetric about the origin.

21.

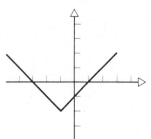

22.

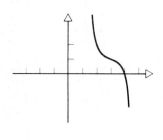

23.

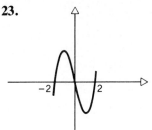

24.

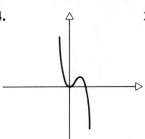

25.

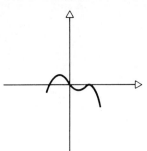

26.

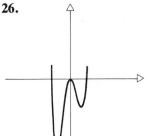

27.

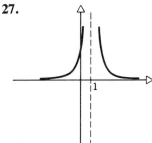

28.

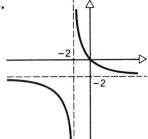

29.

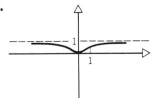

30.

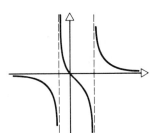

31. $\dfrac{3}{x + 2} - \dfrac{4}{(x + 2)^2}$

32. $\dfrac{4}{x} - \dfrac{3}{x^2 + 1}$

33. $x + \dfrac{3}{x - 1} + \dfrac{1}{x + 2}$

34. $x^2 + 1 - \dfrac{1}{x - 1} + \dfrac{5}{x^2 + 2}$

35. $\dfrac{2}{3}$

36. 7 cm, 9 cm, 21 cm

SECTION 24

The entries for the tables in Exercises 1, 3, and 5 were computed directly with a calculator and then rounded; they will not agree completely with entries computed using the approximations given in the exercises.

1.

x	-3	$-\frac{5}{2}$	-2	$-\frac{3}{2}$	-1	$-\frac{1}{2}$	0	$\frac{1}{2}$	1	$\frac{3}{2}$	2	$\frac{5}{2}$	3
5^x	0.008	0.018	0.040	0.089	0.200	0.447	1	2.24	5	11.2	25	55.9	125

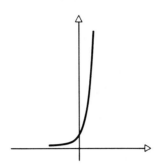

3.

x	-3	$-\frac{5}{2}$	-2	$-\frac{3}{2}$	-1	$-\frac{1}{2}$	0	$\frac{1}{2}$	1	$\frac{3}{2}$	2	$\frac{5}{2}$	3
3.5^x	0.023	0.044	0.082	0.153	0.286	0.535	1	1.87	3.50	6.55	12.25	22.9	42.9

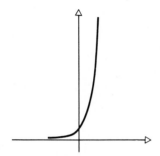

5.

x	-3	$-\frac{5}{2}$	-2	$-\frac{3}{2}$	-1	$-\frac{1}{2}$	0	$\frac{1}{2}$	1	$\frac{3}{2}$	2	$\frac{5}{2}$	3
0.4^x	15.62	9.88	6.25	3.95	2.5	1.58	1	0.63	0.4	0.25	0.16	0.10	0.06

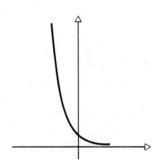

7. 1000

9. 1,000,000

11. 32

13. 3

15. 2

17. 9^{x-y}

19. The graph coincides with Figure 24.3 for $x \geq 0$, and it is symmetric with respect to the y-axis.

21. The graph is symmetric with respect to the *y*-axis and is above the *x*-axis. Some coordinate pairs are (0, 1), (±1, 1.25), (±2, 2.125), (±3, 4.0625), and (±4, 8.0313).

23. Carry out the operations indicated on the left and simplify.

25. $f(x_1 + x_2, y) = (x_1 + x_2)b^y = x_1 b^y + x_2 b^y = f(x_1, y) + f(x_2, y)$

27. $f(x_1 x_2, y_1 + y_2) = (x_1 x_2)b^{y_1 + y_2} = x_1 b^{y_1} x_2 b^{y_2} = f(x_1, y_1)f(x_2, y_2)$

29. 33.35963198, 36.33783888, 36.41584470, 36.45491473, 36.46195209. Yes: $\pi^\pi \approx 36.46215964$

SECTION 25

1. 1   **3.** 1   **5.** 1   **7.** -2 **9.** $\dfrac{1}{3}$

11. 0   **13.** 0   **15.** -4 **17.** $\dfrac{3}{2}$ **19.** -3

21. $\dfrac{3}{4}$ **23.** $x = \log_u v$ **25.** $x = \dfrac{1}{3}\log_4 2 = \dfrac{1}{6}$ **27.** $x = \dfrac{1}{4}\log_y z$

29. $x = \log_5 2$ **31.** $x = 3^\pi$ **33.** $x = 10^{\sqrt{2}}$ **35.** $x = \dfrac{1}{2} \cdot 3^{2y}$

37. $x = a^4$ **39.** $x = 2^{y/3}$

41.

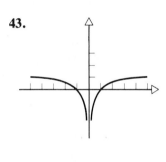

$y = \left(\frac{1}{3}\right)^x$

$y = \log_{\frac{1}{3}} x$

43.

45. For $y = \log_2 x$, reflect the graph of $y = 2^x$ (Figure 24.3) through the line $y = x$. Then for $y = \log_2(x + 1)$ translate to the left by 1 unit. For $y = (\log_2 x) + 1$, translate $y = \log_2 x$ upward by 1 unit.

47. $\{x: \ x \neq 1\}$

49.

51. (a) $x > 1$ (b) $0 < x < 1$

53. $\left\{x: \ \dfrac{1}{3} < x < 3\right\}$

SECTION 26

1. $\log_b x + \log_b y - \log_b z$ 　　　　　 **3.** $-\log_b x - \log_b y$

5. $\log_b x + \dfrac{1}{2}\log_b z$ 　　　　　 **7.** $\dfrac{3}{2}(\log_b x - \log_b y - \log_b z)$

9. $\dfrac{1}{3}\log_b x + \dfrac{1}{4}\log_b y + \dfrac{1}{5}\log_b z$ 　　 **11.** $\dfrac{1}{2}\log_b x - \dfrac{3}{2}\log_b y - \dfrac{3}{2}\log_b z$

13. $\log_b(uv)^2$ 　　 **15.** $\log_b(1/v)$ 　　 **17.** $\log_b v^3 u$ 　 **19.** $\log_b 4$ 　 **21.** 0

23. 2 　　 **25.** 0.9208 　　 **27.** 1.1292 　　 **29.** 0.1781 　 **31.** -1.0686

33. -0.0937 　　　 **35.** 3 　　　 **37.** π 　　　 **39.** 1.73 　　 **41.** $-x$

43. 0 　　　　 **45.** 0 　　　　 **47.** -1

49. 32

51. Start with $\log_a(1/b) = -\log_a b$. Also use Equation (26.10).

53. If $u = \log_b x$, then $b^u = x$ so $b^{ur} = x^r$ and $\log_b x^r = ur = r \cdot \log_b x$.

SECTION 27

1. (a) 0.9299 　 (b) 2.9299 　 (c) $-2 + 0.9299 = -1.0701$

3. (a) 0.5315 　 (b) 3.5315 　 (c) $-1 + 0.5315 = -0.4685$

5. (a) 4.2095 　 (b) 0.2095 　 (c) $-2 + 0.2095 = -1.7905$

7. 0.92992956, 0.82607480, 0.53147892, 2.75587486, 4.20951502, -3.65364703

9. -1.07007044, 1.82607480, -0.46852108, -0.24412514, -1.79048499, 4.34635297

11. $(\log 7)/(\log 5) \approx 1.209$

13. $(\log 4)/\log(64/5) \approx 0.5438$

15. $-(\log 125)/\log(49.5) \approx -2.116$

17. $-(\log 5.5)/(\log 5) \approx -1.059$

19. ± 9 　　 **21.** 25 　　 **23.** ± 100 　　 **25.** 256

27. 1/100,100 　　 **29.** $\{x: \ 0 < x < 500\}$

31. 7 years 　　 **33.** 9 years 　　 **35.** 8 years

37. (a) 11.9 years 　 (b) 12 years 　 (c) 0.84%

39. (a) 6.12 years 　 (b) 6 years 　 (c) 2.0%

41. (a) 3.22 years 　 (b) 3 years 　 (c) 6.8%

43. 19 years

45. (a)

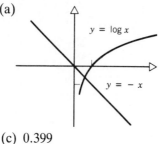

(b) Since $\log 0.40 = -0.3979$ and $\log 0.39 = -0.4089$, the solution is between 0.39 and 0.40.

(c) 0.399

47. 754 million   **49.** 1997   **51.** 2001

53. 38.9 years   **55.** 6.27 billion

57. (a) 41.1 years   (b) 41.2 years   (c) 42.4 years

SECTION 28

1. $24.41, $26.13, $27.15

3. $2613.04, $2714.57, $2718.28

5. $8243.50 (with Table III)   **7.** $210.26 (with Table III)

9. $6749.50 (with Table III)

11. (a) 4.5218   (b) 4.5218   (c) 4.5218

13. (a) -3.2188 (b) -3.2189 (c) -3.2189

15. (a) -7.9576 (b) -7.9576 (c) -7.9576

17. (a) -7.2645 (b) -7.2645 (c) -7.2644

19. $Q = 0.1e^{-[(\ln 2)/11]t} = 0.1e^{-0.063t}$ gm

21. $Q = 0.001e^{-[(\ln 2)/(2.4 \times 10^4)]t} = 0.001e^{-0.0000289t}$ gm

23. (a) 0.0625 mg   (b) 0.751 mg   (c) At the end of 14.9 years.

25. 5.4 billion   **27.** 7.9 billion

29. 784,000   **31.** $t = \dfrac{1}{k}\ln(Q/Q_0)$ **35.**

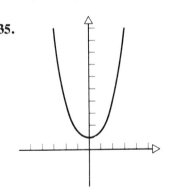

REVIEW EXERCISES FOR CHAPTER V

1. 3   **2.** -2 **3.** 1.5   **4.** -1

5. 49   **6.** 5   **7.** 75   **8.** $(b^{d/a})/c$

9. $\log_b x + \log_b z - \log_b y$ **10.** $\dfrac{1}{5}\log_b y + \dfrac{1}{5}\log_b z$

11. $\log_b(v^3/u)$ **12.** $\log_b \sqrt{15}$

13. $(\log 4)/(\log 5) \approx 0.861$ **14.** $(\log 0.5)/(\log 9) \approx -0.315$

15. $(\log 9)/\left(\log \dfrac{5}{3}\right) \approx 4.30$ **16.** $-(\log 1.25)/(\log 7) \approx -0.115$

17. $-8, 7$ **18.** e^{10}

19. (a) 3.4341   (b) 3.4340   (c) 3.4340

20. (a) 5.8861   (b) 5.8861   (c) 5.8861

21. Justify each of the intermediate steps $\frac{1}{2} = e^{-kT}$, $\ln\frac{1}{2} = -kT$, and $-\ln 2 = -kT$.

22. 108

23. (a) $6x + 2$ (b) $[(\ln x) - 1]/3$

24. $(\ln 4, 4)$ and $(-\ln 4, 4)$

25. $e^{-10} < k < e^{-4}$

26. (a) $(-\infty, -1) \cup (1, \infty)$ (b) $\pm 101, \pm 101/100$

27. The graph of $y = \log x$ is similar to the graph of $y = \log_4 x$, shown in Figure 25.2. Translate 1 unit to the right to get the graph of $y = \log(x - 1)$. Then, using the suggestion, reflect the graph of $y = \log(x - 1)$ through the x-axis to get the graph of $y = \log[1/(x - 1)]$.

28. 1, 4

29. Substitute and simplify.

30. $4.87e^{0.018(10)} \approx 5.83$ billion

31. (a) 7.27 years   (b) 7.2 years   (c) 0.96%

32. $1271

33. The suggestion will lead to the equation $(e^y)^2 - 2x(e^y) - 1 = 0$.

SECTION 29

1.

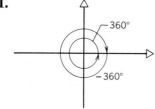

3.

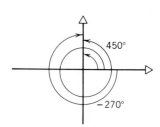

5.

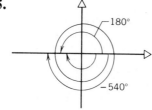

7.

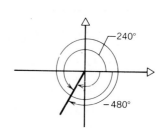

9.

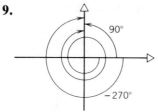

11.

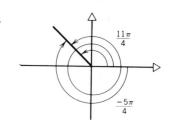

13.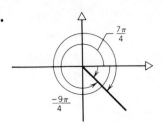

15.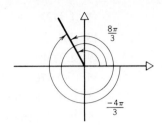

17. $-16.\overline{3}°$ **19.** 14° 45'

21. $-82° 9'$ **23.** 900°

25. $-57.3°$ **27.** 330°

29. $3\pi/4$ **31.** $-3\pi/10$

33. $-5/9$ **35.** 47° 8' 15″

37. $-7\pi/2, -3\pi, -5\pi/2, -2\pi, -3\pi/2, -\pi, -\pi/2$

39. $25\pi/18$ cm   **41.** 2900 mi

43. 55°   **45.** (a) 67.5° (b) 1600/3 mils

47. (a) 1.152 (b) 3956   **49.** 11.78 ft/sec

51. 9.00

SECTION 30

sine	cosine	tangent	cotangent	secant	cosecant
1. $2\sqrt{5}/5$	$\sqrt{5}/5$	2	1/2	$\sqrt{5}$	$\sqrt{5}/2$
3. 3/5	4/5	3/4	4/3	5/4	5/3
5. $-3\sqrt{13}/13$	$-2\sqrt{13}/13$	3/2	2/3	$-\sqrt{13}/2$	$-\sqrt{13}/3$

7. $\pi/2, 3\pi/2$ **9.** $0, \pi$ **11.** IV

13. $\sqrt{2}/2, \sqrt{2}$ **15.** $-1, -1$ **17.** $\sqrt{3}/2$

19. 2   **21.** 2   **23.** $2\sqrt{3}/3$

25. $\dfrac{3}{5}, \dfrac{4}{3}$ **27.** $\dfrac{12}{13}, \dfrac{5}{12}$ **29.** $\dfrac{17}{8}, \dfrac{15}{8}$

31. 0.8165, 1.7321   **33.** $0.8, 1.\overline{6}$ **35.** $\dfrac{5}{3}, \dfrac{5}{4}$

37. 1.1547   **39.** 2.7475   **41.** 0.3640

43. 0.3420   **45.** 1.1918

47. $\sec \theta > 0$ in Quadrants I and IV.
$\sec \theta < 0$ in Quadrants II and III.

49. $\sqrt{5}/5, 1/2$ **51.** $-\sqrt{5}/2, 2$

SECTION 31

Some of the answers from a calculator can be rounded further than they have been here.

1. 0.3057   **3.** 0.6947   **5.** 1.3073

7. 1.4422   **9.** 23.880   **11.** 0.3096

13. 20°   **15.** 85°   **17.** 70°

19. 2.9238   **21.** 0.9962   **23.** 0.3640

25. −0.9781   **27.** −1.6643   **29.** 0.629320391

31. 0.656986599   **33.** 11.43005230

35. 0.207911691   **37.** 2.785230695

39. −13.68279740   **41.** 61.0°   **43.** 79.3°

45. 75.5°   **47.** 22.3°   **49.** 22.19°

51. 86.32°   **53.** 45.22°   **55.** 2.01°

57. 21.30°   **59.** 45.6°   **61.** 16.6°

63. 78.5°   **65.** 36.4   **67.** 1.15

69. 2.20

71. If $a = 2$, then $b = \sqrt{21}$, $\alpha = 23.6°$, and $\beta = 66.4°$.

73. 73.1°   **75.** 783 ft   **77.** 146 m

79. 14.9°   **81.** 11.1 m   **83.** 4.6 ft

CHAPTER VI REVIEW

1. 5.7°   **2.** 21° 49′ 12″   **3.** 72°

4. −480°   **5.** $-10\pi/9$ **6.** $\pi/10$

7. $24/\pi \approx 7.64$ cm   **8.** 45°

9. $2\sqrt{5}/5$, -2, $-\sqrt{5}$

10. $\sqrt{26}/26$, $-1/5$, $-\sqrt{26}/5$

11. $\sqrt{3}$ **12.** $\sqrt{2}/2$ **13.** 1/2

14. $-\sqrt{2}$ **15.** -2 **16.** $\sqrt{3}/3$

17. $-1/2$ **18.** $-\sqrt{3}$ **19.** $-\sqrt{2}/2$

20. 0   **21.** -1 **22.** Undefined

23. $5\sqrt{29}/29$, $2/5$, $\sqrt{29}/2$

24. $3\sqrt{58}/58$, $7/3$, $\sqrt{58}/7$

25. $4\sqrt{41}/41$ **26.** 5/12   **27.** t

28. u **29.** 160°   **30.** $\pi/3$

31. −1.2349   **32.** −0.6428   **33.** −1.0711

34. 1.8871   **35.** 0.374606593

36. 11.47371325   **37.** 1.191753593

38. −0.984807753

39. 0.9659, −0.9659, −0.9659, and −0.9659, respectively.

40. 58.78 cm

41. arctan $2 \approx 63.43°$

42. $1\frac{3}{4}$ mi   **43.** 17.7°   **44.** 105 m

SECTION 32

1. $(-1, 0)$ **3.** $(0, 1)$

	sin	cos	tan	cot	sec	csc
5.	-0.6	-0.8	0.75	$1.\overline{3}$	-1.25	$-1.\overline{6}$
7.	$-1/2$	$\sqrt{3}/2$	$-\sqrt{3}/3$	$-\sqrt{3}$	$2\sqrt{3}/3$	-2
9.	$-\sqrt{3}/2$	$-1/2$	$\sqrt{3}$	$\sqrt{3}/3$	-2	$-2\sqrt{3}/3$

11. 0.4176 **13.** -0.9959 **15.** 0.0707

17. -0.943818209 **19.** -19.98333055

21. 8.390660036 **23.** -1.127214188

25. $-\sqrt{21}/5 \approx -0.9165$ **27.** $(-2\sqrt{5}/5, \sqrt{5}/5)$

29. The cosine and secant are positive and the other functions are negative.

31. IV

33. $\sin t = \tan t = 0$, $\cos t = \sec t = -1$, $\cot t$ and $\csc t$ are undefined.

35. (a) Because $\sec t = 1/\cos t$.
  (b) $\pm\pi/2, \pm3\pi/2, \pm5\pi/2, \ldots = \{t: t = (2n + 1)\pi/2 \text{ for } n \text{ an integer}\}$

37. $\pi/4, 3\pi/4, 5\pi/4, 7\pi/4$

39. Because $\tan t$ and $\sin t$ have opposite signs in Quadrants II and III.

41. $\sin t = \dfrac{\tan t}{\pm\sqrt{1 + \tan^2 t}}$, $\cos t = \dfrac{1}{\pm\sqrt{1 + \tan^2 t}}$,

  $\cot t = \dfrac{1}{\tan t}$, $\sec t = \pm\sqrt{1 + \tan^2 t}$,

  $\csc t = \dfrac{\pm\sqrt{1 + \tan^2 t}}{\tan t}$

43. Any value of t such that $0 < t < 2\pi$ and $t \neq \pi$ will do.

45. Use $t = \pi/2$, for example.

47. $c + (1/d)$

SECTION 33

49. I and IV   **51.** I and III   **53.** III and IV

SECTION 34

1. $2 + \sqrt{3}$ **3.** $(\sqrt{6} - \sqrt{2})/4$ **5.** $2 - \sqrt{3}$

7. $(\sqrt{6} + \sqrt{2})/4$ **9.** $-2 - \sqrt{3}$ **11.** $\tan 5t$

13. $\tan (A/6)$ **15.** $\cos 4B$ **17.** $7/22$

19. $-(3 + 8\sqrt{2})/15$ **21.** $-(960 + 169\sqrt{15})/231$

37. Apply (34.5) and (34.6) to the left-hand side and simplify.

39. The left-hand side equals $\cos^2(\alpha + \beta) + \sin^2(\alpha + \beta)$.

41. Use (34.6) to show that $\sin\left(\theta + \dfrac{\pi}{3}\right) = \dfrac{1}{2}\sin\theta + \dfrac{\sqrt{3}}{2}\cos\theta$, (34.2) to show

$\cos\left(\theta + \dfrac{\pi}{6}\right) = \dfrac{\sqrt{3}}{2}\cos\theta - \dfrac{1}{2}\sin\theta$, and then subtract.

43. Use $\sec(u - v) = 1/\cos(u - v)$, then (34.1), and then divide both numerator and denominator by $\cos u \cos v$ and simplify.

45. Use (34.1) and (34.2) on the left-hand side, and then divide both numerator and denominator by $\cos A \cos B$ and simplify.

47. Use (34.1).

49. Use $\cot(u + v) = 1/\tan(u + v)$ and (34.8)

53. Use $u = v = 0$, for example.

55. Begin by using (34.6) on $\sin(x + h)$.

57. $\dfrac{2}{3}$ **59.** Use (34.7) and (34.10).

SECTION 35

1. $\sqrt{2 - \sqrt{2}}/2$ **3.** $\sqrt{2} - 1$ **5.** $\sqrt{3} - 2$

7. $2 - \sqrt{3}$ **9.** $\sqrt{2 - \sqrt{3}}/2$

11. $\sin 2u = -\dfrac{3}{5}$, $\cos 2u = \dfrac{4}{5}$, $\tan 2u = -\dfrac{3}{4}$

13. $\sin\dfrac{\theta}{2} = 2\sqrt{5}/5$, $\cos\dfrac{\theta}{2} = \sqrt{5}/5$, $\tan\dfrac{\theta}{2} = 2$

15. $\sin\dfrac{\theta}{2} = \sqrt{(3 - \sqrt{5})/6}$, $\cos\dfrac{\theta}{2} = -\sqrt{(3 + \sqrt{5})/6}$, $\tan\dfrac{\theta}{2} = (\sqrt{5} - 2)/2$

31. $3\sin u - 4\sin^3 u$

33. $8\cos^4 u - 8\cos^2 u + 1$

35. -0.7190 **37.** -0.4657

39. -0.7817

SECTION 36

1. $\dfrac{1}{2}(\cos 5t + \cos 3r)$

3. $\dfrac{1}{2}(\cos t - \cos 11t)$

5. $\dfrac{1}{2}[\cos(m - n)x - \cos(m + n)x]$

7. $2\cos(7\alpha/2)\cos(\alpha/2)$

9. $2\cos 2\theta \cos\theta$

11. $-2\sin[(2x + h)/2]\sin(h/2)$

13. $\sqrt{6}/2$ **15.** $\sqrt{2}/2$ **17.** $1/4$

35. Write (36.3) with x in place of u and y in place of v: $\cos x \cos y = \frac{1}{2}[\cos(x + y) + \cos(x - y)]$. Now let $u = x + y$ and $v = x - y$, so that $\frac{1}{2}(u + v) = x$ and $\frac{1}{2}(u - v) = y$. Then $\cos \frac{1}{2}(u + v) \cos \frac{1}{2}(u - v) = \frac{1}{2}(\cos u + \cos v)$, which implies (36.7).

37. $2 \cos(t - \pi/6)$ **39.** $2 \cos(t - \pi/3)$

41. $13 \cos(2t - c)$, where $\cos c = 12/13$ and $\sin c = 5/13$.

43. $\cos[(m + n)\pi x/d] + \cos[(m - n)\pi x/d]$

45. $2 \sin \frac{x}{2}(\sin x + \sin 2x + \sin 3x + \sin 4x) = \cos \frac{x}{2} - \cos \frac{9x}{2}$

SECTION 37

1. $3\pi/2$ **3.** $\pi/4, 5\pi/4$

5. $2\pi/3, 4\pi/3$ **7.** $0, \pi$

9. No solution.   **11.** $\pi/6, 5\pi/6, 3\pi/2$

13. $1.32, 1.82, 4.46, 4.97$

15. $\pi/2 (\approx 1.57), 3.87, 5.55$

17. $3.52, 5.91$

19. $\pi/12, 5\pi/12, 13\pi/12, 17\pi/12$

21. $\pi/16, 5\pi/16, 9\pi/16, 13\pi/16, 17\pi/16, 21\pi/16, 25\pi/16, 29\pi/16$

23. $5\pi/3$

25. $\pi/6, \pi/2, 5\pi/6, 3\pi/2$

27. $\pi/6, 5\pi/6, 3\pi/2$

29. No solution.

31. $\pi/3, 2\pi/3, 4\pi/3, 5\pi/3$

33. No solution.

35. $\pi/4, 3\pi/4, 5\pi/4, 7\pi/4$

37. $\pi/6, \pi/3, 2\pi/3, 5\pi/6, 7\pi/6, 4\pi/3, 5\pi/3, 11\pi/6$

39. $7\pi/24, 11\pi/24, 19\pi/24, 23\pi/24, 31\pi/24, 35\pi/24, 43\pi/24, 47\pi/24$

41. $\{(5\pi + 30k\pi)/3: \ k \text{ is an integer}\} \cup \{(25\pi + 30k\pi)/3: \ k \text{ is an integer}\}$

43. $\{(3\pi + 8k\pi)/4: \ k \text{ is an integer}\} \cup \{(7\pi + 8k\pi)/4: \ k \text{ is an integer}\}$

45. $\{k\pi/2: \ k \text{ is an integer}\}$

47. No solution.

49. $\{k\pi/3: \ k \text{ is an integer}\} \cup \{(\pi + 12k\pi)/2: \ k \text{ is an integer}\} \cup \{(5\pi + 12k\pi)/2: \ k \text{ is an integer}\}$

51. $\{k\pi/4: \ k \text{ is an integer}\} \cup \{(4\pi + 24k\pi)/3: \ k \text{ is an integer}\} \cup \{(20\pi + 24k\pi)/3: \ k \text{ is an integer}\}$

53. $\{(\pi + 4k\pi)/6: \;\; k$ is an integer$\} \cup \{(3\pi + 4k\pi)/6: \;\; k$ is an integer$\}$

55. $\{0.46 + k\pi: \;\; k$ is an integer$\}$

57. $\{(13\pi + 24k\pi)/36: \;\; k$ is an integer$\} \cup \{(29\pi + 24k\pi)/36: \;\; k$ is an integer$\}$

59. $\{k\pi: \;\; k$ is an integer$\}$

61. $22.6°$ **63.** $12.0°$

65. $d = \dfrac{\sqrt{5}}{4} \cos(8t - 1.11)$

CHAPTER VII REVIEW

	sin	cos	tan	cot	sec	csc
1.	$-5/13$	$-12/13$	$5/12$	$12/5$	$-13/12$	$-13/5$
2.	$\sqrt{15}/4$	$-1/4$	$-\sqrt{15}$	$-\sqrt{15}/15$	-4	$4\sqrt{15}/15$
3.	$\sqrt{3}/2$	$-1/2$	$-\sqrt{3}$	$-\sqrt{3}/3$	-2	$2\sqrt{3}/3$
4.	$-\sqrt{2}/2$	$-\sqrt{2}/2$	1	1	$-\sqrt{2}$	$-\sqrt{2}$

5. $\sin 5\pi/6 = 1/2$. $\cos 5\pi/6 = -\sqrt{3}/2$. $\tan 5\pi/6 = -\sqrt{3}/3$. Now use (30.2).

6. $\sin -5\pi/2 = -1$. $\cos -5\pi/2 = 0$. $\csc -5\pi/2 = -1$. $\cot -5\pi/2 = 0$. $\tan -5\pi/2$ and $\sec -5\pi/2$ are undefined.

7. (b) 0.257850033 **8.** (b) -0.684136808

9. (b) 0.325549335 **10.** (b) 0.560404665

11. (b) -1.331060937 **12.** (b) -1.529885656

13. $\sin t$ and $\csc t$ are positive. The other functions are negative.

14. IV

15. $\sin 5\pi = \tan 5\pi = 0$. $\cos 5\pi = \sec 5\pi = -1$. $\cot 5\pi$ and $\csc 5\pi$ are undefined.

16. $\pm\pi/2, \pm 3\pi/2$

17. $\tan t = \sqrt{1 - \cos^2 t}/\cos t$, $\csc t = 1/\sqrt{1 - \cos^2 t}$.

18. $\cot t = -\sqrt{1 - \sin^2 t}/\sin t$, $\sec t = -1/\sqrt{1 - \sin^2 t}$.

19. $\{t: t \neq (k + 1)\pi/2, k$ an integer$\}$ or $\{t: t \neq k\pi/2, k$ an integer$\}$

20. 1.816 or $104.04°$

39. $\cos 15° = \sqrt{2 + \sqrt{3}}/2 = (\sqrt{6} + \sqrt{2})/4 \approx 0.9659$

40. $\tan 105° = -2 - \sqrt{3}$

41. $63/65$ **43.** $\sqrt{6}/3$

45. $\sqrt{5} \cos(3t + 0.46)$

46. $2 \cos(5t - \pi/6)$

47. $7\pi/6, 3\pi/2, 11\pi/6$

48. $0, 4\pi/3$

49. $\pi/12, 5\pi/12, 9\pi/12, 13\pi/12, 17\pi/12, 21\pi/12$

50. $\pi/18, 5\pi/18, 13\pi/18, 17\pi/18, 25\pi/18, 29\pi/18$

51. None.

52. $5\pi/3$
53. $\{(3\pi + 4k\pi)/2:$ k is an integer$\}$
54. $\{(\pi + 2k\pi)/12:$ k is an integer$\}$
55. $\{(3\pi + 4k\pi)/4:$ k is an integer$\}$
56. $\{5(\pi + 6k\pi)/6:$ k is an integer$\}$
57. 89.05 ft

SECTION 38

1. 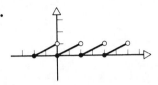   **3.**

5. (a) 1   (b) 4   (c) 4
7. $\{k\pi: k$ is an integer and $-8 \le k < 8\}$
9. $-7\pi/2, -3\pi/2, \pi/2, 5\pi/2$
11. $\pm\pi/2, \pm3\pi/2, \pm5\pi/2, \pm7\pi/2$

13.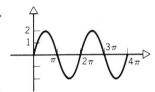

15.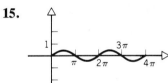

17.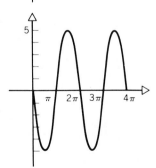

19.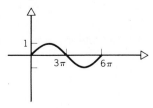

21.

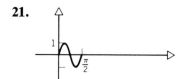

23.

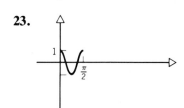

25. $a = 6, b = 4$ **27.** $a = 5, b = \pi/3$

29. $a = 10, b = \pi/2$

31.

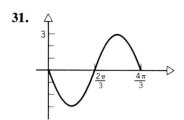

33.

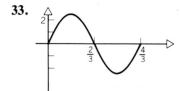

35.

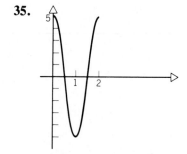

37. (a) If $P(t)$ has coordinates (x, y), then $P(t + \pi)$ has coordinates $(-x, -y)$. Thus $\sin(t + \pi) = -y/r = -\sin t$. [For another proof, use (34.6) with $u = t$ and $v = \pi$.]

(b) Similar to part (a). Use part (a) and $\csc t = 1/\sin t$ for the alternate proof.

39. p/b

41.

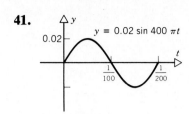

SECTION 39

1. See Figure 39.3 for the graph.
3. See Figure 39.7 for the graph.
5. Replace $y = a \cos bx$ by $y = a \sin bx$, and replace Figure 39.5 by Figure 39.7.
7. $-3\pi/2$ and $\pi/2$. 9. $-\pi$ and π.

11.

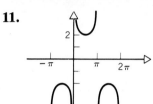

13.

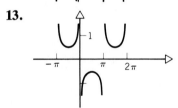

15.

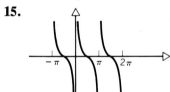

17.

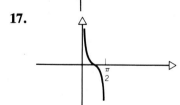

19.

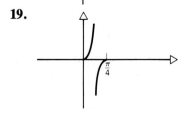

21.

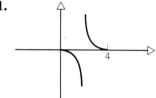

23. $\pi/5$

25. $a = 3$ or $a = -3$, $b = \dfrac{1}{4}$.

27. $x = 2 \tan \dfrac{\pi}{3} t$

SECTION 40

1.

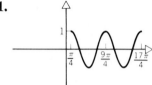

3.

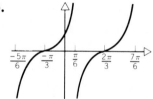

5.

7.

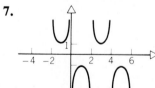

9.

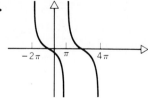

11.

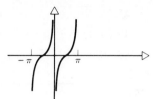

13.

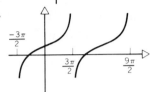

15.

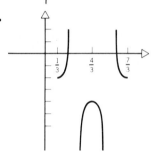

23. The period is $\pi/|b|$. The phase shift is $|c/b|$ units to the right if $c/b > 0$ and $|c/b|$ units to the left if $c/b < 0$.

25. Amplitude $= 2$, period $= \pi/2$, phase shift $= \pi/4$ to the right.

27. Amplitude $= \sqrt{2}$, period $= 4\pi$, phase shift $= 2\pi$ to the right.

29. Period $= \pi/3$, phase shift $= \pi$ to the left.

31. $y = 2 \cos(t - \pi/3)$

33. $y = 5\sqrt{2} \cos(t + \pi/4)$

35. $y = 2 \cos(4t + \pi/4)$

37. (a) The reciprocal of cycles/second, the frequency, is seconds/cycle, the period.

(b) The period is $2\pi/b$ $(b > 0)$.

(c) Amplitude $= 10$, period $= 1/60$, frequency $= 60$, and phase shift $= 1/4$.

39. $I = 20 \sin(240\pi t - 40\pi)$

SECTION 41

1. $\pi/2$ **3.** 0 **5.** $-\pi/3$ **7.** $\pi/4$

9. $\pi/6$ **11.** $\pi/6$ **13.** 1.10 **15.** 0.75

17. 1.55 **19.** 0.800063026 (or 0.80)   **21.** 0.709983514 (or 0.71)

23. 0.099966416 (or 0.10)   **25.** 0.3

27. 5 **29.** 0.75 **31.** $\pi/6$

33. $\pi/6$ **35.** π **37.** $\pi/6$

39. $-\pi/3$ **41.** $4/5$ **43.** $\sqrt{1-x^2}$

45. $x/\sqrt{1-x^2}$ **47.** $-4\sqrt{5}$ **49.** $\sqrt{15}/5$

51. $-3\sqrt{5}/10$ **53.** 0 **55.** 0.939881775

57. 0.981473097 **59.** 0.047400426

61. 0.954781516 **63.** -1.843254503

67. (a) All real numbers.   (b) $-\pi/2 < x < \pi/2$

69. (a) $x \le -1$ or $x \ge 1$ (b) $0 \le x < \pi/2$ or $\pi/2 < x \le \pi$

71.

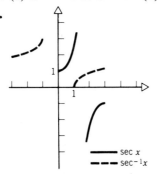

73. $x = \dfrac{1}{b}\left[c + \sin^{-1}\dfrac{(y-d)}{a} \right]$

75. (a) The larger angle at the eye is $\arctan (a+b)/x$. The lower of the smaller angles at the eye is $\arctan b/x$.

(b) $ax/(x^2 + ab + b^2)$

CHAPTER VIII REVIEW

1. 10 **2.** 0.5

3.

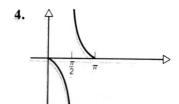

4.

5.

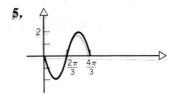

6.

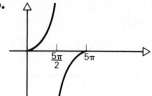

7.

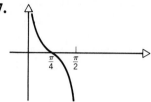

8.

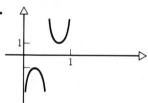

9.

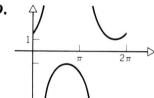

10.

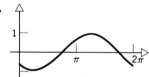

11.

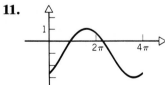

12.

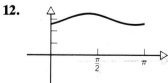

13. $-\pi/3$ **14.** $\pi/6$ **15.** $2\pi/3$

16. $\pi/3$ **17.** $-\pi/2$ **18.** $-\pi/4$

19. 24/25   **20.** $-3/5$

21. (b) -0.7200 **22.** (b) -0.9500

SECTION 42

1. 5.413   **3.** 15.51   **5.** $20\sqrt{3} \approx 34.64$

7. 38.72   **9.** 12.14   **11.** $137.87°$

13. $\alpha = 48.19°$, $\beta = 58.41°$, $\gamma = 73.40°$

15. $c = 42.65$, $\alpha = 1.74$, $\beta = 0.41$

17. $\alpha = 33.56°$, $\beta = 112.89°$, $\gamma = 33.56°$

19. $\alpha = 77.96°$

21. 63.72 ft

23. The angle at C: $48.19°$.

25. $63.35°$

27. Fourth side: 135.4 m. Other angles: $54.23°$ (adjacent to $130°$ angle) and $70.77°$.

33. Use the quadratic formula, treating a, c, and $\cos \alpha$ as constants.

SECTION 43

1. $\beta = 57°$, $b = 19.62$, $c = 22.79$

3. $\alpha = 16°$, $a = 14.39$, $b = 47.70$

5. No solution.

7. No solution.

9. $\beta_1 = 118.72°$, $\gamma_1 = 36.28°$, $b_1 = 10.38$ or $\beta_2 = 11.28°$, $\gamma_2 = 143.72°$, $b_2 = 2.31$.

11. $\beta_1 = 38.68°$, $\gamma_1 = 111.32°$, $c_1 = 7.45$ or $\beta_2 = 141.32°$, $\gamma_2 = 8.68°$, $c_2 = 1.21$.

13. $\beta_1 = 97.83°$, $\gamma_1 = 49.87°$, $b_1 = 46.91$ or $\beta_2 = 17.57°$, $\gamma_2 = 130.13°$, $b_2 = 14.29$.

15. $\alpha_1 = 73.33°$, $\beta_1 = 79.50°$, $a_1 = 16.70$ or $\alpha_2 = 52.33°$, $\beta_2 = 100.50°$, $a_2 = 13.80$.

17. 274.1 square units

19. 118.8 m

21. 22.15 ft

23. 11.95 m

25. 1.12 mi   **27.** 1.64 mi   **29.** 1.93 mi

SECTION 44

1. (a) $6i - 3j$ (b) $-9i + 12j$
   (c) $-3i - 6j$ (d) 5

3. (a) $8i + 4j$ (b) $-3i + 18j$
   (c) $-13i - 26j$ (d) $\sqrt{37}$

5. (a) $\langle 1, 14 \rangle$ (b) $\langle 9, -30 \rangle$
(c) $\langle -11, 2 \rangle$ (d) $\sqrt{109}$

7. (a) $-i - 5j$ (b) $12i + 6j$
(c) $-10i + 4j$ (d) $2\sqrt{5}$

9. (a) $3i - 6j$ (b) $-9i + 12j$
(c) $3i$ (d) 5

11. $\dfrac{1}{4} i - \dfrac{\sqrt{3}}{4} j$ **13.** $3i - 3j$

15. $-5i + 5\sqrt{3}\, j$ **17.** $u_x = c - a, u_y = d - b$

19. S48.2°W at 190 mi/hr.

21. N48.1°W at 192 mi/hr.

23. N11.5°W at 29.6 km/hr.

25. 2934 lb perpendicular to the plane and 624 lb parallel to the plane.

27. 463.6 lb from the north and 187.3 lb from the west.

29. 97.9 kg

SECTION 45

For some of these exercises there is more than one correct answer.

1. $P[3, 100°] = P[-3, 280°] = P[3, -260°]$

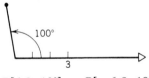

3. $P[6.8, 10°] = P[-6.8, 190°] = P[6.8, -350°]$

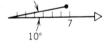

5. $P[7, 11\pi/12] = P[-7, 23\pi/12] = P[7, -13\pi/12]$

7. $(-3/2, -3\sqrt{3}/2)$

9. $(-5, -5\sqrt{3})$

11. $[5\sqrt{2}, 3\pi/4]$

13. $[10, 4.07]$

15. $[25, 2.50]$

17.

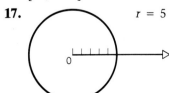

19.

$r = -4.5$

21.

$r^2 = 4$

23.
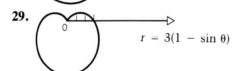
$r = 6 \sin \theta$

25.
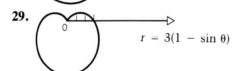
$2r = \cos \theta$

27.

$r = 3(1 - \cos \theta)$

29.
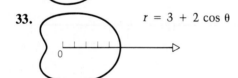
$r = 3(1 - \sin \theta)$

31.
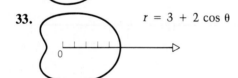
$r = 2 \cos \theta - 1$

33.
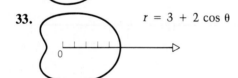
$r = 3 + 2 \cos \theta$

35.

$r = -2\theta$

37.
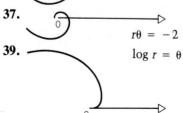
$r\theta = -2$

39.

$\log r = \theta$

41. A rose curve with three loops. **47.** $y = 1$

43. $x^2 + (y - 3)^2 = 9$ **49.** $r = a$

45. $x^2 + y^2 = 4$ **51.** $r = a \cos \theta$

53. $r \cos \theta = 5$ or $r = 5 \sec \theta$

55. (a) The graph of a polar equation is symmetric with respect to the origin if an equivalent equation is obtained when r is replaced by $-r$.

(b) The graph of a polar equation is symmetric with respect to the line $\theta = \pi/2$ if an equivalent equation is obtained when r is replaced by $-r$ and θ is replaced by $-\theta$.

CHAPTER IX REVIEW

1. 10.52   **2.** 14.52

3. 14.24   **4.** 48.92°

5. $\beta = 43.68°$, $\gamma = 126.32°$, $a = 1.509$

6. $\beta = 62.42°$, $\gamma = 37.58°$, $c = 6.19$

7. $\alpha = 0.95$, $\gamma = 0.36$, $a = 25.24$

8. No solution.

9. (a) $14\mathbf{i}$ (b) $-22\mathbf{i} - 5\mathbf{j}$ (c) $\sqrt{17}$

10. (a) $\langle -4, 2 \rangle$ (b) $\langle -48, 19 \rangle$ (c) 13

11. $\mathbf{u} = -5\mathbf{i} + 5\sqrt{3}\mathbf{j}$

12. $\mathbf{u} = -\dfrac{5\sqrt{3}}{2}\mathbf{i} - \dfrac{5}{2}\mathbf{j}$

13. N19.24°E at 259 mi/hr.

14. 69.13 kg directed S4.32°E.

15. $(-7/2, -7\sqrt{3}/2)$

16. $(-15\sqrt{3}/2, 15/2)$

17. $[2, 7\pi/6]$

18. $[15, 5.36]$

19. $(x - 2.5)^2 + y^2 = 6.25$

20. $r = b \sin \theta$

21.

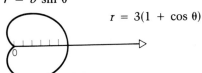

$r = 3(1 + \cos \theta)$

22.

$r^2 = 1$

23.

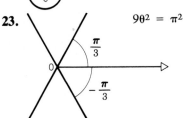

$9\theta^2 = \pi^2$

24.

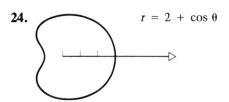

$r = 2 + \cos\theta$

25. Figure 45.11 with $a = 1$.

26. A circle, or a rose curve with one loop having an extreme point at [4, 0].

27. Notice that $(a + b)/b = (a/b) + 1$ and $(\sin\alpha + \sin\beta)/\sin\beta = (\sin\alpha/\sin\beta) + 1$. Now use the Law of Sines.

28. Similar to Exercise 33 of Section 42.

SECTION 46

1. *Real numbers:* $\sqrt{3}$, 0. *Imaginary numbers:* $4 - i$, πi, $2 + \sqrt{-4}$. *Pure imaginary numbers:* πi.

3. *Real numbers:* π, -1. *Imaginary numbers:* $3 + \sqrt{-15}$, $\sqrt{2} + i$, $4i$. *Pure imaginary numbers:* $4i$.

5. 1   **7.** $-4 + 6i$ **9.** $5 - \sqrt{2}i$ **11.** $9 + 3i$

13. $-3 + 2i$ **15.** $3 - 24i$ **17.** $-9 - 2i$ **19.** 2

21. -10 **23.** -5 **25.** $-i$ **27.** 1

29. $\dfrac{3}{10} + \dfrac{1}{10}i$ **31.** $-\dfrac{1}{2} + \dfrac{1}{2}i$ **33.** $\dfrac{4}{5} - \dfrac{3}{5}i$ **35.** $-1 - 5i$

37. $1 - i$ **39.** $1 + i$ **41.** $-11 - 2i$

43. $(1 \pm \sqrt{7}i)/2$ **45.** $(-1 \pm \sqrt{11}i)/2$ **47.** $(5 \pm \sqrt{11}i)/6$

49. $\pm 3i$ **51.** $(3 \pm \sqrt{7}i)/4$ **53.** $x = -2, y = -1$

55. No.   **57.** Yes.

59. Substitute and simplify.

61. (a) $(a + bi) + (a - bi) = 2a$
      (b) $(a + bi) - (a - bi) = 2bi$.

63. $\pm i$ **65.** $\overline{(a + bi)} = \overline{(a - bi)} = a + bi$

67. $\overline{(a + bi) + (c + di)} = \overline{(a + c) + (b + d)i} = (a + c) - (b + di) = (a - bi) + (c - di)$.

69. Use $\overline{(z/w)} = \overline{(z\bar{w})/(w\bar{w})}$, $\bar{z}/\bar{w} = (\bar{z}w)/(\bar{w}w)$, $z = a + bi$, $w = c + di$, and simplify.

71. Replace z by $a + bi$ and simplify both sides.

73.   **75.**

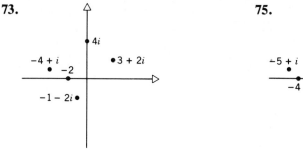

SECTION 47

1. $\sqrt{3} + i$ **3.** $\dfrac{5\sqrt{2}}{2} - \dfrac{5\sqrt{2}}{2}i$

5. $2, 11\pi/6, 2(\cos 11\pi/6 + i \sin 11\pi/6)$

7. $2, 3\pi/2, 2(\cos 3\pi/2 + i \sin 3\pi/2)$

9. $4, \pi/6, 4(\cos \pi/6 + i \sin \pi/6)$

11. $4, \pi/2, 4(\cos \pi/2 + i \sin \pi/2)$

13. $\sqrt{13}, \arctan -1.5 \approx -0.98, \sqrt{13}(\cos -0.98 + i \sin -0.98)$

15. $\sqrt{65}, \arctan -7/4 \approx -1.05, \sqrt{65}(\cos -1.05 + i \sin -1.05)$

17. $-16\sqrt{3} + 16i$

19. -1

21. -4096

23.

25. $\cos \dfrac{\pi + 4k\pi}{8} + i \sin \dfrac{\pi + 4k\pi}{8}$ for $k = 0, 1, 2, 3$.

27. $\sqrt[3]{2}\left(\cos \dfrac{\pi + 2k\pi}{3} + i \sin \dfrac{\pi + 2k\pi}{3}\right)$ for $k = 0, 1, 2$.

29. $\sqrt[4]{2}\left(\cos \dfrac{11\pi + 12k\pi}{24} + i \sin \dfrac{11\pi + 12k\pi}{24}\right)$ for $k = 0, 1, 2, 3$.

31. Use De Moivre's Theorem to prove that each number is an nth root. Also explain why the n numbers are all different.

33. From $z = i \sin \theta, z^n - z^{-n} = 2i \sin n\, \theta$.

35. Use $\sqrt{a^2 + b^2} = \sqrt{a^2 + (-b)^2}$.

37. Use $z\bar{z} = a^2 + b^2$.

39. One method: Use $z = a + bi, w = c + di, |z/w| = |z\bar{w}/w\bar{w}|$, compute, and simplify.

SECTION 48

1. $q(z) = z^2 - iz + 2, r(z) = 3i$

3. $q(z) = 2z^3 + (-1 + 4i)z^2 + (-8 - i)z + (1 - 16i), r(z) = 33 + 3i$

5. $f(i) = -1 - 2i, f(1 + i) = -4 + 6i$

7. $q(z) = z^3 + 2iz^2 - 3z - 6i$

9. $q(z) = z^3 - \sqrt{5}iz^2 - z + \sqrt{5}i$

11. $z^4 + (-1 + 3i)z^3 - 3iz^2 + 4iz - 4i$

13. $x^4 - 4x^3 + 8x^2 - 16x + 16$

15. $(x^2 - 2x + 2)(x^2 + 9)^2$

17. (a) $5, (-1 + \sqrt{3}i)/2$, and $(-1 - \sqrt{3}i)/2$, each of multiplicity one.

(b) $(z - 5)\left(z + \dfrac{1}{2} - \dfrac{\sqrt{3}}{2}i\right)\left(z + \dfrac{1}{2} + \dfrac{\sqrt{3}}{2}i\right)$

(c) $(z - 5)(z^2 + z + 1)$

19. (a) $i, -i, 2i$, and $-2i$, each of multiplicity one.

(b) $(z - i)(z + i)(z - 2i)(z + 2i)$

(c) $(z^2 + 1)(z^2 + 4)$

21. (a) $1 + i, 1 - i$, and -3, each of multiplicity one.

(b) $(z - 1 - i)(z - 1 + i)(z + 3)$

(c) $(z^2 - 2z + 2)(z + 3)$

23. $-1/2, \sqrt{3}i, -\sqrt{3}i$

25. $1/2, 2i, -2i$

27. $0, 5, 2i, -2i$

CHAPTER X REVIEW

1. $7 - i$ **2.** $-1 - 5i$ **3.** 3

4. $14 - 4i$ **5.** $-8 + i$ **6.** $-10 + 10i$

7. $1 + 2i$ **8.** $-\dfrac{15}{26} - \dfrac{3}{26}i$ **9.** $4 + 2i$

10. $11 + 2i$

11. $-\dfrac{3}{2} + \dfrac{3\sqrt{3}}{2}i$ **12.** $-\dfrac{9\sqrt{2}}{4} + \dfrac{9\sqrt{2}}{4}i$

13. $5, 3\pi/2, 5(\cos 3\pi/2 + i \sin 3\pi/2)$

14. $10, 7\pi/6, 10(\cos 7\pi/6 + i \sin 7\pi/6)$

15. $6\sqrt{2}, 3\pi/4, 6\sqrt{2}(\cos 3\pi/4 + i \sin 3\pi/4)$

16. $24, \pi, 24(\cos \pi + i \sin \pi)$

17. $64i$

18. $-512 - 512\sqrt{3}\,i$

19.

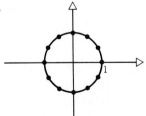

$$\cos \frac{2k\pi}{12} + i \sin \frac{2k\pi}{12} \text{ for } k = 0, 1, 2, \ldots, 11.$$

20. $\dfrac{\sqrt{3}}{2} + \dfrac{1}{2}i, -\dfrac{\sqrt{3}}{2} + \dfrac{1}{2}i, -i$

21. $(1 \pm i)/2$ **22.** $(-3 \pm \sqrt{3}i)/2$

23. $z^4 - iz^3 + 2z^2$

24. $z^4 + (-3 + i)z^3 + (1 - 3i)z^2 + (-3 + i)z - 3i$

25. $z^4 - 4z^3 + 8z^2$ **26.** $z^6 + 11z^4 + 19z^2 + 9$

27. (a) $2i$, $-2i$, and $\dfrac{3}{2}$, each of multiplicity one.

    (b) $(z - 2i)(z + 2i)(2z - 3)$

    (c) $(z^2 + 4)(2z - 3)$

28. (a) $i/2$, $-i/2$, $\sqrt{2}i$, and $-\sqrt{2}i$, each of multiplicity one.

    (b) $(2z - i)(2z + i)(z - \sqrt{2}i)(z + \sqrt{2}i)$

    (c) $(4z^2 + 1)(z^2 + 2)$

SECTION 49

1. $(2, 1)$ **3.** $(0, 3)$ **5.** $(-2, -3)$

7. $\{(x, 3x - 1): \ x \text{ is a real number}\}$

9. $\{(x, -\tfrac{1}{2}(x + 3)): \ x \text{ is a real number}\}$

11. If you add -2 times the second equation to 3 times the first equation, the result will be $-23 = 0$.

13. $(1, 2)$ **15.** $\{(x, 1 - 2x): \ x \text{ is a real number}\}$

17. $\left(\dfrac{1}{3}, \dfrac{1}{3}\right)$ **19.** No intersection.   **21.** $(-12, 6)$

23. $(-3, 5)$ and $(5, 3)$

25. No intersection.

27. No intersection.

29. $(-\sqrt{7}/2, 1/2)$ and $(\sqrt{7}/2, 1/2)$

31. Use 0.5 liter of 20% alcohol and 1.5 liter of 60% alcohol.

33. 65 for children and 270 for adults.

35. Air speed 290 mph, wind speed 10 mph.

37. Runner 10 mph, walker 5 mph.

39. 12 gm of 14-karat gold, 8 gm of 24-karat gold.

41. $\dfrac{27}{41}$

43. 2.5 and 1.5

45. 15 cm and 9 cm

47. 360 mph, 1 hr 10 min

49. Two solutions for $k^2 < 18$, one solution for $k^2 = 18$, and no solution for $k^2 > 18$.

51. a and b are given by $c[1 \pm \sqrt{2/k - 1}]/2$.

53. Use the fact that the graphs coincide iff they have equal slopes and equal y-intercepts.

SECTION 50

1. $(5, -1, -2)$ **3.** $(3, \frac{1}{2}, 3)$

5. $\{(5t + 1, -2t, t): \quad t \text{ is a real number}\}$ **7.** Inconsistent.

9. $\{(t + 4, -2t - 10, t): \quad t \text{ is a real number}\}$ **11.** Inconsistent.

13. $\{(\frac{7}{11}t + \frac{6}{11}, \frac{1}{11}t + \frac{4}{11}, t): \quad t \text{ is a real number}\}$

15. Inconsistent.

17. $\{(\frac{3}{4}t + \frac{1}{8}, \frac{1}{4}t + \frac{11}{8}, t): \quad t \text{ is a real number}\}$

19. $x = -2, y = 4, z = 3, w = 7$ **21.** Inconsistent.

23. 500 pounds each of B and D.

25. National League teams 20, Yankees 19, other American League teams 11.

27. Cleveland 35 inches, Mobile 67 inches, Phoenix 7 inches.

29. $y = 2x^2 + x$ **31.** $x^3 - 2x^2 + x + 2$ **33.** $v_0 = 24$ ft/sec, $s_0 = 216$ ft

SECTION 51

1. Inconsistent.   **3.** $\{(4, -3t - 1, -2t + 7, t): \quad t \text{ is a real number}\}$

5. Inconsistent.   **7.** $(-3, 8, 6, 6)$

9. Inconsistent.   **11.** $(-4, 6)$

13. $(0, 2, 4)$ **15.** $\{(5t - \frac{4}{3}, -3t + \frac{2}{3}, t): \quad t \text{ is a real number}\}$

17. Inconsistent.   **19.** $\{(\frac{4}{7}t + \frac{2}{7}, -\frac{1}{7}t + \frac{10}{7}, t): \quad t \text{ is a real number}\}$

21. Inconsistent.   **23.** $(-4, 0, 2, 1)$

25. $\left(\dfrac{1}{2}, \dfrac{3}{2}\right)$ **27.** $(4, 0)$

29. $\{(v - 3, -2v + 1, v, 2): \quad v \text{ is a real number}\}$

31. 7 cm, 17 cm, 21 cm   **33.** $\dfrac{1}{3}, 1, \dfrac{5}{3}$

35. 9 four-game series, 11 five-game series, 8 six-game series, 22 seven-game series.

37. $-x^3 + x + 4$ **39.** Use the suggestion.

SECTION 52

1.

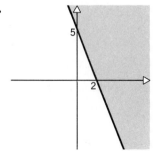

3.

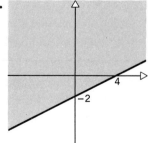

5.

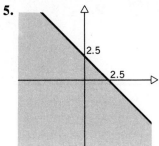

7.

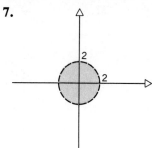

9.

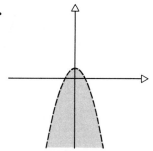

11.

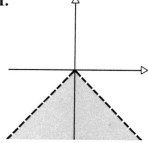

13.

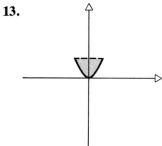

15.

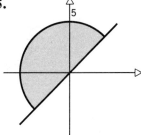

17.

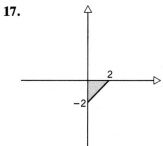

19.

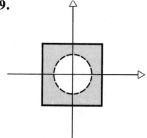

21.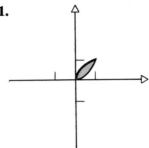

23.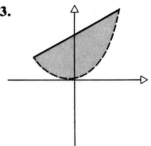

25.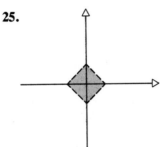

27.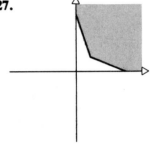

29.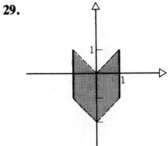

31. $x > 0, y > 0$

33. $x > 0, y < 0$

35. $y > 0, x^2 + y^2 < 25$

SECTION 53

In Exercises 1–6, the vertices of the feasible set are at $(0, 0)$, $(0, 4)$, and $(2, 0)$.

1. Maximum $f(0, 4) = 11$, minimum $f(0, 0) = -1$.

3. Maximum $f(0, 0) = -2$, minimum $f(2, 0) = -8$.

5. Minimum $f(0, 0) = 0$, maximum $f(x, y) = 8$ for $\{(x, 4 - 2x): 0 \le x \le 2\}$.

In Exercises 7–12, the vertices of the feasible set are at $(0, 0)$, $(6, 0)$, $(5, 4)$, and $(0, 3)$.

7. Maximum, $f(6, 0) = 13$, minimum $f(0, 3) = -5$.

9. Maximum $f(6, 0) = 28$, minimum $f(0, 0) = -2$.

11. Maximum $f(x, y) = 48$ for $\{(x, 24 - 4x): 5 \le x \le 6\}$, minimum $f(0, 0) = 0$.

13. $\dfrac{5}{6}$ units of chicken and no tuna.

15. Use 3 bags of brand A and 2 bags of brand B. The total cost will be $100.

17. No orange juice and 6 cups of tomato juice. 13.2 mg.

19. Use the suggestions.

21. Any combination with $a = 3b$ and $b > 0$.

CHAPTER XI REVIEW

1. $\{(x, \frac{1}{3}(x - 2)): \ x \text{ is a real number}\}$

2. $(4, 7)$　　**3.** $(\frac{1}{3}, -\frac{1}{2})$　　　　**4.** None.

5. None.　　　　　　**6.** $\{(x, \frac{1}{2}(4x - 3)): \ x \text{ is a real number}\}$

7. None.

8. $(-1 + \frac{1}{2}\sqrt{14}), 1 - \frac{1}{2}\sqrt{14})$ and $(-1 - \frac{1}{2}\sqrt{14}, 1 + \frac{1}{2}\sqrt{14})$.

9. None.

10. $(-1, 1)$.

11. $\{(x, \frac{1}{3}(x - 2)): \ x \text{ is a real number}\}$

12. $(4, 7)$　　**13.** $(\frac{1}{3}, -\frac{1}{2})$　　　　**14.** Inconsistent.

15. Inconsistent.　　**16.** $\{(x, \frac{1}{2}, (4x - 3)): \ x \text{ is a real number}\}$

17. $(0, -2, 4)$　　　　　　**18.** $(\frac{1}{2}, \frac{1}{4}, \frac{3}{4})$

19. $\{(\frac{4}{3}, t - \frac{5}{3}, t): \ t \text{ is a real number}\}$

20. $\frac{23}{3}, \frac{26}{3}, \frac{5}{3})$　　　　　　**21.** $(8, 2, 0)$

22. Inconsistent.　　　　**23.** $(4, -2, 5)$

24. $(\frac{3}{13}, -\frac{5}{13}, \frac{53}{13})$　　　　**25.** $(\frac{5}{3}, -1, -\frac{2}{3})$

26. $\{(\frac{1}{2}t + \frac{5}{4}, \frac{1}{2}t + \frac{3}{4}, t): \ t \text{ is a real number}\}$

27. $(5, 6, -2)$　　　　**28.** $(17, 6, 10)$

29. $(-1, 4)$ and $(2, 13)$　　**30.** $(-1, 2)$ and $(2, 8)$

31. $\left(\dfrac{3\sqrt{42}}{4}, \dfrac{\sqrt{22}}{4}\right), \left(\dfrac{3\sqrt{42}}{4}, -\dfrac{\sqrt{22}}{4}\right), \left(-\dfrac{3\sqrt{42}}{4}, \dfrac{\sqrt{22}}{4}\right), \left(-\dfrac{3\sqrt{42}}{4}, -\dfrac{\sqrt{22}}{4}\right)$

32. $(3, 2), (3, -2), (-3, 2), (-3, -2)$

33.

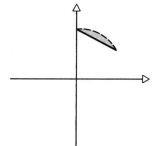

34.

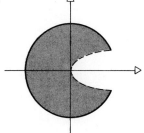

35. Maximum $f(3, 0) = 9$, minimum $f(0, 1) = -9$.

36. Maximum $f(x, y) = 2$ at $\{(x, \frac{1}{2}(5 - x)): \ x \text{ is a real number}\}$, minimum $f(0, 0) = -3$.

SECTION 54

1. $x = -5, y = -4$

3. $c = \dfrac{3}{4}, d = 7, x = 1, y = -\dfrac{1}{2}, z = \dfrac{7}{8}$

5. $\begin{bmatrix} -1 & 28 \\ 14 & -8 \end{bmatrix}$

7. $\begin{bmatrix} 2 & 6 & -12 \\ -4 & -4 & 16 \end{bmatrix}$

9. $\begin{bmatrix} 13 \\ 3 \\ -5 \end{bmatrix}$

11. $[10 \quad 1 \quad 15]$

13. 3×1

15. $m = 6, p = 7$

17. $\begin{bmatrix} -23 & -52 \\ 9 & -18 \end{bmatrix}$

19. $\begin{bmatrix} -16 & -69 & 34 \\ 2 & 27 & -41 \end{bmatrix}$

21. $\begin{bmatrix} -51 & 37 \\ 0 & 0 \\ 46 & 31 \end{bmatrix}$

23. $\begin{bmatrix} -41 & 15 & 34 \\ -3 & 5 & 12 \\ -1 & -1 & 10 \end{bmatrix}$

25. $\begin{bmatrix} 52 & 18 \\ 26 & 9 \\ 4 & 22 \end{bmatrix}$

27. $\begin{bmatrix} -20 & 24 \\ -59 & -23 \\ 21 & 15 \end{bmatrix}$

29. $\begin{bmatrix} -14 & -10 \\ 0 & 0 \\ 10 & -12 \end{bmatrix}$

31. $\begin{bmatrix} -25 & 59 \\ 0 & 0 \\ 37 & -23 \end{bmatrix}$

33. $\begin{bmatrix} -49 & 26 & 3 \\ -42 & 9 & -9 \end{bmatrix}$

SECTION 55

1. $AB = I_2$ **3.** $AB = I_3$

5. $d = -4$, invertible, $\begin{bmatrix} -\dfrac{1}{2} & \dfrac{7}{4} \\ \dfrac{1}{2} & -\dfrac{5}{4} \end{bmatrix}$

7. $d = \dfrac{1}{12}$, invertible, $\begin{bmatrix} 4 & -3 \\ -4 & 6 \end{bmatrix}$

9. $d = 0.1$, invertible, $\begin{bmatrix} 30 & 5 \\ 100 & 17 \end{bmatrix}$

11. $d = 140$, invertible, $\begin{bmatrix} \dfrac{1}{28} & \dfrac{1}{70} \\ -\dfrac{1}{4} & \dfrac{1}{10} \end{bmatrix}$

13. $\begin{bmatrix} -\dfrac{14}{3} & 1 & \dfrac{2}{3} \\ -6 & 1 & 1 \\ \dfrac{17}{3} & -1 & -\dfrac{2}{3} \end{bmatrix}$

15. $\begin{bmatrix} -\dfrac{69}{187} & \dfrac{13}{187} & \dfrac{21}{187} \\ -\dfrac{7}{187} & \dfrac{23}{187} & -\dfrac{6}{187} \\ \dfrac{4}{17} & -\dfrac{1}{17} & \dfrac{1}{17} \end{bmatrix}$

17.
$$\begin{bmatrix} \dfrac{7}{53} & \dfrac{15}{106} & -\dfrac{21}{106} \\[2mm] -\dfrac{8}{53} & \dfrac{51}{106} & -\dfrac{29}{106} \\[2mm] -\dfrac{1}{53} & \dfrac{13}{106} & \dfrac{3}{106} \end{bmatrix}$$

19.
$$\begin{bmatrix} \dfrac{17}{5} & \dfrac{8}{15} & \dfrac{16}{15} & -\dfrac{4}{15} \\[2mm] 1 & -\dfrac{1}{3} & \dfrac{1}{3} & -\dfrac{1}{3} \\[2mm] -\dfrac{3}{5} & -\dfrac{2}{15} & -\dfrac{4}{15} & \dfrac{1}{15} \\[2mm] -\dfrac{1}{5} & \dfrac{1}{15} & \dfrac{2}{15} & \dfrac{7}{15} \end{bmatrix}$$

21. Not invertible.

23. $x = \dfrac{13}{2}, y = -\dfrac{19}{2}$

25. $x = \dfrac{4}{3}, y = \dfrac{1}{12}, z = -\dfrac{5}{12}$

27. $x = \dfrac{7}{9}, y = \dfrac{1}{9}, z = \dfrac{10}{3}$

31. 23 million ounces in South Africa, 16 million ounces outside.

33. $C = I = 15.5$ quads. $G = D = 20.5$ quads.

SECTION 56

1. -4 **3.** 8   **5.** -30 **7.** -4

9. 8   **11.** 80   **13.** 215   **15.** -261

17. 238   **19.** -70 **21.** $-\dfrac{63}{4}$ **23.** $x = \dfrac{2}{7}, y = -\dfrac{6}{7}$

25. $u = \dfrac{17}{22}, v = \dfrac{1}{11}$ **27.** $r = \dfrac{13}{7}, s = \dfrac{17}{7}$ **29.** $x = \dfrac{24}{7}, y = \dfrac{6}{7}, z = \dfrac{11}{7}$

31. Inconsistent.   **33.** Inconsistent.   **35.** Inconsistent.

41. \$3 billion in 1970, \$42 billion in 1978, \$80 billion in 1980.

SECTION 57

1. -42 **3.** -30 **5.** 89   **7.** 228

9. -150 **11.** -204 **13.** -584 **15.** 4

17. Property IV   **19.** Property I   **21.** Property II

23. $\left(-1, 0, \dfrac{2}{3}, -\dfrac{1}{2}\right)$

CHAPTER XII REVIEW

1. $\begin{bmatrix} 2 & 23 \\ -2 & 3 \end{bmatrix}$ **2.** $\begin{bmatrix} 0 & 3 \\ -3 & -1 \\ 13 & 18 \end{bmatrix}$ **3.** $\begin{bmatrix} 1 & -2 \\ 56 & 55 \end{bmatrix}$

4. $\begin{bmatrix} 0 & -5 & 8 \\ 22 & 16 & 2 \\ -1 & -17 & 1 \end{bmatrix}$ **5.** $\begin{bmatrix} \dfrac{2}{13} & -\dfrac{1}{13} \\[2mm] \dfrac{5}{13} & \dfrac{4}{13} \end{bmatrix}$ **6.** $\begin{bmatrix} 12 & -20 \\ -15 & 30 \end{bmatrix}$

7. $\dfrac{1}{14}\begin{bmatrix} -2 & 4 & -4 \\ -25 & 8 & -1 \\ 8 & -2 & 2 \end{bmatrix}$ **8.** $\dfrac{1}{84}\begin{bmatrix} 19 & -3 & 1 \\ 8 & 12 & -4 \\ 14 & 42 & 14 \end{bmatrix}$ **9.** Not invertible.

10. $\begin{bmatrix} 0 & \dfrac{1}{23} & -\dfrac{3}{23} \\ \dfrac{1}{2} & \dfrac{1}{2} & \dfrac{1}{2} \\ 0 & -\dfrac{6}{23} & -\dfrac{5}{23} \end{bmatrix}$ **11.** $x = \dfrac{45}{16},\ y = \dfrac{19}{32}$ **12.** $x = \dfrac{4}{3},\ y = \dfrac{1}{3},\ z = \dfrac{7}{3}$

13. 170   **14.** 0

15. -14 **16.** $(ac - b^2)(bd - c^2)$

17. $x = \dfrac{2}{5},\ y = \dfrac{7}{5}$ **18.** Inconsistent.

19. Dependent.   **20.** $x = \dfrac{6}{5},\ y = \dfrac{13}{5},\ z = -\dfrac{12}{5}$

SECTION 58

1. 5, 10, 15, 20, 25, $5n + 5$ **3.** $\dfrac{1}{2}, \dfrac{2}{3}, \dfrac{3}{4}, \dfrac{4}{5}, \dfrac{5}{6}, \dfrac{n + 1}{n + 2}$

5. $-2, 4, -8, 16, -32, (-2)^{n+1}$

For Exercises 7–23, only the induction step (P_k implies P_{k+1}) is given.

7. $1 + 2 + 2^2 + \cdots + 2^{k+1} = 2^{k+1} - 1 + 2^{k+1} = (2 \cdot 2^{k+1}) - 1 = 2^{k+2} - 1$

9. $5 + 10 + 15 + \cdots + 5(k + 1) = \dfrac{5k(k + 1)}{2} + 5(k + 1)$

$$= \dfrac{5k(k + 1) + 10(k + 1)}{2}$$

$$= \dfrac{5(k + 1)(k + 2)}{2}$$

11. $1^3 + 2^3 + \cdots + (k + 1)^3 = \left[\dfrac{k(k + 1)}{2}\right]^2 + (k + 1)^3$

$$= \dfrac{k^2(k + 1)^2 + 4(k + 1)^3}{4}$$

$$= \left[\dfrac{(k + 1)(k + 2)}{2}\right]^2$$

13. $(1 \cdot 2) + (2 \cdot 3) + (3 \cdot 4) + \cdots + (k + 1)(k + 2)$

$$= \dfrac{k(k + 1)(k + 2)}{3} + (k + 1)(k + 2)$$

$$= \dfrac{(k + 1)(k + 2)(k + 3)}{3}$$

15. $\dfrac{1}{1 \cdot 2 \cdot 3} + \dfrac{1}{2 \cdot 3 \cdot 4} + \dfrac{1}{3 \cdot 4 \cdot 5} + \cdots + \dfrac{1}{(k + 1)(k + 2)(k + 3)}$

$\quad = \dfrac{k(k + 3)}{4(k + 1)(k + 2)} + \dfrac{1}{(k + 1)(k + 2)(k + 3)}$

$\quad = \dfrac{k(k + 3)^2 + 4}{4(k + 1)(k + 2)(k + 3)} = \dfrac{(k + 1)(k + 4)}{4(k + 2)(k + 3)}$

17. $\left(1 - \dfrac{1}{4}\right)\left(1 - \dfrac{1}{9}\right)\left(1 - \dfrac{1}{16}\right) \cdots \left(1 - \dfrac{1}{(k + 1)^2}\right)$

$\quad = \left(\dfrac{k + 1}{2k}\right)\left(1 - \dfrac{1}{(k + 1)^2}\right)$

$\quad = \dfrac{k + 1}{2k} - \dfrac{1}{2k(k + 1)}$

$\quad = \dfrac{(k + 1)^2 - 1}{2k(k + 1)} = \dfrac{k + 2}{2(k + 1)}$

19. $2k \le 2^k$ implies $2(2k) \le 2^{k+1}$, which implies $2(k + 1) \le 2^{k+1}$ because $k + 1 \le 2k$.

21. $\left(1 + \dfrac{1}{2}\right)^k \ge 1 + \dfrac{1}{2^k}$ implies $\left(1 + \dfrac{1}{2}\right)^{k+1} \ge \left(1 + \dfrac{1}{2^k}\right)\left(1 + \dfrac{1}{2}\right) \ge 1 + \dfrac{1}{2^{k+1}}$,

$\quad$ which implies $\left(1 + \dfrac{1}{2}\right)^{k+1} \ge 1 + \dfrac{1}{2^{k+1}}$.

23. $a^k < 1$ and $0 < a < 1$ imply $a^{k+1} < a < 1$, which implies $a^{k+1} < 1$.

25. By connecting appropriately chosen vertices, a $(k + 1)$-sided polygon can be divided into a k-sided polygon and a triangle. From this, the sum of the interior angles will be $(k - 2)180° + 180° = (k - 1)180°$.

27. Induction step: The number of diagonals of a $(k + 1)$-sided polygon is $\dfrac{k(k - 3)}{2} + (k - 1) = \dfrac{(k + 1)(k - 2)}{2}$.

SECTION 59

1. 5, 8, 11, 14, 17   **3.** 40, 28, 16, 4, -8 **5.** $\frac{3}{2}, \frac{7}{4}, 2, \ldots, \frac{15}{4}$

7. 1, 8, 15, 396   **9.** $\frac{1}{4}, \frac{3}{8}, \frac{1}{2}, \frac{115}{4}$ **11.** 120, 125, 130, 16,050

13. (a) $a_1 = a_n - (n - 1)d$ (b) 28

15. (a) $n = \dfrac{a_n - a_1}{d} 1$ (b) 351   **17.** -455

19. $b_1 + b_2 + b_3 + b_4 + b_5$ **21.** $4 + 5 + 6 + 7 + 8 + 9$

23. $\frac{1}{6} + \frac{1}{7} + \frac{1}{8} + \frac{1}{9} + \frac{1}{10} + \frac{1}{11}$ **25.** $\sum\limits_{k=1}^{10} 3^k$

27. $\sum\limits_{k=1}^{20} k \cdot 5^k$ **29.** $\dfrac{5n}{2}(3n + 1)$ **33.** $S_n = \dfrac{n}{2}(a_1 + a_n)$ **35.** 490

37. (a) 482.5 gal   (b) 47th day   **39.** \$40

SECTION 60

1. $\frac{1}{2}$, 2, 8, 32, 128   **3.** 10^{-1}, $10b^{-4}$, 10^{-7}, 10^{-10}, 10^{-13}

5. 100, 10, 1, 10^{-7} **7.** $\dfrac{2^{k-1}}{5^{k-2}}$, $\dfrac{1031}{125}$

9. $3(-\sqrt{3})^{k-1}$, $39 - 12\sqrt{3}$ **11.** $(1 - \sqrt{2})(1 + \sqrt{2})^{k-1}$, $-1 - 2\sqrt{2}$

13. $\frac{665}{128}$ **15.** $1.5(1 - 0.6^{10}) \approx 1.49$ **17.** $22.\overline{2}$

19. (a) $a_1 = \dfrac{a_n}{r^{n-1}}$ (b) 486

21. (a) $a_1 = \dfrac{S_n(1 - r)}{(1 - r^n)}$ (b) $1 + \sqrt{2}$

23. \$15,431.20   **25.** \$7958.55   **27.** \$19,124.88

29. \$3000/32.38 $\approx$ \$92.94   **31.** $a_{k+1} = (a_1 r^{k-1})r = a_1 r^k$

35. $\dfrac{a_1 - a_1(-1)^n}{2}$. S_n alternates between 0 and a_1.

37. $\frac{31}{9}$ **39.** $\frac{697}{33}$ **41.** $\frac{34}{333}$

43. Suggestion: If $b = ar$ and $c = br$, then $b/a = c/b$.

45. Suggestion: If $a_{k+1} = a_k r$, what is $\log a_{k+1} - \log a_k$?

47. 262 cycles per second

SECTION 61

1. 120   **4.** 126   **5.** 1140

7. $(n + 2)(n + 1)$ **9.** $\dfrac{n + 1}{n}$ **11.** $2n$

13.

1		7		21		35		35		21		7		1
1		8		28		56		70		56		28		8		1

15. A common denominator for the right side is $k!(n - k + 2)!$. Or use (61.3) three times.

17. $u^6 + 6u^5v + 15u^4v^2 + 20u^3v^3 + 15u^2v^4 + 6uv^5 + v^6$

19. $a^6 - 6a^5b + 15a^4b^2 - 20a^3b^3 + 15a^2b^4 - 6ab^5 + b^6$

21. $16u^4 - 32u^3v^2 + 24u^2v^4 - 8uv^6 + v^8$

23. 300   **25.** $a^{20} + 20a^{18}b^3 + 180a^{16}b^6 + 960a^{14}b^9$

27. $x^{100} + 100x^{99}y + 4950x^{98}y^2 + 161,700x^{97}y^3$ **29.** 210

31. With $a = b = 1$, the left side of (61.4) equals 2^n, and the right side is the desired sum.

33. Suggestion: Multiply and divide the left side by the product of the first n positive even integers.

CHAPTER XIII REVIEW

1. 3, 27, 162, 810, $3^n(n + 1)n/2$ **2.** $\displaystyle\sum_{k=1}^{\infty} k(-10)^{k-1}$

3. $3 + 7 + 11 + \cdots + (4k + 3) = k(2k + 1) + (4k + 3) = 2k^2 + 5k + 3 = (k + 1)(2k + 3)$

4. $1^2 + 3^2 + 5^2 + \cdots + (2k + 1)^2 = \dfrac{k(2k - 1)(2k + 1)}{3} + (2k + 1)^2$

$= \dfrac{4k^3 + 12k^2 + 11k + 3}{3} = \dfrac{(k + 1)(2k + 1)(2k + 3)}{3}$

5. $10 - 45\pi$ **6.** Use $a_1 = 3$ and $a_n = 4n - 1$ in Equation (59.4).

7. $2\sqrt{3}, -6, 6\sqrt{3}, -486$ **8.** $0.1y^2$ **9.** -1705

10. $\dfrac{a - a^{n+1}b^{2n}}{1 - ab^2}$ **11.** $-\dfrac{1}{5} < t < \dfrac{1}{5}, \dfrac{1}{1 - 5t}$ **12.** 6

13. \$1103.97 **14.** \$75 **15.** $\dfrac{n + 1}{k}$

17. $a^{10} - 5a^8b + 10a^6b^2 - 10a^4b^3 + 5a^2b^4 - b^5$

18. $1,000,000x^6 + 60,000x^5y + 1500x^4y^2 + 20x^3y^3 + 0.15x^2y^4 + 0.0006xy^5 + 0.000001y^6$

SECTION 62

1. 5040 **3.** 117,600 **5.** 380 **7.** 1

9. n **11.** 36, 216, 6^n **13.** 15 **15.** $10! = 3,628,800$

17. (a) 24 (b) 50 **19.** 2880 **21.** 336 **23.** 3, 4, 5, $n + 1$

SECTION 63

1. 10 **3.** 10 **5.** $\dfrac{n(n - 1)(n - 2)}{6}$

7. $k!$ **9.** $n/2$ **11.** 126

13. $C(39, 13)$ **15.** $C(48, 5)$ **17.** $C(13, 4)C(13, 3)^3$

19. (a) 168 (b) $\dfrac{k(n - 1)n}{2}$

21. (a) 35 (b) 63 (c) 203 **23.** $C(n, 3)$

25. $C(n, 4)$ **27.** $\dfrac{n(n - 3)}{2}$ **29.** 31

SECTION 64

1. $\dfrac{3}{8}$ **3.** $\dfrac{1}{2}$

5. (a) $\dfrac{1}{2}$ (b) $\dfrac{2}{3}$ (c) $\dfrac{5}{6}$ (d) 0

7. (a) $\dfrac{1}{36}$ (b) $\dfrac{25}{36}$ (c) $\dfrac{4}{9}$ (d) $\dfrac{1}{4}$ (e) $\dfrac{3}{4}$ (f) $\dfrac{1}{2}$

9. (a) $\dfrac{3}{8}$ (b) $\dfrac{7}{8}$ (c) $\dfrac{1}{2}$ (d) $\dfrac{1}{8}$ (e) $\dfrac{1}{2}$ (f) $\dfrac{1}{2}$

11. 0.72675 **13.** $\dfrac{1}{11}$ **15.** $\dfrac{1}{6}$

17. (a) $\dfrac{1}{2}$ (b) $\dfrac{1}{2}$ (c) $\dfrac{1}{4}$ (d) $\dfrac{3}{4}$

19. (a) $C(39, 13)/C(52, 13)$ (b) $1/C(52, 13)$ (c) $C(48, 13)/C(52, 13)$
(d) $C(48, 9)/C(52, 13)$

21. (a) \$0.50 (b) \$2.50 (c) \$0.20 (d) \$20,000

CHAPTER XIV REVIEW

1. 1320   **2.** 1680   **3.** 21   **4.** 462   **5.** 60   **6.** 23

7. 60   **8.** 95,040   **9.** 210   **10.** 150   **11.** $\dfrac{1}{6}$ **12.** $\dfrac{3}{20}$

13. (a) $\dfrac{3}{7}$ (b) $\dfrac{18}{35}$ (c) $\dfrac{5}{7}$ **14.** $C(13, 5)/C(52, 5)$, $4 \cdot C(13, 5)/C(52, 5)$

15. $\dfrac{1}{10}, \dfrac{3}{10}$ **16.** $\dfrac{11}{12}$

SECTION 65

1. Directrix $y = \dfrac{1}{8}$. Focus $\left(0, -\dfrac{1}{8}\right)$. Vertex $(0, 0)$. Axis $x = 0$.

3. Directrix $x = 3$. Focus $(-3, 0)$. Vertex $(0, 0)$. Axis $y = 0$.

5. Directrix $x = \dfrac{1}{4}$. Focus $\left(-\dfrac{1}{4}, 0\right)$. Vertex $(0, 0)$. Axis $y = 0$.

7. Directrix $x = -\dfrac{7}{4}$. Focus $\left(-\dfrac{9}{4}, -1\right)$. Vertex $(-2, 1)$. Axis $y = 1$.

9. Directrix $y = \dfrac{11}{2}$. Focus $\left(3, \dfrac{1}{2}\right)$. Vertex $(3, 3)$. Axis $x = 3$.

11. Directrix $y = -\dfrac{23}{8}$. Focus $\left(0, -\dfrac{25}{8}\right)$. Vertex $(0, -3)$. Axis $x = 0$.

13. Directrix $y = -\dfrac{1}{4}$. Focus $\left(-\dfrac{3}{2}, \dfrac{1}{4}\right)$. Vertex $\left(-\dfrac{3}{2}, 0\right)$. Axis $x = -\dfrac{3}{2}$.

15. Directrix $y = -\dfrac{1}{4}$. Focus $\left(\dfrac{1}{2}, \dfrac{1}{4}\right)$. Vertex $\left(\dfrac{1}{2}, 0\right)$. Axis $x = \dfrac{1}{2}$.

17. $x^2 = -20y$.

19. $(x + 1)^2 = 16(y - 4)$

21. $x^2 = 16(y + 4)$

23. $(y - 1)^2 = -\dfrac{1}{5}(x - 5)$

25. $\dfrac{45}{8}$ ft or 5 ft $7\dfrac{1}{2}$ in.

SECTION 66

1. Center $(0, 0)$. Foci $(0, -\sqrt{7})$ and $(0, \sqrt{7})$. Endpoints of major axis $(0, -4)$ and $(0, 4)$. Endpoints of minor axis $(-3, 0)$ and $(3, 0)$.

3. Center $(0, 0)$. Foci $(0, -\sqrt{21})$ and $(0, \sqrt{21})$. Endpoints of major axis $(0, -5)$ and $(0, 5)$. Endpoints of minor axis $(-2, 0)$ and $(2, 0)$.

5. Center $(0, -1)$. Foci $(-\sqrt{6}, -1)$ and $(\sqrt{6}, -1)$. Endpoints of major axis $(-3, -1)$ and $(3, -1)$. Endpoints of minor axis $(0, -1 - \sqrt{3})$ and $(0, -1 + \sqrt{3})$.

7. Center $(0, -2)$. Foci $(0, -2 - \sqrt{3})$ and $(0, -2 + \sqrt{3})$. Endpoints of major axis $(0, -4)$ and $(0, 0)$. Endpoints of minor axis $(-1, -2)$ and $(1, -2)$.

9. Center $(-5, 0)$. Foci $(-5, -3\sqrt{3})$ and $(-5, 3\sqrt{3})$. Endpoints of major axis $(-5, -6)$ and $(-5, 6)$. Endpoints of minor axis $(-8, 0)$ and $(-2, 0)$.

11. Center $(-1, -3)$. Foci $(-1, -6)$ and $(-1, 0)$. Endpoints of major axis $(-1, -8)$ and $(-1, 2)$. Endpoints of minor axis $(-5, -3)$ and $(3, -3)$.

13. $\dfrac{(x - 2)^2}{2^2} + \dfrac{(y - 4)^2}{4^2} = 1$

15. $\dfrac{(x + 3)^2}{3^2} + \dfrac{(y - 2)^2}{4^2} = 1$

17. $\dfrac{x^2}{5^2} + \dfrac{y^2}{4^2} = 1$

19. $35x^2 + y^2 = 36$

21. $25x^2 + 24y^2 = 625$

23. $(2, 1)$ and $(-2, 1)$

25. Use $x = 56.25$ in $\dfrac{x^2}{76.25^2} + \dfrac{y^2}{37.5^2} = 1$. (Why?) The answer is 25 ft 4 in.

27. All of the points that are both inside or on the circle $x^2 + y^2 = 25$ except those inside the ellipse $4x^2 + 9y^2 - 36 = 0$. (Why?)

31. $b = 9.68$

33. $e = 0.0509$

35. 67 million miles

37. 29.41 Earth years

39. 164.62 Earth years

SECTION 67

1. Center $(0, 0)$. Foci $(-\sqrt{13}, 0)$ and $(\sqrt{13}, 0)$. Endpoints of transverse axis $(-3, 0)$ and $(3, 0)$. Endpoints of conjugate axis $(0, 2)$ and $(0, -2)$.

3. Center $(0, 0)$. Foci $(0, -2\sqrt{5})$ and $(0, 2\sqrt{5})$. Endpoints of transverse axis $(0, -4)$ and $(0, 4)$. Endpoints of conjugate axis $(-2, 0)$ and $(2, 0)$.

5. Center $(2, -1)$. Foci $(2, -1 - \sqrt{34})$ and $(2, -1 + \sqrt{34})$. Endpoints of transverse axis $(2, -4)$ and $(2, 2)$. Endpoints of conjugate axis $(-3, -1)$ and $(7, -1)$.

7. Center $(0, -2)$. Foci $(-2\sqrt{3}, -2)$ and $(2\sqrt{3}, -2)$. Endpoints of transverse axis $(-3, -2)$ and $(3, -2)$. Endpoints of conjugate axis $(0, -2 - \sqrt{3})$ and $(0, -2 + \sqrt{3})$.

9. Center $(1, -2)$. Foci $(-4, -2)$ and $(6, -2)$. Endpoints of transverse axis $(-3, -2)$ and $(5, -2)$. Endpoints of conjugate axis $(1, -5)$ and $(1, 1)$.

11. Center $(2, -6)$. Foci $(2, -6 - \sqrt{14})$ and $(2, -6 + \sqrt{14})$. Endpoints of transverse axis $(2, -9)$ and $(2, -3)$. Endpoints of conjugate axis $(2 - \sqrt{5}, -6)$ and $(2 + \sqrt{5}, -6)$.

13. $9x^2 - 4y^2 + 36 = 0$

15. $21x^2 - 4y^2 + 84 = 0$

17. $5x^2 - 16y^2 - 80 = 0$

19. $25x^2 - 9y^2 + 225 = 0$

21. $\dfrac{x^2}{3^2} - \dfrac{(y - 3)^2}{4^2} = 1$

23. $(0, -1), (-\sqrt{3}, 2)$ and $(\sqrt{3}, 2)$.

25. $\dfrac{y^2}{b^2} - \dfrac{x^2}{a^2} = 1$

27. $y = \pm x$

CHAPTER XV REVIEW

1. Parabola

2. Ellipse

3. Hyperbola

4. Parabola

5. Two intersecting lines

6. Circle

7. Ellipse

8. Ellipse

9. Parabola

10. Empty set

11. Hyperbola

12. Parabola

13. Hyperbola

14. Two parallel lines

15. $x^2 = -20y$

16. $25x^2 + 4y^2 = 100$

17. $\dfrac{(y - 3)^2}{4} - \dfrac{(x + 2)^2}{5} = 1$

18. $y^2 = -12(x - 2)$

19. $4(x + 1)^2 + 3y^2 = 48$

20. $36x^2 - 9y^2 = 324$

INDEX

TRIGONOMETRIC FUNCTIONS

ACUTE ANGLES

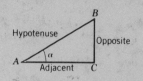

$$\sin \alpha = \frac{\text{opposite}}{\text{hypotenuse}} \qquad \csc \alpha = \frac{\text{hypotenuse}}{\text{opposite}}$$

$$\cos \alpha = \frac{\text{adjacent}}{\text{hypotenuse}} \qquad \sec \alpha = \frac{\text{hypotenuse}}{\text{adjacent}}$$

$$\tan \alpha = \frac{\text{opposite}}{\text{adjacent}} \qquad \cot \alpha = \frac{\text{adjacent}}{\text{opposite}}$$

GENERAL ANGLES

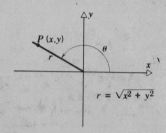

$r = \sqrt{x^2 + y^2}$

$\sin \theta = y/r$	$\csc \theta = r/y$
$\cos \theta = x/r$	$\sec \theta = r/x$
$\tan \theta = y/x$	$\cot \theta = x/y$

REAL NUMBERS

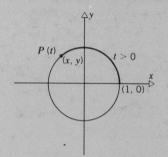

$\sin t = y$	$\csc t = 1/y$
$\cos t = x$	$\sec t = 1/x$
$\tan t = y/x$	$\cot t = x/y$

RECIPROCAL IDENTITIES

$$\csc t = \frac{1}{\sin t}$$

$$\sec t = \frac{1}{\cos t}$$

$$\cot t = \frac{1}{\tan t}$$

COFUNCTION IDENTITIES

$$\cos t = \sin\left(\frac{\pi}{2} - t\right)$$

$$\cot t = \tan\left(\frac{\pi}{2} - t\right)$$

$$\csc t = \sec\left(\frac{\pi}{2} - t\right)$$

QUOTIENT IDENTITIES

$$\tan t = \frac{\sin t}{\cos t} \qquad \cot t = \frac{\cos t}{\sin t}$$

EVEN-ODD IDENTITIES

$\sin(-t) = -\sin t$	$\csc(-t) = -\csc t$
$\cos(-t) = \cos t$	$\sec(-t) = \sec t$
$\tan(-t) = -\tan t$	$\cot(-t) = -\cot t$

PYTHAGOREAN IDENTITIES

$$\sin^2 t + \cos^2 t = 1$$

$$\tan^2 t + 1 = \sec^2 t$$

$$1 + \cot^2 t = \csc^2 t$$